Animal Behavior

Sixth Edition

Animal Behavior

AN EVOLUTIONARY APPROACH

JOHN ALCOCK
Arizona State University

SINAUER ASSOCIATES, Inc. *Publishers*
Sunderland, Massachusetts

The Cover

Male *Telostylinus* flies in combat on a fallen mango tree in a Sulawesi rainforest. Males of this species engage in a resource-defense mating strategy (see Chapter 13), defending holes bored by beetle larvae which are attractive to female flies as egg-laying sites. The combatants rear up as high as possible on their long legs, pressing their heads together and pushing strenuously against each other, in a remarkable example of convergent evolution of threat displays; similar behavior has evolved independently in many other animals, including large vertebrates such as red deer (see Chapter 8). *Photograph by Ken Preston-Mafham/ Premaphotos Wildlife.*

The Frontispiece

An Adélie penguin rookery in Antarctica. Breeding grounds are carefully chosen by these birds based on several factors (see Chapter 11). *Photograph by Colin Monteath/Hedgehog House.*

Animal Behavior: An Evolutionary Approach, Sixth Edition

© Copyright 1998 by Sinauer Associates, Inc.

All rights reserved.

This book may not be reproduced in whole or in part for any purpose whatever without written permission from the publisher. For information or to order, address: Sinauer Associates, Inc., P.O. Box 407, Sunderland, Massachusetts, 01375-0407 U.S.A. Fax: 413-549-1118. Internet: publish@sinauer.com; http://www.sinauer.com

Library of Congress Cataloging-in-Publication Data

Alcock, John, 1942-

      Animal behavior: an evolutionary approach   / John Alcock. – 6th ed.

          p.   cm.

      Includes bibliographical references and index.

      ISBN 0-87893-009-4 (cloth)

      1. Animal behavior—Evolution.   I. Title.

QL751.A58   1997

591.5—dc21   97-27347CIP

10 9 8 7 6 5 4 3

With thanks to George C. Williams
for explaining what evolutionary theory is

Contents

CHAPTER 13 *The Evolution of Mating Systems 483*

CHAPTER 14 *The Adaptive Tactics of Parents 523*

Preface

Another four years or so has passed since I last confronted my computer and stacks of animal behavior journals and reprints in an attempt to bring my textbook up to date while also making it a more effective educational tool. The job of revising is made both easier and harder by the exponential increase in the number and quality of research articles on behavior over the past four years. In the first edition of the book, I cited fewer than 500 scientific publications. I am now up to 1300, of which only a handful were used the first time around. I am not at risk of running out of new discoveries to report to my readers.

On the other hand, this flood of new information is hard to digest in a timely fashion. For example, a few days before writing this Preface, I came across a report in the journal *Nature* demonstrating that the very same genes that are responsible for the development of respiratory gill plates in wingless crustaceans are also present in insects, in which they are involved in the development of wings. This finding has considerable significance for our understanding of how insects came to fly, a topic that fascinates me. However, I found this paper too late to incorporate it into Chapter 7, having decided that if I did not stop adding new material I would never get done.

The task of picking and choosing among literally thousands of good papers forces me to select just a fraction of those I might use, after which I generally reduce a complex story to a line or two of text. Even so, it is a struggle to keep the book at a manageable length. A textbook need not be, and in fact should not be, an encyclopedia. Therefore, in revising, I have tried to keep the focus on the handful of really big ideas in behavioral biology: the distinction between different kinds of underlying causes of behavior, how natural selection theory is used to develop hypotheses on the evolutionary causes of behavior, what is meant by a cost–benefit approach to behavioral analysis, and most importantly, the procedures that scientists follow when trying to discriminate between competing explanations for something.

In order to achieve this goal, I have rewritten the book extensively, incorporating new evidence and new topics, such as the origins of insect flight, where it seemed appropriate. I have, however, retained the traditional organizational scheme of the book, except for one major change. In this edition, the second chapter no longer classifies behavior into instincts and the different kinds of learning, but instead presents the topic of communication to explain the various kinds of internal mechanisms, ranging from genes to neurons, that underlie animal behavior. The following chapters cover each of these mechanisms individually before the book switches gears with two chapters on the evolutionary bases of communication, again using this subject to preview a series of subsequent chapters. In these chapters, evolutionary theory is applied to topics ranging from antipredator to reproductive to social behavior, and finally to human behavior as well. I would like my readers to come away understanding more about how animal behavior has been studied and why the logic of science is worth appreciating.

Acknowledgments

As in all previous revisions, I received aid from many people. First, all the chapters in their initial revised drafts were read by one or more colleagues from the United States or Canada, including Luis Baptista, Martin Daly, Susan Foster, Steven Gaulin, Ann Hedrick, Ronald Hoy, Walter Koenig, Donald Kroodsma, Steven Lima, Robert Montgomerie, Randy Nelson, David Queller, and Jon Waage. All these persons made a large number of suggestions for changes, most of which were clearly correct, and many of which I actually followed. Sometimes, however, laziness on my part, or incomprehension, or stubbornness prevented me from doing the right thing, and so needless to say I am responsible for any errors or shortcomings that remain in the text.

Many other behavioral biologists have assisted me by doing such things as sending me reprints, answering my questions, and responding to requests for photographs or permission to use their material in figures or diagrams. I have tried to acknowledge the permission givers at the appropriate illustration in the text. Acknowledgments to the publishers who have also generously granted permission to use their copyrighted material appear between the Bibliography and Index.

I also thank my editor Peter Farley at Sinauer Associates, where he (and I) have the help of Kerry Falvey and the other Sinauerians, as well as Norma Roche, a remarkably skillful copy editor who saved me from myself in many places in the text. Peter has effectively steered me through the complicated process of producing a manuscript for two editions now, and I continue to bless my lucky stars that I signed on with Sinauer Associates more than two decades ago.

Most authors need support from their family and friends if they are to remain sane, and here too I am fortunate. My wife Sue is amazingly tolerant, even to the point of becoming my unpaid executive field assistant on bee research projects in remote Western Australia. When we are in Arizona, my son Nick comes over to have dinner with us often, after which he and I play three games of ping-pong, one of which he manages to lose so that I will have reason to go on living. A group of politically dubious white male colleagues, including my department chair, Jim Collins, as well as Dave Brown, Stuart Fisher, Dave Pearson, and Ron Rutowski, often keep me company at lunch, where we discuss teaching, the shortcomings of others, and what it is like to grow old. On Friday afternoons, we sometimes wander over to the faculty club, having several years ago traded in large plastic pitchers of Budweiser and greasy french fries at the Chuck Box for small tapered glasses of Pilsner Urquell and delicate cheese-filled jalapeños at the club. There we discuss teaching, the shortcomings of others, and what it is like to grow old. As I grow older, I am increasingly grateful to my friends and family for the many pleasant distractions they provide in the intervals when I am not looking at a computer screen.

CHAPTER 1

An Evolutionary Approach to Animal Behavior

*F*OR HUNDREDS OF THOUSANDS OF YEARS, humans observed animals for a thoroughly practical reason: their lives depended on a knowledge of animal behavior. Even today, the study of animal behavior has great potential importance for our species. For example, understanding the reproductive behavior of agricultural pests may ultimately lead to their control, while a knowledge of the migratory routes of an endangered whale or shorebird may enable conservationists to design adequate reserves to save the animal from extinction. Even if there were absolutely no practical benefits to be gained from learning about animal behavior, the subject would still be worth exploring because it is so fascinating. Who would have guessed that praying mantises can detect the ultrasonic cries of predatory bats, while Belding's ground squirrels treat their full siblings differently from their half-siblings, and male black-winged damselflies use their penises to scrub other males' sperm out of their mate's sperm storage organ before transferring their own sperm? In the pages ahead, you will learn about many other equally remarkable feats of animals. The point of this text, however, is not only to introduce you to these entertaining discoveries, but also to help you understand how scientists have determined that praying mantises can hear sounds inaudible to humans and have demonstrated that the penis is a weapon in the sperm competition wars among black-winged damselflies. This book is dedicated to the proposition that the process of doing science is every bit as interesting as the findings that are its end product. If I can help my readers understand the beautiful, useful logic of science, as well as appreciate the wonderful diversity of animal behavior, my textbook will have done its job.

Questions about Behavior

I lived for one summer in Monteverde, a tiny community in the mountains of Costa Rica, and while I was there a friend loaned me a black light, which I hung up by a white sheet on the back porch of our home. The ultraviolet rays of the lamp attracted hundreds of moths each night, and many stayed on the sheet until I could inspect them. Some mornings I found a huge bright yellow moth of the genus *Automeris* clinging to the sheet. In the chilly dawn, the sluggish moth did not struggle if I picked it up carefully. But if I jostled it suddenly, or poked it sharply on its thorax, the moth abruptly lifted its forewings and held them up to expose its previously concealed hindwings. The hindwings were marvelously decorated, with

1 *Automeris* **moth from Costa Rica.** (Left) Moth in its resting position with forewings held over the hindwings. (Right) After being jabbed in the thorax, the moth pulls its forewings forward to expose the "eyes" on the hindwings. *Photographs by (left) the author and (right) Michael Fogden.*

large circular patches of blue, black, and white scales on a deep yellow background. The patches looked like two eyes, which seemed to stare back at me (Figure 1).

Anyone seeing *Automeris* abruptly expose its hindwing "eyes" for the first time will surely wonder what is going on. I know I did. My questions about *Automeris* wing-flipping grew into a substantial list, but no matter how long the list, I could assign each question to one of just two fundamentally different categories: "how questions," about how **proximate** mechanisms inside the moth cause the behavior, and "why questions," about the **ultimate** reasons for the behavior [782, 879]. How questions ask *how* an individual manages to carry out an activity; they require mechanistic explanations about how structures *within* an animal operate, enabling the creature to behave in a certain way. Why questions ask *why* the animal has evolved the proximate mechanisms that cause it to perform an activity.

How Questions about Proximate Causes

Consider the following series of questions about the wing-flipping reaction of an *Automeris* moth to a sharp poke:

What is the causal relationship between the animal's genes and its behavior?

Is the trait to some extent inherited from the moth's parents?

How has the development of the moth from a single cell to a multimillion-celled adult affected its behavioral abilities?

What stimuli trigger the response, and how are these stimuli detected?

What these four questions have in common, despite their diversity, is an interest in the operation of mechanisms *within* the moth that cause it to pull its forewings forward, revealing the amazing hindwings. The diversity of proximate questions is great enough, however, that we can subdivide them into two complementary groups, one dealing with the effects of heredity and development on the construction of the mechanisms underlying wing-flipping, and the other dealing with how the fully developed physiological mechanisms actually work to cause the reaction.

With respect to the influence of heredity on development, we can ask how the special touch receptors, brain cells, and muscle controllers used in the wing-flipping reaction arise in an individual. The adult moth began life as a fertilized egg, a single cell that contained genetic instructions donated by its father and mother. These instructions affected the development of the moth, channeling the proliferation and specialization of cells along certain pathways that produced a nervous system with special features. This is a wonderfully complex process, still poorly understood for any organism, let alone for *Automeris* moths.

Even if the development of the response were completely understood, we could still explore how the fully developed neural mechanisms within the adult moth detect certain kinds of stimulation, and how messages are then relayed to activate muscular reactions. No one knows much about the neurophysiology of *Automeris* behavior, but perhaps someone will eventually learn how its neural mechanisms help it respond to a poke on the thorax.

Why Questions about Ultimate Causes

Even if we knew everything there was to know about the genetic–developmental and sensory–motor causes of the wing-flipping behavior of *Automeris* moths, we could still ask many more questions, such as:

Has the behavior evolved over time?
If so, why did the changes take place?
What was the original step in the historical process that led to the current behavior?
What is the purpose, the function, of the behavior?
Does the behavior help individuals overcome obstacles to survival and reproduction?

These are all questions about the evolutionary, or ultimate, reasons why an animal does something. Why does the moth suddenly lift its wings and expose its eye spots when it is molested? The British scientist David Blest suggested that this behavior exists because in the past wing-flipping startled and frightened away some bird predators when they mistook moths' eye spots for the eyes of *their* enemies, predatory owls [99].

If Blest is right and wing-flipping behavior has saved the lives of moths in the past, then the evolutionary process has contributed to the persistence of the behavior in today's moths. Behavioral abilities depend in part on genetic information. The genes present in a contemporary moth species are a tiny subset of all the genes that have ever existed in this species over time. Those genes that have persisted to today have been replicated and passed on from generation to generation more often than others, usually because they conferred a reproductive advantage on the individuals whose cells they occupied. If a gene influenced the appearance or behavior of a moth in ways that helped the moth frighten away predatory birds, the moth might live long enough to reproduce, passing the gene to its descendants. This process could help explain why *Automeris* moths living in Monteverde have inherited genetic mechanisms that promote the development of wing-flipping behavior. The developmental plan, and therefore the behavioral abilities, of each member of the species alive today has been defined by differences in reproductive success that occurred during the history of the species.

The current function of a behavior offers insight into its possible usefulness in the past, which could help explain why the trait spread and replaced others during evolution. However, an evolutionary or ultimate analysis of behavior must also take into account the possibility that a new trait first became common because it served a different function, but has been modified subsequently to serve its current purpose. The present abrupt wing-flipping behavior surely was not always practiced by the ancestors of modern-day *Automeris*. Probably the initial phase involved certain wing movements associated with taking flight, movements that have been altered during the moth's history, just as the color pattern of the hind-

Table 1 Levels of analysis in the study of animal behavior

PROXIMATE CAUSES
1. Genetic–developmental mechanisms
   Effects of heredity on behavior
   Gene–environment interactions underlying the development of sensory–motor mechanisms
2. Sensory–motor mechanisms
   Nervous systems for the detection of environmental stimuli
   Hormone systems for adjusting responsiveness to environmental stimuli
   Skeletal–muscular systems for carrying out responses

ULTIMATE CAUSES
1. Historical pathways leading to a current behavior
   Events occurring over evolution from the origin of the trait to the present
2. Selective processes shaping the history of a behavioral trait
   Past and current usefulness of the behavior in reproductive terms

Sources: Holekamp and Sherman [558]; Sherman [1085]; Tinbergen [1191].

wings has certainly changed over time [98]. A full understanding of the ultimate causes of wing-flipping requires investigation into the initial form and subsequent evolution of the behavior.

You should now be able to discriminate proximate (mechanistic) and ultimate (evolutionary) questions (Table 1). If someone were to wonder if *Automeris* moths expose their hindwings when pecked because their nervous systems control the wing-flipping response, that person would be interested in the proximate basis of behavior, as is anyone concerned with how the genetic, developmental, neural, or hormonal mechanisms work within an animal's body. On the other hand, if you were interested in whether the action evolved because of past predation pressure, you would be dealing with an ultimate issue, as is anyone who wants the answers to questions about the reproductive or **adaptive value** of a trait or its historical basis, its evolutionary foundation (Figure 2).

Moreover, if someone were to claim that work on the evolutionary basis of the wing-flipping behavior eliminated the need to answer questions about the physiological foundations of the behavior, you would know that this is a mistake. Unfortunately, this kind of confusion is common. For example, I recently read that capuchin monkeys may rub citrus fruits on their fur for medicinal benefits (the chemicals may help heal skin wounds). The author then added, "Of course, the monkeys may simply enjoy the sensation," suggesting that this is an alternative to the medicinal hypothesis. It is not. At a proximate level, monkeys may derive pleasure from applying certain scents to their bodies because, at an ultimate level, the behavior in the past had medicinal benefits that translated into increased reproductive success. The full analysis of any behavior involves *both* proximate and ultimate features, with the one complementing the other, not excluding it.

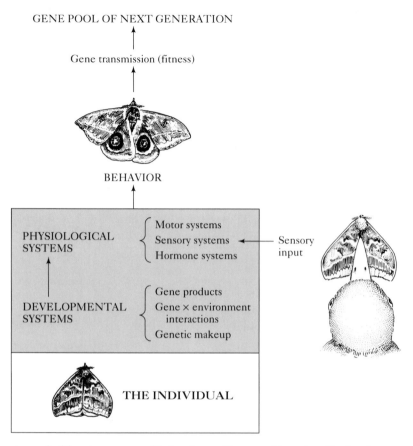

GENE POOL OF NEXT GENERATION

Gene transmission (fitness)

BEHAVIOR

PHYSIOLOGICAL SYSTEMS {
Motor systems
Sensory systems ← Sensory input
Hormone systems

DEVELOPMENTAL SYSTEMS {
Gene products
Gene × environment interactions
Genetic makeup

THE INDIVIDUAL

2 Proximate and ultimate causes of behavior as illustrated by wing-flipping in an *Automeris* moth. At the proximate level, various internal mechanisms enable the moth to execute its wing-flipping behavior. At the ultimate level, the moth's reaction to bird predators determines its reproductive success, as measured by how many copies of its genes reach the next generation. Hereditary differences among individuals in their proximate mechanisms, which result in differences in gene transmission among them, determine which genes are available to influence the development of individuals in the next generation.

Answering Proximate and Ultimate Questions about Behavior

It is one thing to be curious about a mechanism of behavior or its evolutionary foundation, and another thing to satisfy one's curiosity. The scientific approach to answering biological questions will play an important role in every chapter of this book. Understanding the logic of science will enable you to better evaluate the validity of scientific conclusions and can help you solve the puzzles in the natural world that interest you. Our exploration of the "scientific method" begins here with a review of two classic studies by Niko Tinbergen, one on a proximate question and the other on an ultimate question about behavior.

Beewolves and Homing Behavior

Tinbergen helped make the study of animal behavior a part of modern biology. Although Charles Darwin was deeply interested in behavior and explored such diverse topics as earthworm burrowing, bumblebee mating, bowerbird displays, and the facial expressions of dogs and humans, the evolutionary approach to animal behavior did not emerge as a distinct field of scientific study until the mid-1930s. At this time the discipline of **ethology** was developed under the guidance of Tinbergen, a native of the Netherlands, and his friend Konrad Lorenz, an Austrian. They and their colleagues investigated both proximate and ultimate questions about behavior [546, 1191] while studying gulls, jackdaws, butterflies, snow buntings, greylag geese, moth caterpillars, and many other species in their natural environments. The pioneering ethologists ultimately received the Nobel Prize in 1973, which Tinbergen and Lorenz shared with Otto von Frisch (Figure 3), an Austrian researcher famous for his work on honeybee dance communication (see Chapter 10).

Niko Tinbergen explored the causes of behavior at all levels. One of his earliest ethological studies began in 1929, more than 40 years before he was awarded the Nobel Prize, when he discovered a large number of digger wasps nesting in the sand dunes near Hulshorst, Holland. These wasps so fascinated Tinbergen that he and his fellow researchers spent weeks living in a primitive campsite and bicycling up to 70 miles a day in order to learn more about them [1186]. The species of digger wasp that caught Tinbergen's eye was *Philanthus triangulum*, the beewolf, so named because it catches and paralyzes honeybees by stinging them. The female beewolf

3 The founders of ethology. From left to right, Niko Tinbergen, Konrad Lorenz, and Karl von Frisch. *Photographs by (left) B. Tschanz, (middle) Sybille Kalas, and (right) O. von Frisch.*

transports the captured bees, sometimes over a long distance, to her nest, which is a long underground tunnel [1189]. There she places the bees in brood cells off the main tunnel, and lays an egg in each cell. The bees in a brood cell will eventually be eaten by the wasp's offspring when the little grub hatches out from the egg.

Some sand dunes were dotted with hundreds of burrows, each marked with a low mound of yellow sand that the female had transported to the surface when excavating her nest (Figure 4). Tinbergen noted that when a beewolf left her burrow to go bee hunting, she covered up the opening by raking sand over it, hiding it from view, and yet when she came back a half hour or an hour later carrying a paralyzed honeybee, she had no trouble distinguishing her nest from among the many in her neighborhood. By marking females with unique paint marks, Tinbergen verified that Two Red Dots, Yellow, and all the others each built and provisioned only one nest at a time.

The skill with which the marked beewolves darted back to their hidden tunnels intrigued and puzzled Tinbergen. How could they find their own homes among so many nests in a barren sand dune? The wasps provided a hint to a possible answer: when a female left her nest, particularly on the first flight of the day, she often took off slowly and looped over the nest, flying back and forth in arcs of ever-increasing length and height, a characteristic behavior of many ground-

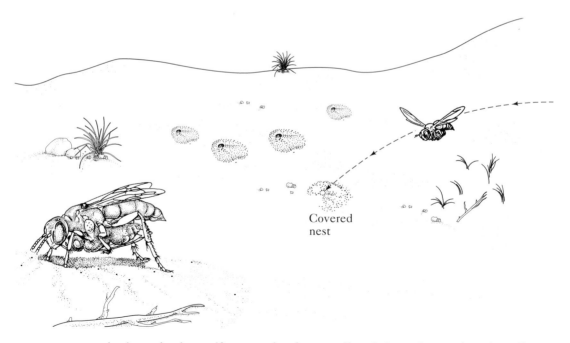

Covered
nest

4 A nesting beewolf wasp at her burrow. Female beewolves navigate long distances between their nests and hunting grounds, carrying back captured bees to place them in underground cells where their offspring will feed on the paralyzed prey.

nesting wasps and bees [1330]. After just a few seconds of this behavior, she abruptly turned and zipped off in a straight line (to the bee-hunting grounds a kilometer away, as it turned out). Tinbergen suspected that the wasps "actually took in the features of the burrow's surroundings while circling above" the nest entrance, and that by memorizing local landmarks, such as the sticks and pebbles scattered in the sand, females were able to find the nest entrance upon their return.

Tinbergen realized that if this **hypothesis**—this possible explanation—was correct, he ought to be able to make it hard for a female to relocate her nest by changing the local landmarks around the burrow. To find out whether he was correct, he waited until some females left to hunt for bees and then swept the areas around their burrows, being careful not to touch the mound of sand around the entrance while displacing the scattered tufts of grass, pebbles, and sticks the beewolves might use to orient themselves. His test showed that he was right about the wasps' dependence on visual cues. When prey-laden females came zooming back close to their nests within the landmark-free zone, they appeared confused, stopping in midair about 4 feet away from their nest mounds before circling out to repeat the approach, again and again. Only by dropping their prey and searching more or less at random on the ground were some females eventually able to find their nest entrances.

Tinbergen's simple experiment confirmed his suspicion that beewolves formed a visual image of the area immediately around their nests, which they used to pinpoint the covered entrances when coming back with prey. But he wanted to test his hypothesis in another way, to be sure that he had it right. As he wrote in his book, *Curious Naturalists* [1191], "The test I did next was again quite simple. If a wasp used landmarks it should be possible to do more than merely disturb her by throwing her beacons all over the place; I ought to be able to mislead her, to make her go to the wrong place, by moving the whole constellation of her landmarks over a certain distance." In other words, Tinbergen knew what he ought to see *if* he performed a particular kind of manipulation. When he did his experiment, he found to his delight that it worked like a charm. By carefully moving the entire set of local landmarks around a nest one foot to the southeast, he induced the returning female to land one foot to the southeast of her real nest entrance. When he shooed the wasp away, and then shifted all the "runway beacons" back to their original locations, the beewolf circled around and came down right at her nest entrance.

Tinbergen then did yet another experiment, this time to see whether he could make the wasps train themselves to landmarks that he provided. He put rings of pine cones around some nests while the nest owners were inside their burrows. When the wasps came out, they looped back and forth over the nest area before heading off to hunt, but then carried on with their work. Two days later, Tinbergen returned to displace the circles of pine cones short distances while the females were off hunting. If the wasps had learned the experimental landmarks, as Tinbergen expected they would, they should have been fooled into landing at the wrong place, within the displaced circle of pine cones, rather than at their nests. The experiment worked as Tinbergen thought it would (Figure 5)[1189].

5 Spatial learning by a beewolf. Tinbergen tested his hypothesis that female beewolves (A) learned the local visual landmarks around a burrow and (B) used this information to find the nest entrance when returning with prey from distant hunting grounds. *After Tinbergen [1189].*

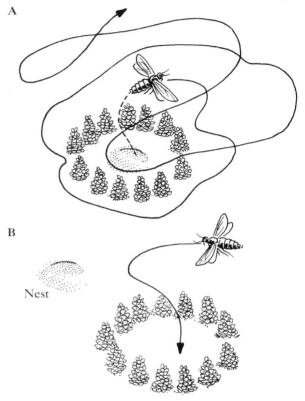

Tinbergen's research provides a beautiful example of the scientific method in action. First, he asked a question: How do beewolves home to their nest entrance? Then, based on his observations, he proposed a tentative answer, a hypothesis, to explain their homing abilities: beewolves possess a visual system that gathers and stores information about local landmarks in the nest area for use when returning to the nest entrance. To test this tentative answer, he tested logical expectations, or **predictions,** derived from the hypothesis: for example, experimental displacement of the local landmarks around a nest should cause a homing female to shift her landing site accordingly. This prediction, when tested, proved to be correct, which gave Tinbergen confidence that his original hypothesis was indeed right.

Thus, we can dissect the procedure Tinbergen followed into a series of steps. He started with (1) a **causal question** about a natural phenomenon, and then developed (2) a hypothesis to explain what he saw. Based on the hypothesis, he generated (3) predictions or expected results, which in turn were (4) subjected to tests, in which actual results were gathered for comparison with the expected ones, in order to (5) reach a conclusion about whether the hypothesis was right or wrong.

Tinbergen's conclusion would have been even stronger if he had systematically tested *several* plausible hypotheses, each with its own mutually exclusive predictions, but the point here is that he made real progress in understanding beewolf homing behavior, thanks to the simple, effective logic of scientific study.

Tinbergen's research is typical of most scientific investigations—not that most scientists formally label their questions, hypotheses, predictions, and planned tests before doing their research. The "scientific method" is familiar enough so that it can be used intuitively. In fact, all people, not just scientists, use it regularly. Just listen to Click and Clack, the mechanics on *Car Talk* on National Public Radio, as they try to figure out what is causing a 1987 Volvo station wagon to stall unexpectedly as its driver, Bill from Bedford, Massachusetts, motors down the highway.

Whatever you call it, the scientific method works; it improves our understanding of the causes of things. By formally identifying the difference between a hypothesis and a prediction, or a test and a conclusion, I think we can better understand how the process works. Clarity of thinking has much to recommend it. In my own research, when I began to write out my hypotheses in order to develop the predictions that followed from them, I found it easier to organize the tests that I would do and to defend the conclusions that I eventually reached.

Gulls and Eggshells

Tinbergen's study of beewolf homing is an example of research on the proximate causes of behavior. He answered a question about how a wasp's visual system helps her get back to her nest. Tinbergen showed that female beewolves respond to certain visual cues in the environment and that they store landmark information for use when homing. Without dissecting a single wasp, Tinbergen was able to tell us something about what goes on inside these creatures, something about the neural physiology of the insect. The focus on internal mechanisms defines his study as a proximate one.

The scientific process works just as well on questions about the ultimate causes of behavior, as we can see by looking at another of Tinbergen's famous studies, one on the black-headed gull's habit of removing the broken eggshells from its nest a short time after its youngsters have hatched. Although this action might seem trivial, since it takes only a few minutes for a gull to fly off and drop the eggshell a short distance away from the nest, Tinbergen was still curious about it. After all, while the gull is away from the nest, it leaves its babies unguarded, and the world of black-headed gulls is full of predators, including other bigger gulls, that would like nothing better than to dine on unprotected, newly hatched chicks. Tinbergen suspected that there must be some major reproductive benefit from removing the eggshells [1192], an argument reinforced by the finding that some other birds also exhibit the eggshell removal trait (Figure 6).

The underlying logic of this argument is worth examining more closely. In essence, Tinbergen assumed that once upon a time black-headed gulls (or the now

6   **Eggshell removal** by a bittern. This species, like the black-headed gull, removes eggshells from the nest soon after the camouflaged young have hatched. *Photograph by Eric Hosking.*

extinct species that eventually evolved into black-headed gulls) did *not* remove eggshells from their nests. Then, the first eggshell remover appeared, perhaps as a result of a genetic mutation that in some way affected the development of the gull's nervous system so that it reacted to eggshells in a particular way. We are not suggesting that this hypothetical new mutation coded for an entire behavior pattern, but rather that this gene played a role in the development of the mutant gull's nervous system (see Chapters 3 and 4). The new gene, with its distinctive developmental effect, could not have persisted in the species to the present *unless* the individual that first carried it reproduced successfully. If its descendants continued to have more surviving chicks on average than the then typical gulls, the mutant gene would have made itself more and more common. But if the reproductive success of eggshell removers was usually lower than that of gulls that ignored eggshells in their nests, the mutation underlying the new response would have disappeared from the species because its owners would have failed to pass it on.

In other words, Tinbergen assumed that eggshell removal evolved because of past differences among black-headed gulls in their behavior and reproductive success. The ultimate question he asked was, What might have caused the spread of eggshell removing in an ancestral population in which the behavior was once rare? He could not go back in time to study what happened in long-gone generations of black-headed gulls, but he could determine whether the behavior offered some reproductive advantage in current generations. If so, he could more plausibly claim that this advantage was also responsible for the spread of the trait in the past. If not, he would know that one possible explanation for the evolution of eggshell removal was probably wrong.

Tinbergen suggested that eggshell removal by today's black-headed gulls might be reproductively advantageous because it eliminated a visual cue that could give the nest away to certain predators. Black-headed gulls nest more or less out in the open, but they appear to try to camouflage the nest, hiding it in whatever vegetation is available. The eggs and chicks are camouflaged as well in coats of mottled brown and gray. In contrast, the white inner part of an opened gull eggshell is highly conspicuous, and might serve as a nest-identifying beacon to carrion crows and other predators if not removed by a parent gull.

To test the hypothesis that eggshell removal evolved to foil predators, Tinbergen first developed a prediction from the hypothesis: if eggshell removal is an antipredator device, then the presence of broken eggshells should help predators locate food. He checked the validity of this prediction by doing a simple experiment. He stole some intact gull eggs from nests in a colony of black-headed gulls and scattered them through sand dunes that were regularly patrolled by carrion crows, which eat gull eggs and chicks. He placed broken eggshells a short distance away from some of the unhatched, cryptically colored eggs; he dropped the eggshells farther away from other unhatched eggs. The eggs that were closest to white eggshell bits were more likely to be found and eaten by foraging crows than those that were farther away from the giveaway cue (Table 2). Since this finding matched the predicted result, Tinbergen concluded that eggshell removal by nesting black-headed gulls could have evolved because birds that happened to behave this way lost fewer offspring to predators than those that did not, and so spread the genetic basis for the behavior [1192].

Note that although this study concerned an evolutionary, or ultimate, question, the procedure for answering it did not differ fundamentally from the method Tinbergen used when he studied the proximate basis of homing behavior in the beewolf wasp. He began with (1) a question: why had black-headed gulls evolved a special response to the eggshells in their nests? He then generated (2) a speculative answer, a working hypothesis: maybe eggshell removal had spread in the past because it helped parent gulls hide their offspring from crows and other visu-

Table 2 Effect of the presence of an eggshell near a gull egg on the chance that the egg will be discovered and eaten by crows

Distance from eggshell to egg (cm)	Eggs taken by crows	Eggs not taken by crows	Percentage eaten
15	63	87	42
100	48	102	32
200	32	118	21

Source: Tinbergen [1192].

ally hunting predators. The hypothesis led logically to (3) a prediction: we would expect to see crows and other predators using the conspicuous cues offered by broken eggshells to narrow their search for unopened eggs and newly hatched chicks. Tinbergen performed (4) a test of the prediction, using an experiment to find out what crows really did and then matching his result against the predicted one. From his test, he reached (5) a **scientific conclusion,** namely, that his original working hypothesis was probably right.

Darwinian Theory and Ultimate Hypotheses

When Tinbergen developed his hypothesis about the evolved function of eggshell removal, he was strongly influenced by the theory of evolution by natural selection, which has had an enormous effect on the way all biologists have explored ultimate questions since 1859. In that year, Charles Darwin (Figure 7) published *On the Origin of Species* [263], with its sweeping explanation of how evolutionary change might occur. Darwin's great idea rests on three commonly observed features of living things:

1. *Variation:* Members of a species may differ in their characteristics (Figure 8).
2. *Heredity:* Parents can sometimes pass on their distinctive characteristics to their offspring.
3. *Differential reproduction:* Because of their special inherited characteristics, some individuals within a population leave more offspring than others.

It was Darwin's genius to perceive that evolutionary change is *inevitable* when these three conditions occur in a species. If some black-headed gulls reproduce

7 Charles Darwin
shortly after returning
from his around-the-world
voyage on the *Beagle* (left),
and at end of his life, long
after having written *On the
Origin of Species* (right).
*Courtesy of the Darwin
Museum at Down House.*

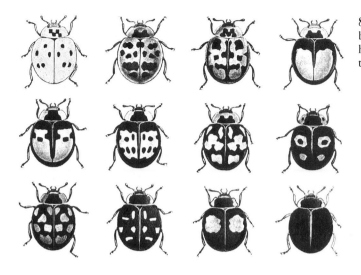

8 A variable species. The ladybird beetle *Harmonia axyridis* exhibits hereditary variation in the color patterns of its wing covers.

more than others, and if their offspring inherit a trait (such as eggshell removal) that advances successful reproduction, then they will also spread the reproduction-enhancing trait (Figure 9). The other side of the coin is equally clear: if some gulls leave fewer offspring than others because of their inherited characteristics, their surviving offspring will inherit the disadvantage and leave few progeny themselves. As a result, traits that compromise reproduction will become progressively rarer over evolutionary time. Darwin called this process **natural selection**

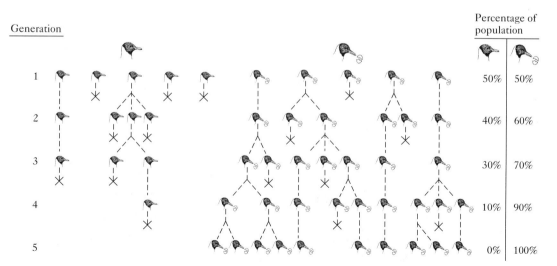

9 Natural selection. If gulls that ignore eggshells in their nests leave fewer surviving offspring *on average* than gulls that remove eggshells, and if these behavioral traits are inherited, eventually only eggshell-removing gulls will appear in the population.

because he saw the elimination of unfavorable traits and the spread of beneficial ones as the natural consequence of individual variation in reproductive success, as measured by the number of reproducing offspring left by an individual. Thus, Darwinian logic leads us to expect that evolutionary change will be in the direction that promotes individual reproductive success.

Darwin developed the theory of natural selection before critical discoveries about the nature of heredity had been made. Genes are now known to be nucleic acids that faithfully encode the information needed for the synthesis of proteins, the building blocks of all living things. Genes are copied and transmitted to offspring when organisms reproduce. In some sense, the genes are what is reproducing, with the organism merely acting as a mortal vehicle for the process. Therefore, we can restate Darwinian logic in genetic terms in much the same way that Darwin described it in relation to individuals, which will help us better understand the modern evolutionary approach to animal behavior [277]:

1. *Genetic variation:* Genes can occur in two or more alternative forms, or **alleles,** within a species. Different alleles lead to the production of slightly different forms of the same protein.
2. *Heredity:* By definition, alleles can be transmitted from parent to offspring.
3. *Differential reproduction:* Some alleles produce effects that usually cause their bearers to replicate them more often than other individuals bearing different alleles of the gene in question.

If these three conditions apply, then reproductively "successful" alleles will become more common in the population. Other forms of these genes may be completely replaced if the relationship between genetic differences and reproductive success remains constant long enough. (We assume that populations cannot grow exponentially forever, so that only a finite number of copies of a gene can exist at any one time.)

The logical conclusion is that selection on individuals will favor alleles that help build bodies that are unusually good at promoting the propagation of those alleles, or as E. O. Wilson puts it, a chicken is really the chicken genes' way of making more copies of themselves [1294].

Darwinian logic and the study of behavior

No matter how the logic of natural selection is presented, it is a blockbuster of an idea. If the concept is valid, it means that humans and all other living things have been designed by past selection to reproduce and pass on their genes; this is the biological significance of life. An understanding of this point *helps us identify questions worth answering while at the same time shaping the kinds of hypotheses that we will test.*

We will illustrate the utility of Darwinian theory with the case of Hanuman langurs. These graceful primates live in bands, which usually consist of one large, reproductively active male and a group of smaller adult females and their offspring (Figure 10). From time to time, the resident male is pushed out of the group

10 A band of Hanuman langur females and their offspring. Males fight to monopolize sexual access to the females in groups like this. *Photograph by Sarah Hrdy-Blaffer/Anthro-Photo.*

by a newcomer, usually after a series of violent clashes. After such a takeover, infants tend to die. Although the cause of death is often unclear [1134, 1157], the new male is the prime suspect in many instances, and has been seen doing the deed several times, even though females try to protect their infants (Figure 11) [578, 579].

The phenomenon of infanticide is precisely the sort of thing likely to attract the attention of a Darwinian biologist. Why should a male langur that has just spent days in a dangerous running battle with the previous resident turn on the very females he has finally succeeded in joining? Having avoided incapacitating injury during his battles with the other male, why should he now run the risk of being bitten by one of his new female companions, with all the attendant risks of infection? In other words, how can infanticide possibly advance a male's reproductive success? This question jumps out at an evolutionary biologist precisely because infanticide by the male seems so unlikely to have evolved by natural selection.

Of course, it is possible that the behavior is not an evolved trait and therefore does not contribute to the male's reproductive success, but is instead a social pathology brought on by overcrowding. Under high-density conditions, males may encounter each other so often that fighting becomes commonplace, with the hyperaggressive males then assaulting not just rivals of the same sex, but females and their offspring as well. In fact, this nonadaptive or incidental effect hypothesis was the first explanation offered by some langur watchers, who knew that langurs were often fed by Indian villagers and so could have reached unnaturally high population densities.

But another student of langurs, Sarah Hrdy, still wondered whether infanticidal behavior by males could have evolved by natural selection, despite its appar-

11 An infant-killing male langur flees from a female whose baby he has fatally wounded. *Photograph by Volker Sommer, from Sommer [1134].*

ent lack of reproductive value for the male. She thought that perhaps infanticide had evolved as a result of sexual competition between males [578]. A killer male might increase his reproductive success if he eliminated nursing offspring sired by a rival male, leaving the mothers of those infants no option other than to mate with him and have his babies.

For the moment, let us not worry about whether the sexual competition hypothesis is right or wrong. The point is that this answer to the puzzle of infanticide derives from Darwinian theory. Hrdy is suggesting that males tend to gain descendants through the selective practice of infanticide. This suggestion differs fundamentally from the social pathology hypothesis, which states that killer males gain nothing from infanticide because it is an abnormal behavior induced by unnatural conditions.

Now consider yet another hypothesis for infanticide: Killer males have evolved in order to help prevent overpopulation, which would threaten the survival of groups of Hanuman langurs by destroying the food base on which they depend. Although this hypothesis also claims that the behavior spread in the past because of its beneficial consequences, it is *not* based on Darwinian theory. Here the beneficiary of infanticide is not the male that kills infants, but the group to which he belongs. In other words, the evolutionary mechanism for the spread of infanticide has nothing to do with differences among individuals in their reproductive success (Darwinian natural selection), but rather with differences among groups in their survival (via a process that has been called **group selection**).

The "for-the-good-of-the-group" argument received its formal presentation in *Animal Dispersion in Relation to Social Behaviour* [1311], written in 1962 by V. C. Wynne-Edwards. According to Wynne-Edwards, only those groups—and a "group" can be an entire species—that possessed population-regulating mechanisms could have survived to the present; others that lacked these mechanisms had surely become extinct through overexploitation of the critical resources on which they depended. With groups thus competing unconsciously to survive, only those whose members sacrificed potential reproductive output for the benefit of their population would be likely to persist. Thus, Wynne-Edwards argued that evolutionary change regularly occurs because of differences among groups in their possession of self-sacrificing individuals, which in turn affects their survival chances.

This argument was challenged by G. C. Williams in his book *Adaptation and Natural Selection* [1287]. Williams showed that the survival of alternative alleles was much more likely to be determined by reproductive differences among genetically different *individuals* than by survival differences among groups. The basis for this claim can be illustrated with reference to langurs. Imagine that in the past there really were male langurs prepared to risk their own lives to reduce the population for the long-term benefit of their group. In such a case, group selection, as defined by Wynne-Edwards, would be said to favor the allele(s) for male risk-taking because the group as a whole would benefit from the removal of excess infants. However, in a population of langurs with some risk-taking infanticidal males, Darwinian natural selection would also be at work, provided that there were two genetically distinct types of males: the infanticidal type, which spent its energy and sometimes shortened its life for the good for the group, and another type, which lived longer and reproduced more because it let the infanticidal males carry the burden of population reduction. Now which of these two types would constitute more of the next generation? If the two types of langur males were genetically distinct, whose hereditary material would be more often transmitted to following generations? What would happen over evolutionary time to infanticidal tendencies in our hypothetical population of langurs?

The general point that Williams made is that selection acting on differences among variant individuals within a population will usually have a much stronger evolutionary effect than selection acting on differences among entire groups. Group selection can occur, provided that groups retain their integrity for long periods and differ in genetic constitution in ways that affect their survival chances. But if group selection favors a trait, such as reproductive self-sacrifice, while natural selection acts against it, natural selection seems likely to win, as we have just seen in our hypothetical langur example. Although research continues on forms of group selection more complex than that proposed by Wynne-Edwards [36], almost all behavioral biologists have accepted the arguments of Williams. Researchers are now very careful to distinguish between group benefit and individual (or gene) benefit hypotheses. Today most persons interested in ultimate causation use Darwinian theory to develop testable hypotheses about the possible reproductive value of a trait to individuals.

However, we can still test non-Darwinian hypotheses using the logic of the scientific method that we illustrated with Tinbergen's studies. The social pathology hypothesis (which is neither Darwinian nor group selectionist) and the population regulation hypothesis (which is based on the theory of group selection) for infanticide by male langurs both happen to yield the same key prediction: if high population density really does cause abnormal behavior (the social pathology hypothesis), or if it truly threatens the survival of langur groups and so stimulates self-sacrificing infanticide (the population regulation hypothesis), then we would expect to see infanticide by males *only* in areas in which the Hanuman langur populations are abnormally or unusually high. Contrary to this prediction, infanticide regularly occurs in troops living at moderate densities in natural areas where they are not fed by people [861, 1157], a finding that weakens our confidence in both the social pathology and population regulation hypotheses in this particular case.

Testing Alternative Hypotheses

Does our tentative rejection of the two non-Darwinian hypotheses mean that Hrdy's Darwinian explanation is correct? Clearly not. First, studies of langurs living under presumably natural conditions are very few, so the evidence against the social pathology and population regulation hypotheses is not utterly compelling.

Second, the sexual competition hypothesis is not the only possible explanation based on Darwinian theory. For example, perhaps males commit infanticide after takeovers in order to replenish dangerously depleted energy reserves by cannibalizing infants. If so, killer males could derive benefits from their actions that could keep them alive and help them reproduce more than males that did not kill and consume youngsters in their new bands.

In order to feel confident that we have identified the true ultimate cause of infanticide, we will have to do two things: (1) generate a full list of plausible alternative hypotheses and (2) test them in a way that insures we can reject the incorrect ones and retain the right explanation if we have included it in our list. When it comes to testing, note that more than one hypothesis can lead to the same prediction, as we just saw when examining the two non-Darwinian hypotheses on infanticide, both of which yielded the same prediction: infanticide should be limited to high-density populations. If we had found that males killed infants only when populations had increased markedly, we would not have been able to accept one hypothesis over the other.

Likewise, the sexual competition hypothesis and the cannibalism hypothesis both produce the prediction that infanticide by males will tend to occur shortly after takeovers. In the first case, we expect males to stop killing infants by the time the first of their own offspring are born to the resident females. In the second case, we expect males to stop killing infants after they have recovered from their energetically expensive takeovers. The fact that infanticide is indeed limited to the period soon after a takeover does not help us decide between these two alternative Darwinian explanations for the behavior. However, if the cannibalism hypothesis is true, we should sometimes see male newcomers eating an infant. No such records exist, but remember that observations of males in the act of infanticide are rare. If cannibalism really does not occur in this species, the adaptive cannibalism argument cannot apply [578, 1135].

The sexual competition hypothesis produces the prediction that females deprived of their young infants will quickly become sexually receptive to the males that killed their offspring. In many mammals, females that are producing milk for a young infant no longer ovulate and so cannot become pregnant. Once they stop lactating, however, they soon resume the ovulatory cycle and can conceive. As expected from the sexual competition hypothesis, langur mothers that have lost their infants to the new male quickly start cycling again and regain their sexual receptivity; as a result, they soon replace their dead babies with new ones fathered by the male responsible for the deaths of their previous infants. This finding increases confidence in the sexual competition hypothesis [1134, 1135].

But the more tests, the better. If sexual competition has led to the evolution of infanticide in Hanuman langurs because males that kill infants gain sexual access to females sooner, then we would expect to observe infanticide in many other species, provided that the same set of conditions apply. This prediction has now been confirmed through studies of a host of other animals in which newcomer males kill infants fathered by other males and then mate with the females that have lost their young, reaping a reproductive benefit from their murderous behavior [519, 689, 1214].

For example, infanticide often occurs when a group of male lions ousts the resident males from a pride containing a number of females with young cubs [966]. The incoming males hunt down cubs less than 9 months old and try to kill them (Figure 12), although, as is true for langurs, lionesses try (sometimes successfully) to protect their cubs. Lionesses that keep their cubs alive give birth at 2-year intervals, but females unable to prevent infanticide resume sexual cycling and mate with the killers of their offspring. Since a male can expect to remain in a pride and have access to its females for just 2 years on average, the reproductive benefits of infanticide from the male's perspective are evident. Male lions probably kill a quarter of all the cubs that die in their first year in some populations [966]. That several animal species commit infanticide under certain predictable conditions makes the sexual competition hypothesis more likely to be true.

If these conditions favor the evolution of infanticide by males, then infanticide should be practiced by females when they can use it to gain access to males that will care for their offspring [353, 588, 643]. This prediction has been confirmed for a giant water bug whose males take care of egg masses; these egg masses are sometimes attacked by egg-stabbing females. After the destruction of a clutch of eggs,

12 **Infanticide by a male lion.** *Photograph by George Schaller.*

the male egg nurse goes off with the infanticidal female to mate with her and care for her eggs.

Likewise, in the jacana, a tropical bird species, only males care for the eggs and young. Territory-defending females sometimes attack the chicks of neighbors, forcing the brooding male to abandon these offspring. He may then accept a new clutch of eggs from the infanticidal female. When Stephen Emlen and Natalie Demong experimentally removed some territorial females, nearby females quickly invaded the vacated territories and, in three of four cases, either killed the baby jacanas there or forced them to flee the area. Within 48 hours, the males that had lost their offspring were involved in sexual liaisons with the infanticidal females. By committing infanticide, these females had gained caretakers for their future clutches of eggs sooner than if they had waited for the males to finish rearing their current broods [353].

Certainty and Science

My summary of the tests on infanticide by male langurs surely suggests that the sexual competition hypothesis is right. I think it is—but I could be wrong. Some researchers have had different views about the evolution of infanticide in Hanuman langurs [101]. In fact, scientific conclusions in any discipline are always tentative. Room has to be left for a reversal of opinion because a previously unconsidered hypothesis may come down the pike, or new data may surface that destroys an established hypothesis. When I was a student at Amherst College, my paleontology professor convinced me that the continents of the earth have always been where they are now located. New evidence gathered since 1964, however, eventually persuaded almost everybody, including me, that the old view was wrong, and that the continents actually do "float" around the planet on moveable plates.

The rejection of established wisdom happens in science because scientists tend to be a skeptical lot, perhaps because special rewards go to those who show that what others believe to be correct, is not. Researchers constantly criticize one another's ideas, in good humor or otherwise, often causing their fellow scientists to have to change their minds. The uncertainty about Truth that scientists accept (at least when talking about other people's ideas) often makes nonscientists nervous, in part because scientific results are usually presented to the public as if they were the Ten Commandments. But anyone who has taken a look at the history of any scientific endeavor will learn that new ideas continually surface and old ones are regularly replaced or modified. Complete certainty, I repeat, is *never* achieved in science. The strength of science stems from the willingness of scientists to keep an open mind, to incorporate new insights, to test an idea repeatedly, and to throw out those hypotheses that fail their tests.

I hope that you will keep this point in mind as we review the findings of scientists and their interpretations of evidence in the chapters ahead. We will first examine the different components of a proximate analysis of behavior (in Chapters 2–6) before turning to ultimate questions about history and adaptive value (in

Chapters 7–15). The book concludes by applying an evolutionary approach to human behavior. Thanks to a small army of behavioral researchers, there is much to say on these topics, so let us get started.

SUMMARY

1. Basic questions about animal behavior fall into two categories. "How" questions ask about the proximate causes of behavior: How do genetic–developmental and sensory–motor–physiological mechanisms cause an individual to behave in particular ways? "Why" questions ask about the ultimate causes of behavior: Why have certain genes and certain proximate mechanisms persisted in species to the present, and why has evolution followed one path instead of another?

2. Both proximate and ultimate questions can be investigated scientifically by following these steps:

1. Pose a question about what causes something to happen: the causal question.
2. Devise a possible answer to the question: the hypothesis.
3. Develop a test of the hypothesis.
   a. Determine what one would expect to observe in nature if the hypothesis is true.
   b. Collect the appropriate observations to match against the prediction.
   c. Match the actual observations, data, or results against the predicted ones.
4. Reach a scientific conclusion: The hypothesis is tentatively rejected or tentatively accepted on the basis of the test(s).

3. The theory that a person uses affects the kinds of hypotheses he or she is likely to propose and test. Two major evolutionary theories exist. Charles Darwin proposed that evolutionary change occurs by natural selection if the species contains genetically different individuals whose special characteristics cause them to have different numbers of surviving offspring. In contrast, Wynne-Edwards proposed a group selection theory, arguing that evolutionary change occurs if genetic differences among groups cause them to differ significantly in their long-term survival chances.

4. Users of Darwinian theory formulate hypotheses on how traits might promote the survival of the genes of individuals with those traits; users of Wynne-Edwardsian group selection theory produce hypotheses on how traits might advance the survival of the group or species to which the individual belongs.

5. Today almost all behavioral biologists use Darwinian rather than Wynne-Edwardsian theory as the foundation for their hypotheses because selection at the level of individuals should be a more powerful force for evolutionary change than selection at the level of groups. Why? Because a trait favored by group selection may lead individuals to sacrifice their genes for the good of the group. If other members of the group have alternative traits that better propagate their genes, those genes will replace the ones that are being sacrificed for the benefit of the group.

6. The beauty of science lies in its logical approach to testing alternative hypotheses, whether proximate or ultimate, whether based on theory X or theory Y. Persons who use the scientific approach help eliminate explanations that fail their tests while retaining other hypotheses that have passed their tests.

SUGGESTED READING

Books that capture the sense of curiosity and excitement that biologists feel as they study the proximate basis of animal behavior include Vincent Dethier's *To Know A Fly* [287] and Kenneth Roeder's *Nerve Cells and Insect Behavior* [1008]. Niko Tinbergen's *Curious Naturalists* [1189] and Konrad Lorenz's *King Solomon's Ring* [737] bridge the gap between proximate and ultimate approaches, as do Howard Evans's *Life on a Little Known Planet* [366] and Michael Ryan's *The Túngara Frog* [1031].

For books that capture the delight of field studies of animal behavior, consider Evans's *Wasp Farm* [367], Bernd Heinrich's *In a Patch of Fireweed* [527], George Schaller's *The Year of the Gorilla* [1050], and Tinbergen's *The Herring Gull's World* [1190], as well as Shirley Strum's *Almost Human* [1154], Cynthia Moss's *Elephant Memories* [841], and *Journey to the Ants* by Bert Hölldobler and E. O. Wilson [562]. A realistic account of a high-profile field research project is given in Craig Packer's *Into Africa* [895]. The study of langur infanticide is described in *The Langurs of Abu* by Sarah Hrdy [579], while infanticide in general is reviewed in *Infanticide and Parental Care* [909]. Bernd Heinrich's superb *Ravens in Winter* offers an unusually clear picture of how scientists test alternative hypotheses [529]. For a provocative essay on the nature of science, read James Woodward and David Goodstein's essay [1306].

Charles Darwin had something useful to say about the logic of natural selection in *On the Origin of Species* [263]. G. C. Williams's classic *Adaptation and Natural Selection* [1287] thoroughly demolishes "for-the-good-of-the-species" arguments, but other forms of group selection are still vigorously debated [1292]. Richard Dawkins offers highly entertaining analyses of behavioral evolution in *The Selfish Gene* [277] and *The Blind Watchmaker* [280].

DISCUSSION QUESTIONS

1. Two people are talking about why humans eat so much candy and drink so many soft drinks. Which of the following explanations are proximate hypotheses, and which are ultimate hypotheses?

a. Candies contain sugar, which tastes sweet to people.

b. Sweet taste is remembered as good; the memory of pleasure leads people to seek more of the same.

c. The sugar in candy provides energy that helps keep people alive.

d. Our primate ancestors depended on sugar-rich fruits; we have inherited from them the same kind of taste receptors they had.

e. The genetic information in our bodies shapes the development of nerve cells that provide perceptions of sweetness and pleasure.

f. In the past, those individuals who liked sugar left more descendants than those who were indifferent to sweet-tasting foods.

g. The sensory input from taste receptors in the tongue to selected brain cells leads to a positively reinforcing sensation of sweetness.

2. Lemmings are small mouselike rodents that live in the Arctic tundra. They are known for extreme fluctuations in population size. At high population densities, large numbers leave their homes to travel long distances. In the course of their journey, many die, some by drowning in attempts to swim across lakes and rivers. One widely circulated explanation for their behavior is that these travelers are attempting to commit suicide to relieve pressure on their population. If some die, the survivors will have something left to eat. What theory is the foundation for this hypothesis? What would G. C. Williams have to say about it? How would he use Gary Larson's cartoon to make his point?

3. In Dawson's burrowing bee, males come in two sizes, large and small, with no intermediates [10]. The large males fight to mate with virgin females in places where the females are emerging from underground nests; the small males avoid fighting, and instead patrol flowers, where they sometimes find and copulate with females that have emerged without mating. You want to answer the question, Why do small male Dawson's burrowing bees behave differently than large ones when it comes to finding mates? The following list of statements contains both hypotheses and predictions. Identify which is which.

a. Small males should lose fights when they try to compete aggressively with larger individuals.

b. Small males search for and locate mates in areas that larger males ignore.

c. Small males use a mate-finding method that larger males are incapable of using.

d. Small males are avoiding direct competition with larger fighting rivals.

e. Small males will secure at least some receptive females in the areas in which they search.

f. Large males are expected to be unable to fly with the rapidity and agility of smaller males.

4. Organize the following statements into the following sequence: observation—hypothesis—prediction.

Example 1:

a. In the past, men were hunters who often traveled widely in search of prey, creating stronger selection on men than on women to be able to navigate long distances without getting lost.

b. In tests of spatial orientation abilities, men score higher than women on average in our society.

c. In species other than humans in which the sexes differ in distance traveled from a home base, the sex that travels farther should exhibit greater spatial orientation abilities.

Example 2:

a. The difference between males in tail length ought to be hereditary, at least in part.

b. Female preference for unusually long-tailed males enables the male offspring of choosy females to acquire genes that will make them attractive mates when they are mature.

c. In some bird species, females prefer the males that have the longest tails.

Example 3:
Construct another case of this sort involving bird song and hormones.

5. Donald Dewsbury [293] has rejected the proximate–ultimate classification scheme in favor of a three-part system: genesis, control, and consequences. "Genesis" refers to the origins of behavior, including the evolutionary origins and pathways leading to the behavior, as well as the effects of cultural transmission in shaping human behavior and all the other developmental effects on behavior generally. "Control" refers to the internal mechanisms that enable individuals to respond to their environment as well as the external stimuli that, once detected, cause certain actions to occur. "Consequences" refers to the changes for individuals that result from their behavior (as in learning), as well as its reproductive consequences, which can lead to evolutionary change. Can you place each of the components of this three-part system within a proximate–ultimate framework? Can you think of reasons to favor or reject this three-part system?

6. Construct a table in which you contrast the conditions required for evolutionary change via Darwinian natural selection and Wynne-Edwardsian group selection.

CHAPTER 2

The Proximate Causes of Behavior: Analyzing Communication

I BEGAN BIRD-WATCHING AVIDLY AT AGE SIX. The hobby is an addictive one, thanks to the challenge of identifying different species that can be very similar in appearance. Over the years, my world "life list" has grown to well over a thousand species, with the recent addition of an eared trogon in the Superstition Mountains of Arizona. The pleasure of the event was almost as great as that of putting the first bird on my list, a mallard duck on the mill pond near my childhood home in southeastern Pennsylvania. Early on, I learned from my father and other bird-watchers that some birds can be identified by their distinctive songs. To help me remember which species sang what, I was told that the song of the yellow warbler went something like "sweet-sweet-sweeter-than-sweet," in contrast to the song sparrow's "Madge-Madge-Madge, put on the tea kettle-ettle." These rough approximations helped me identify these birds when I heard them singing.

What is there about the yellow warbler that makes it sing its distinctive song? Biologists have produced both proximate and ultimate answers to this kind of question. This chapter will focus on studies of species-specific communication, especially bird vocalizations, to illustrate the different components of a proximate approach to behavior. (Later, in Chapters 7 and 8, we will turn to the ultimate aspects of the topic.) Here we shall argue that, on the one hand, a yellow warbler sings "sweet-sweet-sweeter-than-sweet" because of the way its genes and its environment interact in the development of its nervous system. But complementary to this type of analysis is research into the operating rules and design features of the neural mechanisms that control singing by an adult yellow warbler. Descriptions of both the developmental and neurophysiological causes of behavior are required for a complete picture of the proximate causes of behavior.

Species Differences in Behavior: Behavioral Development

Why do bird species differ in their communication signals? The ability to recognize bird species by their songs and calls comes in handy when bird-watchers hear a bird that they cannot see but would still like to identify. However, yellow warblers are not really communicating with bird-watchers. The evolved function of bird song offers fascinating evolutionary problems that we shall explore later in the book. For the moment, let's investigate the proximate mechanisms that control singing. From this perspective, if two bird species differ in their songs, there must be some difference in their internal physiology. We now know that the physiological machinery most involved in bird singing behavior includes a part of the bird's nervous system, which acts in concert with its vocal apparatus (the syrinx) and allied muscles to generate a complex pattern of sound waves that constitutes the bird's song [187].

The bird's neural song system and its syrinx develop as the fertilized egg, a single cell, gives rise to many millions of daughter cells, some of which are transformed into the specialized components of the adult bird's singing apparatus. From the start, the developmental process is regulated by a complex interaction between (1) the genetic information (the DNA) that the fertilized egg inherited from its parents and (2) the environment in which that information expresses itself, an environment that initially is composed of the proteins and other materials in the egg that help produce the body of a yellow warbler.

If this is how development works, then the differences in the songs of two species, such as the yellow warbler and the song sparrow, could arise because of differences in the proximate neural–muscular mechanisms that they possess. These differences could in turn be caused by differences in the genetic information possessed by the two species, or by differences in the environments in which their genetic information was expressed, with either factor leading to differences in the developmental pathways followed by members of the two species (Figure 1).

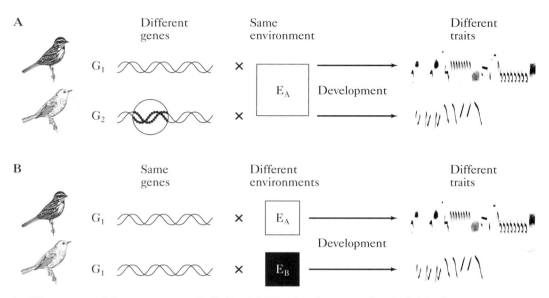

1   What causes differences among individuals? The development of an individual's traits depends on the interaction between the genetic information inherited from its parents and environmentally supplied materials. Therefore, differences among individuals can be caused by differences in either (A) their genes or (B) the environment.

The Development of Song Differences in Different Species of Birds

If this sketch is correct, we can start at ground level in exploring the possible proximate causes of communication signals by formally presenting two alternative hypotheses for the development of song differences among bird species.

> Hypothesis 1: The differences are caused by differences in the birds' DNA, their *genes*, which cause differences in the development of the internal mechanisms controlling their songs.
>
> Hypothesis 2: The differences are caused by differences in the birds' *environments*, the proteins in their cells, the foods in their diets, their acoustical experiences, everything external to the birds' DNA that might affect the development of their song systems.

Of course, it is entirely possible that a particular difference between two species could be caused by some combination of genetic and environmental differences, but for the sake of clarity, we shall first consider the two polar opposites. We can test these hypotheses in many different ways. For example, imagine that the differences between the communication signals of two species are caused solely by differences in the foods they receive and the acoustical environments in which they are reared. If we then reared the young of the two species on identical diets and gave them access to the same acoustical experiences, we ought to find that the differences between their signaling behaviors would evaporate. If we eliminate the differences in the environment, broadly defined, of young birds, the gene–*environment* interactions taking place as the birds develop should become more similar, and so should lead to similar outcomes. On the other hand, if the two species communicate differently only because they differ genetically, then members of the two species should develop different songs even in identical environments.

Two Australian ornithologists, Ian Rowley and Graeme Chapman, have collected evidence that enables us to test these competing predictions for two species of parrots, the galah, *Cacatua roseicapilla*, and the pink cockatoo, *Cacatua leadbeateri* [1024]. These big, noisy birds both squawk out a full spectrum of vocalizations, which differ substantially between the two species. For example, the begging calls that baby galahs use to communicate their hunger to their parents do not sound like the begging calls of pink cockatoos. Likewise, each species has its own alarm calls that adults produce as warnings when they are surprised by a brown falcon or some other predator. And the same goes for the contact calls that the parrots use to keep in acoustical touch with the other members of their flocks as foraging groups move through woodlands.

Because the ranges of the two species overlap, and because both parrots nest in holes in trees, it sometimes happens that the same tree hole attracts a nesting pair of galahs and a nesting pair of pink cockatoos. The two pairs may share the same tree hole for a short time without interacting because the birds do not get serious about incubating their eggs until the entire clutch is laid, a task that takes 3 or 4 days. Eventually, however, conflict erupts when both cockatoos and galahs try to incubate their eggs, which are both in the same nest. Invariably, the larger pink cockatoos drive the smaller galahs away; the victors then unwittingly incubate both their own eggs and those of the evicted galahs, whose babies are taken care of by their foster parents after the eggs hatch.

Rowley and Chapman realized that baby galahs reared by pink cockatoos constitute a wonderful natural experiment with the potential to determine the role of genetic versus environmental differences in shaping the different communication signals of the two species. The adopted baby galahs obviously retain their species' genes, but they share much the same environment as their pink cockatoo nestmates, both with respect to the food they receive and the communication signals they hear.

As it happened, the adopted baby galahs gave the characteristic begging calls of galahs when they were hungry; their pink cockatoo foster parents fed them anyway. After fledging, the cross-fostered galahs remained with their foster parents and nestmates in a pink cockatoo flock. The galahs gave typical galah alarm calls when coming across a dangerous predator, not the alarm calls of their cockatoo foster parents. But when it came to the contact calls that flocking parrots use to maintain flock cohesion, the adopted galahs sounded exactly like pink cockatoos (Figure 2).

Based on these findings, Rowley and Chapman concluded that the differences between the foster-reared galahs and pink cockatoos with respect to their begging and alarm calls arose largely because of differences in the genetic information con-

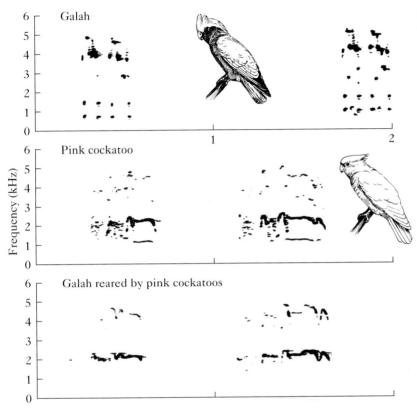

2 Differences in the contact calls of galahs reared by pink cockatoos and galahs reared by their own species are due to differences in their acoustical environment. The sonograms represent the contact calls of a galah, a pink cockatoo, and a galah reared by pink cockatoos (top to bottom). *Source: Rowley and Chapman [1024].*

tained in galah and pink cockatoo cells. However, because foster-reared galahs and naturally reared galahs produce very different contact calls, we can conclude that a difference in their environments caused the two kinds of galahs to develop different signals. Perhaps the key environmental factor was the response (or lack of it) by pink cockatoos to the initial contact calls given by the adopted galahs. When a galah gave a galah contact call, meaning roughly, "Here I am; where are you?," perhaps its pink cockatoo flockmates did not respond. But when the adopted galahs produced modified sounds that resembled the contact calls they were hearing from their pink cockatoo companions, they were answered, and were therefore better able to monitor the movements of the other birds in their flock. As a result, they came to mimic their companions' contact calls and to differ from fellow galahs that were reared by galah parents.

Please note that Rowley and Chapman's work [1024] does not support the conclusion that the alarm call of the galah is "genetic" while its contact call is "environmentally determined." Both genetic information and environment are involved in the developmental interactions that underlie the construction of every single element of a galah's (or pink cockatoo's) nervous and muscular systems. Without genetic information, the galah would lack the ability to make the battery of enzymes needed to regulate the myriad chemical reactions underlying its development. Likewise, without the environment, those enzymes could not be made, because they are constructed from amino acid building blocks that come from substances in the egg and later from foods consumed by the bird. Thus, both genes and environment play different but essential roles in the building of enzymes, which then become part of the cellular environment, where they influence the development of a galah's (or pink cockatoo's) nervous system.

The ability of galahs to give pink cockatoo contact calls gives us a hint about the operation of the galah's nervous system; namely, that it can incorporate information about how its companions react to its acoustical signals. Typically, young galahs travel with other galahs, hearing their flock members give the galah contact call and receiving feedback in the form of responses to their own contact calls. Galahs that have been reared by pink cockatoos, however, associate with pink cockatoos, in whose company they receive different auditory and social experiences, which apparently affect the gene–environment interactions underlying the development and operation of the call production system. As a result, foster-reared galahs learn to produce pink cockatoo contact calls.

The possibility that acoustical experience can have this kind of effect by selectively altering gene activity has been tested in some birds other than galahs. When a zebra finch, an attractive little Australian bird, hears the songs of its own species, a specific regulatory gene in a particular part of its brain becomes much more active. Over twice as much messenger RNA is transcribed from that gene following a 45-minute exposure to zebra finch song than following exposure to the songs of other species. (Messenger RNA is a molecule that is essential for converting the coded information in a gene into a specific protein.) In essence, the song of its own species is much more effective in "turning on" the gene in specific brain cells than are other acoustical stimuli [804]. Activity in this gene could change cell chemistry, changing the way in which the nerve cell operates, thereby producing changes in the mechanisms involved in song perception and song production (Color Plate 1).

In summary, if galahs grow up in a social environment dominated by their fellow galahs, gene–environment interactions occur that contribute to the development of the typical differences between galahs and cockatoos in their various calls. But change the social environment by having a galah reared by pink cockatoos, and the customary difference between the two species in their contact calls disappears. However, the experiences that occur in this environment do not override the role that genetic differences play in the development of distinctive begging and alarm calls in the two species. Thus, although the ability to produce each kind of signal depends on the production of specific physiological mechanisms, which develop during an interaction between the individual's genes and its environment, the differences between individuals in signaling behavior can sometimes be traced to either environmental or genetic differences among them (see Figure 1).

The Development of Song Differences in Different Species of Fruit Flies

Just as yellow warblers and song sparrows sing different songs, so too do many species of fruit flies. Laboratory geneticists are extremely fond of these very small flies, so much is now known about their reproductive behavior. When males court females, whether in a laboratory vial or on a piece of rotting fruit in nature, they vibrate their wings from time to time. In so doing, they create sound waves that females can hear, and that researchers can record with highly sensitive equipment. The recordings reveal complex trains of sound pulses, with patterns that vary among species.

For example, when the fruit fly *Drosophila melanogaster* "sings," it produces a long train of sound pulses, each pulse corresponding to a burst of wing vibration. An interval or pause, called the interpulse interval or IPI, occurs between any two pulses of wing vibration. The IPI averages about 30 milliseconds in *D. melanogaster,* but over a series of wing vibrations the pauses steadily grow longer, then shorter, then longer, then shorter, in a regular cycle lasting about 60 seconds (the so-called IPI period). Both the IPI and the IPI period differ from those of other species of *Drosophila,* with *D. simulans,* for example, having a mean IPI of roughly 50 milliseconds and a mean IPI period of about 35 seconds [1265].

In order to explain this difference between the two fly species, we can begin with our two familiar developmental hypotheses: The development of whatever mechanisms are responsible for song production in the two flies varies because the gene–environment interactions in the two species differ due to (1) different genes or (2) different environments.

Before we test these two extreme alternatives, we shall digress for a moment to focus on the timing of when fruit flies fly and walk around. In *D. melanogaster,* adults typically have an activity cycle that repeats itself every 24 hours. However, a very few individuals move about at random intervals, with no cyclic pattern at all, while others follow a much shorter than typical schedule of about 19 hours, and still others exhibit a lengthened 29-hour cycle.

Breeding experiments with flies of known genetic constitution have revealed that these behavioral differences stem from differences in a single gene, which comes in a variety of different forms or alleles. Individuals with the typical, or wild-type, *period* gene (*per*$^+$) have the 24-hour cycle, while flies endowed with two copies of the allele *per*^{o} exhibit no apparent activity rhythm. Flies with the *per*^{s} allele have the shortened activity cycle, while flies with *per*^{L} exhibit the longer-than-customary cycle (Figure 3) [68].

Using the techniques of molecular genetics, two research teams have shown that each of the mutant alleles differs from the typical form of the gene by just one pair of nucleotides in a chain of DNA composed of more than 3500 such pairs [67, 1320]. In the *per*^{o} form of the gene, the single alteration in the DNA molecule

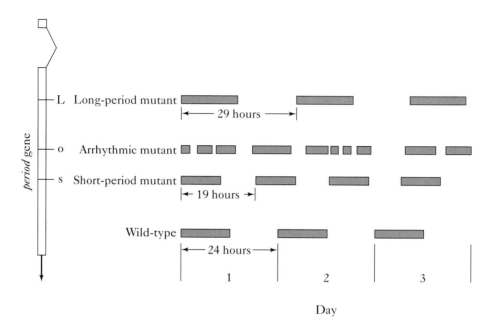

3   Mutations and behavior. The *period* gene of *Drosophila* fruit flies occurs in many forms. On the left is the strand of DNA that constitutes the *period* gene. Each mutant allele has a characteristic effect on the activity patterns of fruit flies, shown on the right (dark bars indicate when the flies are active). The normal pattern of nonmutant (wild-type) flies appears at the bottom of the diagram. *Source: Baylies et al. [68].*

results in the production of a protein chain that contains about 400 amino acids, instead of the 1200 appearing in the wild-type protein. This mutation causes the cell's machinery to stop "reading" the gene's information prematurely.

The information contained in the *per*ˢ and *per*ᴸ alleles is used by the fly's cells to make a protein chain with the full 1200 amino acids. However, the mutant chains differ from the wild-type protein by a single amino acid (a different substitution for each mutation). Even this small alteration causes dramatic differences in the activity patterns of the flies. If the *per*⁺ form of the *period* gene carries the information needed for the development of normal activity rhythms, then if we could insert this allele in arrhythmic individuals that had the *per*° allele, we should be able to restore the normal activity pattern.

Molecular geneticists are an ingenious lot, and they have discovered procedures for carrying out certain gene transplants of this sort. One trick is to transfer a fragment of a fruit fly chromosome that contains the *per*⁺ allele into a tiny viruslike entity called a *plasmid*, which has the capacity to insert this DNA fragment into the DNA contained in a cell of a recipient fly. By microinjecting a solution containing many such plasmids into an embryonic fruit fly with the *per*° genotype, one can endow the developing fly with genetic information that it originally lacked. When tested as adults, those individuals that have received the right bit of DNA are genetically transformed; they exhibit a wild-type activity rhythm, a result that supports the contention that the *per⁺* gene is critical for the development of this trait in fruit flies (Figure 4) [1329].

Now we can return to the songs of fruit flies, because as it turns out, males with different forms of the *per* gene sing different songs. Thus, the *per* gene has multiple effects on development, a common occurrence given that each gene codes for an enzyme that catalyzes a particular biochemical reaction; the same reaction often participates in the development of several traits (Figure 5). Males of *D. melanogaster* with the wild-type *per* gene sing the typical song of their species, with

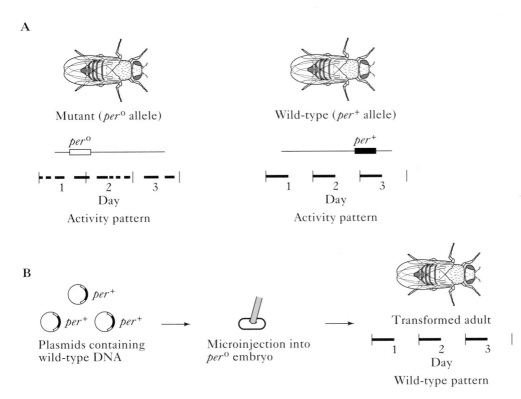

4 Genetic transformation experiment. (A) Mutant and wild-type flies that differ in the *period* gene differ in their activity patterns. (B) Wild-type DNA from fruit flies can be taken up in tiny plasmids, some of which will then carry the *per⁺* allele. The plasmids can be microinjected into the embryos of fruit flies with the *perᵒ* allele. These embryos would have developed into arrhythmic adults, but instead they are genetically transformed by receiving the *per⁺* allele and will exhibit the normal activity pattern of wild-type adults.

5 A gene with multiple effects. A single gene may have many effects on the development of an individual if the gene's information is used to produce an enzyme that participates in a biochemical reaction fundamental to the development of several traits.

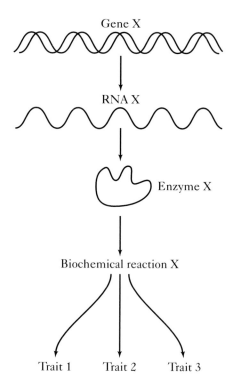

its characteristic IPI and IPI period. If differences in the *per* gene are responsible for differences in the songs sung by males within *D. melanogaster*, perhaps the distinctive songs of different species of fruit flies arise because the wild-type or typical form of the *per* gene differs from species to species.

This hypothesis has been tested in spectacular fashion by persons using the techniques of molecular biology to insert a copy of the wild-type *per* gene from *D. simulans* into embryos of another species, *D. melanogaster*, which carried the *per*° allele. Adult males of *D. melanogaster* endowed with the *per*° allele develop arrhythmic songs. If the wild-type *per* allele from *D. simulans* carries the information needed to restore song pattern, the transformed *D. melanogaster* males should have exhibited the typical *D. simulans* song cycle—and they did [1265]. When the reciprocal experiment was done, the expected result occurred: males of *D. simulans* that would have developed arrhythmic songs were induced to sing the typical song of *D. melanogaster* (Figure 6). These results confirm the hypothesis that the *differences* among species in male song occur because of the developmental effects of the various forms of the *per* gene.

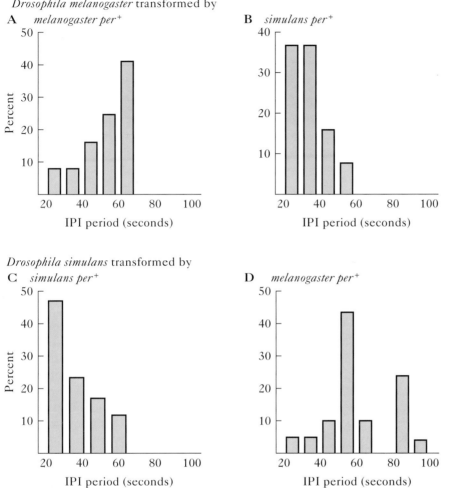

Drosophila melanogaster transformed by

A   *melanogaster per*⁺

B   *simulans per*⁺

Drosophila simulans transformed by

C   *simulans per*⁺

D   *melanogaster per*⁺

6   Genetic transformation across species. The graphs show the mean periods of the interpulse intervals (IPI) for four groups of genetically transformed flies, all of which had the *per*⁰ genotype prior to their genetic transformation. (A) Fruit flies of the species *Drosophila melanogaster* transformed by receipt of the wild-type *period* allele from *melanogaster* DNA. Note that 40 percent of the recorded IPI periods lasted more than 60 seconds. (B) Fruit flies of *D. melanogaster* that received a wild-type *period* allele from another species, *D. simulans.* Note that most IPI periods lasted less than 60 seconds for this group of transformed flies. (C) *D. simulans* transformed by the *simulans* wild-type *period* gene; these flies have the short IPI periods characteristic of their species. (D) In contrast, *D. simulans* that receive the wild-type *period* allele from *D. melanogaster* have much longer IPI periods on average. *Source: Wheeler et al. [1265].*

Sex Differences in Behavior: Neural Mechanisms

In answering the question, why do bird species produce different vocalizations? we first focused on the developmental differences in gene–environment interactions that cause the galah and pink cockatoo to produce different calls. But it is not just different species that differ in their calls; often male and female birds have very different singing abilities. Why? In dealing with this question, we will expand our proximate approach to look at how physiological mechanisms produced by gene–environment interactions cause the sexes to differ in their ability to produce a complex song.

We return to the zebra finch as an example of a bird species whose sexually mature males, but not females, produce a courtship song, which has been described as a series of "tya" notes connected by a chattering trill [589]. At the developmental level, one hypothesis for the difference between the sexes is that they differ in the genetic information they possess, and thus in the nature of the gene–environment interactions that occur in their developing bodies. Of course, it is also possible that the sexes differ because of differences in their dietary and acoustical environments as they are growing up.

Male and female zebra finches both start life as a single fertilized cell, which is surrounded by yolk and albumin and enclosed in an eggshell. The materials inside the shell provide what it takes for the fertilized cell to divide into many millions of descendant cells, each of which contains genes vital for structuring the growth and specialization of the tissues within the chick. The genetic information contained in the cells of male and female zebra finches (and all other birds) is not the same because the two sexes differ in the sex chromosomes they possess. Male birds have two Z chromosomes, whereas females have a Z and a W (unlike mammals, in which males have two different sex chromosomes, an X and a Y, while females have two X chromosomes).

Because chromosomes are where the genes are located, and because the W chromosome is known to have many fewer genes than the Z chromosome in birds, male and female birds are obviously genetically distinct. Differences in sex chromosomes cause a key developmental difference between embryonic male and female birds [813]. Female embryos with a W and a Z chromosome develop gonadal cells destined to become ovaries. Male embryos have two Z chromosomes, and thus a different genetic makeup, with the result that the very same cells develop differently, eventually forming sperm-producing testes [483].

Thus, from a very early stage, male and female gonadal cells develop differently, as these cells represent the product of different gene–environment interactions. Not only do gonadal cells in the two sexes follow different developmental pathways, toward sperm- versus egg-producing tissues, but they also differ early on in the kinds of chemicals they produce for transport to other cells. Most significantly, the pre-testicular cells manufacture the hormone estrogen, whereas the pre-ovarian cells do not. Therefore, estrogen is available for export to other cells in the very young male zebra finch, but not in the female. The estrogen that only male gonadal cells can produce becomes part of the chemical environment of other cells, notably those that will give rise to the nervous system. When estrogen reaches and enters some nerve cells, enzymes present in those cells, thanks to other gene–environment interactions, convert the hormone into a chemical similar to testosterone, which in turn generates yet another gene–environment interaction specific to males.

The Development of the Song System in Male and Female Birds

The cascading series of gene–environment interactions in a zebra finch culminates in the development of a special chain of neural elements that runs from the front of the brain to the spinal cord, where it connects with neural pathways to the syrinx, the organ that produces vocalizations. This network, the song system, grows rapidly in the male's brain during the first 40 days after hatching, while the number of cells in the corresponding parts of the female's brain declines through selective cell death (Figure 7) [291, 653]. As a result, the mature male has a brain that is structurally and functionally different from that of the female, enabling the male to sing a courtship song while the female zebra finch does not.

These conclusions about the development of physiological differences between the sexes were once hypotheses that had to be tested before they could be accepted. For example, if estrogen really is the critical self-manufactured environmental signal for masculinization of the young male's brain, then if we were to insert small pellets containing estrogen under the skin of a nestling female, her song system should grow, rather than shrink. Experimental application of estrogen to immature females does have the predicted effect, confirming its critical role in setting off the chain of events that leads to the development of a male song system in a zebra finch's brain [483].

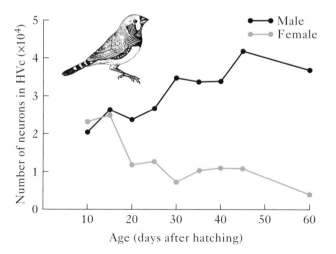

7 Changes in the song system of young zebra finches. The number of nerve cells in the female's HVc, a component of the song system, declines rapidly, whereas the number of HVc cells in males increases greatly between 10 and 40 days after hatching. *Source: Kirn and DeVoogd [653].*

Furthermore, the timing of estrogen implants has to match that of estrogen production in the male if the female's song system is to become completely masculinized. Brain development in nestling females is far less affected in individuals that receive experimental hormone implants 4 days after hatching than in those that receive estrogen treatment sooner [946]. The song system's development evidently follows a tight schedule. Only in recently hatched birds are brain cells highly sensitive to the testosterone derived from estrogen. With the passage of time, irreversible cell differentiation and nervous system organization take place, closing off possible avenues of development.

Even if a female zebra finch has been given estrogen in the proper dose at the proper time, she will not spontaneously sing the male courtship song—unless she receives an additional hormone implant as an adult. Although her brain has been masculinized, the song system requires testosterone to activate song production. The mature male's testes can, under the right set of environmental influences, manufacture more testosterone, which affects cells in the song system and motivates him to sing. The mature female lacks testes, and so cannot produce the chemical needed to turn on her song system, even if she has been earlier induced (via hormone treatment) to develop a male song system in her brain. But with experimentally supplied testosterone at the appropriate time, a female with a masculinized brain will sing like a male, having learned the song by listening to singing zebra finches earlier in her life (the role of learning in bird song will be examined shortly).

Based on this sketch of zebra finch developmental biology, how would you respond to someone who said that the courtship song of a male zebra finch was "genetically determined?" I hope that you would react calmly but firmly, pointing out that song depends on a neural song system, which could not have developed without both genetic control and environmental inputs. Furthermore, you might continue, any one gene–environment interaction may lead to a new chemical product that then becomes a part of the cellular environment and is available to interact directly or indirectly with genetic material in the cell, resulting in yet another chemical and possibly another novel gene–environment interaction, with still more developmental consequences. To say that courtship song is genetically determined, when it is an end product of an extraordinarily large number of these developmental events, all of them environment-dependent, simply does not make sense.

And how would you respond to someone who said that the song *differences* between normal adult male and normal adult female zebra finches were genetically determined? I hope that you would agree that they were, in the sense that differences in the genetic information contained within the sexes create differences in the gene–environment interactions that occur early on in the lives of males and females. The different substances that typical males and females can make then play their roles in the internal environment of the developing body, with estrogen having a particularly significant organizational effect on the developmental pathway followed by an individual.

So are the differences between the sexes in their singing abilities also environmentally determined? By now you would be sure to say, of course. A difference in the genes of males and females gets the process started, but as their bodies grow, the presence or absence of key hormones is an environmental cause of the growth or decline of the song system, which affects the singing abilities of males and females.

Differences in the Response of Male and Female Birds to Song

In addition to the general rule that male songbirds sing and females do not, the sexes also respond differently to a song that they hear a member of their species singing. Not surprisingly, it is the female, not the male, that sometimes reacts by inviting the singing individual, a male, to copulate with her.

Consider an adult female canary in breeding condition as she listens to a host of sounds, among them the songs of her own and other species of birds. Sexually receptive females react to canary song with a copulation solicitation display in which they crouch and raise the tail. They never respond in this way to the songs of male birds of other species, suggesting that they possess a special song-perceiving mechanism that enables them to recognize the song of male canaries.

Eliot Brenowitz tested the hypothesis that a special neural mechanism plays a critical role in song perception by female canaries [122]. Although, as we have seen, the song systems of males and females differ, females have most of the same brain nuclei (anatomically distinct aggregates of nerve cells) that make up the song system in males. These include the caudal nucleus of the ventral hyperstriatum (HVc), which is essential for song production by male canaries and also contains cells that fire when a male hears a recording of his own song. Brenowitz wondered, therefore, if the HVc might promote song *perception* in female canaries. If it did, then selective destruction of the HVc should produce females that could not "tell the difference" between the songs of their own species and those of other birds.

Brenowitz was able to cause small lesions in the brains of some females by directing a damaging amount of electricity through an electrode implanted in the HVc. After this operation, the females were given opportunities to respond to taped songs of canaries and white-crowned sparrows. The females could still hear sounds, but they reacted with copulation solicitation displays to white-crowned sparrow songs as avidly as to canary songs (Figure 8). Brenowitz concluded that an intact HVc enables the female canary to identify and react appropriately to canary song.

Individual Differences in Behavior: Proximate Levels of Analysis

Differences in songs occur not only among bird species, but also within some species whose members sing variants of their species-typical song in different parts of their range. In the white-crowned sparrow, for example, males living even a short distance apart may put their own distinctive twist on the basic song pattern, creating a geographic mosaic of what are called "dialects." In fact, any moderately

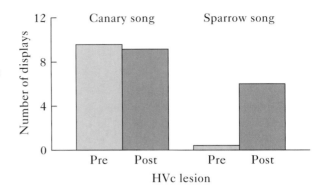

8 Song discrimination by female canaries. The HVc region of the female's song system helps her discriminate between the song of male canaries and the song of other species. After destruction of the HVc, female canaries will perform copulatory displays to tapes of white-crowned sparrow song. *Source: Brenowitz [122].*

experienced bird-watcher from San Francisco can learn to tell the difference between the dialect of white-crowns from Marin, north of San Francisco Bay, and the dialect sung by males from Sunset Beach, 50 miles farther south (Figure 9). By examining the proximate causes of dialects in this sparrow, we can consolidate our understanding of how developmental processes can lead to different physiological systems, and thus different behavioral outcomes [772].

We begin with our two standard alternative developmental hypotheses for behavioral differences: these song differences might be caused (1) by differences in the males' hereditary makeup or (2) by differences in the environments that the males experience. Both are legitimate possibilities, because some bird species are known to form subpopulations with hereditarily distinctive song types, whereas other species are geographically subdivided into groups whose members are able to learn the song variant of a group other than their own [54]. But please note that the claim that a male white-crown sings the Sunset Beach song type because the trait is genetically determined (or environmentally determined) cannot be correct. The song of a male depends on both his genes and his environment, as the zebra finch story illustrated. However, the *differences* between two males' songs are the product of gene–environment interactions, which could differ because the males have different genes, or because their environments are not identical. Once again, a combination of genetic and environmental differences could also produce different behavioral outcomes.

How might we test the gene versus environment alternatives for song differences among male white-crowns? If the Sunset Beach males belong to a genetically distinct population whose distinctive heredity promotes the development of their particular kind of song, then if we hand-reared nestling males from this population in a laboratory, we would expect them to sing the Sunset Beach dialect when they grew up, whereas Marin white-crowns treated exactly the same way should grow up to sing the Marin dialect.

Peter Marler has done many experiments of this type, taking eggs from white-crowned sparrow nests into the laboratory, where individuals can be raised to adult-

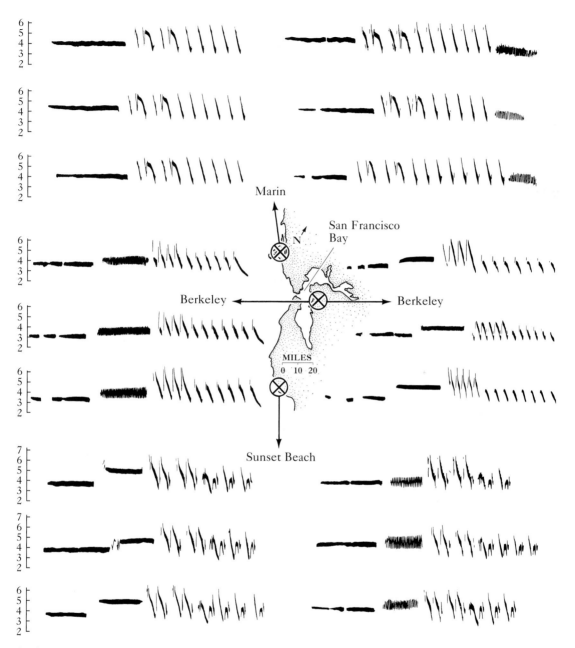

9   Song dialects in white-crowned sparrows from Marin, Berkeley, and Sunset Beach, California. Each dialect is different, and males in each location also have their own distinctive song, as revealed in these sound spectrograms of six singers from each location. *Sonograms courtesy of Peter Marler.*

hood in soundproof chambers. The song development of these birds follows a revealing pattern. A young male is usually several months old before he produces his first "song," which is usually a twittering subsong that has only a vague similarity to a mature male's full song. The bird continues to sing as it grows older, but its song never becomes a typical white-crown male's courtship–territorial song, let alone a particular dialect [770].

Clearly something critical is missing from the hand-reared bird's environment. Marler hypothesized that the missing ingredient was the chance to hear adult male white-crowns sing their songs. Without this acoustical experience, the isolated young male might lack the information needed to complete the development of his territorial song. If this hypothesis was correct, then a young male isolated in his soundproof chamber but exposed to tapes of white-crowned sparrow song ought to be able to sing a proper full song eventually. And this is exactly what happened with 10- to 50-day-old subjects that listened to tutor tapes of white-crowned sparrow song. The isolated birds subsequently reproduced perfectly the song they had heard on tape after going through the subsong phase. If a young male had listened to a Sunset Beach dialect, he would mimic that song; if he had instead been treated to a steady diet of Marin song on his tapes, he would sing that dialect by the time he was 200 days old.

These results offer powerful support for the environmental differences hypothesis for why male white-crowns from different areas sing their own dialects. Young birds that grow up near Marin hear only the Marin dialect as sung by older males in their neighborhood. They store the acoustical information they acquire from their "tutors" and later match their own song output against their memories of what the song sounded like, eventually coming to duplicate their tutors' song type.

Additional laboratory experiments with isolated birds and tutor tapes led Marler to the following conclusions about the role of the environment in song acquisition in the white-crowned sparrow:

1. A male white-crowned sparrow cannot acquire the song of any bird species except his own. Isolated white-crowns subjected to tapes of song sparrows develop aberrant songs that generally resemble those of birds that have had no songs of any sort played to them. If the bird hears a tape with both song sparrow and white-crowned sparrow songs on it, he will develop normal white-crowned sparrow song, learning the white-crown dialect from the tape while ignoring the song sparrow song [671].

2. A male must hear white-crown song from 10 to 50 days after hatching if he is ever to sing a normal full song. This is a **critical period** for song learning by the socially isolated, tape-tutored male.

3. During the interval from 150 to 200 days after hatching, when the young bird begins to produce subsong, he must be able to hear himself sing if he is ever to produce normal full song [670]. If he is deafened before subsong begins, he cannot match his vocal output with the memory of his species

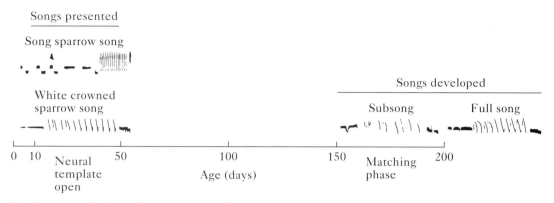

10 Song learning hypothesis based on laboratory experiments with white-crowned sparrows. According to this hypothesis, young white-crowns have a critical period 10 to 50 days post-hatching when their neural systems can acquire information by listening to white-crown song, but not from any other species' song. Later in life, the bird matches its own subsong with its memory of the tutor's song, and eventually imitates it perfectly—unless the young male is deafened. *Based on a diagram by Peter Marler.*

song acquired when he was 10 to 50 days old. As a result, development of the song is unguided, and the end product remains a highly abnormal twittering "subsong."

Marler interpreted the results of his experiments to mean that the gene–environment interactions that underlie the development of a male white-crowned sparrow's brain produce a highly specialized neural mechanism that can acquire acoustical information during an early critical or sensitive period in the juvenile male's life, but only from songsters of his own species (Figure 10) [770].

Social Experience and Song Acquisition in Songbirds

If Marler's interpretation is correct, then we can predict that male white-crowned sparrows will never be able to sing another species' song. Occasionally, however, ornithologists have heard white-crowned sparrows *in nature* singing like song sparrows. Observations of this sort caused Luis Baptista to wonder whether some other factor in addition to acoustical experience might influence the song development of white-crowns. Marler's famous tape-tutor experiments were all done with birds deprived of social interactions with adults. Perhaps social stimuli also have an effect on what a young white-crown learns about song.

Baptista and his colleague Lewis Petrinovich tested the social learning hypothesis by placing fledgling hand-reared white-crowns in cages near social tutors in the form of living adult song sparrows or strawberry finches (which are not even closely related to the sparrows). These young white-crowns learned their social tutor's song even though they could hear, but not see, other white-crowns (Figure 11)! Moreover, white-crowns that were older than 50 days when they first encoun-

11   Social tutors affect song development. A white-crowned sparrow that has been caged next to a strawberry finch will learn the song of its social tutor. (A) The song of a tutor strawberry finch; (B) the song of a sparrow caged nearby. The letters beneath the sonograms label the song syllables of the finch song and their counterparts in the song learned by the sparrow. *Sonograms courtesy of Luis Baptista.*

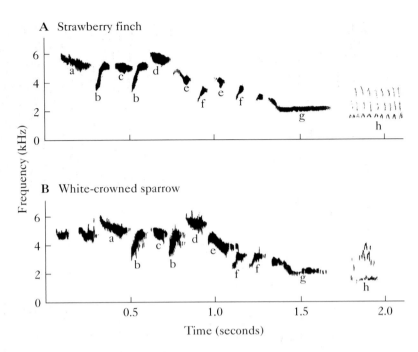

tered a social tutor could still learn the alien song, showing that the proposed critical period for song learning applied only to tape-tutored birds [58].

Researchers now know that the power of social experience in shaping singing behavior applies to many birds in addition to white-crowned sparrows. For example, captive starlings are adept at mimicking human speech, producing such phrases as "see you soon baboon" or "basic research" as well as the sounds of laughter or kissing or coughing. But they will do so only if they are hand-reared in a human household where they participate in the social life of their caregivers (Figure 12). Hand-reared starlings that lack a human social tutor but instead are housed in an aviary with other birds fail to mimic human sounds [1256], but will learn the songs of adult male starlings living with them [332].

In nature, young male starlings, zebra finches, and white-crowned sparrows receive both social and acoustical information from adult males of their own species, with both factors shaping the development of their songs. In a few species, the most effective social tutor may be the young bird's father [812]. Does the same hold true for white-crowned sparrows? To find out, Luis Baptista and Martin Morton recorded the songs of ten white-crown fathers in the Sierra Nevada of California, which is home to many geographically separate populations, each with its own dialect. They then compared the fathers' dialects with those of twelve of their male offspring that had been banded as nestlings and returned to breed in these mountains.

Only three adult sons sang the same dialect as their fathers. The others sang dialects characteristic of the males in neighboring territories. Baptista and Morton

12   Social effects on song learning. Kuro the starling learned to include words in his vocalizations because he had a close relationship with the family of Keigo Iizuka. *Photograph by Brigitte Nielsen, courtesy of K. Iizuka.*

concluded that although young male white-crowns may learn some song themes in their first few weeks of life, year-old males retain the flexibility to be influenced by their singing neighbors when they attempt to establish their first territories [57]. These neighbors provide critical social experience through their interactions with their young competitors. White-crowns with song acquisition mechanisms that can be influenced by these experiences will be able to match their songs with those of their immediate neighbors.

So, in proximate terms, why do male white-crowns sing different dialects? At the developmental level, the research of Marler and others shows that the differences in white-crown dialects are caused by an environmental difference in the sounds that males hear from other territorial males singing nearby. But these sounds have their developmental effects only because the brains of young males possess neural networks that are extraordinarily specialized for the task at hand. Thus, at the physiological level, behavioral differences in singing arise in white-crowns because they have specialized batteries of nerve cells that are "designed" to record information only from neighboring male white-crowns and, rarely, a few other social competitors as well. The learned information is used for a particular purpose, altering the vocal output of the listener, who must hear his own songs in order to match what he sings with his memories of the songs of territorial rivals and neighbors. This special kind of learning demands special physiological mechanisms in the bird's brain, which develops because of a complex and highly ordered array of gene–environment interactions.

Social Influences on Song Learning in a Brood Parasite

It is not just male white-crowned sparrows that sing geographically distinct types of songs. Male brown-headed cowbirds do too, and they represent an especially interesting case because cowbirds are **brood parasites** whose young are reared by adult red-eyed vireos, yellow warblers, bluebirds, and over 100 other host species, but never by members of their own species. If fledgling male cowbirds had brains capable of being influenced by the social and acoustical stimuli present in their immediate surroundings in their first months of life, one would expect that they would develop vireo- or warbler-influenced songs. But they do not. No matter what the foster species, male cowbirds sing only cowbird songs.

One of these songs, the flight whistle, varies from place to place. Adrian O'Loghlen and Stephen Rothstein have shown that young males rarely sing the local dialect at first, but by their third breeding season, the birds perfectly match their flight whistle "song" with that produced by the older males in the area [874]. In their first year, juvenile males rarely associate with adult male cowbirds, so perhaps they must wait before acquiring the acoustical experience they need to memorize the local dialect [873]. An alternative possibility is that young birds do learn the song, but don't dare sing it because to do so generates violent aggression from older male cowbirds [429].

Learning is clearly involved in acquisition of the flight whistle, but it is not so obvious that it plays a role in the development of the cowbird's "perched song." The perched song of a young male growing up in a laboratory cage with no social or acoustical companions of any sort is close to identical to this song as sung by wild cowbirds. The perched song, which is used both to intimidate rivals and to stimulate females [1019], appears resistant to the kinds of environmental influences that shape the songs of white-crown sparrows and starlings.

But two psychologists, Meredith West and Andrew King, found that certain kinds of experience subtly shape the development of this cowbird signal as well [649, 1257]. West and King suspected that something was odd upon discovering that adult females were *more* sexually stimulated by the perched song of males that had been reared in isolation than by the song of naturally reared males (Figure 13). This result indicated that the songs of wild birds must have been modified by experience of some sort.

The experience in question is aggression among the members of a flock of cowbirds, which forces free-living males to accept their place in a **dominance hierarchy.** In nature or in an aviary, one male in a cowbird flock physically dominates all the others, who give way to him when he approaches. The other males rank second, third, fourth, and so on in their ability to displace others in their group. The dominant male in a captive flock sings high-potency perched songs that effectively stimulate females to adopt the precopulatory display, enabling him to monopolize the sexually receptive females. The other males learn from the attacks and threats of the dominant male not to sing their most stimulating song, but to modify it so as to avoid being set upon or chased away. A socially isolated male does not have the chance to acquire the experience that teaches him to restrain his high-

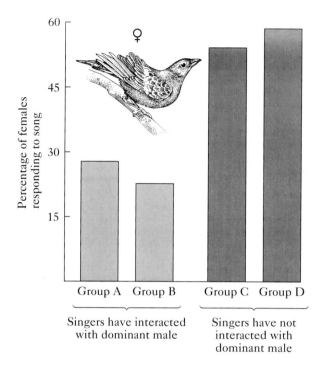

13   Song potency in cowbirds. Female cow-birds find the songs of males that have not been reared with other males more sexually stimulating than the songs of males reared under more typical conditions. Males in groups A and B had at least visual contact with a dominant male in another cage; males in groups C and D lived in partial or complete social isolation from other cowbirds. *Source: West et al. [1257].*

potency perched song until he becomes dominant, when it is safe to sing it [1257]. If naive, socially deprived males are introduced into aviary flocks with an established dominance hierarchy, these isolates persist in singing their high-potency songs, triggering violent assaults by the dominant male.

Social learning, male to male, clearly affects song development in young male cowbirds. But males might also modify their songs in response to the reaction they get from females. To test this hypothesis, West and King used two subspecies of cowbirds, geographically separated from each other and with slightly but consistently different perched song types. We shall call them subspecies A and B. When an A male sings his subspecies' song to a B female, she is far less likely to respond with the precopulatory crouch than when she hears the song of her own subspecies. When A males are given opportunities to interact only with B females, they gradually change their songs toward a pattern characteristic of B males. They do this even though they never hear the song of the B subspecies. Females do not sing, but they do react more positively to some variant songs than others (with a subtle wing flick display), and this social feedback modifies male singing behavior (Table 1) [650, 1255].

We have now briefly examined why certain bird species differ in their vocalizations, why the sexes of some birds differ in their singing behavior, and why males from different populations of white-crowned sparrows and cowbirds may sing somewhat different songs. We have, however, only touched upon the rich

Table 1 Effects of social experience with female cowbirds of two subspecies on the song development and potency of males belonging to subspecies A

Female companion	Song Development of Males		
	Number of notes in phrase 1 of song	Percentage of subspecies B song	Copulatory response by subspecies A females[a]
Subspecies A	4.8	0	54
Subspecies B	2.6	64	23

[a]Percentage of females giving precopulatory display to a recording of male song.
Sources: King and West [649]; West and King [1255].

diversity of developmental patterns that exist in the world of bird communication. For example, various flycatchers can acquire their vocal repertoire in the laboratory without any opportunity for learning from others [677]. Even within the white-crowned sparrow, different subspecies exhibit quite different patterns of imitation of tutor songs as well as different schedules of learning [856]. Thus, we have by no means covered all aspects of song development in birds, but rather have shown selected examples of how the topic can be explored at the developmental (gene–environment) level and the physiological (brain mechanisms) level. For a final case of this approach, let's turn to the development of human speech.

The Development of Human Speech

Humans, like white-crowned sparrows and cowbirds, produce different acoustical signals in different parts of their range, speaking on the order of 4000 languages in various parts of the globe [940]. At the developmental level, the language *differences* among people are largely or entirely environmental, the result of hearing different languages spoken in different places. Take a baby from English-speaking parents and rear that child in a place where it hears only Zulu or Spanish or Japanese, and the child will speak only Zulu or Spanish or Japanese. However, just as the ability of a white-crowned sparrow male to acquire a song dialect requires a very specialized song system, the ability of a human infant to acquire a language depends on some highly specialized elements of the human brain, which develop via the essential gene–environment interplay.

We can learn a great deal about the nature of our language mechanisms without inspecting a single brain (although anatomical studies have also revealed a great deal about the structures underlying language ability [432, 915, 1080]). For example, only about 40 speech sounds, or phonemes, are used to construct the thousands upon thousands of words in all human languages, and many of the same phonemes are employed in different languages. This finding suggested to some researchers that the brains of very young human infants might be "tuned in" to phonemes, enabling babies to detect and discriminate among the phonemes in the language sounds they were hearing around them. Such an ability ought to be very helpful in facilitating language learning. Imagine the difficulties you would have if you could not tell the difference between the sounds of the consonants "b" and "p," or "d" and "t." These are very similar sounds, but we have no trouble telling them apart—and, as it turns out, neither do babies.

Peter Eimas made this discovery by playing tapes of artificially synthesized strings of sounds that would be perceived by English-speaking adults as "ba . . . , ba . . . , ba . . ." to babies from English-speaking households, after first giving each infant subject a pacifier wired to record its sucking rate (Figure 14). The infants' sucking rates typically increased sharply when the tape started playing, but with repetition of the sound, the infants habituated to it, and their sucking rates fell. However, when a new syllable was introduced to the string, such as "pa," or even one featuring a novel, but very similar, consonant sound that Thai speakers use but that does not occur in English, the infants became aroused and their sucking rates promptly accelerated, indicating that they had perceived the difference between "ba" and "pa." Thus, the perceptual mechanisms needed for the subtle discrimination of speech sounds have already developed in the infant's brain by the time it is born [336].

Soon after birth, all young humans go through a stage in which they babble, producing a variety of unstructured sounds [770]. They seem to match the sounds they make against their stored memories of language sounds they have heard. Their ability to make fine discriminations among similar consonant and vowel sounds surely aids in this matching process. If a young child is congenitally deaf and cannot hear spoken language, it cannot acquire information about speech sounds. It will go through a babbling phase, but eventually will stop producing sounds altogeth-

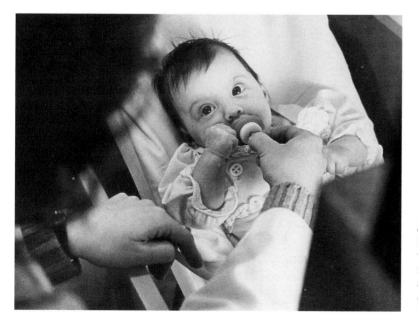

14 A language learning experiment. The infant has received a specially wired pacifier that will measure its sucking rate. *Photograph courtesy of Peter Eimas and Northwestern University.*

er. However, if such children are exposed to sign language users, they quickly develop facility in the use of this visual language, following the same developmental schedule shown by hearing children as they acquire a vocal language [802].

Children learn a language, whether English, Russian, Zulu, or American Sign Language, in a predictable manner [840]. The child first uses one-word sentences ("No." "Go." "Mama."), then always enters a two-word stage ("Dada come." "Bring me."). This stage endures for some time before the child enters the third phase: *telegraphic speech*, in which sentences consist of nouns without plurals and verbs without tense endings—speech stripped to its essentials. Finally, young children begin to elaborate their control over the grammar of speech (or sign language), acquiring the rules that give language its meaning.

Without a knowledge of these rules, language communication would be impossible. (Language knowledge these a communication without impossible be of rules.) My ability to communicate with you exists because English speakers use subjects, verbs, predicates, verb tenses, and prepositions grammatically. Young children can acquire the necessary grammar skills without ever receiving formal instruction in them. As a result, some persons have hypothesized that the human infant's brain is designed to extract grammatical rules just by listening to the speech of others.

This hypothesis generates a number of testable predictions. First, in a language with irregular verbs, children are expected to make consistent errors of a sort they never hear adults make but which are derived from rules that properly apply to regular verbs. Let me illustrate what I mean with an anecdote from an American friend who reared her children in Peru, where they became bilingual. One of her young children asked, "Mommy are my clothes planched?" The verb *planchar* means "to iron" in Spanish. The child had apparently modified this verb using the English rule of adding the suffix "-ed" to signify the past tense. The child had extracted this rule by listening to English-speaking persons; she did not have to have the rule explained to her, nor did she have a conscious revelation about past tense.

Errors of this sort, such as saying "goed" instead of "went," typically do not appear until a child is at least 20 months old. In fact, before this age, children slavishly imitate adult speech, rather than generalizing grammatical rules that they have somehow picked out of the speech sounds they hear around them. Thus, during the period between 20 to 36 months, the young child's brain apparently develops the special neuronal mechanisms that enable us to use language creatively, grammatically, rather than repeating memorized phrases or sentences [731]. The rare occurrence of individuals with *specific language impairment syndrome* (SLI) is consistent with this hypothesis. Persons with SLI have little difficulty remembering the past tense of irregular verbs that are often used (come–came, feel–felt), but they have great difficulty using the general -ed rule to convert regular verbs to their past tense forms. For example, when asked to convert the nonsense verb *zoop* to talk about something in the past, 83 percent of nonimpaired controls say *zooped*, but less than half of individuals diagnosed with SLI make the appropriate conversion [939].

Some persons have suggested that our "grammar-extracting" brain mechanism actively promotes adoption of certain grammatical rules over others [940]. If this is true, then we again expect special "lawful" errors to occur in children's speech that reflect this bias or preference for specific grammars. One possible grammatical predisposition involves a word order bias in favor of the subject–verb–object construction. This sequence is standard in English for declarative statements (e.g., "You like juice."), as it is in many other, but not all, languages. However, English is unusual in that word order is often modified when a question is asked instead of a statement made. For example, the question "Do you like juice?" places an auxiliary verb ("to do") in front of the subject "you." In most languages, questions involve a change in inflection only ("You like juice?"), not a change in word order.

Young children in English-speaking households have trouble with the word order shifts that occur in English-language questions. They often produce sentences such as "Where it is?" and "What you did with it?", sentences that they never heard an adult produce, but which maintain the standard subject–verb–object sequence. Errors of this sort suggest that the human brain is designed to facilitate adoption of a standard grammar [296]. The bias can be overridden through experience, but its existence tells us something about the operation of the brain's "language system," the batteries of nerve cells that contribute to our astonishing capacity to communicate via human speech. All the elements of this system develop through gene–environment interactions of staggering complexity, which means that although the differences between a person speaking Spanish and a person speaking Russian are caused by environmental differences, genetic information is vital in order for the environmental differences to have their effects on the development of this or any other behavior.

SUMMARY

1. At the proximate level, an animal's actions are caused by internal physiological mechanisms. In multicellular species, organized clusters of nerve cells are involved in every behavioral act, whether it is the song of a white-crowned sparrow or a sentence spoken by a human. Differences in physiological systems translate into behavioral differences. Female zebra finches do not have the same kind of brain mechanisms as males, which is why they do not sing like males.

2. The internal mechanisms controlling behavior are the product of a complex developmental process dependent on both genetic information and environmental inputs. Development of a bird's song cannot occur without genes coding for the many enzymes regulating the myriad biochemical reactions needed to produce its song system. But development of these mechanisms also requires environmentally supplied building blocks that are incorporated into the proteins and other chemicals used in the construction of hormones, nerve cells, bones, and muscles.

3. In the developmental process, initial gene–environment interactions result in the production of materials that then become part of the cellular environment, leading to new gene–environment interactions that were not possible prior to those that preceded them. When cells in the testes of an infant male zebra finch produce the hormone estrogen, this chemical is then available for transport to cells in the growing nervous system, where it affects the genetic activity and therefore the chemical reactions taking place there, with profound organizational consequences.

4. Because of the interactive nature of development, differences among individuals in either their environments or their genes can generate different gene–environment interactions, and thus different patterns of development. Differences in the genetic information contained in the cells of galahs and pink cockatoos mean that they have different enzymes available to them, which

in turn results in differences in the call control mechanisms that develop within growing galahs and pink cockatoos. On the other side of the coin, differences in the environments of individuals also can shift development one way or the other. Thus, the presence of estrogen in nestling males, but not females, leads to differences in the development of the song system in male and female zebra finches. Likewise, the acoustical and social experiences that affect male white-crowned sparrows and humans can result in differences in the song dialects or languages they will exhibit.

5. Two complementary components make up the proximate approach to understanding behavior. The developmental component requires an understanding of how genes and environment contribute to the construction of the mechanisms that make behavior possible. The physiological component requires an understanding of how the mechanisms, once developed, cause animals to behave.

SUGGESTED READING

Bird Song: Biological Themes and Variations [187], by Clive Catchpole and Peter Slater, offers a thorough and readable review of research on the proximate and ultimate causes of bird song. See also the more technical *Ecology and Evolution of Acoustic Communication in Birds*, edited by Donald Kroodsma and Edward Miller [680]. Work by Peter Marler [770], Luis Baptista and colleagues [57, 58, 59], and Meredith West and An-

drew King [1256, 1257] on song learning by songbirds has greatly improved our understanding of the role of environment and heredity in the development of behavior. In Jared Diamond's *The Third Chimpanzee* [296], you will find a delightful essay on the learning biases underlying human language, a topic explored in depth by Steven Pinker in *The Language Instinct* [940].

DISCUSSION QUESTIONS

1. Birds are not the only animals to use vocalizations as apparent mate-attraction signals. The acoustical signals of some male frogs and toads also have this function. Different species typically produce different calls. The figure on the following page contains data from a study of two closely related frog species from southeastern Australia

whose ranges overlap partially [725]. The data are pertinent to what hypothesis about the evolution of species differences in mate-attracting signals? What prediction can be evaluated on the basis of these data? What scientific conclusion can you derive from the test?

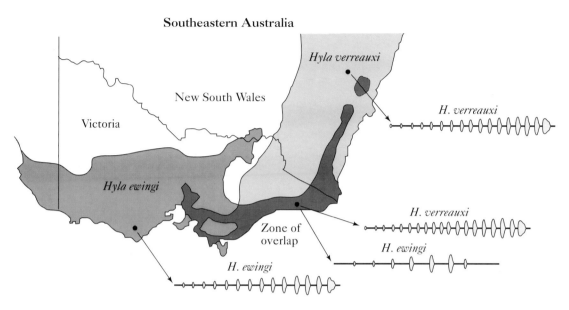

Southeastern Australia

2. Humans living in Australia and Arizona employ the same language, but decidedly different dialects. Here are two statements about this phenomenon: (1) "The dialect difference between the Australian and the Arizonan is purely environmentally caused." (2) "The dialect spoken by an Australian (or by an Arizonan) is purely environmentally determined." Which of these two statements can be readily defended, and which can be severely criticized, and why?

3. We observe that a white-crowned sparrow's species-specific song is more similar to that of his neighbor than to his father's dialect. Categorize the following hypotheses as "proximate" or "ultimate."

   a. Young adult white-crowned sparrow males try to match their song as closely as possible with that of their neighbors.
   b. White-crowned sparrow males possess genetic information that steers the development of a biased song learning mechanism.
   c. By singing the same song as his neighbors, the young adult male more effectively deters invasion of his territory by those neighbors.
   d. The hormonal condition of the young adult male predisposes him to listen to and mimic the songs of his territorial neighbors.
   e. Those males with the song flexibility to mimic the songs of their neighbors have in the past attracted females more reliably than males without this ability.

CHAPTER 3

The Development of Behavior: The Role of Genes

W E HAVE USED STUDIES OF ANIMAL COMMUNICATION (Chapter 2) to help define the different proximate components of behavior. Now we begin four chapters that analyze those components in more depth, starting with the role that genes play in behavioral development. What does it mean to say that genes are involved in a certain behavior? For example, do blackcap warblers differ in their choice of wintering grounds because of differences in the genes they carry? Some of these tiny European songbirds began to winter in Great Britain in the 1950s instead of the continuing on to their traditional wintering grounds in northern Africa, 1000 to 1500 kilometers to the south. Why the change? Did some British birds lose their hereditary ability to migrate to North Africa? Or are the blackcaps that winter in Britain birds from Scandinavia that now stop in Britain instead of flying much farther south? Behavior geneticists have learned a great deal about the connection between genetic differences among blackcaps and the migratory behavior of different populations, one of several research stories that we will review in this chapter. My goal is to describe some of the techniques that geneticists have developed for testing whether the differences in behavior among individuals stem in part from differences in their genes. We shall examine a spectrum of studies on the genetic differences hypothesis, some of which show that an alteration in even a single gene may change the way individuals behave. The chapter concludes with a study that unifies proximate and ultimate approaches to genetic research on behavioral development, showing how the two levels of analysis complement one another, producing a fuller, richer understanding of animal behavior.

How Genetic Differences Affect Behavioral Development

We have already examined the evidence that differences among animals in their communication signals can be traced in some cases to their chromosomes. For example, the presence or absence of a W chromosome has a lot to do with why male and female zebra finches differ in their behavior. In other cases, signal differences occur because of differences in a specific gene. For example, the distinctive *per* genes of various fruit fly species affect the development of their songs. Thus, differences in **genotype** (defined as the genetic constitution of an individual) sometimes affect the kinds of gene–environment interactions that determine the development of the **phenotype** (defined as anything observable about an individual, including its nervous system and behavior). So, whenever we find a phenotypic difference between individuals, it is possible that we are seeing the effects of heredity. However, this possibility needs to be checked in each case, because it is also true that phenotypic differences can arise because of differences in the environments in which the animals develop, as shown by the phenotypic effects of hearing different dialects on young white-crowned sparrows.

Therefore, when Ann Hedrick and Susan Riechert found differences in the speed with which funnel web spiders in the southwestern United States attacked prey that wandered onto their webs, they did not know in advance whether the spiders differed because of their genes or their environments. Spiders living in desert grasslands were quicker to respond to prey than were spiders living nearby along the banks of streams. Differences in prey abundance, an environmental variable, might well have made the desert spiders hungry and quick to respond to prey, whereas those living in the relatively lush streamside environments could afford to react more slowly.

One way to eliminate any environmental differences among the spiders would be to take members of the two wild populations into the laboratory, where the grassland and streamside spiders could be bred separately, each producing a new generation to be reared under exactly the same conditions [522]. When Hedrick and Riechert performed this experiment, the offspring of slow-to-attack streamside parents were also slow to respond to prey, taking about a minute on average to come after a cricket placed on the web apron. The offspring of quick-to-kill grassland parents reacted in less than three seconds on average, even though they had received just as much food as the laboratory descendants of streamside spiders. These results suggest that much of the *difference* in attack behavior between grassland and streamside funnel web spiders is hereditary [997]. The simple logic of the spider experiment was that if adult spiders from the two areas behaved differently because of their distinctive genotypes, then their offspring would inherit their parents' special genes, and would behave like their parents as well. Geneticists often use breeding experiments in uniform environments to find out whether genetic differences contribute to phenotypic differences among individuals.

Comparing Parents and Offspring: Migratory Behavior of Blackcap Warblers

You will recall that some blackcap warblers winter in Britain, whereas others go all the way to Africa. In order to study the basis for the "winter in Britain" phenotype, Peter Berthold and his colleagues removed some wild blackcaps from their British wintering area and brought them to Berthold's laboratory in Germany. The birds were held indoors until spring, at which time pairs were released into outdoor aviaries to breed and produce offspring [87].

To check on the migratory behavior of blackcaps reared in captivity, it is not necessary to release the birds in the fall and then try to follow them wherever they go—a nearly hopeless task. Instead, Berthold placed some blackcaps in special cages that had been electronically wired to record the number of times a bird hopped from one perch to another, while others were housed in cages shaped like funnels and lined with typewriter correction paper, which recorded the scratch marks made by birds leaping up from the base of the funnel (Figure 1). These cages enabled the researchers to record whether their subjects became restless at night during the fall and spring migratory periods, and if so, in what direction they were attempting to fly.

When tested in the first fall of their lives, the offspring of birds taken from the British wintering grounds exhibited the kind of heightened activity characteristic of migratory blackcaps, as did their parents, which were tested as well. These results enabled Berthold to reject the hypothesis that the British wintering population was composed of birds that had lost their ability to migrate.

If the birds wintering in Britain were migrants, they obviously had a different migratory pattern than other European members of their species (Figure 2). Where did they come from? The marks left on the paper in the funnel cages provided an answer to this question. When caged birds try to begin their nighttime migratory flight, they orient in the direction they wish to fly in and jump off their perch.

1   Funnel cage for recording the migratory orientation of captive birds. The cage is shown cut in half. The bird can see the night sky through the wire mesh ceiling of the cage, and as it jumps up onto the surface of the funnel, it leaves revealing marks on the typewriter correction paper lining the funnel.

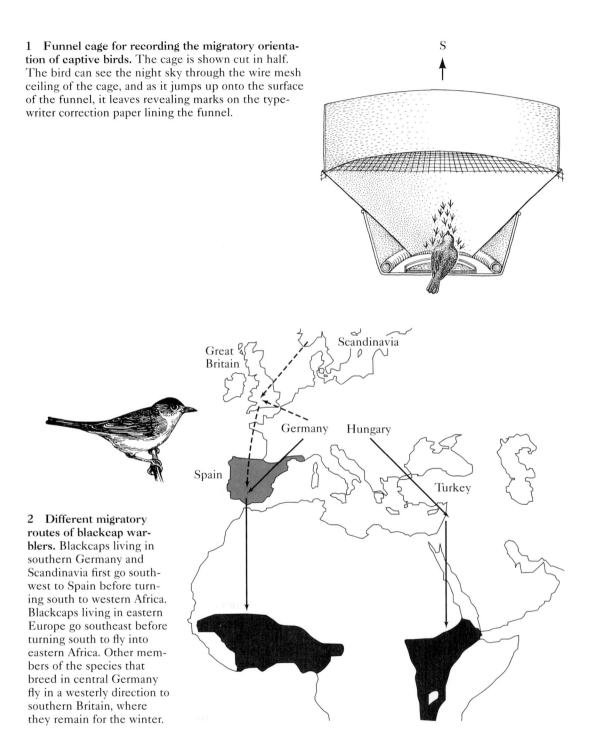

2   Different migratory routes of blackcap warblers. Blackcaps living in southern Germany and Scandinavia first go southwest to Spain before turning south to western Africa. Blackcaps living in eastern Europe go southeast before turning south to fly into eastern Africa. Other members of the species that breed in central Germany fly in a westerly direction to southern Britain, where they remain for the winter.

They do not get far in a funnel cage, but they try and try again, leaving a scratchy record of their orientation preference. Berthold's subjects, both experienced adults and novice youngsters, oriented in a westerly direction. This finding eliminated the possibility that these blackcaps came from northern Britain or still farther north in Scandinavia. Birds from those populations oriented in a southerly or southwesterly direction. Instead, the birds that Berthold and his coworkers tested must have come from central Germany or Belgium, due east of the British wintering grounds, a possibility that has been confirmed by the discovery of wintering blackcaps in Britain that were initially banded in central Germany.

The differences between the German blackcaps that migrate to Britain and those that fly to Africa in their first fall could conceivably have been caused by differences in their environments. In order to test this possibility, Berthold and colleagues captured some adult blackcaps from southwestern Germany, arranged for them to breed in the same kind of outdoor aviaries used by the blackcaps that wintered in Britain, collected the young, and tested them in the fall. Since these youngsters were reared under the same conditions as those with British-wintering parents, any migratory differences between the two groups could not have been caused by environmental differences. And yet the birds behaved differently, with the southwestern German novice migrators generally orienting toward the southwest, rather than taking the westerly tack of the birds with British-wintering parents (Figure 3). This experiment clearly demonstrates that genetic differences, not environmental ones, are largely responsible for the differences between the two populations of German birds in their migratory orientation [86, 87].

Andreas Helbig has extended the blackcap story by examining whether hereditary differences also cause birds from Austria to follow yet another migratory route. Unlike the blackcaps from southwestern Germany, which fly in a southwesterly direction to Spain and then on to Africa, Austrian blackcaps follow a flight path that takes them *southeast* across Turkey to Lebanon and Israel before turning due south to Ethiopia and Kenya (see Figure 2). Helbig created "hybrids" by crossbreeding captive southwestern German and Austrian blackcaps. He then measured the flight orientation of the parental and hybrid birds by placing them in funnel cages at night. The mean orientation of the marks left on the typewriter correction paper was calculated for each bird. When the directions chosen by hybrid offspring were compared with those of their German and Austrian parents, they proved to be intermediate (Figure 4). This result supports the hypothesis that southwestern German and Austrian birds differ in their genetic makeup in ways that affect the initial orientation of their migrations. Hybrids between the two populations receive a mix of the distinctive alleles of particular genes, with compromising effects on the development of migratory orientation [537].

Foraging in Fruit Fly Larvae

By holding the environment constant, behavior geneticists interested in spider feeding behavior and blackcap migratory behavior have been able to show that

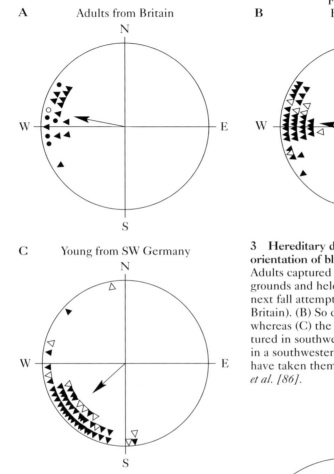

A   Adults from Britain

B F_1 offspring of British adults

C   Young from SW Germany

3   Hereditary differences in migratory orientation of blackcap warblers. (A) Adults captured on the British wintering grounds and held in Germany until the next fall attempted to fly west (toward Britain). (B) So did their offspring, whereas (C) the offspring of adults captured in southwestern Germany oriented in a southwesterly direction that would have taken them to Spain. *Source: Berthold et al. [86].*

4   A test of the genetic differences hypothesis for why migratory blackcap warblers from western and eastern Europe take different flight paths. The inner ring shows the migratory orientations of birds from southwestern Germany (the solid triangles) and from Austria (the open triangles). The outer ring shows the orientations of hybrid offspring whose parents differed in migratory orientation (solid circles). The large arrows show the mean directions taken by the three groups of birds. As predicted, the hybrids tended to orient due south, an intermediate direction compared with those of their parents. *Source: Helbig [537].*

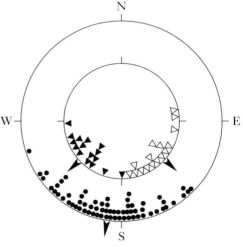

certain phenotypic differences must have been caused by differences in the genetic information possessed by animals from different populations. But what about behavioral variation within a single population?

An instructive case takes us again to fruit flies. Working with a single species of *Drosophila*, Marie Sokolowski and her colleagues created two genetically uniform strains, one whose larvae covered a lot of distance while feeding (labeled "rovers") and one whose larvae did not (labeled "sitters"). When larvae of these strains were reared under identical conditions, individuals with the rover phenotype traveled about four times farther on a yeast-coated petri dish during a 5-minute test than members of the sitter strain [283]. When adults of the two strains were permitted to reproduce together, they created offspring (the F_1 generation) all of which were rovers during the larval phase. When these larvae matured and interbred, they created an F_2 generation with a ratio of rovers to sitters of 3:1 (Figure 5).

Persons familiar with Mendelian genetics will recognize that these results indicate that rovers carry one or two copies of the dominant form of a gene affecting larval foraging behavior, whereas sitters have two copies of the recessive form of this gene. So here is a case in which the difference between two behavioral types stems from a difference in the information contained in a single gene, now known to be located on the second of the four chromosomes that fruit flies possess [282].

As noted in Chapter 2, other single-gene effects have been found in fruit flies, although rarely in natural populations, probably because most mutations are highly deleterious. For this reason, researchers often expose laboratory populations of adult flies to conditions that cause genetic mutations, after which their offspring can be scanned for behavioral oddities, and the apparent mutants tested in breeding experiments [335]. Among the many mutant alleles discovered in such experiments are *stuck* (males with the mutant gene fail to dismount after the normal 20 minutes of copulation) and *coitus interruptus* (males with this allele disengage after just 10, not 20, minutes of copulating) [79]. A whole series of mutations have now been identified that affect the sexual behavior of fruit flies by decreasing male courtship vigor, eliminating female receptivity, altering the courtship song pattern, or causing males to treat both males and females as suitable objects of courtship [498].

Are studies of this sort designed to show that a particular gene codes for a particular behavior? No. When a single gene difference causes a difference in phenotype, it does so by affecting the production of one enzyme. This one enzymatic difference can set in motion a whole series of changes in other gene–environment interactions dependent on or influenced by that enzyme. The result can be a shift in the developmental pathways followed by individuals with different genotypes, leading eventually to major phenotypic differences between them.

Consider the differences between two strains of laboratory mice. Individuals with a mutant allele that affects the production of a single enzyme, α-calcium-

5 Genetic differences cause foraging differences in fruit fly larvae. Representative tracks made by sitter and rover phenotypes feeding in a petri dish appear at the top of the figure. When adult male flies of the sitter strain mate with adult females of the rover strain, their larval offspring (the F_1 generation) almost all exhibit the rover phenotype (move more than 7.6 cm in 5 minutes: shaded areas). When flies from the F_1 generation interbreed, their offspring (the bottom graph) are composed of rovers and sitters in the ratio of 3:1. *Source: de Belle and Sokolowksi [283].*

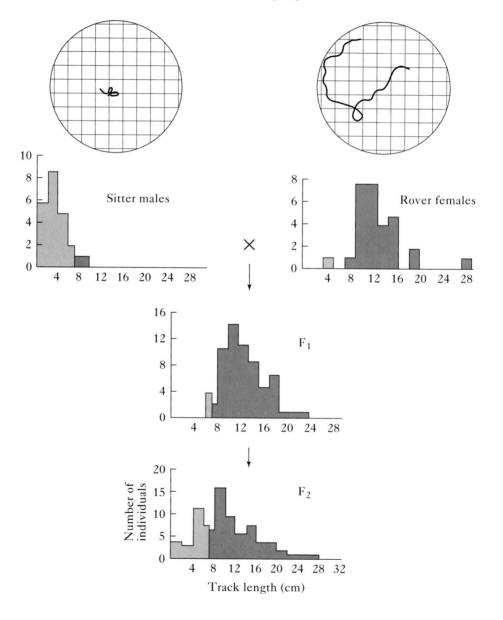

calmodulin kinase, behave differently from other mice whose only distinctive hereditary feature is a different form of the kinase gene. When the spatial learning ability of the two strains of mice is tested, the mutant mice cannot retain information that their slightly different fellow mice easily remember (Figure 6) [1106]. The enzyme difference between the mice apparently alters biochemical pathways involved in the development of the hippocampus, a region of the vertebrate brain that plays a major role in spatial learning. Thus, a single genetic difference between the two types of mice leads to a single enzymatic difference that cascades into major biochemical, developmental, neurophysiological, and eventually, behavioral differences.

Thanks to technological advances in molecular manipulations, today's researchers can even target a particular gene for alterations that prevent it from being transcribed. One such tailor-made mutant mouse with inactivated *fosB* genes appears to be normal in almost every respect, including size, hormonal makeup, learning ability, and sensory equipment. However, adult females with two inactivated *fosB* genes treat their newborn pups with spectacular indifference. After giving birth, these mutant females inspect their pups, but then leave them alone. Wild-type females homozygous for active *fosB*, which are genetically identical to the mutants in every other respect, typically gather their pups together and crouch over them, keeping them warm and permitting them to nurse. Should a pup wriggle away or be displaced, wild-type females quickly retrieve their errant offspring; mutant females ignore them altogether (Figure 7) [152].

These stunning behavioral differences between the two kinds of mice have been tentatively explained in the following way. Both mutant and wild-type mice

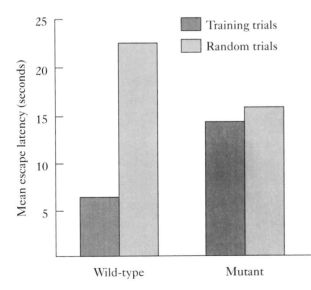

6 A single gene affects learning ability in laboratory mice. Mice were placed in a circular water-filled arena containing one small, barely submerged platform. The mice had to find the platform in order to stop swimming. Both wild-type and mutant mice were able to learn the location of the platform. Two days later the mice were tested again. When the platform was in the same position as during the training trials (dark bars), the wild-type mice found it significantly more rapidly than those with the α-calcium-calmodulin kinase mutation. When the location of the platform was set randomly (lighter bars), the wild-type mice did not perform any better than the mutants, showing that the difference between the two groups during the first part of the test was due to the superior retention of learning by the wild-type mice. *Source: Silva et al. [1106].*

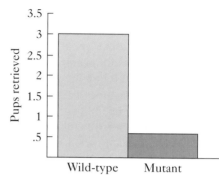

7 A single gene affects maternal behavior in laboratory mice. Wild-type female mice gather their pups together and crouch over them (above), but females with inactivated *fosB* genes do not exhibit these behaviors (below; the pups are seen scattered in the foreground). A quantitative measure of the difference between the two is based on experiments in which three mouse pups were placed in the corners of their mother's cage; the number of pups retrieved in 20 minutes is shown in the graph. *Source: Brown et al. [152].*

that have just given birth almost always have the neuronal circuits in place that cause them to inspect their newborns. The resulting sensory experiences, especially olfactory stimuli from the pups, generate signals in receptor cells that travel to higher regions of the female's nervous system. The multiple inputs that come from sniffing, touching, and hearing newborns are integrated in a particular region of the hypothalamus, called the preoptic area (POA), where the volleys of neuronal signals (see Chapter 5) serve to activate *fosB* alleles in certain cells. As a result, wild-type females quickly produce a functional protein (FosB). However, these signals have no such effect on the altered *fosB* alleles of mutant females. Thus, the two types of mice differ in the effects of selected experiences on gene activation. The production of FosB eventually leads to additional genetic and enzymatic changes that produce structural alterations in particular neural circuits within the POA. The modified neural machinery of wild-type females motivates them to pull their pups underneath their bodies and care for them solicitously. In contrast, the essential neural changes required to initiate and maintain nurturing behavior never occur in mutant females for lack of the gene (and allied protein) needed to initiate the series of biochemical events that reshape certain neurons in the brain [152].

Comparing Other Relatives

Controlled experiments involving comparisons of parents and offspring have helped test the hypothesis that some phenotypic differences among individuals arise from genetic differences among them. But ethical considerations generally preclude such experimental manipulations of members of our own species. Nevertheless, we can still test whether people differ behaviorally because of genetic differences by making the right kinds of comparisons. Among the most interesting of these comparisons are those involving identical and fraternal twins. Identical, or monozygotic, twins are derived from the same fertilized egg, which divides and gives rise to two genetically identical embryos. In contrast, fraternal, or dizygotic, twins are born at the same time, but each is formed by the union of a different egg and sperm. As a result, dizygotic twins are no more similar genetically than any other pair of siblings. The probability that two siblings—other than identical twins—will inherit the same allele from a parent when two different alleles are represented in the parent's gametes is always 1/2. As a result, if behavioral development is influenced by the presence of a particular allele, the behavior of fraternal twins should often differ (because they will often have different forms of a gene), whereas that of identical twins should be the same (because they possess exactly the same genotype). In actual fact, fraternal twins differ from each other more than identical twins do with respect to a host of traits, including personality measures.

However, an alternative hypothesis exists for the high degree of behavioral similarity between identical twins. Behavioral development is influenced not only by the genes an individual inherits, but also by his or her environment and experiences. These factors may be more similar for identical than for fraternal twins, perhaps because parents and other people treat identical twins the same way because of their nearly identical appearance (but see [945]). This shared environmental influence, so the argument goes, could be the reason why identical twins are so similar behaviorally.

To control for this potentially variable factor, it would be helpful to observe the behavior of identical twins who happened to grow up in separate families. This is why Jack Yufe and Oskar Stohr and other identical twins who were reared apart provide such valuable information. Oskar Stohr was raised as a Catholic in Nazi Germany by his grandmother. Jack Yufe grew up on various Caribbean islands with his Jewish father. One can hardly imagine more diverse environments for the development of two human beings, and yet these men are remarkably similar in appearance (Figure 8) and behavior. Although they were separated at birth and lived apart for 47 years, they both "like sweet liqueurs, . . . store rubber bands on their wrists, read magazines from back to front, dip buttered toast in their coffee and have highly similar personalities" [556].

A study of more than 50 pairs of identical twins reared apart demonstrates beyond reasonable doubt that genetic differences are responsible for a significant portion of the differences among humans in personality, temperament, and social attitudes [113]. For example, when identical twins filled out elaborate question-

8   Identical twins separated at birth: Jack Yufe (left) and Oskar Stohr (right). *Photograph by Bob Burroughs.*

naires designed to provide quantitative measures of personality traits (e.g., degree of cooperativeness, sociability, and so on), the scores of identical twins reared apart were nearly as similar as those of identical twins reared together. Both groups were much more alike than were fraternal twins, whether reared together or apart.

From these studies, one can calculate what proportion of the variance (a statistical measure of variation among individuals) in the personality traits of these groups was due to genetic differences among them. In the lingo of behavior genetics, this is the **heritability** of the trait in question. Heritabilities can range from 0 (none of the variance in phenotypes is caused by genetic differences) to 1 (all of the variance is due to genetic variation in the population studied). The twin studies produced a heritability of about 0.5 for personality scores; in other words, about 50 percent of the *differences* in personality scores among all the individuals tested were due to genetic differences among them. The other 50 percent of the *differences* were environmental in origin.

Genetic Differences and IQ Differences

Identical twins are also more similar than fraternal twins in their scores on IQ tests [114] (Table 1). Moreover, identical twins typically have very similar IQ scores

Table 1 Familial correlations for IQ scores: Predicted values based on the
genetic differences hypothesis versus actual correlations

Category	Predicted correlation	Actual median correlation	Number of studies
Identical twins reared together	1.0	0.85	34
Identical twins reared apart	1.0	0.67	3
Fraternal twins reared together	0.5	0.58	41
Siblings reared apart	0.5	0.45	69
Genetic parent–child	0.5	0.39	32
Adoptive parent–child	0.0	0.18	6

Source: Bouchard and McGue [114].

even when they have grown up in different households. By comparing the correlations between identical twins and between fraternal twins in various environments, Thomas Bouchard and his colleagues calculated a heritability of about 0.7
for IQ. That is, about 70 percent of the differences in IQ scores in the population
they studied stemmed from the genetic variation that existed among the individuals whose IQs they measured. Only the remaining 30 percent can be attributed
to the diverse environments that the participants in the study experienced as they
were growing up.

As an alternative hypothesis on the close IQ similarity between identical twins
reared apart, consider the argument that their adoptive homes were more similar
than those selected for fraternal twins that were reared apart. If that were true, the
more similar home environments could have channeled the intellectual development of the separated identical twins along similar lines, leading ultimately to very
similar IQ scores. But when Bouchard's team measured various aspects of the adoptive home environment, such as socioeconomic status, education level of both
adoptive parents, intellectual orientation of the household, and the like, they found
that there was little or no connection between the similarity in environments of
twins reared apart and their similarity in IQ scores [113].

People sometimes interpret the high heritability of IQ scores to mean that intelligence is "genetically determined." Leaving aside the serious problem of establishing the relation between IQ and intelligence, about which much ink has been
spilt, this interpretation is incorrect (Figure 9). Once again, all behavioral phenotypes, including IQ scores, develop as a result of gene–environment interactions.
Taking IQ tests requires a brain, which in turn is the product of an almost incom-
prehensibly complex interaction between the genetic information in a fertilized
egg and its "environment," which includes the nongenetic materials in the egg
and the egg's surroundings. These materials are utilized by the growing embryo
and are absolutely essential for the construction of brain tissue. Moreover, as the
brain develops and begins to generate electrical activity and receive messages from

9   IQ scores are not genetically determined. An IQ score is a phenotype dependent on neural activity; brains develop via processes in which both genes and environment play essential roles.

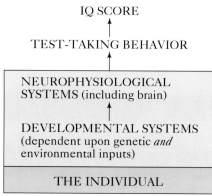

outlying sensory receptors linked to it, it is creating and receiving "experiences" that are vital to its further development.

Thus, the results of the familial comparisons within populations outlined above show just one thing: genetic differences among the people studied to date contribute to the differences in the scores they achieve on IQ tests. But environmental differences also play a role in the development of IQ differences among people, as we can see because the correlation between the IQs of identical twins is less than 1.0 (the figure expected if environmental differences did not affect IQ development at all). Even identical twins develop in somewhat different environments. Each twin experiences slightly different surroundings in the womb, each consumes somewhat different foods, and each has its own unique set of social interactions, as is especially true of twins separated early in life.

Producing Genetic Mosaics

Let's leave human genetics behind for now and consider an elegant experimental technique for exploring how genes affect the development of behavioral differences. In sexually reproducing species, males and females never behave identically. One proximate explanation for these differences focuses on genetic differences between the sexes—specifically, on the possibility that sex-determining genes control the development of the male and female nervous systems. If this hypothesis is correct, then if we were able to create embryonic individuals that were mixtures of cells, some containing the female genotype and others containing the male genotype, we might be able to produce individuals whose bodies were largely of one sex but whose brains, or parts of brains, were composed of cells of the other sex [79, 1196].

Researchers working with fruit flies have been able to do just that. Sex is determined in fruit flies by the number of X chromosomes; females have two X chromosomes in their cells, while males have one. **Genetic mosaic** flies have some cells with two X chromosomes, which therefore develop into female organs, and other cells with a single X chromosome, which become male organs (Figure 10).

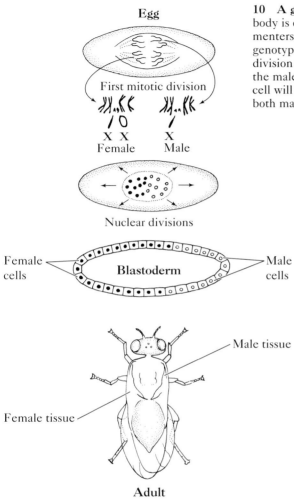

Egg

First mitotic division

X X
Female

X
Male

Nuclear divisions

Female cells

Blastoderm

Male cells

Male tissue

Female tissue

Adult

10 A genetic mosaic fruit fly. To create a fly whose body is composed of both male and female cells, experimenters use fly eggs that contain a special XX (female) genotype. As eggs of this sort begin to divide, an early division results in a cell with only one X chromosome—the male genotype. All the descendants of the one-X cell will be genetically male, creating a mosaic fly with both male and female tissues. *After Hall et al. [499].*

Mosaic flies can be manufactured with sophisticated experimental techniques that generate individuals whose male and female tissues can be identified (e.g., by different colors) so that researchers can tell which parts of the fly have one or two X chromosomes. When different kinds of mosaic flies interact with one another, their behavior can reveal the effects of genetically different male and female tissues on reproductive behavior.

A fly whose upper brain lies within a zone of male tissue will pursue any other fly whose posterior abdomen is chromosomally female [497, 572, 1101]. The fly will initiate wing-waving courtship even if it has female genitalia, eyes, wings, and antennae. Female antennae, like male antennae, must therefore be capable of detecting the olfactory cues that receptive females provide to attract males. But a female's brain operates differently from that of a male, because only brains whose nerve cells carry a single X chromosome will order a fly to court females with a wing-waving song. Mosaic flies can tell us exactly where one or two X chromosomes have their specific effects on the development of nervous systems.

Recent techniques have been used to create a new type of mosaic fly. Thanks to complex maneuvers that we need not outline here, it is possible to insert a sex determination gene called *transformer* into particular parts of the male fruit fly's brain [386]. Against the proper genetic background, this gene expresses itself, resulting in the feminization of just that portion of the brain in which olfactory information is processed. Males with feminized "olfactory regions" are prone to court females and males indiscriminately, perhaps because the altered brain tissues lack the inhibitory networks possessed by normal males, which are involved in turning off the courtship response when typical males encounter their fellow males. Thus, these experiments confirm that genetic differences between male and female fruit flies lead to differences in their nervous systems, thereby causing sexual differences in behavior.

Artificial Selection Experiments

The various techniques of behavior genetics that we have described thus far have been dedicated largely to testing the hypothesis that genetic differences can cause behavioral differences among individuals, including members of our own species. But this hypothesis generates yet another prediction, which we have not yet examined. If individuals vary in their behavior because of genetic differences, then researchers should be able to select for or against a particular trait, increasing or decreasing the frequency of the particular allele(s) associated with that trait over time [335]. In contrast, if variation among individuals has no genetic basis, then selective breeding cannot cause evolutionary change within the population—because those individuals that reproduce will not endow their offspring with a distinctive allele or alleles "for" a particular attribute.

Consider the work of Carol Lynch with house mice, which build nests of soft grasses and other plant materials in nature, but will happily accept cotton for their nests in a laboratory [743]. The amount of cotton a mouse collects can be quantified as the number of grams pulled into a nest cage over a 4-day period. In the starting generation in Lynch's experiment, individuals moved between 13 and 18 grams of cotton into their cages from an external cotton supply. But was the phenotypic variation in nest building caused by genetic or environmental factors?

To get an answer, Lynch attempted to create a "high line" by interbreeding males and females that collected a relatively large amount of cotton, as well as a "low line" (by crossing males and females that gathered relatively little cotton) and a "control line" (by crossing males and females chosen at random from each generation). The offspring produced by these crosses were reared under conditions identical to those of their parents, eliminating environmental variation as a cause for differences in their behavior. When the offspring of the three groups became adults, they were tested for the amount of nest material they collected in 4 days. The most avid cotton collectors in the high line were permitted to breed, creating a second selected generation, as were the least eager collectors in the low line.

Lynch repeated these procedures over 15 generations, with the eventual result that the high-line mice gathered about 40 grams of cotton for their nests on average, while the low-line mice brought in only 5 grams on average. The 15th-generation control mice brought in about 15 grams, the same amount as their ancestors (Figure 11). This result offers evidence that there were genetic differences among the mice in the original population related to nest building, which enabled Lynch to cause evolutionary change in the populations she controlled.

Fundamentally similar results have been achieved in selection experiments with many other animals. Ary Hoffmann used variation among male fruit flies in their ability to defend a food resource as a basis for a selection experiment. He suc-

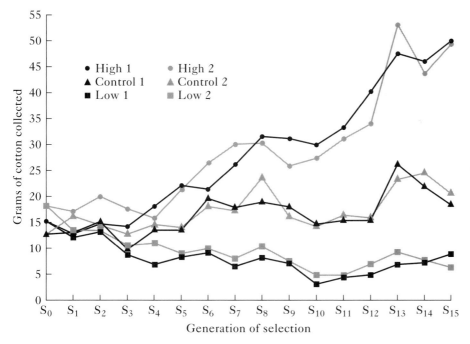

11   Response to selection on nest-building behavior by mice. Artificial selection favoring mice that collect large amounts of cotton for their nests eventually led to the evolution of populations (the high lines) whose members collected much more cotton on average than the control lines, whose behavior was not selected and so did not evolve. Likewise, selective breeding of mice that gathered relatively little cotton resulted in the evolution of low lines, whose members made very small nests. The symbols represent the amount of cotton collected over 4 days. *Source: Lynch [743].*

ceeded in creating a line of males highly successful in territorial disputes [549]. William Cade did much the same with respect to calling time in crickets [169]. In any one population, some individuals chirp away for many hours each night, while others call for just a few hours, and still others almost never sing. The differences between the male crickets might arise because their environments differ (for example, the more food a cricket has, the longer it might be able to call at night) or because they differ genetically. By selectively breeding only the most persistent chirpers and the least dedicated callers from each generation, Cade eventually produced two lines of crickets, one whose males called for hours on end and one whose males were usually silent (Figure 12). Cade could not have changed the behavior of his cricket populations unless the males in the first generation varied genetically in ways that affected their chirping behavior.

Such **artificial selection** studies are of great interest, not just as demonstrations that genetic differences cause some of the behavioral variation among individu-

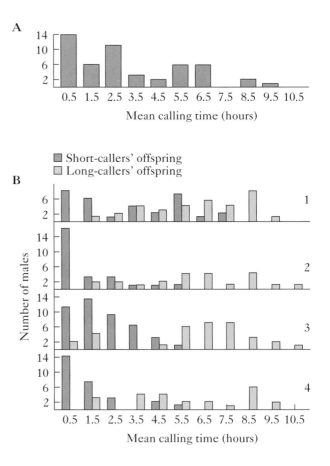

12 Response to selection on calling duration in crickets. (A) The distribution of short-, intermediate-, and long-calling males in a population of crickets. (B) By eliminating intermediate callers from breeding populations over four generations (1–4), William Cade selected for and created populations of males that differed ever more strongly in the time they spent calling each evening. *Source: Cade [169].*

als in a population or species, but because they also tell us a great deal about the potential for behavioral evolution. In making the case for evolution, Charles Darwin carefully studied domesticated animals because he realized that the artificial selection involved in producing breeds of dogs or pigeons was fundamentally the same as natural selection [286]. Artificial selection experiments almost always reveal that modern populations—and, by implication, past ones as well—contain a great deal of genetic variation. Therefore, one of the primary conditions required for natural selection almost certainly applied widely in the past, so that behavioral evolution could have occurred in most, if not all, species.

Genetic Differences and Alternative Phenotypes

One of the most interesting products of behavioral evolution is the coexistence of hereditarily discrete phenotypes within the same species, despite the tendency of natural selection to reduce genetic variation within populations [479]. We have reviewed several cases of this phenomenon, including the different migratory forms of the blackcap warbler and the different predatory types of funnel web spiders found in different locations. We return to this topic now to consider a species in which two different types of individuals live together in the same place. The species is a cichlid fish, *Perissodus microlepis*, that makes its living, believe it or not, by snatching scales from the bodies of other fish in Lake Tanganyika of Africa. Two structurally different forms of the scale-eater exist, one with its jaw twisted somewhat toward the right, the other with its jaw bent toward the left. This structural dimorphism is linked to the attack behavior of the fish, which always comes from behind to snatch scales from the rear side of a victim. Scale-eaters with a jaw turned to the right always take scales from the prey's left flank, while the other form invariably goes for the prey's right flank (Figure 13). The frequency of these two forms in the population fluctuates narrowly around 1:1.

The phenotypes "jaw bends to right" and "jaw bends to left" are heritable, according to Michio Hori, who compared parents and their fry [570]. Presumably the differences between the behavioral phenotypes "attack left flank" and "attack right flank" are also hereditary, given the perfect correlation between jaw structure and behavior in adult fish. If the success of an attacking scale-eater increased when its phenotype was less common than the other form in the population, then we could explain why the frequencies of the two forms remain about the same. Hori found that in a year in which the "left-jawed" phenotype made up somewhat more than 50 percent of the population, fish with the other jaw type and attack mode apparently had greater success in grabbing scales from their victims, judging from the fact that scale-donating prey had more scales missing from the left flank on average than from the right flank (remember that " right-jawed" fish attack the *left* flank). When one phenotype was more common, the prey were approached by scale-eaters more on one side than the other. The prey

13 Alternative phenotypes in the cichlid fish *Perissodus microlepis* from Lake Tanganyika. The "right-jawed" and "left-jawed" forms of the fish attack their prey from opposite sides.

"Left-jawed" *Perissodus* attack prey from the right rear side

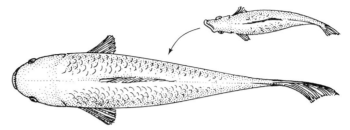

"Right-jawed" *Perissodus* attack prey from the left rear side

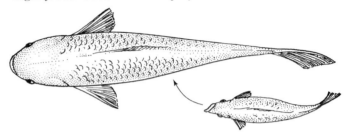

then began to guard against attack on the more frequently approached flank—which enabled the less frequent phenotype to sneak through the prey's defenses more often.

Higher foraging success for the rarer of the two phenotypes appears to translate into higher reproductive success as well, resulting in an increase in its frequency in the next generation. Whenever right-jawed individuals constituted a slight minority, they had higher fitness than their left-jawed competitors, increasing the frequency of right-jawed fish in the next generation. If right-jawed fish then came to be in the majority, the other type experienced higher fitness, keeping both types oscillating around the 50 percent mark. Because this kind of selection depends on the relative frequencies of alternative phenotypes, it is labeled **frequency-dependent selection** [38]. The consequence of frequency-dependent selection in this case is the long-term coexistence of the two different genotypes underlying the two different phenotypes in *P. microlepis* in Lake Tanganyika (Figure 14).

Another case of a structural and behavioral **polymorphism** is found in the sand cricket, whose males can be divided into a short-winged, flightless cohort and a long-winged group that can fly. Flight ability comes in handy when the local environment deteriorates, so it is easy to envision why alleles that promote the development of long wings could spread in a cricket whose local environments regularly change in quality. But the ability to fly comes at a price. Peter Crnokrak and

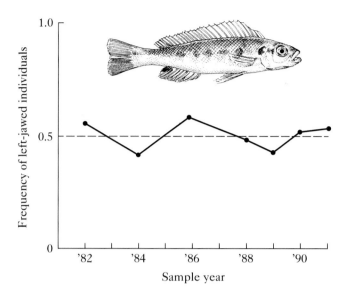

14   Frequency-dependent selection and the maintenance of "right-jawed" and "left-jawed" forms of *Perissodus microlepis*. When one form becomes slightly more abundant relative to the other, a reduction in fitness of the more common type causes the other to become more common in subsequent generations. As a result, the frequency of both types oscillates around 50 percent. *Source: Hori [570].*

Derek Roff have shown that long-winged males do not attract females as well as their short-winged flightless competitors. The reduced mating success of long-winged males stems from their failure to sing for as many minutes per night as their rivals [245]. Apparently, short-winged males have more energy available for calling than long-winged individuals, perhaps because they have not had to build or maintain large wings. Thus, in a stable environment, short-winged males may leave more descendants than long-winged ones. However, if conditions change, favoring a move away from the natal area, long-winged males are probably better at finding a new location in which they can survive and reproduce.

Different Populations, Different Genes, Different Behavioral Traits

The examples of alternative behavioral phenotypes we have just discussed involve species whose members live in the same region. Cases of this sort are probably less common than those in which the alternative phenotypes occur in geographically separate populations of the same species. A classic example involves the garter snake *Thamnophis elegans*, which lives throughout much of western North America, including both foggy, wet coastal California and the drier, elevated inland areas of that state [32]. The diets of snakes living in the two areas, referred to hereafter as "coastal" and "inland," differ markedly.

Coastal snakes search about in humid areas where they find their major prey, banana slugs. Their ability to consume these creatures (Figure 15) arouses my bewildered admiration. I once made the mistake of picking up a banana slug; 10 minutes later I was still at the kitchen sink scrubbing frantically, trying to remove the repulsively sticky mucus that the slug applied liberally to my hand. Slugs do

15 A coastal Californian garter snake about to consume a banana slug, a favorite food of snakes in this region. *Photograph by Stevan Arnold.*

not live in inland northern California, and not surprisingly, the inland snakes find other things to eat, primarily fish and frogs, which they capture while swimming in lakes and streams.

Stevan Arnold was intrigued by these differences in the behavior of the snakes from the two regions [32]. He first attempted to learn the proximate causes of the behavioral variation within this species. To determine whether there were genetic differences involved, he took pregnant female snakes from the two populations into the laboratory, where they were held under identical conditions. When the females gave birth to a litter of babies (garter snakes produce live young rather than laying eggs), each baby was placed in a separate cage, away from its littermates and its mother, to remove these possible environmental influences on its behavior. Some days later Arnold offered each baby snake a chance to eat a small chunk of freshly thawed banana slug by placing it on the floor of the young snake's cage. Naive young coastal snakes usually ate all the slug hors d'oeuvres they received; the inland snakes usually did not (Figure 16). In both populations, slug-refusing snakes did not even make contact with the slug food, but ignored it completely.

Arnold took another group of isolated newborn snakes that had never fed on anything and offered them a chance to respond to the *odors* of different prey items. He took advantage of the readiness of newborn snakes to flick their tongues at, and even attack, cotton swabs that have been dipped in fluids from some species of prey (Figure 17). Chemical scents are carried by the tongue to the vomeronasal

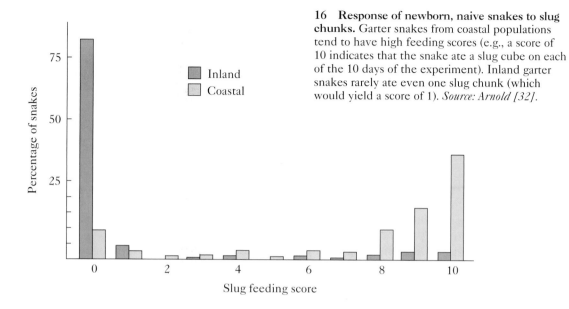

16 Response of newborn, naive snakes to slug chunks. Garter snakes from coastal populations tend to have high feeding scores (e.g., a score of 10 indicates that the snake ate a slug cube on each of the 10 days of the experiment). Inland garter snakes rarely ate even one slug chunk (which would yield a score of 1). *Source: Arnold [32].*

organ in the roof of the snake's mouth, where the odor molecules are analyzed as part of the process of detecting prey. By counting the number of tongue flicks that hit the swab during a 1-minute trial, Arnold measured the relative responsiveness of inexperienced baby snakes to different odors.

17 A tongue-flicking newborn garter snake senses odors from a cotton swab that has been dipped in slug extract. *Photograph by Stevan Arnold.*

Populations of inland and coastal snakes reacted about the same to swabs dipped in toad tadpole solution (a prey of both groups), but behaved very differently toward swabs daubed with slug scent (Figure 18). Within each group, not every snake responded identically, but almost all inland snakes ignored the slug odor, whereas almost all coastal snakes flicked their tongues enthusiastically. Because all the young snakes had been reared in the same environment, the differences in their willingness to eat slugs and to tongue-flick in reaction to slug odor must have been caused by genetic differences among them.

Arnold then did a heritability study of the sort we discussed earlier in the context of human twin studies. By comparing the tongue-flick scores of siblings, Arnold determined that *within* each population, only about 17 percent of the *differences* in chemoreceptive responsiveness to slug odor stemmed from genetic differences among individuals. The low heritability of tongue flicking within the coastal or the inland population simply means that what little phenotypic variation exists *within* each population is largely caused by subtle environmental differences.

If the feeding differences *between* the two populations arise because most coastal snakes have a different allele or alleles than most inland snakes, then crossing

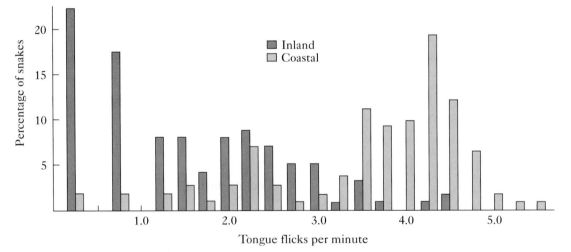

18   Odor preferences of inland and coastal garter snakes as measured by the frequency of tongue flicking in response to cotton swabs dipped in slug extract. Coastal snakes tongue-flicked much more than inland snakes. *Source: Arnold [32].*

adults from the two populations should generate a great deal of variation in the resulting group of "hybrid" offspring. Arnold conducted the appropriate experiment and found evidence in support of the expected result, confirming again that the differences *between* populations have a strong genetic component.

Having identified a genetic and physiological basis for the differences in food preference between coastal and inland garter snakes, Arnold turned his attention to the evolutionary basis for these differences. He proposed that among the original colonizers of the coastal habitat were a very few individuals that carried the then rare allele(s) for slug acceptance. (The snake almost certainly colonized coastal California more recently than inland western North America.) These slug-eating individuals were able to take advantage of an abundant food resource in their new habitat. If, as a result, their reproductive success was as little as 1 percent higher than that of their slug-rejecting fellows, the coastal population could have reached its present state of divergence from the inland population in less than 10,000 years.

It is easy to imagine why slug-accepting alleles might enjoy an advantage and spread rapidly in coastal populations. But why would they be actively selected against in (and nearly eliminated from) inland populations? Arnold found that slug-eating snakes will also consume aquatic leeches. These blood-sucking animals are absent in coastal California, but plentiful in inland lakes. It is possible, though unproven, that snakes that try to eat leeches will be damaged by their "victims." Leeches can survive even after being swallowed by a garter snake, and if they attached themselves to the wall of the snake's digestive tract, they might seriously injure their consumer. Therefore, snakes with leech-accepting alleles might reproduce less well in inland California, eliminating from this population the alleles that also led to the development of the ability to detect and attack slugs.

Thus, the geographic differences in the feeding behavior of the snakes can be explained in terms of their proximate basis: different alleles predominate in the two populations, which cause the development of differences in the chemoreceptors that detect certain molecules present in slugs and leeches. At the ultimate level, certain genes and chemosensors are associated with different fitnesses in different areas: inland snakes must contend with potentially dangerous leeches, whereas coastal snakes are exposed only to sticky but edible slugs. This difference in the snakes' environments influences the survival and reproductive success of behaviorally different phenotypes in the two areas [32]. Because natural selection is acting in different directions in the two populations, behavioral and hereditary differences now characterize snakes from the two regions. Thus, Arnold's work integrates both proximate and ultimate levels of analysis, while also suggesting why genetic variation persists in the garter snakes he studied.

SUMMARY

1. Genetic differences among individuals may result in behavioral differences among them. To demonstrate that a particular allele contributes to the development of a behavioral characteristic is not to say that the trait is "genetically determined." The statement, "There is an allele for IQ score, or slug acceptance, or copulation duration" is shorthand for the following: "A particular allele in an individual's genotype codes for a distinctive protein whose contribution to the biochemical reactions within cells may influence the development of the physiological mechanisms underlying a particular behavioral ability."

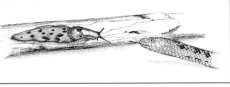

2. Researchers in behavior genetics have developed several ways to test the hypothesis that specific behavioral differences between individuals are caused by genetic differences. These include comparing parents and offspring (and other relatives), twin studies, "hybridizing" members of different populations, creating genetic mosaics, and conducting artificial selection experiments. The genetic difference hypothesis has been supported in many, but not all, cases.

3. Because behavioral differences among living animals can often be traced to genetic differences among them, we can conclude that behavior readily evolves by natural selection, a process that requires hereditary variation within a population. But if natural selection has shaped the evolution of a species, why hasn't it eliminated most genetic variation affecting behavioral traits? Factors that may help maintain hereditary differences within species include frequency-dependent selection and changes in the direction of natural selection.

4. Natural selection appears to be operating in favor of different genes and different prey preference phenotypes in a species of garter snake that occupies two geographic regions in California. A detection and feeding preference for slugs does not confer the same fitness benefits in coastal and in inland regions. Therefore, alleles that have increased in frequency in one area have not done so in the other. Differences in the reproductive success of different behavioral phenotypes have resulted in rapid evolutionary divergence in the feeding behavior of snakes in the two locations.

SUGGESTED READING

Stevan Arnold's comprehensive study on the genetics, physiology, and ecology of garter snake feeding behavior [32] should be read by everyone interested in behavior genetics. Jeffrey Hall has recently written a major review of the genetic basis of courtship and mating in *Drosophila* fruit flies [498]. Research on the behavioral similarities of twins is presented by Thomas Bouchard and his colleagues [113]. The entire field of behavior genetics is reviewed in books by Lee Ehrman and Peter Parsons [335], and by Jeffrey Hall, Ralph Greenspan, and William Harris [499]. The subject of the maintenance of hereditary polymorphism is nicely covered by Skuli Skulason and Thomas Smith [1118].

DISCUSSION QUESTIONS

1. The calls of frogs differ distinctively from species to species. You wish to test the hypothesis that hereditary differences contribute to the calling differences between two closely related species, the gray tree frog and the pinewoods tree frog. How could you take advantage of the ability to produce hybrids in the laboratory simply by taking a male from the back of an egg-laying female of his own species and switching him to an egg-laying female of the other species? Make predictions about the calling behavior of hybrid males and the response to these mate-attracting calls by hybrid females. See [305] after developing your predictions and tests.

2. A few blackcap warblers live year-round in southern France, although 75 percent of the breeding population migrates to and from this area. Let's propose that the difference between the two types of warblers is environmentally induced. Make a prediction about the outcome of an artificial selection experiment run over several generations in which the experimenter tries to select for both nonmigratory and migratory behavior. Describe the procedure and present your predicted results graphically. Check your predictions against the actual results as shown in [85].

3. Someone repeats Arnold's experiment with garter snakes, but instead of checking on the behavior of newborn garter snakes from coastal and inland parents, he compares rates of slug acceptance in snakes captured as adults from the two regions. He finds that coastal snakes are far more likely to eat slugs than are inland snakes. He concludes that the genetic difference hypothesis is supported by these results. You challenge his conclusion on what grounds?

4. Analyze the science behind Carol Lynch's study of nest-building behavior in mice described in this chapter. What question was she trying to answer? What hypothesis or hypotheses did she propose? What prediction(s) and test(s) of the hypotheses did she produce? What conclusion did she reach?

5. You hypothesize that the difference among fruit flies in their activity rhythms is caused by differences in a single gene, the *per* gene, which you believe is on the X chromosome. Fruit flies are diploid, and sex is determined chromosomally (typical males are XY, females XX). You cross a male without an activity rhythm with a female with the wild-type activity pattern. If your hypothesis is correct, what prediction must be met about the behavior of the offspring of this cross?

CHAPTER 4

The Development of Behavior: The Role of the Environment

EVELOPMENT REQUIRES MORE THAN THE GENES that individuals inherit from their parents. If you were to remove the yolk from a fertilized white-crowned sparrow egg, even without harming a single gene in the nucleus of the cell, the development of the white-crowned sparrow would never get off the ground. The environmentally supplied nutrients in the yolk are essential ingredients for the construction of white-crowned sparrow bodies. The full development of white-crown behavior also depends on the hormones produced by specialized cells within the birds' bodies, as well as some kinds of sensory stimulation, especially the acoustical experiences provided by other singing white-crowned sparrows (see Chapter 2). All of these things—nutrients, hormones, social experience—are elements of the environment that have much to do with development. Nor is there anything fundamentally special about white-crowned sparrows in the broad range of environmental factors that affect their behavioral development. As we shall see shortly, the position of an embryonic mouse in its mother's uterus can affect its aggressive behavior as an adult. Or the experience of storing food can change the course of brain development in a marsh tit. Or a scarcity of pollen in a honeybee hive can cause a worker bee to switch much sooner than usual from being a nurse bee to becoming a food collector for her colony. The way in which the environment of a developing animal can have these effects is the subject of this chapter, which concludes with a look an important evolutionary question: Why has natural selection sometimes favored mechanisms that enable animals to be developmentally flexible in response to their environments, while at other times favoring systems that promote a standard behavioral outcome for most individuals, despite differences in their environments?

The Interactive Theory of Development

Most researchers studying behavioral development base their work on the theory that the development of an organism's attributes is caused by a complex interaction between genotype and environment. Consider the differences in the behavior of male and female rats. Males tend to be more aggressive toward members of their own sex and to wander over larger areas than females. Females have their own behavioral attributes, including a special precopulatory posture and a variety of parental traits, such as retrieving pups that have wiggled away from the nest. Why these differences?

The interactive theory of development suggests that male or female behavior arises from an interplay between a rat's genetic makeup and its environment. The basic pattern is reminiscent of the one we have already described for zebra finches (see Chapter 2), with a sex-determining chromosomal mechanism leading to the production of testes in embryonic males, while females acquire ovaries. A male embryo's testes manufacture testosterone, while the female's ovarian cells do not. This hormonal difference activates a developmental switch mechanism that endows males with a masculinized brain capable of male behavior, while supplying females with a feminized brain that promotes female behavior (Figure 1) [793].

The hypothesis that testosterone is an internal signal that changes the developmental trajectory of target cells elsewhere in the body, especially in the nervous system, has been tested in mammals using the same kinds of experimental manipulations of hormones that we described for zebra finches. Inject a small amount of testosterone into a newborn female rat, and when she is an adult, she will attempt to copulate with other females—provided that she receives another injection of testosterone at this time. Her brain became masculinized as a result of exposure to testosterone early in life, and the masculinized components of her brain that control male copulatory behavior can be activated by additional testosterone, which

A   Female development   B   Male development

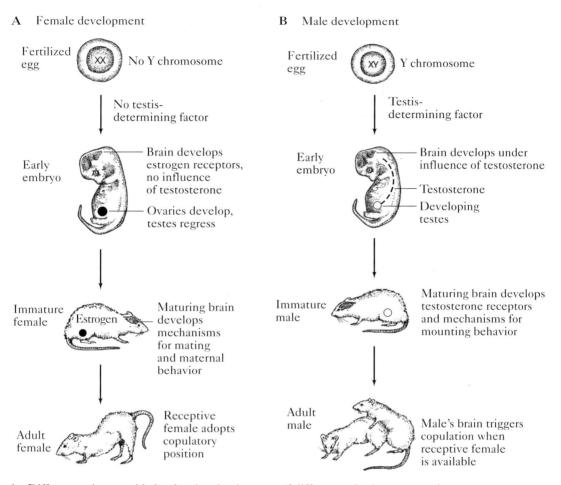

1   Differences in sexual behavior develop because of differences in the gene–environment interactions occurring within the bodies of the females (A) and males (B). In the white rat, as in mammals generally, chromosomal (genetic) differences between females and males are present in fertilized eggs. These differences initiate a cascading series of differences in the development of neural mechanisms underlying the sexual behavior typical of females and males.

in an adult male would be produced in abundance by his testes, organs that the experimentally altered female lacks.

Conversely, if one removes the testosterone-producing gonads from newborn male rats, they develop female-like brains, which order them to adopt the precopulatory posture of females (upon receipt of the proper stimulation) if they have also received estrogen treatment as adults [750].

Hormones in the Uterine Environment of Mouse Embryos

The hypothesis that mammalian brains can be masculinized or feminized by the hormonal environment of very young animals has been examined in a novel way with laboratory mice [1224]. In a litter of mouse embryos, some embryos will by chance be sandwiched between two sisters, while others will rub shoulders with one male and one female sibling; still others will happen to lie between two brothers. As fetuses develop, their cells release any number of biochemical products, among them sex hormones, which diffuse into fetal siblings via the amniotic fluid that surrounds the embryos. Therefore, a 2M embryo (one between two males) and a 0M embryo (one between two females) are exposed to different concentrations of male and female sex hormones.

In order to determine whether these slight differences might have an effect on the behavior of males, Frederick vom Saal and his colleagues delivered mouse pups via cesarean section to document their positions in the uterus. Males were then castrated and later given hormonal implants of the male hormone, testosterone, to ensure that any differences between individuals could be attributed solely to the effects of early uterine position on brain development. 2M males, whose embryonic brains had received a little extra testosterone from their brothers, behaved more aggressively than 0M males when they were given replacement testosterone (Figure 2). Apparently the brains of 2M males were more strongly masculinized early in life than the brains of 0M males, which had experienced the feminizing influence of estradiol (a form of estrogen) derived from their embryonic sisters [1224].

Likewise, females that have been exposed embryonically to different levels of hormones behave differently as adults. In populations of wild house mice living on unmowed grassy patches formed by highway cloverleafs, 2M females occupied significantly larger areas than did 0M females (Figure 3) [1331]. Since male

2   Organizational effect of hormones on the development of male behavior in mice. (A) The levels of estradiol, a female hormone, are higher in the amniotic fluid surrounding embryonic 0M males (sandwiched between two sisters) than in that surrounding 2M males (between two brothers). (B) At three months of age, 2M males are more likely than 0M males to attack another mouse when placed in an arena with the stranger for eight 10-minute trials spread over 16 days. Tests of aggression were conducted with males that had been castrated as newborns, but later received testosterone implants. *Source: vom Saal et al. [1224].*

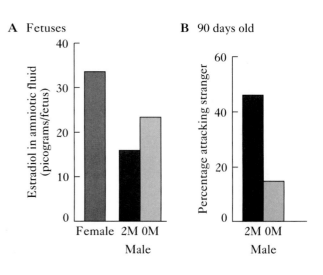

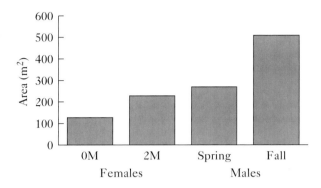

3 **Home range sizes of female mice** that developed embryonically between two sisters (0M females) and between two brothers (2M females) compared with the home range of males in the spring and fall. *Source: Zielinski et al [1331].*

house mice typically have larger home ranges than females, these results suggest that females bathed in relatively high concentrations of testosterone in their mother's uterus become behaviorally masculinized as a result. Further support for this conclusion comes from observations that 2M females are more aggressive and less sexually attractive to males than female littermates that did not develop between two embryonic brothers [197].

Hormones and the Division of Labor in Honeybee Colonies

Hormones are a critically important part of the cellular environment that affects behavioral development in mammals—and in some insects as well, including the honeybee. The typical honeybee colony contains tens of thousands of sterile worker bees, which labor on behalf of the queen, caring for her larval offspring and gathering pollen or nectar from flowers for transport back to the hive. At any one time, some bees dedicate themselves to "nurse duties," feeding larvae in their brood cells, capping cells that contain full-grown larvae, and the like, while others are full-time foragers. What causes different workers to do different things?

Studies of marked workers of known age have shown that the division of labor is usually based on age (Figure 4) [1074]. The younger workers stay inside the hive for their first three weeks of life, where they care for the brood; the same individuals leave the hive in their fourth week to track down food for their nestmates during their last week or so of life. As workers age, they undergo hormonal changes that are believed to be responsible for their shifts in behavior. Young nurse workers have very low levels of juvenile hormone circulating in their blood (or haemolymph), whereas older foragers have much higher levels of this hormone. An increase in juvenile hormone appears to trigger changes in the brain of the honeybee, altering the size of some parts of the "mushroom bodies" (Figure 5). Bees that have been exposed to a chemical that mimics juvenile hormone, but that have been prevented from leaving the hive to forage, undergo alterations in their mushroom bodies, demonstrating that foraging experience is not necessary for these neural changes to occur. Instead, it appears that juvenile hormone triggers modi-

4 Development of worker behavior in honeybees. The tasks adopted by worker bees are linked to their age. *After Seeley [1073].*

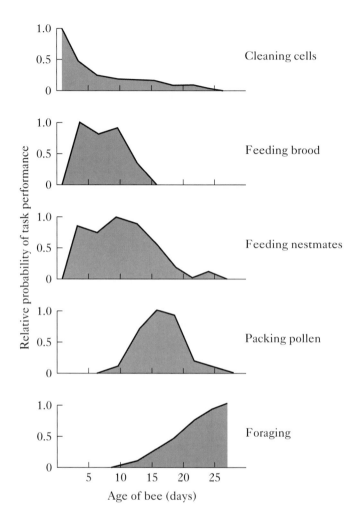

fications in the brain in anticipation of the new needs of foragers, perhaps helping them learn spatial landmarks so they can travel back and forth from the hive to distant patches of flowers [1302].

But what causes the changes in juvenile hormone that take place within the bodies of honeybee workers? As it turns out, these hormonal changes are not absolutely fixed with worker age. This conclusion is based on experiments with colonies that have been manipulated so that all the workers are the same, relatively young, age. Under these conditions, a division of labor still manifests itself, with some individuals remaining nurses much longer than usual while others start foraging as much as two weeks sooner than average. As a result, the brood is cared for continuously while the colony also receives food supplies.

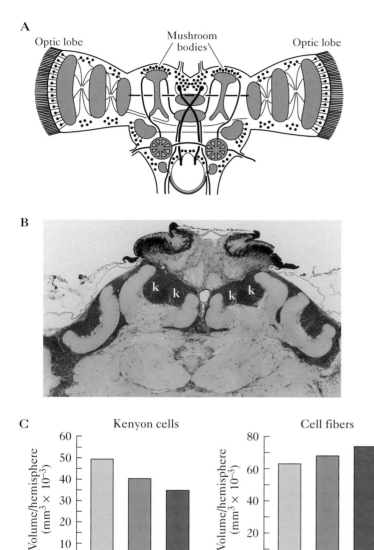

5 Developmental changes in worker bee behavior are correlated with changes in the bee's brain over time. (A) The location of the mushroom bodies in the brain of the honeybee. (B) A photomicrograph of a bee's mushroom body, showing the dark Kenyon cell region (k) and the fibers leading from the cell bodies to other regions of the brain (the lighter stalk of the mushroom body). (C) Changes in the volume of the Kenyon cell bodies (left) and cell fibers (right) of the mushroom body are linked with changes in the tasks performed by the worker bees. *Sources: (A, B) Fahrbach and Robinson [375]; (C) Withers et al. [1302].*

What enables the bees to make these adjustments? Workers inside the colony move about, presumably monitoring the presence of brood and food stores as well as interacting with their fellow workers. These environmental factors provide stimuli that the bees detect and evaluate. For example, in colonies with many cells containing larvae, or even larval scents, workers collect more pollen than in colonies without these stimuli [603]. Larvae need food, and the proximate mechanisms controlling worker behavior reflect that demand. Likewise, in colonies from which cells containing stored pollen have been removed by researchers, lowering the amount of pollen available, the bees boost their pollen-collecting effort greatly, restoring the quantity of pollen in the hive to its original level in a week or so [387].

In addition to a shortage of food, a deficit in social encounters with older foragers may stimulate the developmental transition from nurse to forager behavior. This possibility has been tested by adding groups of older foragers to experimental colonies made up of only young workers. Under these circumstances, the young nurse bees do not undergo an early transformation into foragers, even when the entrance of the hive is sealed so that the transplanted older bees cannot collect pollen for the colony (Figure 6) [580]. The behavioral interactions between the young residents and the *older* transplants must inhibit the development of foraging behavior, because transplants of *young* workers have no such effect on young resident bees.

One way in which cues in the hive environment might alter the development of worker bees is by changing their hormonal status. Those young bees whose juvenile hormone levels rise in reaction to certain food or social signals would tend to become foragers earlier than otherwise. Do early foragers have higher juvenile hormone levels than nurse bees, including nurses that remain in the business of brood care longer than is customary when there are no very young workers to fill this role? Yes, precocious foragers do have higher concentrations of juvenile hor-

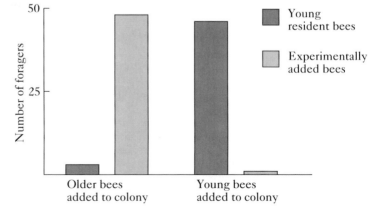

6   Social environment and task specialization by worker honeybees. In experimental colonies composed exclusively of young (resident) workers, the young bees do not forage if older forager bees are added to their hive. But if young (nonforaging) bees are added in-stead, the young residents develop into foragers very rapidly. *Source: Huang and Robinson [580].*

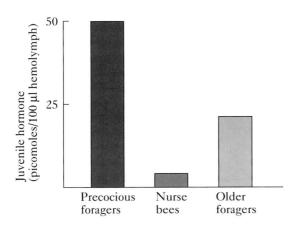

7   Juvenile hormone levels in precocious and older foragers and in nurse bees the same age as the precocious foragers. Data from two honey-bee colonies have been averaged here. *Source: Robinson et al. [1005].*

mone than their fellow workers who stay in the nurse role (Figure 7) [1005]. Thus, the within-colony environment of honeybee workers can affect their behavioral development, with workers adjusting the standard pattern of age-related changes in task specialization in response to special conditions in their hive.

Early Experience and Behavioral Development

In animals as different as honeybees and birds, sensory experiences participate in the gene–environment interactions underlying behavioral development. In some animals, the timing of these experiences has a profound effect on development, as we have seen in song learning by sparrows. Likewise, the normal structural development of the visual system in mammals is highly dependent on the infant's early exposure to visual stimuli [581]. The development of certain behaviors in geese and ducks also requires certain kinds of sensory inputs soon after the birds have hatched, when the youngsters normally follow their mother away from the nest [738]. Konrad Lorenz studied this phenomenon by rearing a clutch of greylag goose eggs and permitting the goslings to waddle after him when they were old enough to leave the hatching box. By virtue of this experience, the baby birds formed an attachment to Lorenz. They trailed after him wherever he went, instead of following an adult female of their own species. Because he was the first object they had followed after hatching, they had evidently learned to recognize Lorenz as an individual, just as they would normally have learned to identify their mother.

Imprinting

Social learning of this sort that is based on early experience is called **imprinting,** and it occurs in some mammals as well as in birds. For example, young shrews of a European species will follow their mother by holding onto the fur of another shrew (either the mother or a sibling). The mother then sets off with a conga line of babies trailing behind her (Figure 8). Experiments have shown that between 5 and 14 days after birth, the baby shrews become imprinted on the odor of the individual that is nursing them [1332]. Usually, of course, their caretaker is their mother, and she alone induces caravan formation by her youngsters. However, if 5-day-old shrews are given to a substitute mother of another species, they will become imprinted upon her, and when returned to their biological mother at 15 days of age, will not follow her or any siblings they rejoin. They *will* follow a cloth impregnated with the odor of the foster mother, a response that demonstrates that they learned the scent of the female that nursed them when they were younger.

In addition to these short-term effects of imprinting, some remarkable long-term developmental effects occur as well. As adults, the male greylag geese that imprinted on Konrad Lorenz courted human beings—including, but not specifically limited to, Lorenz—in preference to members of their own species. The experience of following a particular individual early in life must somehow alter those regions of the goose's nervous system responsible for sexual recognition and courtship.

The relation between imprinting and adult sexual behavior in zebra finches has been studied by Dave Vos [1227]. He permitted some birds with all-white plumage to breed and produce young. Typical zebra finches have a complex and variable plumage color pattern, but Vos eliminated this complication by using albinos. When the nestlings were about 8 days old, just before their eyes opened, he painted the parents' bills different colors. If he applied orange nail polish to the mother's bill, he gave the father's bill a coat of red nail polish, and vice versa. He periodically freshened up the bill paints for the next 7 weeks, after which the youngsters were removed to cages where they were held in visual isolation from other

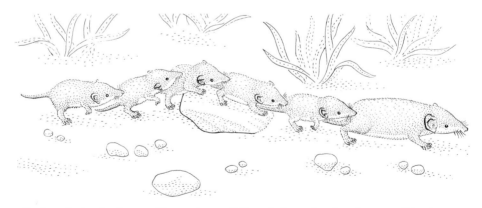

8 Imprinting in shrews. Young shrews follow their mother in a "caravan," having learned her odor very early in life. *After Zippelius [1332].*

zebra finches until they became sexually mature 10 weeks later. Then he ushered each subject into the central compartment of a special cage that had two small side cages attached to it, each containing a white zebra finch, the "stimulus" finch.

In one of Vos's experiments, one of the two "stimulus" finches had its bill painted red, and the other had a bill coated with orange nail polish. The young zebra finch being tested therefore had a choice of associating with a bird with a bill the same color as its mother's or spending its time near the side cage containing a bird with a bill the same color as its father's. Vos recorded the total time per 20-minute trial that the young adults spent near each of the two cages. In one of his experiments, he found that 12 of 14 male subjects preferred to approach and sing to the stimulus bird that had the same bill color as their mother, whether or not that bird was a female; all that counted was its bill color. In fact, in another experiment in which the young birds were given a choice between males with the same bill color as their mother versus females with same bill color as their father, the male subjects spent more time courting stimulus males than stimulus females (Figure 9). Thus, nestling male zebra finches somehow distinguish their mother from their father, and are designed to incorporate information about their mother's appearance for much later use in recognizing potential mates.

No such mechanism exists in young female zebra finches, which in Vos's experiments consistently approached and observed the male when the side cages contained a male and a female stimulus bird. The bill colors of the stimulus birds were not a factor in female preferences, which were apparently much more influenced by the behavior of males than by their appearance. In similar choice studies with naturally plumaged birds, females spent their time next to cages with males that sang frequently, rather than preferring red-billed or orange-billed males [215].

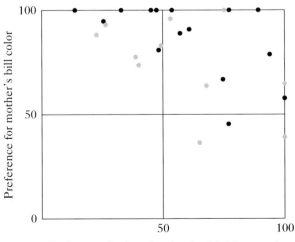

9 Imprinting in zebra finches. Male zebra finches reared in captivity with mothers whose bills had been painted red or orange were given the opportunity to court individuals (males or females) with (or without) the color of their mother's bill. When they had equal opportunity to court a *male* with their mother's bill color or a *female* with a different-colored bill, the imprinted males usually courted the male more often than the female. Shaded circles represent birds given a choice between a female with an orange bill and a male with a red bill; black circles, between a female with a red bill and male with an orange bill. Half the subjects in the experiment had been reared with a mother with a red bill, and half with a mother with an orange bill. *Source: Vos [1227].*

Thus, the effect of early visual experience on the development of adult sexual behavior can vary markedly between the sexes, presumably because male and female zebra finches differ genetically, and this difference affects the nature of the imprinting mechanisms in their bodies.

Early Experience and Recognition of Kin

When a baby shrew or male zebra finch imprints on its mother, it has learned to recognize a close relative. Imprinting is thus one proximate mechanism of **kin discrimination,** the differential treatment of members of the same species in a way that depends on their genetic relatedness to the discriminating individual [1087]. The ability to respond differently to individuals of varying degrees of genetic similarity occurs in many animals that exhibit parental care (see Chapter 14) and in social species in which relatives cooperate (see Chapter 15). As we shall discover in the chapters ahead, kin discrimination facilitates actions that may help propagate the discriminating individual's genes. Here our focus is on the special proximate learning mechanisms that promote the development of kin recognition.

One such mechanism for recognizing kin may cause the young animal to attend to the distinctive sensory cues associated with its early companions, which are usually its siblings. Thus, for example, young captive spiny mice prefer to huddle with individuals with whom they associated in their litter as opposed to individuals with whom they have no prior experience. In nature, this preference would ordinarily lead siblings to remain together, but if one experimentally creates a litter composed of nonsiblings that are cared for by the same female, these unrelated littermates will treat each other as if they were siblings [953]. This "discrimination error" shows that the young of this species learn who their littermates are.

Similar errors can be induced in some insects as well. David Pfennig and his co-workers suspected that paper wasp females absorb odors from their nest [930], which would mean that, under natural circumstances, siblings would share a similar nest odor. This shared scent could provide a cue that would enable females to treat sisters one way and nonsisters in another fashion. If this hypothesis is correct, it should be possible to fool paper wasps into tolerating nonkin by transferring newly emerged queens to a foreign nest, where they can learn the odor of this nest and their new nestmates. As predicted, queens that participated in this experiment were especially tolerant of unfamiliar females that had been exposed to separate fragments of the nest in which they had been placed. Thus, paper wasps employ a rule of thumb that reads, "Treat as relatives those individuals that smell like the nest in which you have been reared."

Another proximate mechanism based on odor may promote kin discrimination in Belding's ground squirrels. To study this phenomenon, Paul Sherman captured some pregnant females and shipped them to Warren Holmes's laboratory. When two females gave birth close in time, Holmes switched some of the pups to create four classes of juveniles: (1) siblings reared apart, (2) siblings reared together, (3) nonsiblings reared apart, and (4) nonsiblings reared together. Foster pups were readily accepted by the adult females.

When the juveniles had reached the postweaning stage, pairs were placed in an arena and given a chance to interact. Animals that were reared together, whether siblings or not, generally treated each other nicely, whereas animals that had been reared apart were likely to react aggressively to each other. This experiment shows that the little squirrels learn something from the experience of growing up together, and that they use this information in their social relations [567], as do other ground squirrels [566].

But perhaps the most remarkable finding from the study of Belding's ground squirrels was that biological sisters *reared apart* engaged in a significantly lower rate of aggressive interactions than nonsiblings reared apart (Figure 10). In other words, sisters have some way of recognizing one another—perhaps an odor similarity—that is not dependent on the experience of sharing a mother and a burrow [565, 567].

Holmes and Sherman have given the label *phenotype matching* to the mechanism underlying kin discrimination by female Belding's ground squirrels and other animals like them. These creatures apparently learn something about their own phenotype—their appearance or odor or some other cue—and then discriminate among others on the basis of how similar those individuals are to them [568]. Such a mechanism can have the effect of causing individuals to behave differently toward others in a manner correlated with their degree of relatedness.

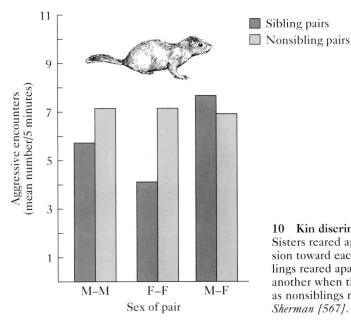

10 Kin discrimination in Belding's ground squirrels. Sisters reared apart display significantly less aggression toward each other than other combinations of siblings reared apart, which are as aggressive to one another when they meet in an experimental chamber as nonsiblings reared apart. *Source: Holmes and Sherman [567].*

Jerram Brown and Amy Eklund [149] have reviewed a fascinating possible contributor to this process: the genes that code for a special class of proteins, the major histocompatibility complex (MHC) glycoproteins. These proteins occupy positions in cell membranes, where they monitor the environment outside the cell for invader organisms. MHC glycoprotein chemistry is such that these proteins react differently to other genetically identical cells of the body in which they reside than to genetically different cells from another individual—typically a disease organism of some sort, such as a virus or bacteria. In effect, they discriminate between self and nonself, and order an immune response to objects that activate the nonself reaction.

Although the MHC genes presumably evolved in the context of fighting disease-causing organisms, they may now also be involved in learned kin recognition in some animals. The MHC genes come in many alleles (as many as 50 per gene in human beings and house mice), and no one allele predominates. As a result, different individuals tend to have different MHC genotypes. Furthermore, the MHC alleles that a mouse has somehow affect the odors it produces, as has been demonstrated by the ability of inbred strains of mice to distinguish among individuals that differ only in their MHC genes. Mice of these strains prefer to mate with partners that have different MHC genes, as do mice living under seminatural conditions [954]. Perhaps distinctive MHC-coded odors contribute to the learned kin discrimination in animals capable of phenotype matching.

Learning as Behavioral Development

The male zebra finch imprinted on its mother's bill color and the Belding's ground squirrel that knows what its nestmates smell like have both learned certain things from their environment. The information these animals acquire leads to long-lasting changes in their behavior, presumably through permanent modifications of the neural mechanisms underlying their behavior. Therefore, imprinting and learned kin discrimination can both be considered a kind of behavioral development based on cellular changes dependent on preceding gene–environment interactions [375].

The fundamental similarity between developmental processes and learning applies to spatial learning as well. Many species, including beewolf wasps (see Chapter 1), memorize visual landmarks in order to move efficiently from place to place. Similarly, some birds possess excellent spatial memory. In a single day, a black-capped chickadee may hide several hundred seeds or small insects in bark crevices or patches of moss scattered over a wide area. In nature, the birds store only one food item in each hiding spot and never use the same location twice, and yet chickadees remember precisely where they have hidden their food, for as long as 28 days after the event [548].

David Sherry provided captive chickadees with a chance to store food in holes drilled in small trees placed in an aviary. After the birds had placed a sunflower seed in 4 or 5 of 72 possible storage sites, they were shooed into a holding cage for 24 hours. Sherry removed the seeds in the interim and closed each of the 72 storage sites with a Velcro cover. When the birds were released back into the aviary to search for the food, they spent much more time inspecting and pulling at the covers at their hoard sites than at spots where they had not stored food 24 hours before (Figure 11). Because the storage holes were empty and covered, there were no olfactory or visual cues provided by stored food to guide the birds in their search; they were relying on their memory of where they had hidden food [1089].

Clark's nutcrackers may have an even more impressive memory, for these birds scatter as many as 9000 caches of pine seeds (1–10 seeds per cache) over entire hillsides. The bird digs a little hole for each store of seeds, and then completely covers the cache. A nutcracker does this work in the fall, and then relies on its stores through the winter and into the spring, so that it may be months before it comes back to retrieve the seeds from a particular cache [48].

It could be that nutcrackers do not really remember where each cache is, but instead rely on a simple rule of thumb, such as "caches will be made near little tufts of grass." Or they might only remember the general location where food was stored, and once there, look around until they saw signs of caching. But experiments similar to those performed with chickadees show that the birds do remember exactly where they hid their food. In one such test, a nutcracker was given a chance to store seeds in a large outdoor aviary, after which it was moved to another cage. The

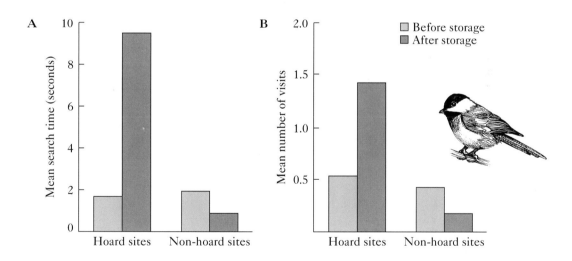

11 Spatial learning by birds. (A) Black-capped chickadees spent much more time searching parts of an aviary where they had stored food 24 hours previously (hoard sites) than they did during their initial exposure to those sites, even though experimenters had removed the stored food. (B) The chickadees also made many more visits to hoard sites than to other sites, evidently because they remembered having stored food there. *Source: Sherry [1089].*

observer, Russell Balda, mapped the location of each cache, then removed the buried seeds and swept the cage floor, removing any signs of caches. Thus there were no visual or olfactory cues available to the bird when it was permitted to go back to the aviary a week later and hunt for food. Balda mapped the locations where the nutcracker probed with its bill, searching for the nonexistent caches. The bird's spatial memory served it well, for it dug into as many as 80 percent of its ex-cache sites, while only very rarely digging in other places [48]. Additional experiments of this sort have demonstrated that nutcrackers can retain information about where they have hidden food for at least 6, and perhaps as long as 9, months [49].

The ability of nutcrackers and chickadees to store spatial information in their brains is surely related to the ability of certain brain mechanisms to change biochemically, and probably structurally, in response to certain kinds of sensory stimulation. One anatomically distinct region in the brain, the hippocampus, appears to be especially important in memory storage. The hippocampus may not only be capable of storing information about hoard sites, but may actually require the sensory experiences associated with food storing behavior in order to develop properly. To test this possibility, Nicky Clayton and John Krebs hand-reared some marsh tits, close relatives of the black-capped chickadee, in the laboratory. They gave some birds opportunities to store whole sunflower seeds at three stages after hatching, while others were always fed powdered sunflower seeds, which the birds cannot and will not store [201]. The experience of storing food had a strong effect on the number of cells in the hippocampus of the marsh tit's brain (Figure 12). Birds with food storing experience at any time of life tended to have a larger hippocampus with more cells than birds that lacked opportunities to hide food. Not caching and retrieving sunflower seeds appears to result in a loss of cells from the hippocampus, a case of "use it or lose it."

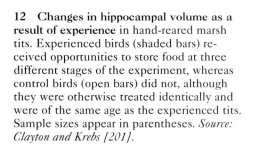

12 Changes in hippocampal volume as a result of experience in hand-reared marsh tits. Experienced birds (shaded bars) received opportunities to store food at three different stages of the experiment, whereas control birds (open bars) did not, although they were otherwise treated identically and were of the same age as the experienced tits. Sample sizes appear in parentheses. *Source: Clayton and Krebs [201].*

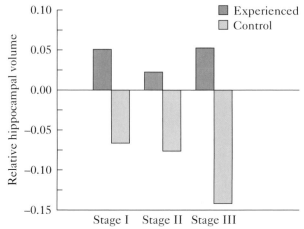

The Evolution of Behavioral Flexibility

The studies that we have reviewed here and in earlier chapters demonstrate the remarkable range of environmental effects on behavioral development. First, the "material" environment influences the construction of the physiological foundations, especially nervous systems, that make behavior possible. Change the availability of amino acids or fats or carbohydrates in the diets of a young animal, and the kinds of gene–environment interactions that occur will change as well, potentially affecting the structural development of its brain or its hormone secretions, with behavioral consequences for the affected individual.

As we have seen, however, the "experiential" environment is in many cases just as important for behavioral development as the "material" environment. Certain sounds a male sparrow hears as a fledgling or as a young adult will be recorded in his brain and later matched against his own song type, enabling the sparrow to change his singing behavior as he comes to mimic the songs of other individuals of his species. The experience of hoarding food means life instead of death for certain cells in a marsh tit's hippocampus, with consequent effects on its ability to remember where it has stored food. The experience of being in a hive without a sufficient number of older foragers triggers an early shift to the foraging role in young worker bees.

The discovery that behavioral development can shift this way or that in response to particular proximate inputs and cues raises interesting evolutionary questions. What reproductive advantage do individual marsh tits gain by having a hippocampus that can grow or decline depending on the experience of storing food? Why do honeybees have a hormonal system that reacts to conditions within the colony in ways that can alter its program for age-related shifts in activities performed for the colony? What possible benefit accrues to zebra finch males from learning what a female of their species looks like as a result of memorizing their mother's bill color?

At the proximate level, the ability of an individual to change its behavior requires mechanisms that can shift developmental pathways in response to certain kinds of environmental inputs or "signals." The genetic information that makes this possible must have survived the process of natural selection thus far, outlasting other genes that helped produce other kinds of mechanisms with no flexibility or with a different kind of developmental agility. If this argument is correct, then the particular kind of developmental flexibility currently exhibited by members of a species ought to promote the reproductive success of individuals. The question is, how?

The Flexibility to Become a Cannibal

Let's first consider the kind of developmental flexibility that results in large and obvious structural and behavioral changes. For an example of a developmental mechanism that results in two alternative forms, or *morphs*, within a species, we shall focus on the tiger salamander, in which at least some individuals can develop into either (1) the "typical" aquatic larva, which eats small pond invertebrates such as dragonfly nymphs, or (2) the cannibal type, which grows larger, has much more powerful teeth, and feeds on other tiger salamander larvae unfortunate enough to live in ponds with it (Figure 13; Color Plate 2).

The development of the cannibal morph, with its distinctive form and behavior, depends on certain environmental factors, as has been shown by experimenters who have varied the densities and genetic relationships of young salamanders in aquaria. Cannibal morphs appear only when many salamanders live together [214]. Moreover, they are much more likely to develop when the larvae differ greatly in size; typically cannibals arise from among the largest individuals in an aquarium [763]. In addition, the cannibal morph is more likely to develop when the population consists largely of unrelated individuals than when the tank holds the offspring of a single mating—in other words, a group of siblings (Table 1) [929]. An immature tiger salamander receives various sensory stimuli from the other salamanders around it, including their odor, which is the proximate cue that this species uses to recognize kin. If a larger-than-average salamander lives with many other young salamanders that do not smell like its close relatives [932], it may shift its development from the typical track to the one that turns it into a giant, fierce-toothed cannibal.

13 Tiger salamanders occur in two forms. The typical form (left) feeds on small invertebrates and grows more slowly than the cannibal form (right), which feeds on smaller tiger salamanders. Cannibals have broader heads and larger teeth than their insect-eating companions. *Photograph by James P. Collins.*

Table 1 The effect of genetic relatedness on the development of cannibal morphs of the tiger salamander in aquaria with equal larval densities

Composition of aquarium population	Cannibal develops in tank	No cannibal develops in tank	Total experiments
Siblings only	31 (40%)	46 (60%)	77
Nonsiblings present	67 (84%)	12 (16%)	79

Source: Pfennig and Collins [929].

What benefit do tiger salamanders derive from having two potential developmental systems and a switch mechanism that enables them to change their growth pattern and behavior? If numerous salamander larvae occupy a pond, and if most are smaller than the individual that becomes a cannibal, then the cannibal gains access to an abundant food source that is not being exploited by its fellows, and so grows quickly. Rapid growth enables cannibals to metamorphose into adults before their fellow pond dwellers, an important advantage in ponds that evaporate in the summer.

Flexibility in making the switch is a plus because in some ponds, would-be cannibals would starve because their prey was too scarce, or too similar to their own size to be captured. Also, for reasons we will discuss later (see Chapter 15), an individual can reduce the transmission of its own genes if it harms its relatives. Salamanders in the past that indiscriminately cannibalized their siblings would have inadvertently reduced the frequency of the genes they shared with their relatives, reducing the likelihood that their shared genes and traits would survive. Since salamanders have no way of "knowing" in advance whether their environment will be loaded with nonrelatives, selection has presumably favored individuals that happened to have the ability to secure the appropriate information after they have hatched, which determines whether or not they will adopt the developmental route leading to a cannibal's lifestyle.

Social Unpredictability and Brain Development

The kind of flexibility exhibited by tiger salamanders, which involves major structural and behavioral changes, is merely one end of a spectrum of mechanisms that involve the collection of information from the environment, leading to adjustments to unpredictable local conditions. As we shall see in Chapter 12, individuals sometimes monitor their competitive ability in relation to others, picking among a range of options in ways that maximize their reproductive chances. No major structural changes necessarily accompany these "decisions," but they are similar to the developmental switch mechanisms of tiger salamanders in helping individuals make adaptive choices in socially variable environments.

A stunning example of this phenomenon is provided by a cichlid fish that lives in Lake Tanganyika in Africa. Males of *Haplochromis burtoni* come in two types: (1)

a brilliantly colored black, yellow, blue, and red male that uses his body colors in territorial displays that repel male rivals and attract females to a nest in the lake bed, and (2) a nondescript pale brown male that is relatively inactive and definitely not territorial. The second type has been labeled the **satellite** morph because males of this type hang around the territories of the reproductively active, brightly colored males, but flee from these individuals, which charge at and collide with them (Color Plate 3).

The external differences in the structure and behavior of the two forms of males are caused by internal differences in their brains, testes, and hormones. In particular, a set of distinctive brain cells in one region of the hypothalamus (the so-called GnRH neurons) is six to eight times larger in the aggressive territorial males than in the submissive satellite males (Figure 14). These cells release a hormone that stimulates development of the testes, which in turn produce male sex hormones that cycle back to the brain, promoting aggressive behavior by modulating the activity of certain cells in that organ.

Russell Fernald has found, however, that males are not locked into one behavioral role by their current hypothalamic–testicular condition. If a territory owner is ousted after a battle with an intruder, he changes color, almost instantly adopting the drab hues of a satellite, and moves in with the satellite crowd on the edges of the nesting area. His brain changes as well, with the GnRH neurons shrinking dramatically in size [385]. But should there be a new territorial opening, as when (for example) a territorial male is killed by a predator, a satellite can quickly switch from that mode to the intensely aggressive territorial morph, quickly acquiring big GnRH neurons. Thus, brain development in *Haplochromis burtoni* is supremely flexible, with changes in neuronal structure and function arising in response to cues from the physical and social environment. The availability of suitable breeding sites and the nature of the competition from aggressive territory owners are unpredictable variables in the fish's environment. Individuals with the flexibility to adjust their internal mechanisms and behavioral output can claim a territory when possible, but adopt a camouflaged color and an energy-saving, risk-reducing, wait-and-see behavioral tactic when suitable territories are occupied by formidable defenders [403].

For all intents and purposes, the cichlid is using something very like a learning mechanism to adjust its behavior. In response to particular kinds of experience with rivals and potential nesting sites, it modifies its behavior and underlying brain structures. The same sort of thing characterizes all forms of animal learning, in which individuals use mechanisms that detect certain kinds of information in the environment to make neuronal changes that translate into adaptive behavioral adjustments.

Unpredictable Environments and Learning

To illustrate the benefits of such mechanisms, consider an unlikely example, certain Australian thynnine wasps that are famous for their apparent behavioral *inflexibility*. The males of these generally nondescript insects spend their adult lives

A

B

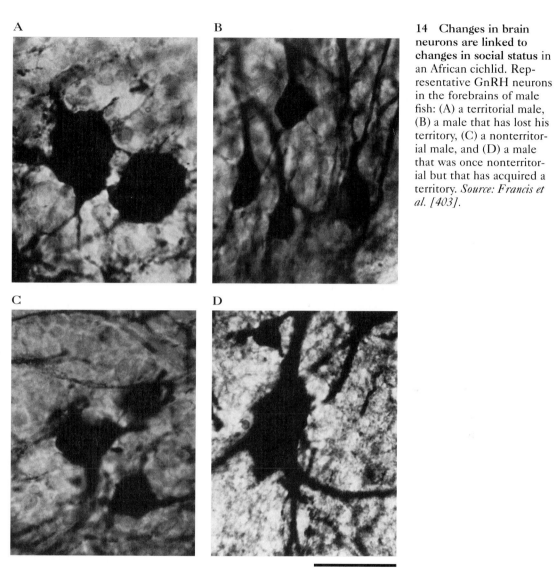

14   **Changes in brain neurons are linked to changes in social status** in an African cichlid. Representative GnRH neurons in the forebrains of male fish: (A) a territorial male, (B) a male that has lost his territory, (C) a nonterritorial male, and (D) a male that was once nonterritorial but that has acquired a territory. *Source: Francis et al. [403].*

C

D

20 μm

searching for receptive wingless females, which announce their readiness to mate by releasing a sex **pheromone,** or scent, while perched on a stem or twig (Figure 15A). Males fly to scent-releasing females and carry them away to copulate with them elsewhere [1151].

However, male wasps will also attempt to copulate with the bizarre flowers of certain small orchids that occur where they live. The flowers of these wasp-attracting

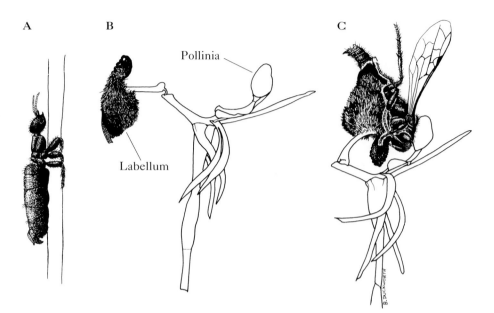

A   B   C

Pollinia

Labellum

15   Orchid deception. (A) A wingless female thynnine wasp on her perch, where she releases a sex pheromone to attract males of her species. (B) An Australian orchid whose labellum (a highly modified petal) resembles and smells like the female wasp. (C) A male wasp deceived by the orchid attempts to grasp the female decoy and fly away with it; instead he is catapulted into the pollinia of the orchid. *Drawings by G. B. Duckworth, from Peakall [914].*

orchids produce a sex pheromone that smells like the odor released by a receptive thynnine female (although the details of the chemical mimicry have not been documented in this group as well as they have for some equally devious European orchids [105]). Male wasps track this odor to an orchid, which has a special petal that only vaguely resembles the body of a wingless female wasp. The male pounces upon and tries (unsuccessfully) to pick up the petal (Figure 15B and C). In the course of the attempt, the male's body comes into contact with the pollinia, or pollen-bearing sacs, of the flower. The sticky pollinia adhere to the male, and when he finally gives up the futile task of trying to carry the "female" away, the pollinia go with him. Should he be drawn to another individual of the orchid species that deceived him, he will transfer pollen to the plant, fertilizing it rather than a female of his species (Color Plate 4).

Thus, thynnine males appear to be remarkably obtuse automatons programmed by their internal mechanisms to rush about responding to simple cues in their environment—so simple that plants can mimic those cues and cause the insects to waste their time and energy. But since one almost never sees wasps actually visiting orchids in nature, perhaps thynnines are not quite as thick-headed as they seem. In order

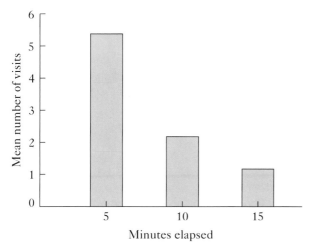

16 Learning by male thynnine wasps, which soon stop visiting deceptive orchids after they have been placed in a new location. *Source: Peakall [914].*

to observe wasp–orchid interactions, Australian botanists have to cut orchids, put them in water-filled jars, and move them to a new location. Only then will they have a chance of seeing males come rushing in to pounce upon the labellum of the orchid. After a brief frenzy of "pseudo-copulations," however, the number of males coming to a sample of cut orchid specimens falls off sharply (Figure 16).

These experiments suggest that male thynnines apparently are attracted to an orchid shortly after its flower opens and the mimetic scent is released for the first time. But after being deceived and grappling with a particular orchid's decoy, the male stores information about the experience somewhere in his brain and thereafter avoids responding sexually to the scent *coming from that particular orchid* [914]. The reproductive benefits of the male wasp's behavioral flexibility are obvious. Male wasps cannot be programmed in advance to know where their females and deceptive orchids will appear on any given day. By using experience to learn to avoid specific orchids while remaining responsive to novel sources of sex pheromone, the male wasp saves time and energy and improves his chance of encountering a receptive female that has begun to release sex pheromone from a perch somewhere in his searching zone.

The Benefits and Costs of Behavioral Flexibility

We can use the salamander and wasp examples to illustrate a theory on the evolution of developmental switch mechanisms underlying structural and behavioral changes in individuals. If a novel flexibility mechanism is to spread through a population, its benefits must exceed its reproductive costs to the individual. If the salamander and the wasp are to be our guides, flexibility confers an advantage on individuals that confront biologically important variables that cannot be known in advance, such as the population density of companions or the location of female wasps and orchids. Likewise, a nutcracker cannot know before its birth where it is going to hide several hundred or more food caches. The flexibility that learning confers enables it to recover its food stores efficiently, thereby presumably gaining energy for its reproductive efforts.

But any such mechanism comes with a price tag. Calories are expended to produce and maintain the systems that underlie the potential development of two body forms, or the neurons that record experience and make learning possible. The view that learning ability requires costly neuronal tissue is supported by a study of male long-billed marsh wrens from the western and eastern United States. Young West Coast marsh wrens have more song types in their song repertoire than their East Coast counterparts. When permitted to listen to a tutor tape in captivity, the West Coast wrens learned nearly 100 songs, whereas their East Coast counterparts incorporated only about 40 songs into their repertoire [679]. In West Coast wrens, the neural structures underlying singing behavior (the song system) weighed on average 25 percent more than the equivalent regions in the brain of an East Coast wren.

If the large brain of humans is related to our great ability to learn, as many persons have argued, then the cost of human learning mechanisms may be great indeed. Although the brain makes up only 2 percent of our total body weight, it demands 15 percent of all cardiac output and 20 percent of the body's entire metabolic budget [30—but see 797]. The costs of learning mechanisms include not just the expense of developing and maintaining these devices, but their damaging effects when they malfunction. For example, animals sometimes use experience to modify their behavior in a way that lowers, not raises, their fitness. If a rat or human being eats a novel food then, and becomes ill for some other reason, they may mistakenly form a learned aversion to a perfectly nutritious item [190].

Given the costs of behavioral flexibility, we would expect it to evolve only when an investment in the underlying mechanisms is repaid because of the nature of the animal's lifestyle. Thus, for example, species that regularly store food ought to have a larger memory apparatus and better spatial learning skills than species that cache food less often, or not at all. This prediction has been tested by comparing the spatial learning abilities of four bird species that are all members of the same family, the Corvidae, but that vary in their reliance on food storing. As we have seen, Clark's nutcracker is a food-storing specialist, and it has a large pouch for the transport of pine seeds to storage sites. The pinyon jay also has a special anatomical feature, an expandable esophagus, for carrying large quantities of seeds to hiding places. In contrast, the scrub jay and Mexican jay lack special seed transport devices and appear to hide substantially less food than their relatives.

Individuals from the four species were tested on two different learning tasks in which they had to peck a computer screen to receive rewards. One task required the birds to remember the *color* of a circle on the screen (a nonspatial learning task),

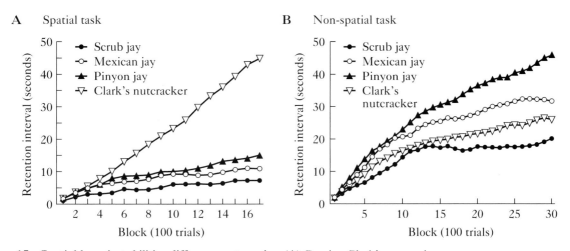

17 Spatial learning abilities differ among species. (A) Captive Clark's nutcrackers performed much better than three other jays in experiments that required the birds to retain information about the location of a circle to peck. (B) But in another experiment that examined the birds' ability to remember the color of a circle to peck, the nutcrackers did not excel in this nonspatial learning test. *Source: Olson et al. [875].*

and the other required memory of the *location* of a circle on the screen (a spatial task). When it came to the nonspatial learning test, pinyon jays and Mexican jays did substantially better than scrub jays and nutcrackers. But in the spatial learning experiment, the nutcracker went to the head of the class, followed by the pinyon jay, then the Mexican jay, and (last) the scrub jay (Figure 17). These results show that the birds have not evolved all-purpose learning abilities that apply equally to all tasks; instead, their learning skills are designed to promote success in solving the special problems that they face in their natural environments [875].

Sex Differences in Spatial Learning Ability

If behavioral flexibility evolves only when special environmental demands favor versatility, and if the sexes within a species differ in the size of their home ranges, then the sex that moves over a wider area ought to exhibit superior spatial learning skills. Steven Gaulin and Randall FitzGerald tested this prediction in studies of three species of small rodents. In one species, the polygynous meadow vole (*Microtus pennsylvanicus*), males have home ranges more than four times as large as those occupied by each of their several mates. In contrast, both prairie and pine voles (*Microtus ochrogaster* and *Microtus pinetorum*) are monogamous; males and females share the same-sized living space. When tested in a variety of mazes, which the animals had to solve in order to receive food rewards, males of the wide-ranging meadow vole consistently made fewer errors than females of their species (Fig-

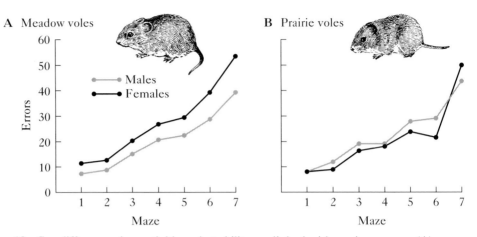

18 Sex differences in spatial learning ability are linked with mating system. (A) Polygynous male meadow voles consistently made fewer errors on average than females when learning how to get through seven different types of mazes of increasing complexity. (B) In contrast, females matched male performance in the monogamous prairie vole. *Source: Gaulin and FitzGerald [426].*

ure 18). Moreover, other researchers have found that males from litters with a large proportion of males do better at spatial tests than males from litters in which most of their siblings are females, suggesting that sex hormones influence the development of the trait [417]. In both the prairie and pine voles, however, males and females did equally well on spatial learning tests; in these species, the two sexes have similar home ranges and so are confronted with equivalent spatial learning problems in their natural lives [426, 427].

Given the role of the hippocampus in spatial learning, it is not surprising that male meadow voles invest more heavily in this structure, as measured by the proportion of brain volume it occupies, than do females of their species. In contrast, no differences in hippocampal size exist in male and female pine voles [597]. Moreover, the differences in spatial learning and hippocampal size between male and female meadow voles appear to be expressed only during the summer breeding season, when males are searching widely for mates. When the animals are not breeding, males gain no reproductive advantage from special learning skills, but would pay the metabolic price of maintaining an enlarged hippocampus. By having a brain that shrinks when the extra tissue is not useful, a meadow vole can reduce the costs of a neural mechanism required for learning [596].

In a species in which females face greater spatial challenges than males, we would expect females to make larger investments in the neural foundations of spatial learning. The brown-headed cowbird is such a species, because females search widely for nests of other birds to parasitize. They must also remember where potential victims have started their nests in order to return to them one to sever-

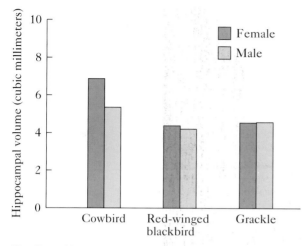

19 Sex differences in the hippocampus. Female cowbirds have a larger hippocampus than males, as predicted if this brain structure promotes spatial learning and if selection for spatial learning ability is greater on female than male cowbirds. Red-winged black-birds and common grackles do not exhibit this difference. *Source: Sherry et al. [1090].*

al days later when the time is ripe for the cowbird to add her egg to those already laid. In contrast, male cowbirds do not confront such spatial problems. As predicted, the hippocampus (but no other brain structure) is considerably larger in female brown-headed cowbirds than in males (Figure 19). No such difference occurs in some nonparasitic relatives of this species [1090]. Similarly, among the South American cowbirds, breeding females of two parasitic species also make larger investments in hippocampal tissue than males [980], but only during the breeding season, when the benefits of spatial learning are high [202].

The Evolution of Other Specialized Learning Skills

We have focused thus far on the kind of flexibility that spatial learning confers, finding that the ability to remember the location of key features of one's environment evolves in response to the special challenges associated with locating widely and unpredictably dispersed resources or mates. But are the many other kinds of learning similarly linked to special ecological problems?

Consider the widespread phenomenon of associative learning, which occurs in several forms [307], including **operant conditioning,** in which an animal learns to associate a voluntary action with the consequences that follow from performing it [1117]. Operant conditioning (or trial-and-error learning) occurs outside psychology laboratories (Figure 20), but it has been studied most extensively in Skinner boxes, named after the psychologist B. F. Skinner. After a white rat has been introduced into a Skinner box, it may accidentally press a bar on the wall of the

20 Trial-and-error learning. After one experience with a foul-tasting millipede, the toad will reject this prey on sight. *Photograph by Thomas Eisner.*

cage (Figure 21), perhaps as it reaches up to look for a way out. When the bar is pressed down, a rat chow pellet pops into a food hopper. Some time may pass before the rat happens upon the pellet. After eating it, the rat may continue to explore its rather limited surroundings for a while before again happening to press the bar. Out comes another pellet. The rat may find it quickly this time, and then turn back to the bar and press it repeatedly, having learned to associate this particular activity with food. It is now operantly conditioned to press the bar.

Operant conditioning techniques have been used with success to guide the learning of responses far more complex than pressing a bar, as demonstrated by the ability of Skinnerians to train pigeons to play table tennis with their beaks or to "communicate" symbolically with one another [364]. We have already mentioned earlier in this chapter that operant conditioning techniques have been used to test the spatial learning abilities of nutcrackers and jays. They have even been employed to condition humans and other animals to regulate internal processes such as heart rate or brain electrical activity, which at one time were thought to be entirely involuntary [97].

Although Skinnerian psychologists once claimed that it was possible to condition almost any operant—any action that an animal could perform—with equal ease, workers in this field soon began to discover contradictory cases. For example, one can readily condition a white rat to do some things, such as running in a running wheel, by sounding a warning noise and giving it an electric shock if it fails to perform the activity. A rat that has had some experience with hearing the sound and receiving a shock while standing in the wheel, but not when running, will make

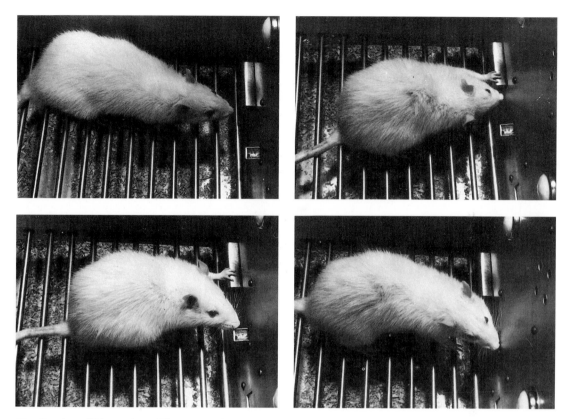

21   Rat in a Skinner box. The operantly conditioned rat approaches the bar (top left) and then presses it (top right). The animal awaits the arrival of a pellet of rat chow (bottom left), which it consumes (bottom right), so that the bar-pressing behavior is reinforced. *Photographs by Larry Stein.*

the appropriate association and start running whenever it hears the warning cue [102]. But Robert Bolles found that a rat cannot be conditioned to stand upright in the running wheel using the same procedures. Rats that happen to rear up just after the sound and are not shocked and they fail to make the association between this behavior and avoidance of electric punishment. In fact, the frequency with which they perform the response when they hear the warning signal actually *declines* over time (Figure 22).

Likewise, John Garcia and his co-workers have found that white rats can learn to avoid some, but not all, sensory cues that are associated with certain other punishing consequences [422, 423]. For example, if a rat feeds on or drinks a novel, distinctively flavored food or liquid and then is exposed to X-rays, the animal becomes ill, because even tiny doses of radiation cause a buildup of toxic chemi-

22 Biased learning. Learning curves for three operants (running, turning, rearing) that were equally rewarded (white rats were not shocked when they performed the appropriate operant). Rats failed to learn to rear on their hind legs to avoid a shock. *Source: Bolles [102].*

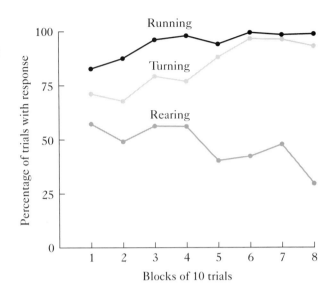

cals in irradiated tissues and body fluids. Subsequently, the rat often refuses to touch whatever it ate or drank before it became sick. The degree to which the food or fluid is avoided is proportional to (1) the intensity of the resulting illness, (2) the intensity of the taste of the substance, (3) the novelty of the substance, and (4) the shortness of the interval between consumption and illness [423]. But even if there is a long delay (up to 7 hours) between eating a distinctive food and exposure to radiation and consequent illness, the rat still links the two events and uses the information to modify its behavior. Most kinds of associative learning will not take place when the interval between a cue and its consequence lasts more than a few seconds.

The specialized nature of this taste aversion learning is further shown by the rat's complete failure to learn that a distinctive sound (a click) precedes internal illness. Rats can learn to associate clicks with shock punishment and will learn to take appropriate action to avoid shocks upon hearing the click. Yet when the punishment is a nausea-inducing treatment, they fail to learn to link the sound with the treatment. In addition, rats have great difficulty in making the association between a distinctive taste and shock punishment. If, after drinking a sweet-tasting fluid, the rat receives a shock on its feet, it often remains as fond of the fluid as it was before, as measured by the amount drunk per unit of time, no matter how often it is shocked after drinking sweet liquids. Thus, the nature of cue and consequence determine whether a rat can learn to modify its feeding and drinking behavior (Figure 23).

Why is it that white rats are so adept at learning to avoid novel foods with distinctive tastes that are associated with illness, even hours after ingesting the food? The white rat is a domesticated laboratory variant of the wild Norway rat. The food-sampling behavior of wild rats exposes them to the risk of poisoning, and this skill

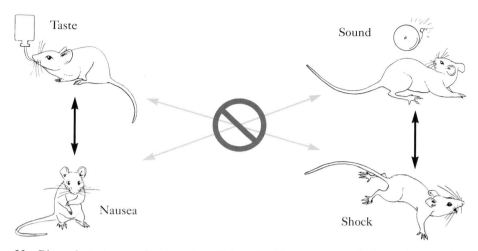

23   Biases in taste aversion learning. Although white rats can easily learn that certain taste cues will be followed by sensations of nausea and that certain sounds will be followed by skin pain caused by shock, they have great difficulty forming learned associations between taste and consequent skin pain or between sound and subsequent nausea. *Source: Garcia et al. [423].*

helps them to minimize that risk. Under natural conditions, a Norway rat becomes completely familiar with the area around its burrow, foraging within that area for a wide variety of foods, plant and animal [736]. New plants and insects are constantly coming into season and then disappearing. Some of these organisms are edible and nutritious; others are toxic and potentially lethal. A rat cannot clear its digestive system of toxic foods by vomiting. Instead, the animal takes only a small bite of anything new. If it gets sick later, it *should* avoid this food or liquid because eating large amounts might kill it [423]. This case demonstrates that even what appears to be a general, all-purpose form of learning is actually a specialized response to the particular kinds of associations that occur in nature, an ability based on natural selection acting on rats with variant nervous systems over evolutionary time.

The Evolution of Developmental Homeostasis

We have to this point emphasized the selective flexibility of the developmental process, with differences in environmental factors causing some individuals to follow different developmental pathways or to learn different things. The flexibility of the developmental process sometimes can be shown to have clearly adaptive outcomes, as in cases in which animals can adjust their behavior to special or unpredictable aspects of their particular environment. But, on the other side of the coin, it is impressive that in nature so few male white-crowned sparrows fail to develop a fully functional species-specific song, despite the complexity of song development, and that so few female mice fail to exhibit fully functional parental behavior, despite the multitude of things that might possibly interfere with the development of normal mothering.

The acquisition of "normal" behavior becomes all the more impressive when one considers that each white-crowned sparrow and each wild house mouse has a unique genotype drawn from the sample of genes contained in its parents' bodies. Moreover, no two sparrows (and no two mice) eat exactly the same foods, experience identical climatic conditions, hear the same sounds, or encounter identical social situations. The development of each individual is therefore the result of an interaction between a unique genotype and a unique environment. Even so, most individuals of a species develop much the same behavioral responses and capacities, including the very mechanisms, such as learning systems, that make some behavioral flexibility possible.

To illustrate that development follows a predictable course in different individuals of the same species, one need only consult an embryology textbook. Developmental biologists can predict with some confidence when certain structures will first appear in a rat or chick embryo and can chart with precision the sequence of changes in those structures over time. Developmental variation occurs among individuals, but it is generally modest. Researchers that follow the fate of identifiable single cells usually find that cellular development follows a standard course. For example, in the grasshopper *Schistocerca nitens*, in essentially every 5-day-old embryo you can find certain distinctive nerve cells occupying a particular position in the thorax. Corey Goodman has shown that these cells undergo a repeatable pattern of development (Figure 24), despite the considerable genetic diversity in the species and the great environmental differences affecting its members [447].

It is almost as if the nerve cells of the grasshopper "know where they are supposed to go," a property insect neurons share with nerve cells in organisms ranging from marine mollusks [785] to various vertebrates [297, 302]. In all these animals, neurons exhibit an impressive capacity to migrate to and make appropriate connections with the proper target tissues, even when embryonic nerve cells are experimentally moved from one spot to another [437]. For example, in some birds, such as zebra finches, entirely new nerve cells are produced in special regions of the brain during song learning, nerve cells that manage to grow in ways that link the song system to the muscles involved in singing behavior [867]. As a result, almost all zebra finch males acquire the essential mechanisms for song production.

Behavioral Development under Abnormal Conditions

The ability of animals to consistently acquire the neural and hormonal substrates for normal behavior, even under less than optimal conditions, has been called **developmental homeostasis.** This phenomenon is especially impressive in animals that have been experimentally deprived of normal social interactions early in their lives and yet still develop reasonably normal social behavior. For example, baby rats that have been separated from their mother and siblings when less than

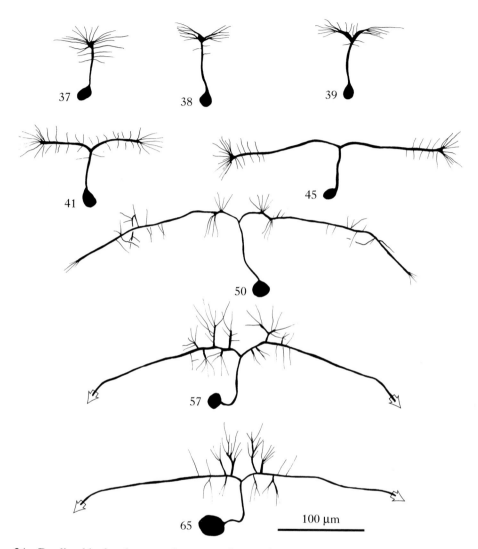

24   Predictable developmental pattern of a grasshopper nerve cell. Here, the growth of one particular cell is shown from the 37th to the 65th day of the insect's life. *From Goodman and Spitzer [447].*

2 days old and fed artificially for 3 weeks still approach scents from the anal excreta of female rats, just as naturally reared rat pups do when temporarily separated from their mothers [418]. Male crickets that have spent all their lives isolated from their fellows will sing a normal species-specific song, despite their severely restricted social and acoustical experiences [78], and the same is true of ring doves

that have been deafened at 5 days of age [870]. Captive, hand-reared female cowbirds that have never heard a male cowbird sing will nevertheless adopt the appropriate precopulatory pose when they hear cowbird song for the first time, if they have mature eggs to be fertilized [649]. These examples illustrate that normal behavioral development sometimes occurs despite the absence of experiences that might be thought essential.

Similar results come from the classic experiments of Margaret and Harry Harlow on the development of social behavior in rhesus monkeys [507, 508]. (These experiments were conducted in an era when animal rights were not as prominent an issue as they are today; readers can decide for themselves whether the harsh treatment of infant monkeys in these experiments yielded information of sufficient importance to justify the research). In one such experiment, the Harlows separated a young rhesus from its mother shortly after birth. The baby was placed in a cage with an artificial "surrogate" mother (Figure 25), which might be a wire cylinder or a terry cloth figure with a nursing bottle. The baby rhesus gained weight normally and developed physically in the same way that nonisolated rhesus infants do. However, it soon began to spend its days crouched in a corner, rocking back and forth, biting itself. If confronted with a strange object or another monkey, the isolated baby withdrew in apparent terror.

The isolation experiment demonstrated that a young rhesus needs social interactions to develop normal social behavior. But what kind of social experience—and how much—is necessary? Interactions with a mother are insufficient for full social development of rhesus monkeys, since infants reared alone with their mothers failed to develop truly normal sexual, play, and aggressive behavior.

25 Surrogate mothers used in social deprivation experiments. This isolated rhesus infant was reared with wire cylinder and terry cloth dummies as substitutes for its mother. *Photograph courtesy of Harry Harlow.*

26   Socially isolated rhesus infants that are permitted to interact with other social isolates for short periods each day at first cling to each other during the contact period. *Photograph courtesy of Harry Harlow.*

Perhaps normal social development in rhesus monkeys requires the young animals to interact with each other. To test this hypothesis, the Harlows isolated some infants from their mothers but gave these infants a chance to interact with three other such infants for just *15 minutes* each day [507]. At first, the young rhesus monkeys simply clung to one another (Figure 26), but later they began to play. In their natural habitat, rhesus babies start to play when they are about 1 month old, and by 6 months they spend practically every waking moment in the company of their peers. Even so, the 15-minute play group developed nearly normal social behavior. As adolescents and adults, they were capable of interacting appropriately with other rhesus, showing typical social and sexual behavior and not exhibiting the intense aggression or withdrawal of animals that had no social contacts as infants.

Socially isolated rhesus infants can develop normal social behavior even if they are reared with mongrel dogs as surrogate mothers (Figure 27). As a result of their social interactions with these highly interactive but extremely atypical companions, the monkeys showed close-to-normal social behavior when they were eventually introduced to members of their own species [779].

27   Rhesus infant with a dog as a surrogate mother. Youngsters reared with these surrogates develop nearly normal social behavior. *Photograph courtesy of W. A. Mason.*

Developmental Homeostasis and Human Behavior

Studies of rhesus monkeys have provided important information on the resilience of the developmental mechanisms of that primate. Naturally one wonders about the relevance of these studies for another primate, human beings. Many people believe that the experiences of very young children greatly affect the development of their intelligence, personality, and other attributes. But the development of human behavior is often buffered against environmental shortfalls, even if these occur early in life [313]. Consider, for example, the results of a study of Dutch teenagers who were born or conceived at a time when their mothers were being starved as a result of the Nazi transport embargo during the winter of 1944–1945, which prevented food from reaching the larger Dutch cities during this time [1142]. Deaths from starvation were common, and for most of the famine period, the average caloric intake was about 750 calories per day. As a result, women living in cities under famine conditions produced babies of very low birth weights. In contrast, rural women were less dependent on food transported to them, and their babies weighed much more than urban infants born or conceived at the same time.

One plausible hypothesis states that normal brain development depends on adequate nutrition during pregnancy, when much of brain growth occurs. However, Dutch boys born in famine areas did not exhibit a higher incidence of mental retardation at age 19 than rural boys (Figure 28). Nor did those born in food-deprived urban areas score more poorly than their relatively well nourished rural counterparts when they took the Dutch intelligence test administered to draft-age men. Children born at low weights apparently suffered no permanent intellectual damage [1142].

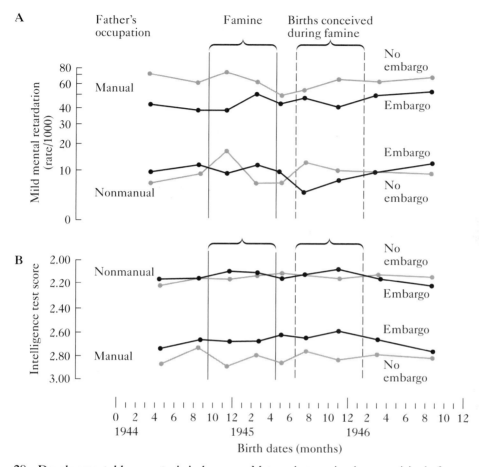

28 Developmental homeostasis in humans. Maternal starvation has surprisingly few effects on intellectual development in humans, judging from a study on (A) the rates of mild mental retardation and (B) the intelligence test scores of 19-year-old Dutch men whose mothers lived under German occupation while pregnant. The subjects were grouped according to the occupations of their fathers (Manual or Nonmanual) and whether their mothers lived or gave birth in a city subjected to food embargo by the Nazis (Embargo) or in a rural area unaffected by the embargo (No embargo). Individuals who were conceived or born under famine conditions exhibited the same rates of retardation and the same level of test scores as men conceived by or born to unstarved rural women. *Source: Stein et al. [1142].*

No one believes that it is good for a fetus if its mother is starving, but the developmental systems of young humans clearly have some resilience. Additional support for this hypothesis comes from the studies of Jerome Kagan, a child psychologist, who expected to find that the child-rearing customs of rural Guatemalans would permanently stunt the intellectual growth of their children. Kagan knew that Guatemalan babies are tightly wrapped in blankets for long periods, and that parents do not coo and babble to their infants during the first year of life. Even so, by age 11, Guatemalan children scored as well on Kagan's intellectual development tests as did North Americans of the same age [623].

Research on normal development in abnormal environments testifies to the adaptively guided, structured nature of development, which can be knocked off the rails only by extremely unusual environmental or genetic problems. A developing organism is not a passive entity capable of being equally affected by every possible environmental or genotypic peculiarity. Such a system would often lead to reproductive disaster. Individuals might fail to develop critically useful traits due to a transitory environmental deficit, such as a temporary shortage of food or lack of social stimulation. The genes of such developmentally "sensitive" individuals would surely be less likely to survive than those of individuals with the capacity to overcome obstacles in the way of acquiring important, fitness-enhancing attributes. Although a great deal remains to be learned about the proximate basis of developmental homeostasis, there can be little doubt that it is a widespread phenomenon, and that in an ultimate, evolutionary sense, individuals benefit by withstanding disruption of their behavioral development.

SUMMARY

1. The development of any trait is an interactive phenomenon involving the genotype of a fertilized egg and the environment of the developing organism. For example, sexual differences in some birds and mammals have their roots in relatively small chromosomal (genetic) differences between males and females. These lead to differences in the hormones produced by embryonic gonadal tissues. The presence of a key hormone acts as an organizational trigger for other cells, altering their genetic and biochemical activity with cascading effects on the physiological and behavioral development of the individual.

2. The environment of a developing organism consists not only of the metabolic products of its cells and the food materials it receives, but also of its sensory and social experiences. All of these factors can act as cues (sometimes only during a restricted critical period) that may have long-term developmental consequences.

3. Various kinds of developmental mechanisms confer behavioral flexibility on individuals that possess them. Developmental switches enable individuals to "select" alternative pathways leading to different structural and behavioral phenotypes, depending upon key cues in their environment. Learning mechanisms are fundamentally similar, with individuals using certain kinds of experience to modify their behavior. The flexibility to make behavioral adjustments provides fitness advantages for individuals in environments where they are likely to encounter unpredictable but biologically important variation during their lifetimes.

4. The developmental process often ignores or overcomes the kind of environmental (and genetic) variation that might disrupt the production of certain traits. This channeled or buffered aspect of development (developmental homeostasis) is seen in (1) the constancy with which individuals pass through certain species-specific stages and (2) the development of normal physiological and behavioral characters in animals placed experimentally in highly abnormal environments. The resilience of the developmental process helps individuals develop key characteristics that promote their reproductive success.

SUGGESTED READING

The relation between genes, hormones, and sexual differentiation in animals is covered in a textbook by Randy Nelson [857] and in the chapters included in *The Differences between the Sexes* [1095]. David Pfennig and Paul Sherman review the fascinating phenomenon of kin recognition [931]. Russell Fernald's article on the environmental basis of neuronal changes in a cichlid fish makes for especially good reading [385].

DISCUSSION QUESTIONS

1. "Some researchers suspect that male homosexuality in humans arises from the hormonal environment experienced by the fetus in its early development. But others refuse to rule out the possibility that childhood interactions with a domineering mother have some effect on the later expression of the trait. Given a connection between fetal hormone exposure and homosexuality, there ought to be some distinctive cells in regions of the brain known to influence sexual preference, cells that have hormone receptor molecules that differ from those in heterosexual men. However, these cells have not been found to date. In addition, some recent studies have uncovered no difference in the nature of mother–son interactions experienced by homosexual and heterosexual men." In this make-believe press report, identify the hypotheses, the predictions, the tests, and the conclusions—if any.

2. Once an adult female green turtle has come ashore and laid her eggs in the sand of a beach, she is very likely to return to that beach again. Thus, different populations of the turtle nest in different locations. A study of females from several different populations reveals that they are genetically distinct [805]. Is this result consistent with the hypothesis that the differences in choice of nesting site by females could stem from genetic differences among them? Is it also possible that females could be returning to the beach where they hatched, and that they have learned the odor associated with their natal area? If that is true, could the differences among females in nesting behavior be purely environmentally caused? If so, can you reconcile the genetic data with this possibility?

3. Recall the case of the cichlid fish whose males can adopt two different reproductive roles, thanks to developmentally flexible brain neurons. Is there any role for developmental homeostasis in this system?

4. Many persons have noted that human boys and girls tend to differ in their involvement in rough-and-tumble play. Provide possible proximate and ultimate hypotheses for the development of these differences between the sexes.

CHAPTER 5

The Control of Behavior: Neural Mechanisms

*I*N EXPLORING WHAT CAUSES AN ADULT MALE white-crowned sparrow to sing, biologists have found that the genetic–developmental mechanisms in a young bird respond to various hormonal, acoustical, and social influences. These and other factors masculinize the bird's song system, which consists of huge numbers of cells in discrete regions of its brain that underlie the sparrow's ability to learn to sing its species' song. All the behavioral abilities of creatures with nervous systems depend on the operation of their nerve cells. The point of this chapter is to illustrate what researchers have been able to learn at this level of proximate analysis, one that focuses on the link between nervous systems and behavior. The job of nervous systems can be divided into three major components:

the acquisition of sensory information from the environment, the processing of this information by central decision-makers, and the motor response to that information. Despite the smallness and astronomical numbers of nerve cells and the mind-boggling complexity of most nervous systems, researchers have been wonderfully ingenious in answering questions such as, how does a moth take action to avoid an approaching bat? How does a homing pigeon released 100 miles from its loft manage to fly back home? As we look at the research findings of neurobiologists, we will see that although all nerve cells share some fundamental properties, what they actually do varies enormously, both within individuals and across species. At the ultimate level, the different outcomes of nerve cell function appear to be related to ecological differences among species in the nature of their predators and other obstacles to their reproductive success. By considering this point, we can integrate the proximate and ultimate aspects of the neural mechanisms of behavior.

How Nerve Cells Control Behavior

Essentially all animals have nervous systems that control behavior. Since we are animals, we too depend on our nerve cells and networks for all our perceptions, decisions, and actions. Remove or injure even small parts of the brain of a human being and behavioral changes and deficits are likely to occur, often highly specific ones. So when some years ago the newspaper headlines shouted "Homosexuality Has Biological Basis," few biologists were surprised. What was newsworthy and controversial was not the idea that the brain played a role in the sexual orientation of humans, but the claim that a particular tiny region of the brain was associated with homosexuality in human males.

In human populations everywhere, some individuals have little or no sexual interest in the opposite sex, but instead prefer their own sex. Homosexuality raises intriguing ultimate questions (How can such a trait persist if it reduces individual fitness?) as well as offering a puzzle for persons interested in its proximate causes. Do the differences between heterosexual and homosexual men have a genetic basis? Twin studies (see Chapter 3) strongly suggest that they do [41, 330, 541], but since identical twins quite often exhibit different sexual preferences, environmental differences must also play a role [167]. Does homosexuality arise as a result of the organizational effects of hormones on embryos (see Chapter 4)? Is a person's sexual preference dependent on the operation of their brain?

It is this last proximate question that neuroscientist Simon LeVay investigated by testing the hypothesis that the brains of homosexual men possess distinctive structural and functional attributes not present in heterosexual men [709]. Based on earlier discoveries, LeVay knew that a structure called the anterior hypothalamus plays a key role in regulating sexual activity in humans and other vertebrates, and that large differences exist between human males and females in the size of certain nuclei in the anterior hypothalamus, suggestive evidence that these cells influence our sexual behavior. Based on this evidence, LeVay predicted that there would be differences in the size of certain nuclei in the anterior hypothalami of homosexual and heterosexual males.

LeVay tested this neural hypothesis by studying the anterior hypothalami of several homosexual men who had died of AIDS. He located the key nuclei and measured their size. He did the same with the brains of a group of men who were presumed to be heterosexual, and who had not died of sexually transmitted AIDS. In the homosexual group, LeVay found that a nucleus called INAH-3 (Figure 1) was about the same size as INAH-3 in women, and much smaller on average than in heterosexual men, where it is typically about the size of a grain of sand.

LeVay's study provides some support for the hypothesis that specific clusters of cells in the human brain have an effect on the sexual orientation of adults. But the issue is far from closed [118], as LeVay himself acknowledges [710], because the hypothesis that a small INAH-3 causes a homosexual orientation in males has not been tested against some alternatives. For example, perhaps homosexual activity causes a reduction in the size of INAH-3, rather than the other way around [1221]. That behavioral interactions can alter the size of certain brain cells has been demonstrated by the studies of cichlid fish [403] discussed in Chapter 4. Or perhaps the AIDS disease itself could have reduced the size of the INAH-3 cells. Persons with AIDS produce less testosterone as the disease progresses, and low testosterone levels in the blood could conceivably cause a reduction in the size of neurons in INAH-3 [167].

Thus, although without doubt human sexual orientation is determined by the way certain cells work somewhere in the human brain, just where and how still remains uncertain. The daunting complexity of the relationship between human brains and human behavior has led many researchers to explore the physiological basis of behavior in animals other than humans. Taking the lead in this respect were the pioneering ethologists Niko Tinbergen and Konrad Lorenz (see Chapter 1), who developed an especially fruitful approach to studying how the nervous systems of fish, birds, and insects activated particular behavior patterns.

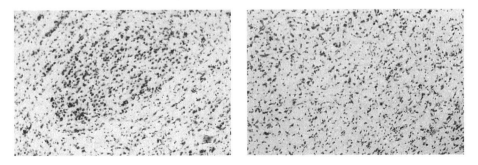

1 The INAH-3 region in human brains. On the left, a section through INAH-3 from the brain of a heterosexual male shows a dense oval collection of cells. In the corresponding area from the brain of a homosexual male, shown on the right, the oval structure is absent. *Source: LeVay [709].*

Fixed Action Patterns and Sign Stimuli

The key to the ethological analysis of the proximate causes of behavior lay in discovering the proximate cues that triggered natural behavior patterns in animals living under natural conditions. For example, one of Tinbergen's classic studies dealt with the begging response of newly hatched herring gull chicks. He knew that the chicks pecked at the red dot toward the end of their parent's bill, which caused the adult to regurgitate a half-digested fish or other tasty morsel [1193]. He subsequently experimented at length with recently hatched chicks in order to figure out what "turned on" the adaptive pecking response of the young bird. Tinbergen established that newborn herring gulls preferred (1) long, pointed objects over short, stubby ones, (2) red objects over all other colored objects, (3) high-contrast dots over low-contrast dots, and (4) moving objects over stationary ones. Furthermore, a three-dimensional model or stuffed head of a herring gull was no more effective in stimulating begging pecks by newborn chicks than a two-dimensional pointed strip of cardboard with a red dot painted on it [1193].

From these experiments, Tinbergen deduced that the chick's visual system must be organized so that the young bird focuses on a few simple cues usually associated with its parent's beak. The gull chick seems to screen out extraneous detail and focus only on the essential stimuli; sensory signals from these stimuli are relayed to the chick's brain, where decision-making neurons generate motor commands carried out by other cells. As a result, the chick pecks at the effective stimulus, whether it is located on its mother's beak or a piece of painted cardboard.

Tinbergen and Lorenz collaborated on another famous experiment of this sort. They found that if they removed an egg from under an incubating greylag goose and put it a few feet away, the goose would retrieve the egg by stretching its neck forward, tucking the egg under its lower bill, and rolling the egg carefully back into its nest. Incubating greylag geese "know" how to retrieve an egg, even with their first clutch of eggs [1188]. Furthermore, when Tinbergen and Lorenz placed an artificial giant egg outside the nest, an egg quite unlike any that the goose could possibly produce itself, the incubating bird automatically stretched out its neck and pulled the egg into its nest. When the researchers took an egg away from the goose when the bird was retrieving it, the goose continued pulling its head back as if an egg were still balanced against the underside of its bill.

From these results, Tinbergen and Lorenz deduced that the goose must have a rather odd perceptual mechanism that reacts to simple visual cues associated with eggs; when it perceives these stimuli, some system in its bird brain automatically activates the motor program for egg retrieval, which runs to the finish once it gets started. Hundreds of similar examples exist. When a human baby's cheek touches its mother's breast, the baby turns toward the breast and opens its mouth to grasp the nipple. Touch the baby's cheek with your finger and it will mechanically perform the same behavior.

Tinbergen and Lorenz and their fellow ethologists specialized in the study of **instincts,** defined as behavior patterns that appear in fully functional form the *first*

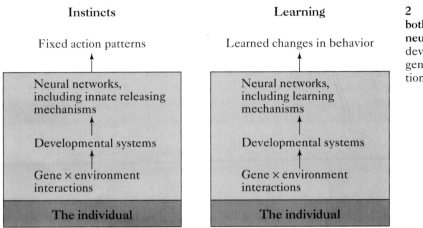

Instincts

Fixed action patterns

↑

Neural networks, including innate releasing mechanisms

↑

Developmental systems

↑

Gene × environment interactions

The individual

Learning

Learned changes in behavior

↑

Neural networks, including learning mechanisms

↑

Developmental systems

↑

Gene × environment interactions

The individual

2 Instincts and learning both require complex neural networks, which develop via complex gene–environment interactions.

time they are performed, even though the animal may have had no previous experience with the cues to which it reacts. Remember that instincts are neither "genetically determined" nor rigid and unmodifiable (see Chapters 3 and 4). Innate responses, like learned ones, depend on an internal mechanism that develops as a result of complex gene–environment interactions. Once development is completed, the nerve cells that make up instinct mechanisms simply do different things than the neurons in learning mechanisms (Figure 2).

The ethologists labeled an instinctive response a **fixed action pattern,** or FAP, while the key component of the object that activates an FAP was called a **sign stimulus** or a **releaser** (if the sign stimulus was a signal from one individual to another). The hypothetical neural mechanism that receives sensory input from sign stimulus detectors and makes the decision to activate the FAP was labeled an **innate releasing mechanism** [1188]. Thus, the red dot at the end of a herring gull's bill is a releaser of an FAP, the begging pecks of baby herring gulls. An especially familiar FAP, a human yawn, is also a releaser (Figure 3). Yawns are very similar in appearance no matter who is doing the yawning, they last about 6 seconds, are difficult to stop in midperformance, and are infectious, releasing yawning in other humans that observe, or even hear, the yawner [961].

The simple relationship between cue and FAP is highlighted by the ability of some parasitic species to exploit the FAPs of other species, a tactic known as code-breaking [1272]. For example, a premier code breaker is the rove beetle *Atemeles pubicollis*, which lays its eggs in the nests of the ant *Formica polyctena* [560]. The resulting larvae produce an attractant scent, or pheromone, that causes ant workers to carry the parasitic beetle grubs into their brood chambers, where the grubs feast on ant eggs and ant larvae. Not content with this larder, the parasites also mimic the food-begging behavior of ant larvae by tapping a worker ant's mandibles with their own mouthparts. This action is a releaser that triggers regurgitation of liquid food

3 Releasers provide key stimuli detected by innate releasing mechanisms, which trigger innate responses. (A) An infant herring gull that sees the pointed tip of its parent's bill with the bright red dot on the lower beak will peck at the dot. Its pecking is in turn a releaser of regurgitation of food by the parent gull. (B) A yawn, a releaser of yawning in other humans. *Photograph by Tim Ford, courtesy of Robert Provine.*

A

Releaser (visual signal)

Innate releasing mechanism

B

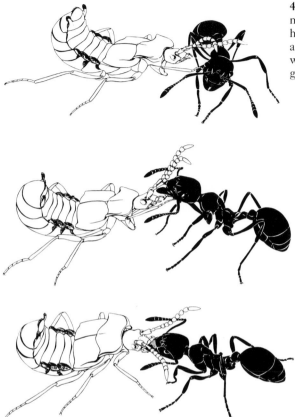

4 Code-breaking rove beetle, whose behavior mimics releasers of feeding behavior in its ant hosts. The beetle first taps a worker ant with its antennae, then touches the ant's mouthparts with its forelegs, and then consumes food regurgitated by the ant. *Drawing by Turid Forsyth.*

(an FAP) by the ant. The larvae eventually metamorphose into adult beetles, which also mimic the releasers that cause adult worker ants to feed them (Figure 4).

Code-breaking species may even provide an exaggerated sign stimulus, or **supernormal stimulus,** that is more effective than that provided by the biologically correct object. When Lorenz and Tinbergen constructed a giant egg for use in their experiments on egg retrieval by greylag geese, they created an artificial supernormal stimulus. Naturally occurring supernormal stimuli have evolved in certain brood parasites, among them the European cuckoo and the North American cowbird [503, 1272]. A female of these species locates the nest of some other bird, generally one smaller than herself. When the owner of the nest leaves during a pause in egg laying or incubation, the lurking parasite slips into the nest, quickly lays an egg, and disappears. When the owner returns, it often accepts the addition to its clutch, incubates the egg, and hatches the parasite's offspring.

The hatchling cowbird or cuckoo may instinctively eject some of the other eggs in the nest (Figure 5), the better to monopolize the care of its foster parents. The

5 Innate egg ejection behavior by a very young cuckoo, which is pushing its host's egg out of the nest. *Photograph by Eric Hosking.*

well-fed parasite grows rapidly and soon becomes larger and more active than its hosts' offspring would have been. Parental feeding behavior in most songbirds depends on how high a nestling reaches, how noisy its begging calls are, and how energetically it bobs its head and body. Host parents confronted by a large, voracious nestling cowbird or cuckoo are subjected to a supernormal begging stimulus, and they react by working overtime to satisfy the parasite's insatiable demands (Figure 6).

6 A supernormal stimulus. The fledgling cuckoo provides exaggerated sensory signals that trigger parental feeding in the host more effectively than the host's own offspring do. *Photograph by Eric Hosking.*

How Do Moths Evade Bats?

Although modern neurobiologists rarely employ the old ethological terminology, many still seek to understand how neural systems provide wild animals with the capacity to reproduce successfully in their natural environment. Consider the classic studies of Kenneth Roeder on the relationship between nerve cells and adaptive behavior patterns in certain noctuid moths. Roeder's work drew its inspiration from observations he and his colleague Asher Treat made of bats and moths flying on warm summer nights in Massachusetts. In many places, you can see what Roeder and Treat saw [1011] if you possess "a minimum amount of illumination, perhaps a 100-watt bulb with a reflector, and a fair amount of patience and mosquito repellent" [1008]. By peering at the weakly illuminated patch of light around the bulb, you will sometimes see a bat burst into view and swoop down upon a moth attracted to the light. But you will also sometimes see a moth turn abruptly or dive straight down just before a bat rushes past, evidence that at least some moths can detect and avoid these predators.

How do the evasive moths succeed in outwitting their enemies? Perhaps they hear the vocalizations that bats produce as they hurtle through the air. Hunting bats produce a more or less steady stream of sound by vibrating the larynx—the voice box. When a bat's muscles cause its larynx to vibrate, pressure waves are generated that travel out of the animal's mouth and through the air. These waves have a *frequency*, or pitch, measured in kilohertz (kHz), and an *amplitude*, or intensity, measured in decibels (dB). A typical nocturnal, insectivorous bat produces very high-pitched (ultrasonic) sound waves with frequencies between 20 and 80 kHz, which we cannot hear, despite the fact that the energy in bat calls is high.

The recognition that bats use ultrasonic calls to navigate did not come until the 1950s, when Donald Griffin proposed that some species emit pulses of high-frequency sound and then listen for the weak echoes reflected back from objects in their flight path [471]. This idea was initially greeted with extreme skepticism. But Griffin tested the sonar hypothesis by placing some little brown bats, a common New England species, in a room filled with wire obstructions strung from the ceiling to the floor [472]. Bats successfully negotiated the obstacle course, uttering their cries and gobbling up the fruit flies that Griffin had provided for them. He then tested whether a bat's ultrasonic calls were critical for its navigational abilities by turning on a machine that generated sounds higher than 20 kHz. As soon as these extraneous sounds began to bombard them, the flying bats began to collide with obstacles and crash to the floor, where they remained until Griffin silenced the jamming device. In contrast, loud (high-intensity) sounds of 1–15 kHz had no effect on the bats because these low-frequency sounds did not mask the high-frequency echoes they require to fly safely and find food in the dark. These results led Griffin, and eventually everyone else, to conclude that the little brown bat employs a sonar system to avoid obstacles and detect prey.

As Roeder watched the nocturnal moth and bat drama outside his home, he guessed that moths might use pulses of bat ultrasound as a basis for evasive action.

7 Noctuid moth ears. (A) The location of the ear. (B) The design of the ear, which features two sensory fibers (A1, A2) linked to a tympanic membrane. *After Roeder [1008].*

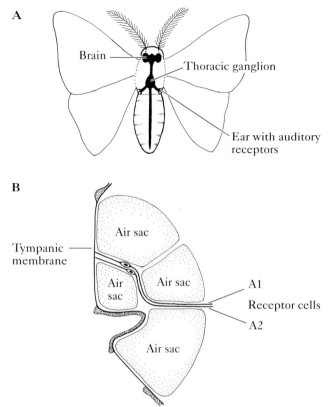

If he was right, he knew he should be able to find ears somewhere on the moths—which he did on the insect's thorax, one ear on each side (Figure 7). The ear consists of a thin, flexible sheet of cuticle—the tympanic membrane—on the surface of the body, which is attached to two nerve cells, the A1 and A2 sensory receptors. When intense pressure waves strike the moth, they cause the tympanic membrane to vibrate, which activates the receptor cells.

How Nerve Cells Work

Nerve cells, or **neurons,** are cells that respond to certain kinds of stimulation by changing the permeability of their membranes to sodium ions. Mechanical stimulation of a noctuid moth's acoustical receptor cells opens stretch-sensitive channels in the membrane, permitting positively charged sodium ions to enter the cell. As these ions enter, they change the electrical charge differential of a portion of the receptor cell's membrane. This abrupt, local change may spread to neighboring portions of the membrane, sweeping around the neuron's cell body and down its axon, the "transmission line" of the cell (Figure 8). This brief, all-or-nothing

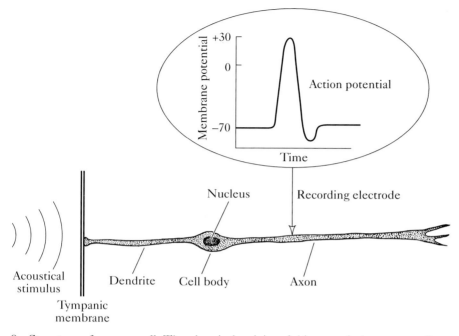

8 **Structure of a nerve cell.** The electrical activity of this acoustical receptor cell depends first on the effect of acoustical stimuli on the dendrite. Electrical changes in the dendrite's membrane can, if sufficiently great, trigger an action potential that begins near the cell body and travels along the axon of the receptor toward the next cell in the network.

change in membrane potential along an axon is called an **action potential,** and is the signal that one neuron uses to communicate with another through a network of cells, sometimes producing a chain reaction of "excited" neurons. In this way messages can be speedily routed from receptor cells to distant portions of the nervous system, where neurons stimulated by the input can continue the chain reaction with action potentials reaching muscles, activating a behavioral response.

When an action potential arrives at the end of an axon, it often causes the release of a neurotransmitter chemical at this point. This chemical signal diffuses across the narrow gap, or **synapse,** separating the axon tip of one cell from the body of the next cell in the network. Neurotransmitters can affect the membrane permeability of the postsynaptic cell in ways that increase or decrease the probability that it will produce its own action potential(s). (See Camhi [173] for a much fuller description of the elements of neural action and interaction.)

In the case of the noctuid moth, the A1 and A2 receptor cells relay information via other cells called interneurons whose action potentials can change the activity of aggregates of neurons, or ganglia, in the thorax ganglia and in the head.

As the chain reaction proceeds, certain patterns of activity by cells in the thoracic ganglia trigger other interneurons whose action potentials in turn reach motor neurons that are connected with the wing muscles of the moth. When a motor neuron fires, the neurotransmitter it releases at the synapse with a muscle fiber changes the membrane permeability of the muscle cells. These changes initiate the contraction or relaxation of the muscle, with consequent effects on the movements of the wings and, thus, the moth's behavior.

Action Potentials and Information

The neurons of noctuid moths have essentially the same properties as the neurons of other animals. Animal behavior is the product of an integrated series of chemical and biophysical changes in cells, initiated by receptor cells and carried on by sensory interneurons, brain cells, motor interneurons and motor cells, and muscles (Figure 9). Because these changes occur with remarkable rapidity, an individual can react to changing stimuli in its environment in fractions of a second.

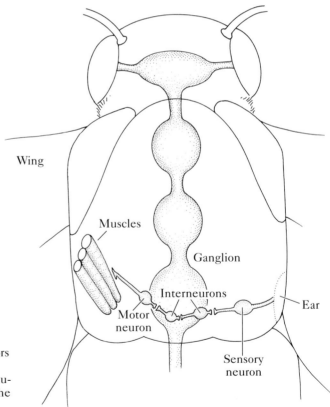

9 Neural network of a moth. Receptors in the ear relay information to interneurons in the thoracic ganglia which communicate with motor neurons that control the wing muscles.

Although the basic design of noctuid moth neurons is not unusual, they perform very special tasks for their owners. For example, the A1 and A2 receptors gather critical information about bats. We know this thanks to the simple, but elegant, experiments of Kenneth Roeder [1008, 1009]. He attached recording electrodes to these receptors in a living, but restrained, moth and projected a variety of sounds at its ear. The electrical activity that resulted was relayed to an oscilloscope, which recorded the activity of the two receptors, revealing the following features of the two cells (Figure 10):

1. The A1 cell is sensitive to sounds of low to moderate intensity. The A2 receptor is not, and begins to produce action potentials only when a sound is loud.

10 Properties of the auditory receptors of a noctuid moth. (A) Sounds of low or moderate intensity do not generate action potentials in the A2 receptor. The A1 fiber fires sooner and more often as sound intensity increases. (B) The A1 receptor reacts strongly to pulses of high-frequency sound, but ceases to fire after a short time if the stimulus is a steady hum or buzz.

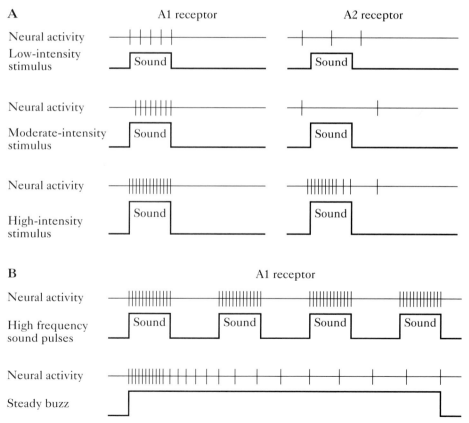

2. As a sound increases in intensity, the A1 neuron fires more often and with a shorter delay between the arrival of the stimulus at the tympanic membrane and the onset of the first action potential.

3. The A1 receptor fires much more frequently to pulses of sound than to steady uninterrupted sounds.

4. Neither neuron responds differently to sounds of different frequencies over a broad ultrasonic range. A burst of sound at 30 kHz elicits much the same pattern of firing as an equally intense sound at 50 kHz.

5. The receptor cells do not respond at all to low-frequency sounds. The moths are deaf to stimuli that we can easily hear.

Although each ear has just two receptors, the amount of information they can provide to the central nervous system about sonar-using bats is impressive. The key property of the A1 receptor is its great sensitivity to pulses of ultrasound, so much so that the cries of a little brown bat cause this highly sensitive cell to respond with action potentials when the predator is 100 feet away, long before a bat can detect the moth. Because the rate of firing in the A1 cell is proportional to the loudness of the sound, the insect has a system for determining whether a bat is coming toward it.

In addition, the moth's ears gather information that could be used to locate the bat in space. For example, if a hunting bat is on the left, the A1 receptor in the left ear will be stimulated sooner and more strongly than the A1 receptor in the right ear, which is shielded from the sound by the moth's body. As a result, the left receptor will fire sooner and more often than the right receptor. The moth's nervous system could also detect whether the bat is above it or below it. If the predator is higher than the moth, then with every up and down movement of the insect's wings, there will be a corresponding fluctuation in the rate of firing of the A1 receptors as they are exposed to, and then shielded from, bat cries by the wings. If the bat is lower than the moth, there will be no such fluctuation (Figure 11).

As chain reactions of neural activity initiated by the receptors sweep through the moth's nervous system, they may ultimately generate motor messages that cause the moth to turn and fly directly away from the ultrasonic stimulus [1011]. When a moth is moving away from a bat, it exposes less echo-reflecting area than if it were flying at right angles to the predator and presenting the full surface of its wings to the bat's vocalizations. If a bat receives no echoes from its calls, it cannot detect a prey. Bats rarely fly in a straight line for long, and therefore the odds are good that a moth will remain undetected if it can stay out of range for a few seconds. By then the bat will have found something else within its 8-foot moth detection range and will have veered off to pursue it.

In order to employ its antidetection response, a moth need only orient so as to synchronize the activity of the two A1 receptors. Differences in the rate of action potential production by the receptors in the two ears are probably monitored by the brain, which relays neural messages to the wing muscles via the thoracic ganglia and allied motor neurons. The resulting changes in muscular action steer the moth

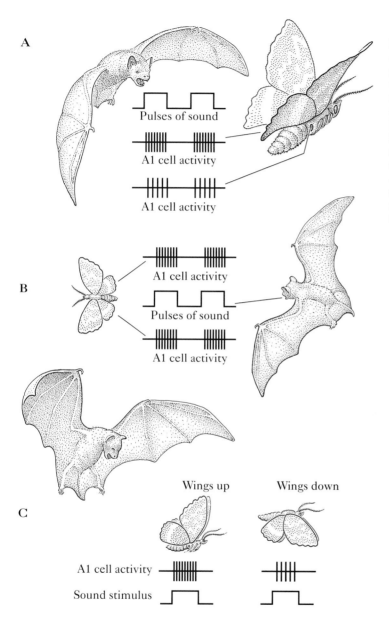

11 How moths might locate bats in space. (A) A bat is to one side of the moth; the A1 receptor on the side closer to the predator fires sooner and more often than the shielded A1 receptor in the other ear. (B) A bat is directly behind the moth; both A1 receptors fire at the same rate and time. (C) A bat is above the moth; activity in the A1 receptors fluctuates in synchrony with the wingbeats of the moth. (Figures not drawn to scale.)

away from the side of its body with the ear that is more strongly stimulated. As the moth turns, it will reach a point where both A1 cells are equally active, at which time it will be facing away from the bat and flying away from danger (see Figure 11B).

Although this reaction is effective if the moth has not been detected, it is useless if a speedy bat has come within the 8-foot detection range. At this time a moth has at most a second, and probably less, before the bat reaches it [625]. Therefore, moths close to bats do not try to outrace them, but employ drastic evasive maneuvers, including wild loops and power dives, that make it relatively difficult for bats to intercept them. A moth that executes a successful power dive and reaches a bush or grassy spot is safe from further attack because echoes from the leaves or grass at the moth's crash landing site mask those coming from the moth itself [1011].

Roeder speculated that the physiological basis for the erratic flight of the moth lies in circuitry leading from the A2 receptors to the brain and back to the thoracic ganglion [1010]. When a bat is about to collide with a moth, the intensity of the sound waves reaching the insect's ears is high. It is under these conditions that the A2 cells fire. Their messages are relayed to the brain, which in turn may shut down the central steering mechanism that regulates the activity of the flight motor neurons (see Figure 10). When the steering mechanism is inhibited, the moth's wings begin beating out of synchrony, irregularly, or not at all. As a result, the insect does not know where it is going—but neither does the pursuing bat, whose inability to plot the path of its prey may permit the insect to escape.

Stimulus Filtering: A Mechanism for Selective Perception

Although Roeder's hypothesis on the control of diving by noctuid moths has evidently not been directly tested with the species he studied, its plausibility is weakened by the discovery that notodontid moths can perform both turns and dives despite an ear endowed with a single acoustical receptor, not two. The ability of these moths to make do with a single receptor in a natural environment filled with a wide array of acoustical stimuli is testimony to the irrelevance of most stimuli to their survival and reproduction. It is typical of animal nervous systems that they respond selectively, extracting relevant information from the sensory barrage while ignoring a great many things at the same time. This process, known as **stimulus filtering,** is at work when a hungry gull chick looks up at its parent and "sees" only a red dot on a pointed object, or when a flying noctuid moth listens to the sounds of the night and "hears" only the staccato buzz of ultrasound produced by an approaching bat.

The noctuid moth's auditory system offers an object lesson on the operation and utility of stimulus filtering. The moth's ear does not relay information about low-frequency sounds audible to humans, because its acoustical sensors respond only to ultrasound. Even when the receptors do fire, they do not produce different patterns of signals in response to sounds of different frequencies, whereas human receptors, for example, provide the signal information that enables us to tell the difference between C and C-sharp. The noctuid moth's ear appears to have just one task of paramount importance: the detection of cues associated with its nocturnal predators. To this end, its auditory capabilities are tuned to pulsed ultrasonic sound at the expense of all else. Likewise, its behavioral repertoire of responses is simple—but adaptive. It turns away from low-intensity ultrasound and dives, flips, or spirals erratically when it hears high-intensity ultrasound.

In addition to illustrating the concept of stimulus filtering, the noctuid moth story also shows us that a knowledge of natural history and the real-world problems confronting an animal can help a researcher formulate hypotheses on a species' sensory mechanisms. Kenneth Roeder knew that noctuid moths live in a world filled with sonar-using moth killers; he searched for, and found, a specialized proximate mechanism that helps moths cope with these enemies. Likewise, persons interested in bat hearing have searched for, and found, neurons that are specialized for the detection of echoes returning to them from their prey [877, 949].

If the noctuid moth's ear is designed to filter out the irrelevant and focus on the critical, then we might expect insects unrelated to moths, but also subject to bat predation, to have independently evolved similarly biased perceptual systems—and they have. Certain praying mantises (Figure 12) [1314] and lacewings (Figure 13) [809–811] possess distinctive bat-detecting mechanisms. In terms of design, these devices have little or nothing in common with those of noctuid moths. For example, the pressure wave sensors of lacewings consist of 25 receptors located within an enlarged vein in each forewing. Ultimately, however, the ultrasound detectors of moths, mantises, and lacewings help all these insects when they are suddenly confronted by a hungry bat.

The relationship between stimulus filtering and a species' special obstacles to reproductive success is evident in every animal whose sensory systems have been carefully examined. For example, some parasitoid flies of the genus *Ormia* deposit larvae on male crickets. When the little maggots burrow into their host and proceed to devour it from the inside out, it is an unhappy experience for the cricket but vital to the reproductive success of female *Ormia*.

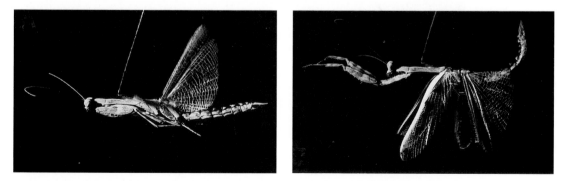

12   Evasive behavior by a praying mantis. In normal flight, a mantis holds its forelegs close to its body (left), but when the insect detects ultrasound, it rapidly extends its forelegs (right), which causes it to loop and dive erratically downward. *Photographs courtesy of D. D. Yager and M. L. May [1314].*

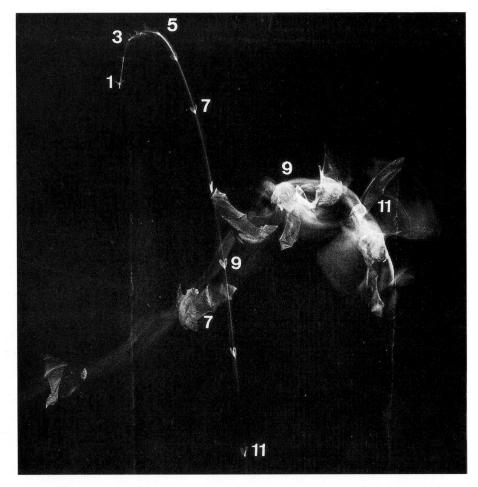

13 Anti-interception response of a lacewing. The numbers superimposed on this multiple-exposure photograph show the relative positions of the bat and lacewing over time. The bat missed the power-diving lacewing, despite performing an aerial somersault. *Photograph by Lee Miller.*

How do these flies find their victims? They listen for singing males, as discovered by researchers who caught *Ormia* at loudspeakers by playing tapes of cricket songs at night. The unique ears of the fly consist of two air-filled structures with tympanic membranes and associated acoustical receptors on the front of the thorax. Vibration of the fly's tympanic "eardrums" activates the sensory cells, just as in noctuid moth ears, and thus provides the fly with information about sound in its environment—but not every sound. As predicted by a trio of evolutionary neu-

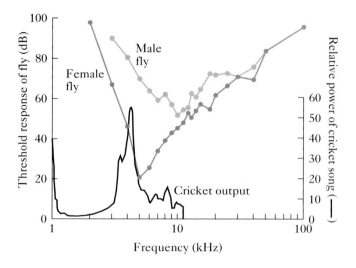

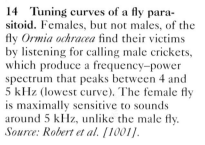

14 Tuning curves of a fly parasitoid. Females, but not males, of the fly *Ormia ochracea* find their victims by listening for calling male crickets, which produce a frequency–power spectrum that peaks between 4 and 5 kHz (lowest curve). The female fly is maximally sensitive to sounds around 5 kHz, unlike the male fly. *Source: Robert et al. [1001].*

robiologists, Daniel Robert, John Amoroso, and Ronald Hoy, the fly's auditory system is tuned to (i.e., most sensitive to) sounds of the frequencies (4–5 kHz) that dominate the host cricket's songs (Figure 14). In other words, the fly can hear sounds of 4–5 kHz much better than sounds of 7–10 kHz. Its acoustical mechanism actively collects information about sounds that male crickets make while ignoring a host of other sounds in its environment, unless they are very loud indeed [1001].

When I say "the fly," I mean the female *Ormia*, not the male, which has an altogether different tuning curve (see Figure 14). The proximate differences between the sexes make ultimate sense because the reproductive chances of male flies are not dependent on their ability to find singing male crickets. As a result, they are much less sensitive to sounds in the 4–5 kHz zone than are female flies.

Another fly parasitoid related to *Ormia* tracks down singing male katydids, *Poecilimon veluchianus*, whose ultrasonic mate-attracting calls fall largely in the 25–35 kHz range [1155]. If fly hearing has evolved to promote success in detecting prey, what frequencies should this katydid hunter, *Therobia leonidei*, be tuned to? As you can see from Figure 15, its ears are matched to the song frequencies of its host, which are much higher than the sounds female *Ormia* flies hear best [1001].

Selective Tactile Detection and Analysis in the Star-Nosed Mole

The star-nosed mole reinforces the point that evolved perceptual systems provide sensations relevant to an animal's lifestyle. This burrowing mole lives in wet, marshy soil, where it tunnels about in search of earthworms and other prey. In its black tunnels, earthworms cannot be seen, and indeed, the mole's eyes are greatly reduced. Instead, the mole appears to rely on touch to find its food, using its strangely wonderful nose to sweep the tunnel walls as it moves forward.

15 Tuning curves of a katydid killer. Females of the fly *Therobia leonidei* parasitize male katydids, whose stridulatory calls contains most of their energy in the range of 20–30 kHz. The female fly is much more sensitive than the male fly to sounds in this range. *Source: Stumpner and Lakes-Harlan [1155].*

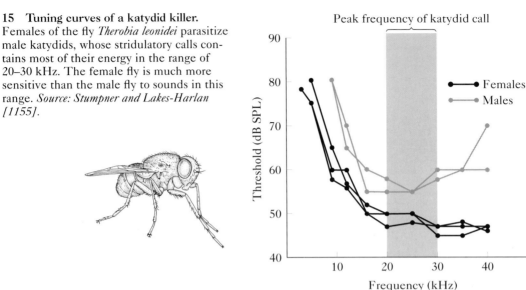

The two nostrils are ringed by 22 fleshy appendages, 11 miniature "fingers" radiating outward from each side of the nose (Color Plate 5; Figure 16). Each "finger" is in turn blanketed with a thousand or so tiny circular structures called Eimer's organs. Each of these organs contains several receptor cells that react to changes in pressure on the upper surface of the organ. These receptor cells are so numerous that they probably relay extremely complex patterns of information to the brain about the surfaces touched by the purely tactile "fingers." Thus, for example, when the nose brushes a worm, the sensory receptors could convey data to the mole's brain on the segmented nature of the object, permitting its recognition as a food item to be grasped by the mole's mouth and eaten.

In order for tactile sensory inputs to be put to good use, however, they must be interpreted by cells capable of this function. Kenneth Catania and Jon Kaas have succeeded in locating the regions of the brain that perform this task [186]. To do so, they exposed the cerebral cortex of an anesthetized mole and then recorded with microelectrodes the response, if any, of individual brain neurons when the skin of the subject, including the skin of the nose, was touched. As it turns out, a large chunk of the mole's brain, called the *somatosensory cortex*, receives and decodes only sensory signals from touch receptors in the skin, but much of the somatosensory cortex responds only to tactile stimulation of the nose appendages. Furthermore, the signals from each nose appendage are received by a specific part of the somatosensory cortex, creating a kind of nose map laid out on the surface of the brain. The kind of analysis that the cortical map of the appendages provides is focused on the spatial location of objects touching the nose relative to the mouth.

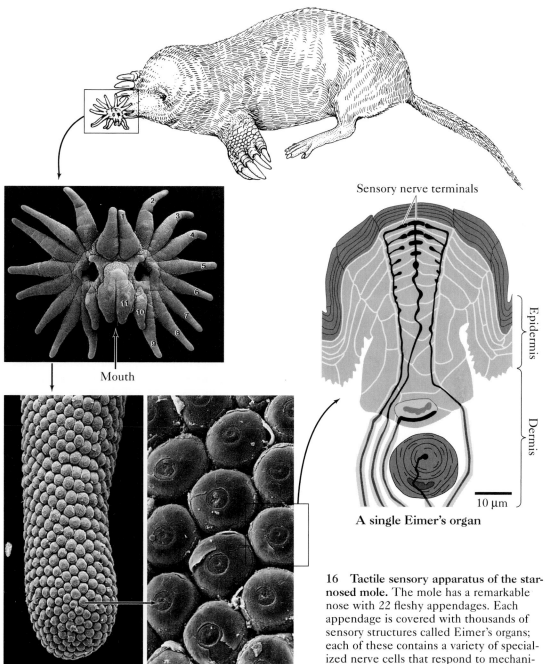

Mouth

Eimer's organs

Sensory nerve terminals

Epidermis

Dermis

10 μm

A single Eimer's organ

16 **Tactile sensory apparatus of the star-nosed mole.** The mole has a remarkable nose with 22 fleshy appendages. Each appendage is covered with thousands of sensory structures called Eimer's organs; each of these contains a variety of specialized nerve cells that respond to mechanical deformation of the skin above them. *Source: Catania and Kaas [186].*

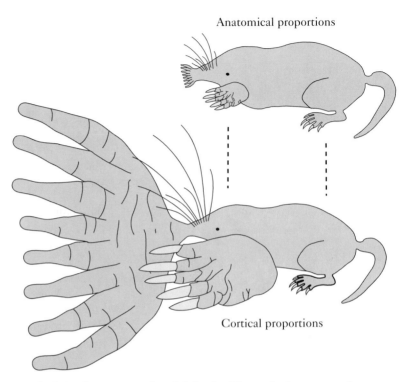

Anatomical proportions

Cortical proportions

17   Sensory analysis by the star-nosed mole's brain. The mole devotes much more brain cortex to processing sensory inputs from its nose and forelimbs than it does to signals coming from receptors on other parts of its body. The upper panel shows the actual body proportions of the mole; the lower panel shows how the body is proportionally represented in the cortex of the mole's brain. *Source: Catania and Kaas [186].*

Thus, the mole exhibits what has been called the cortical magnification of signals from certain body parts, investing a highly disproportionate amount of brain tissue in analysis of signals from the nose appendages (despite the relatively small surface area of the skin covering these appendages relative to the rest of the body). A much smaller proportion of brain tissue is devoted to tactile messages coming from areas other than the nose and forelimbs (Figure 17). The highly biased nature of sensory data collection and analysis in this species is also evident in the amount of cortex devoted to nerve inputs coming from the various nose appendages. The number of Eimer's organs per appendage and the sizes of the nerves coming from each appendage do not determine the amount of cortical tissue invested in analyzing a given appendage's signals. Instead, the brain is "more interested in" information from appendages 10 and 11 than from the other appendages, presumably because these appendages are right above the mouth, and so signals from them are most relevant for making decisions about grasping prey [186].

Stimulus Filtering and Selective Visual Perception

A major conclusion that we can draw from studies of acoustical perception in insects and tactile perception in moles is that nervous systems do not gather information in a neutral fashion. Instead, thanks to natural selection in the past, they zero in on the biologically significant items and events in their environment—the sign stimuli, the cues associated with predators or prey.

We can illustrate this point again by looking at the visual perception system of a fellow vertebrate, the European toad. This animal has a visual network similar to our own, with two large eyes whose retinal surfaces are exposed to the environment. The retina contains a layer of receptor cells that detect the energy in the light reflected from objects. Changes in the membrane permeability of these receptors generate action potentials that are relayed to the next link in the network, the bipolar cells, which in turn feed their output to ganglion cells. The long axonal projections of the ganglion cells run together to form the optic nerve, which carries action potentials to the regions within the optic tectum and thalamus of the toad's brain where information from the visual receptors is decoded (Figure 18).

To determine what a toad's brain "sees," Jörg-Peter Ewert devised a sophisticated apparatus to record the activity of single cells in the animal's optic tectum [372, 373]. The apparatus can be mounted on a living toad that is free to move about and respond to stimuli in its laboratory environment. This device informs us that when a toad is sitting still—something toads are good at—and nothing is moving nearby,

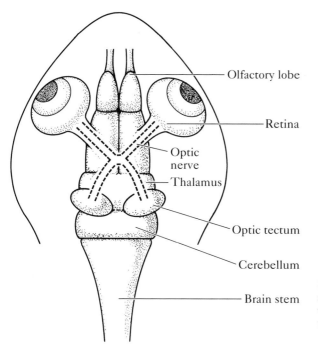

Olfactory lobe

Retina

Optic nerve

Thalamus

Optic tectum

Cerebellum

Brain stem

18 Toad visual system. This view from above shows the optic nerve, which carries visual input from the retina to the optic tectum and thalamus.

the animal apparently does not see a thing! Photons of light are striking its retinal receptors, but the optic nerve does not relay action potentials to the optic tectum.

The ganglion cells in the visual system perform stimulus filtering for the toad. Each ganglion cell has a **receptive field,** defined as that area of the retina whose receptors feed messages back to that ganglion cell via numerous bipolar cells. In the toad and other animals, each ganglion cell monitors a small elliptical portion of the retina. Visual stimulation of this part of the retina causes the receptors there to fire. The resulting action potentials are relayed to the bipolar cells, whose responses affect the probability that the associated ganglion cell will fire. Interestingly, a ganglion cell's receptive field typically is organized into a central excitatory region surrounded by an inhibitory ring (Figure 19). What this means is that messages from bipolar cells monitoring activity in the *outer* ring of the receptive field *reduce* the likelihood that the ganglion cell will fire. On the other hand, if light reaching the retina strikes only the *central* excitatory region of a receptive field, those bipolar cells that cover receptors in this zone will be stimulated, and their activity will in turn generate an excitatory response on the part of the ganglion cell that they report to.

Imagine a brown beetle moving over dark soil in front of a toad. The beetle casts a very small image on the surface of the toad's retina as the light waves reflected from it enter the lens and are focused on the back of the eye. The image moves over clusters of receptors. Some of these clusters will constitute the excitatory central area of the receptive field of at least one ganglion cell. This cell will fire, and information will be sent to the brain.

Now imagine a large object such as a human hand passed close to a toad's eye. The hand will cast a relatively large image on the retina, an image that will cover the entire receptive field of most ganglion cells. In this case, the inputs the ganglion cells receive from the inhibitory surrounding ring will largely or entirely cancel any messages from the excitatory central zone. These cells are not likely to fire in response to this visual stimulus.

Thus, the toad has ganglion cells that respond to *changes* in light intensity that reach the retina, changes typically caused by *moving* objects, whose movement affects how much light enters the retina at a particular point. From a toad's perspective, small moving images on its retina are much more likely to be caused by its prey (nearby beetles and worms) or its enemies (distant herons or hedgehogs) than are large stationary images.

The selective analysis of visual stimuli does not end at the level of the ganglion cells. Neurons within the optic tectum receive messages from many neighboring ganglion cells. Each brain neuron has, therefore, a receptive field of its own, consisting of that area of the retina monitored by the ganglion cells linked to it. The properties of the receptive fields of tectal cells can be studied by applying a recording electrode to a tectal cell in a live, but immobilized, toad and then moving various objects in front of its eye. Some stimuli will elicit a considerable response, others will generate a few action potentials, and still others will have no effect on the activity of the neuron.

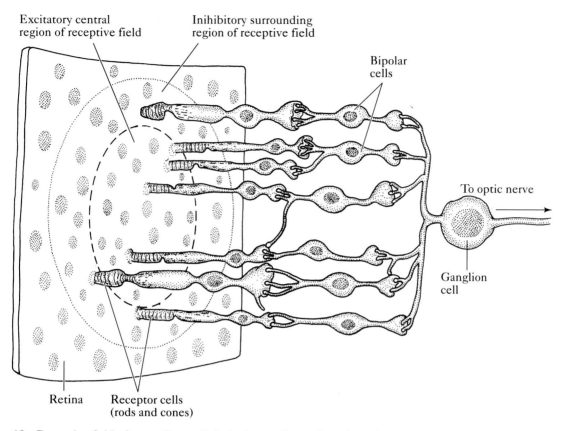

19 Receptive field of a ganglion cell. A single ganglion cell receives signals from many bipolar cells, which in turn are linked to many receptor cells. As a result, the ganglion cell integrates information from a portion of the surface of the retina, called its receptive field. Because of the inhibitory nature of the surrounding region of the receptive field, stimuli that cover the entire receptive field elicit little response in the ganglion cell; however, smaller stimuli that move through the excitatory central region of the field may trigger many action potentials in the cell.

Some cells in the European toad's tectum respond most to long, thin objects that move horizontally across the toad's visual field (Figure 20). These cells have a roughly circular receptive field that consists of an excitatory central strip lying horizontally within an inhibitory surrounding region. Objects that happen to move primarily through the excitatory region of the receptive field will cause the cells to fire. Objects that move through both regions produce few or no action potentials because the stimulation of the excitatory area is canceled by equivalent stimulation of the inhibitory region.

20 Receptive field of a "worm detector" cell in the European toad. A cell in the optic tectum of the toad responds maximally to long, thin objects moved horizontally through the cell's receptive field. *Source: Ewert [373].*

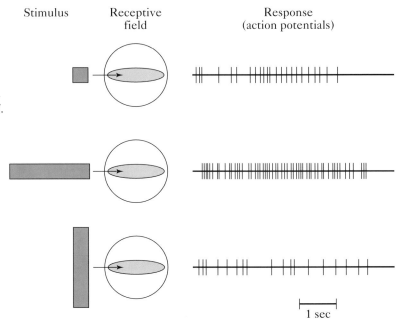

The consequences of the design of these tectal neurons become clear if one imagines how they would respond to a nearby moving worm. The worm creates a small, 5- to 10-millimeter image on the toad's retina. This horizontally oriented image passes through the excitatory central strip of the receptive fields of many tectal cells, causing them to fire rapidly. These messages travel to brain cells that have the capacity to order the toad to turn toward the object. Once the toad is properly oriented, three other kinds of cells sequentially provide the motor commands that cause it (1) to stop turning when both eyes are fixed on the potential victim, (2) to lean closer within tongue range, and (3) to open the mouth, flip out the tongue, and snap up the worm [372].

Optical Illusions and Face Detectors

The fact that humans perceive various optical illusions tells us that our visual mechanisms, like those of toads, have special features that "encourage" us to detect specific stimuli, even at the expense of distorting reality. Certain of our visual cells react exceptionally strongly to edges, actually enhancing the contrast between neighboring dark and light areas. No more light reaches our eyes from the white paper right next to the black squares in Figure 21A than from the central grayish regions that are not bordered directly by black . But our brain tells us that the edges are brighter, creating a useful illusion of the sort that contributes to our ability to see the outlines of objects.

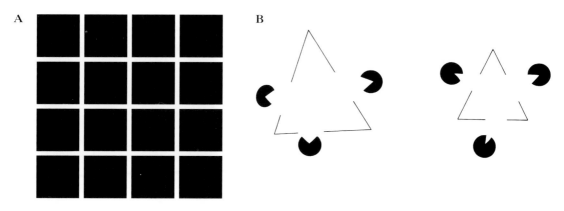

21 Optical illusions and selective perception. (A) We see illusory gray spots at the intersections of the white bars. Our brains exaggerate the contrast wherever there is an edge (black bordered by white), and therefore white areas without edges appear darker than they are. (B) We perceive white forms overlying the triangles—but the forms that we see do not exist. *Source: Marr [773].*

The interpretative capacity of human visual mechanisms leads us to see objects where none actually exist (Figure 21B) because the brain is far more than a passive reporter of environmental stimulation. It assists us in the perception of the partly hidden body of an animal in vegetation, a sharp twig projecting into the path we are traveling, the subtle change of expression in a companion. Indeed, we may have visual mechanisms designed to assist us in the recognition of human faces. Evidence for this hypothesis is provided by a few unfortunate people who have suffered injury to a particular part of the brain's temporal lobe. These people can see things, they can recognize and name many objects, they can identify particular individuals by the sound of their voice or by familiar clothing, but when shown the faces of their friends, their spouses, even themselves, they are at a complete loss. The possibility that a particular neural structure helps us remember faces is also supported by the rare occurrence of brain-damaged individuals with just the opposite problem. They cannot accurately identify ordinary objects when they see them, but have no difficulty recognizing particular faces [74].

To test whether such a face-detecting mechanism might be present in other animals, David Perrett and Edmund Rolls conducted experiments with a primate relative of humans, the rhesus monkey [917]. They used microelectrodes to record the activity of single neurons in the region of the monkey's temporal lobe that corresponds to the area implicated in face recognition in humans. The monkeys were conscious as they were presented with a variety of visual stimuli. In these experiments, Perrett and Rolls discovered a category of temporal lobe cells that fired two to ten times as much when the monkeys looked at images of human or monkey faces as when they looked at any other object. Among the cells in this region

22   **Activity of a "face detector" cell** in the cortex of a rhesus monkey. The cell fires at a high rate whenever the monkey sees a human face or a photograph of a monkey's face. The cell fires often whether the face is nearby or far away, and even when the face appears through colored filters or is upside down. A face in profile, however, does not stimulate the cell. *Source: Perrett and Rolls [917].*

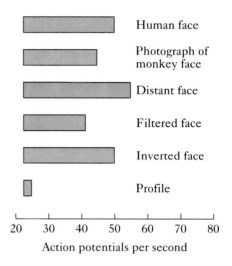

were certain ones that reacted equally strongly to faces close up or distant, upside down or covered with red filters, but did not fire more often in response to a face in profile (Figure 22).

These results are consistent with the hypothesis that the cells in question have something to do with the perception of faces, a task highly pertinent to the lives of these social primates. But finding a *correlation* between cell activity and the presentation of a particular stimulus does not prove that perception of the stimulus is actually *caused* by these particular cells, which might be incidentally stimulated by inputs from other networks of cells that are really responsible for the perception in question.

The Perception of Movement

To demonstrate a causal relationship between neural activity and perception, a research team led by William Newsome [858, 859, 1044] examined movement perception by rhesus monkeys. The experimental subjects watched a monitor on which a picture appeared for just a second, a picture of thousands of flashing, moving dots, most of them moving randomly, but with a certain fraction moving in one direction (Figure 23). The monkeys had been trained by operant conditioning techniques (see Chapter 4) to use eye movements to signal to the experimenter the direction in which the subpopulation of nonrandomly moving dots was going (e.g., up or down, to the right or to the left). The monkeys had little difficulty signaling correctly—until the subset of dots headed in one direction or the other fell to about 5 percent of the total number. The randomly moving 95 percent created visual chaos, making the monkeys' task difficult and increasing the likelihood of an incorrect judgment on their part.

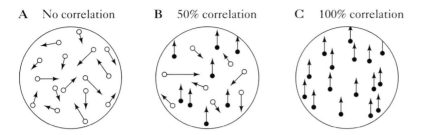

23 Moving dot visual display used to identify the neurons involved in motion detection in the rhesus monkey brain. In (A), no two dots are moving in the same direction; (B) illustrates correlated motion of 50 percent (half the dots are moving in the same direction); in (C), all the dots are moving in the same direction. Rhesus monkeys can identify correlated motion signals of 5 to 10 percent. *Source: Newsome and Paré [859].*

During the experiment, Newsome's research team recorded the activity of small groups of cells in a particular part of the monkeys' visual cortex, a part called area MT. Certain cell clusters in area MT consistently fired only in response to *upward*-moving objects; other, different clusters fired specifically in response to objects that were moving *downward* in a monkey's visual field. When the monkeys made errors, such as signaling that the very small minority of nonrandomly moving dots were moving downward, when they were actually moving upward, the "downward motion detectors" tended to be active, while the "upward motion detectors" tended to be inactive. This result strongly suggested that the perception of the direction of moving objects is directly dependent on the relative activity of these particular cells. But this conclusion too is based on a correlation between neuronal activity and perception (as revealed by the monkeys' behavior), which leaves open the possibility that the active area MT cells coincidentally fire in concert with some other neurons that are truly responsible for the animal's perception of movement.

If it is true that the apparent "downward motion detectors" cause the monkeys to perceive downward movement, then experimental manipulation of these cells should affect the animals' perception. Newsome and his co-workers tested this prediction by using electrodes to stimulate selected cells in area MT while their subjects were examining the moving dot patterns on the monitor. As predicted, stimulation of the upward motion detectors biased the monkeys' perception, so that they were more likely to signal that they had seen upward-moving dots when the screen had actually shown downward-moving dots [1044]. This result provides very strong evidence for a causal relationship between the activity in particular small groups of cells in the monkey's brain and what the monkey thinks it sees.

The Sensory Basis of Navigation

We can draw several conclusions from the neurobiological research reviewed thus far. First, behavior depends on the proximate features of the cells that make up the nervous system. Second, the sensory component of nervous systems supplies different species, and sometimes even the two sexes of the same species, with different kinds of information about the environment on which to base their behavioral "decisions." Third, at the ultimate level of analysis, these differences in sensory inputs are related to specific obstacles to reproductive success. Thus, the mechanisms for gathering information from the environment are species-specific; they do not provide general, all-purpose data about the world, but rather bias the input toward fitness-relevant events and objects.

Let's reinforce these points one more time by considering how some animals secure information that enables them to do something that humans are unable to do without special technology—namely, navigate across unfamiliar terrain to a specific destination. If you or I were dropped off in a strange spot even a few miles from home, we would probably have no idea which way to go. Nor would we be any better off if we were given a compass by the persons studying our homing abilities. We would have to be told in which direction to head; only then would the compass be helpful. Navigation to a destination requires that you know where you are relative to your goal (a map sense) and in what direction to move (a compass sense). These two abilities are possessed by many species, especially those that migrate vast distances and those that forage well away from a home base to which they regularly return (Figure 24).

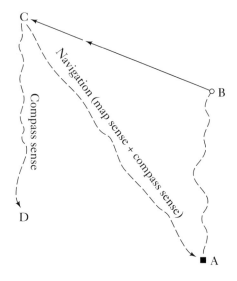

24 A compass sense is necessary, but not sufficient, for navigation. An animal that has moved from its home site, A, to site B is captured and transported to site C. If the animal is able to return directly home from C, it demonstrates an ability to navigate. If instead, the animal moves to site D, it has a compass sense, but not a map sense, and cannot navigate home.

Although we know something about how a few species use visual landmarks to construct mental images that assist them in their travels [e.g., 318, 1247], our understanding of navigational maps is rudimentary. What we know about the compass sense involved in navigation comes in large measure from studies of honeybees and homing pigeons. Both species are skilled navigators, as demonstrated by a honeybee's ability to make a beeline back to its hive after a meandering outward journey in search of food, and by a homing pigeon's ability to make a pigeonline back to its loft after having been released in a distant and strange location.

Honeybees and homing pigeons are active during the daytime, and, as one might suspect, both use the sun's position in the sky as a directional guide [629, 1226]. Even we can do this to some extent, knowing that the sun rises in the east and sets in the west—provided we also know approximately what time of day it is. Every hour the sun moves 15 degrees on its circular arc through the sky. One has to adjust for the sun's movement if one is to use its position as a compass. A bee leaving its hive notes the position of the sun in the sky relative to the hive and flies off on a foraging trip. It might spend 15–30 minutes on its trip and might move into unfamiliar terrain in search of food. If it were to try then to return home, orienting as if the sun had not shifted, the bee would get lost, because the sun's position changes with the passage of time.

Honeybees rarely get lost, in part because they learn visual landmarks in foraging areas, but also because they use their internal clock mechanism (see Chapter 6) to compensate for the sun's movement [723]. This ability can be demonstrated by training some marked bees to fly to a sugar-water feeder some distance from the hive (say, 300 meters due east of the hive). One can then trap the bees inside the hive and move everything to a new location. After 3 hours have passed, the hive is unplugged, and the bees are free to go in search of food. They do not have familiar visual landmarks to guide them, and yet some marked individuals remember that food is found 300 meters due east. They fly 300 meters due east of the new hive location, to the spot where the food source "should be." They have compensated for the 45-degree shift in the position of the sun that has taken place during their 3-hour confinement.

Pigeons too can be tricked into demonstrating how important a clock sense is if they are to orient accurately by the sun [1223]. The birds can be induced to reset their biological clocks by placing them in a closed room with artificial lighting and then shifting the light and dark periods in the room out of phase with sunrise and sunset in the real world. For example, if sunrise is at 6:00 A.M. and sunset at 6:00 P.M., one might set the lights to go on 6 hours earlier (midnight) and off 6 hours sooner (noon). A pigeon exposed to this routine for several days would become *clock-shifted* 6 hours out of phase with the natural day. If taken from the room and released at 6:00 A.M. at a spot some distance from the loft, the bird will behave as if the sun has been up for 6 hours (as if it is noon), which will cause it to orient improperly. For example, let's say that the pigeon is released at a place 50 miles due west of its loft. Its map sense somehow tells it this, and it attempts to orient

25 Clock-shifting and altered navigation in homing pigeons. The results of a hypothetical experiment. The pigeons are released at 6:00 A.M. near Augusta, Maine. The dashed line represents the flight path of a bird kept in natural light; the dotted line shows the flight path of a bird whose clock has been shifted by 6 hours.

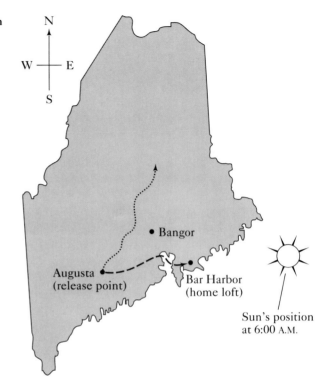

itself to fly east (Figure 25). As Charles Walcott points out: "To fly east at 6:00 A.M., you fly roughly toward the sun, but because your clock tells you it is really noon, you know that the sun is in the south and that to fly east, you must fly 90 degrees to the left of the sun. And this is exactly what the birds do, although they presumably do not go through the reasoning process I have described."

Backup Orientation Mechanisms

If a sun compass were the only mechanism available to bees and pigeons, they would not be able to home on cloudy days. But both species navigate successfully under total overcast [318, 1298]. In fact, some pigeons can home accurately at night, or when they have had frosted contact lenses placed over their eyes, reducing their vision to a murky blur! Thus, these species must have more than one compass mechanism, including one that might operate on clear nights (celestial orientation based on patterns of star locations) and others that might come into play when neither sun nor stars are visible [357].

Among the nonvisual compass mechanisms that have been explored is one based on a bird's ability to sense the weak lines of magnetic force created by the earth's magnetic field, which you and I cannot detect at all. If a bird were to use the cues provided by these north–south-running lines of magnetism, then experimental deflection of the magnetic field around the bird should disrupt accurate orientation in individuals that have access to no other compass system. Charles Walcott altered the magnetic field around some pigeons, either by strapping a magnet to the birds or by outfitting them with helmets consisting of a battery-powered Helmholtz coil (Figure 26). The altered magnetic field around the pigeons disoriented them when they were released far from home—but only on overcast days. If the sun was shining, the birds attended to the cues provided by sun position and ignored the signals detected by their magnetism sensors. But when reliable information from the sun was missing, the pigeons used the next best thing, information about lines of magnetic force, as an aid to getting home [1233].

Pigeons are fine birds, excellent navigators, and superb subjects for persons interested in the mechanisms of navigation. However, they are by no means the only birds capable of getting from point A to point B across unfamiliar terrain. Indeed, the feats performed by some long-distance migratory species far outshine the homing flights of pigeons (see Chapter 11). These birds typically fly at night, leaving around dusk on trips that may cover hundreds of miles. Experiments with several migratory songbird species have shown that the pattern of polarized light at sunset is the predominant cue that these birds use to set off in the right direction on their nocturnal journeys [1]. But here too multiple orientation cues can come into play, as the position of the setting sun, the earth's magnetic field, and the position of the stars also can be used to varying degrees, depending on the species. Thus, just as diurnal travelers have navigational backup devices based on different sensory systems, night-flying birds track more than one cue with their battery of senses.

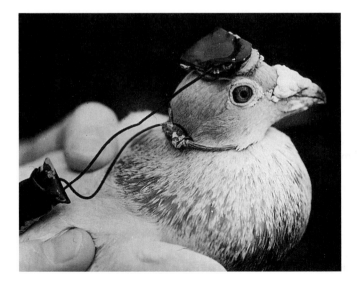

26 **Homing pigeon with a Helmholtz coil fitted on its head** to disrupt its ability to detect the earth's magnetic field. *Photograph by Charles Walcott.*

Olfactory Navigation?

Instead of exploring how researchers have reached conclusions about the ability of some migrating songbirds to detect magnetic fields and polarized sensory signals, we will return to the humble pigeon to focus on the possibility that it smells its way home, a hypothesis that is still controversial [2] long after its presentation in 1971 by the Italian ornithologist Floriano Papi [902]. Papi's experiments convinced him that homing pigeons form a mental map of odors (Figure 27), learning which scents always come on winds from the north, which from the west, and so on. If this hypothesis is correct, then interference with the olfactory system, either by cutting key olfactory nerves or by applying a temporary local anesthetic to olfactory receptors, should produce a smell-blind pigeon that is unable to navigate home. Experimental manipulations of the olfactory network often do reduce the accuracy of the initial orientation of released homing pigeons (Figure 28) [1234].

An alternative explanation for these results is that the experimental manipulations affect the birds' motivation and attention, which in turns alters their orientation and homing ability independently of their changed sense of smell. Birds that have their olfactory nerves severed might be less motivated to return to their lofts. But surgically altered pigeons that are released in familiar places, and so know the local landmarks, orient normally when released and zip home with the same efficiency as birds with intact olfactory nerves. Thus, the operation apparently does not harm the bird's readiness to fly home or to pay attention to the cues

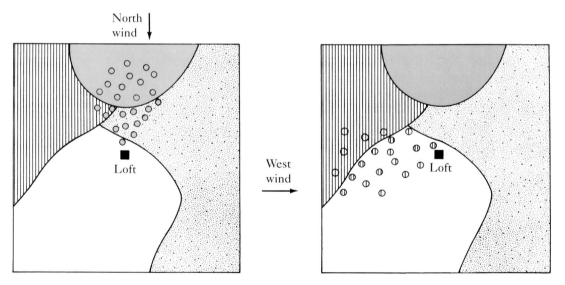

27 An olfactory map? If different locations have different odors, a north wind and a west wind would each have a distinctive odor. Pigeons at a home loft might be able to store this information to construct an olfactory map for later use when navigating back to their loft. *After Papi [902].*

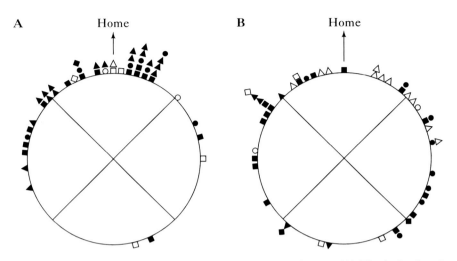

28 Summary of olfactory experiments with homing pigeons. (A) The behavior of control pigeons with unaltered olfaction. Each symbol represents the mean initial orientation of pigeons tested in a given study. Note the clustering of symbols around the home direction. (B) Pigeons with blocked olfaction are much less likely to orient accurately toward the home loft on release. *After Papi [903].*

needed to do the job. Still, it would be good to test the olfactory map hypothesis without assaulting the birds' sensory equipment.

Consider that if one were to deflect the route taken by winds arriving at a loft, the homing pigeons there should construct a "deflected" olfactory map. For example, if winds from the west were deflected by wooden baffles so to enter the loft from the south, the affected pigeons would be deceived into thinking that odors from places in the west were in the south, and this would cause them to shift their orientation when they were released at a site away from the loft and smelled the local air. Experiments involving birds reared in such deflector lofts (Figure 29) yielded the expected result: the initial orientation of birds at the release site was shifted by the angle predicted if they had been fooled into constructing a deflected odor map [50].

However, pigeons held in deflector lofts may misorient on release for non-olfactory reasons. Perhaps the deflector panels also deflect light entering the loft, and this visual effect alters the orientation mechanism of homing pigeons [1057]. To control for this possible effect, some researchers kept two groups of young, naive pigeons in identical cages without deflectors. The experimental group was exposed to a special scent, benzylaldehyde (BA), blown into the cage from one direction, say, the northwest. The other group had access to unaltered air flows. Then both groups of pigeons traveled to the same release point while BA was applied to their carrying cages at intervals during the journey and at the release site. The control

29 Deflector loft. (Top) Baffles mounted on a loft deflected the wind clockwise by 90 degrees, leading pigeons housed in the loft to mistakenly associate odors coming from the north with an easterly source. (Bottom) Thus, when the pigeons were released northeast of the loft, instead of flying southeast to return home ("Home"; 208°), all but one headed directly away from what they now perceived to be north (which was actually to the east) and flew west (open circles). The arrow indicates the mean initial orientation (264°) of the group as a whole. *After Baldaccini et al. [50].*

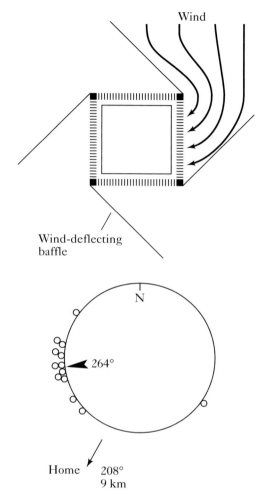

pigeons, which experienced BA for the first time, apparently ignored it and set off toward home accurately. The experimental pigeons, which perceived the scent that had always been coming to them from the northwest at their home cage, were expected (according to the olfactory hypothesis) to assume that they were in the path between the source of BA in the northwest and their home loft in the southeast. Therefore, to return home, these birds should have headed toward the southeast—and they did [590] (Figure 30).

Yet another ingenious test of the olfactory map hypothesis has been performed by Jakob Kiepenheuer [647]. He took air samples from four future release sites (A, B, C and D). Pigeons were then introduced into airtight containers with one of the four air samples and kept there for a time, after which they were removed

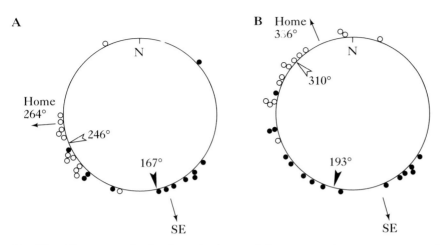

30   The olfactory navigation hypothesis tested. In two experiments, one group of homing pigeons was exposed at intervals to the odor of benzylaldehyde (BA) blown by fans into the home loft from the northwest; a control group was not exposed to the odor. Both groups were then exposed to BA on the trip to a release site, and at the release site, which was to the east of the loft in one experiment (A) and to the north in the other (B). Under the olfactory hypothesis, it was predicted that the experimental birds (solid circles) would "think" that they had been carried northwest, toward the source of the BA they had smelled in their loft, and would thus fly southeast to get home. Many did just that. In contrast, the control pigeons (open circles) ignored the odor of BA and oriented accurately toward their home loft in both experiments. The solid and open arrows respectively indicate the mean initial orientation of the experimental and control groups in the two experiments. *Source: Ioalé et al. [590].*

and given a local anesthetic that blocked their sense of smell. The birds were then packed up and taken to site A, B, C, or D for release. Only the birds that had smelled air from A before being released at A tended to fly off in the direction of home when let go (Figure 31). (Likewise, birds that had smelled B air could orient properly only if released at B, and so on.) It is as if birds with experience with site A air before being made smell-blind said to themselves upon release, "I last smelled A air; site A is located west of the home loft; therefore in order to return to the loft, I must go east." Again, they presumably do this "unconsciously" or automatically, but however they do it, the results are consistent with the hypothesis that they are using an olfactory map.

Kiepenheuer's experimental design was subsequently employed by Jörg Ganzborn after he had first released unaltered birds in various locations around Tübingen, Germany [421]. He measured the directions taken by the birds as they disappeared from view, and found that different release sites could be clustered together in four groups (or patches) in terms of the initial orientations of the released pigeons. The tendency of pigeons to fly off in a particular direction when released

31 Olfactory cues and orientation in homing pigeons. (A) The directions taken by released pigeons that had not been exposed to air from their release site before they were made smell-blind. The mean orientation (arrow) of these experimental birds was not toward the home loft (triangle). (B) In contrast, control pigeons that had experience with air from the release site before their sense of smell was blocked tended to fly toward the home loft. *Source: Kiepenheuer [647].*

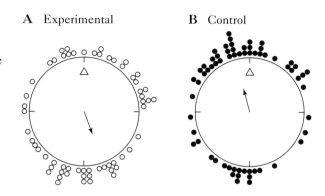

at a particular site, no matter where their home loft is located, is a regular, but still puzzling, feature of pigeon homing behavior [648]. Ganzborn then exposed some pigeons to air from a site belonging to one of the various patches and then made them temporarily unable to smell through application of an anesthetic to their nasal mucosae. The pigeons were subsequently released at another site belonging to a different patch. In some places, but not all, the smell-blind pigeons had no trouble orienting properly toward home despite their inability to smell.

Ganzhorn interprets these apparently conflicting results to mean that the map that homing pigeons form has an airborne olfactory component, but other cues can overrule the olfactory cue in some, but not all, areas. If homing pigeons do employ a multisensory map system with both airborne and nonolfactory components, then no wonder that advocates and opponents of the olfactory map hypothesis often get contradictory results and reach opposing conclusions [2, 1299].

Scientists often search for a single cause of an observed phenomenon, believing that the procedure of testing alternative hypotheses will eventually result in the elimination of all but the one "right" explanation. But when there are several causes, experiments may generate confusion until researchers reach a consensus that multiple causal factors are at work. Perhaps this approach will eventually help us truly understand the proximate mechanisms of animal navigation.

Mechanisms of Motor Control

Thus far we have focused heavily on how specialized sensory systems detect critical stimuli such as ultrasound, moving objects, the visual cues associated with faces, or the olfactory inputs that may enable a pigeon to map the odorous world around it. Just as sensory detection reflects specialized features of nerve cells, so too does an animal's reaction to critical sensory cues. The pigeon uses its map sense to do something special—namely, return to its home loft. The European toad filters visual input in order to locate worms to eat. Noctuid moths sense ultrasound in order to evade approaching bats.

The cricket *Teleogryllus oceanicus* also has ultrasound detectors, whose activity also helps these insects avoid bats [818]. The sensory side of the process begins with the firing of certain ultrasound-sensitive acoustical receptors in the cricket's ears, which are found on its forelegs. Sensory messages from cells that react to sounds of 40–50 kHz travel to a diversity of cells in the cricket's central nervous system. One such cell is a pair of sensory interneurons called *int-1*, one of which is located on each side of the insect's body. Ron Hoy and his co-workers established that *int-1* played a key role in the perception of ultrasound by playing sounds of different frequencies to a cricket while recording the response of its *int-1* cells. These cells became highly excited when the cricket's ears were bathed in ultrasound. The more intense a sound in the 40–50 kHz range, the more action potentials were produced, and the shorter the latency between stimulus and response—two properties that exactly match those of the Al fiber in noctuid moths.

These results suggest that the *int-1* cell is part of a circuit that helps the cricket respond to ultrasound. If this is true, then it follows that if one could experimentally inactivate *int-1*, ultrasonic stimulation should not generate the typical reaction of a tethered cricket, which is to turn away from the source of the sound by bending its abdomen (Figure 32). Note that only flying or suspended crickets respond to ultrasound. As expected, suspended crickets with temporarily inactivated *int-1*s do not attempt to steer away from ultrasound, even though their acoustical receptors are intact and producing action potentials. Thus, the *int-1* cells are necessary for the steering response.

The corollary prediction is that if one could activate *int-1* in a flying, tethered cricket (and one can, with the appropriate stimulating electrode), the cricket should change its body orientation as if it were being exposed to ultrasound, even though it is not. Experimental activation of *int-1* does cause the cricket to bend its abdomen [864]. Therefore, activity in this neuron is sufficient to cause the steering response. These experiments convincingly establish a causal relationship between *int-1* activity and the apparent bat evasion response of flying crickets. But the *int-1* cells are sensory interneurons; they do not innervate the cricket's locomotory muscles directly. Instead, they send their signals to the flying cricket's brain, which integrates inputs from many sources before "making a decision" to send signals to motor neurons that control muscles in the cricket's body.

What is the proximate motor mechanism that enables a flying cricket to carry out the order to steer away from a source of ultrasound? Mike May, then a graduate student in Ron Hoy's laboratory, started work on this problem by simply watching tethered, flying crickets that were periodically exposed to bursts of ultrasound [781]. One day, while he was not conducting any experiment or formally testing any hypothesis, but instead was (in his own words) "toying with the ultrasound stimulus and watching the responses of a tethered cricket," he noticed that the beating of one hindwing seemed to slow down every time he zapped the cricket with an ultrasonic stimulus. Crickets have four wings, but only the two hindwings power the flight of the insect. If the hindwing opposite the source of ultrasound

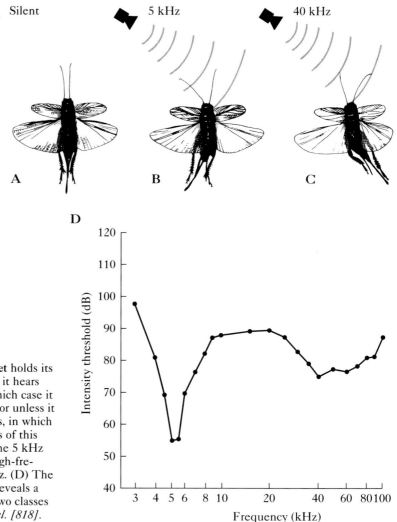

32 **A flying tethered cricket** holds its abdomen straight (A) unless it hears low-frequency sounds, in which case it turns toward the sound (B), or unless it hears high-frequency sounds, in which case it turns away (C). Males of this species produce sounds in the 5 kHz range; some bats produce high-frequency calls of about 40 kHz. (D) The tuning curve of the cricket reveals a special sensitivity to these two classes of sounds. *Source: Moiseff et al. [818].*

really did slow down, the result should be reduced power or thrust on that side of the cricket's body and a corresponding turning (or "yawing") of the cricket away from the stimulus (Figure 33).

May converted his initial interpretation of his observations into a formal hypothesis: that by lifting a hind leg into a hindwing, the cricket altered that wing's beat and thereby changed its flight path. To test this hypothesis, May took a number of high-speed photographs of crickets with and without hind legs. Without the appropriate hind leg to act as a brake, both hindwings continued beating unimpeded when the cricket was exposed to ultrasound. As a result, crickets without hind legs required about 140 milliseconds to begin to turn, whereas intact crickets started their turns in about 100 milliseconds. May concluded that central neurons in the ultrasound detection system order the appropriate motor neurons to induce muscle contractions in the opposite-side hind leg of the cricket. As these muscles contract, they lift the leg into the wing, interfering with its beating movement and contributing to an adaptively speedy turn away from an ultrasound-producing bat [781].

33 How to turn away from a bat—quickly. (Top) A flying cricket typically holds its hind legs so as not to interfere with its beating wings. (Middle) Ultrasound coming from the right causes the leg on the opposite side of the body to lift up into the wing. (Bottom) As a result, the beating of the left hindwing slows, and the cricket's body turns away from the source of ultrasound while the cricket dives to the ground. *Drawing by Virge Kask, from May [781].*

The Song of the Midshipman Fish

We return to communication signals for a second example of how nerve cells regulate the muscle movements required to produce a particular behavior. The plainfin midshipman is not the most handsome of fish, but its elegant neurophysiology more than compensates for its grotesque appearance [66]. The behavior of special interest to Andrew Bass and his co-workers is the humming song of the fish. Large males produce this song at night in the spring and summer while guarding rock shelters that serve as nests. The song attracts females to the rock shelters; the fish spawn at these sites, and the male subsequently guards the eggs his mates attach to his rocks.

How do the male fish produce their songs? Inspection of the anatomy of the midshipman reveals that it has a large, air-filled swim bladder sandwiched between layers of muscles. The bladder serves as a drum; rhythmic contractions of the muscles "beat" the drum, generating vibrations that other fish can hear (Figure 34).

In order to figure out how the contraction of the muscles is controlled, Bass dissected the swim bladder apparatus, exposing the motor nerve cells that were connected to the sonic muscles. He then applied a cellular dye called biocytin to the cut ends of these nerve cells, which absorbed the material, staining themselves brown. And the stain kept moving along, crossing the synapses between the first cells to receive it and the next ones in the circuit, and so on, through the whole network of connected cells. By cutting the brain into fine sections and searching for cells stained brown by biocytin, Bass and his colleagues mapped the sonic control system. In so doing, they discovered two discrete collections of interrelated cells whose job it is to generate the signals that translate into the coordinated muscle contractions required for midshipman humming. These two clusters, containing some 2000 neurons near the base of the brain (Figure 35), constitute the paired sonic motor nuclei. The long axonal processes of these cells travel out from the brain, fusing together to form the occipital nerves, which reach the swim bladder muscles.

34 Song-producing apparatus of the male midshipman fish. The sonic muscles control the movement of the swim bladder, thereby controlling the fish's ability to hum for females. *After an illustration by Margaret Nelson, in Bass [66].*

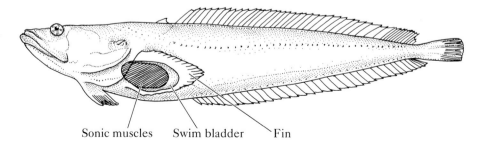

Sonic muscles   Swim bladder   Fin

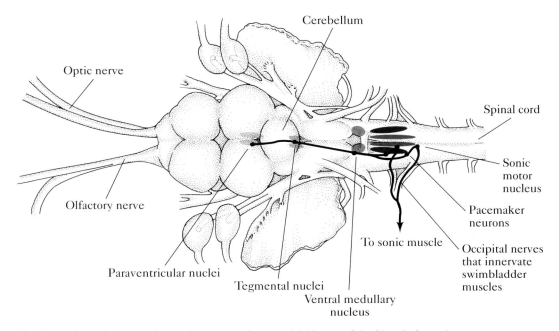

35 Neural regulation of the sonic muscles in the midshipman fish. Signals from the central region of the brain (the mesencephalon) travel by way of the cerebellum and ventral medullary nucleus to the sonic motor nuclei in the upper part of the spinal cord. The firing of the pacemaker neurons regulates the frequency of firing by cells in the sonic motor nuclei; these signals in turn set the rate of contraction of the sonic muscle, and thus, the frequency of the sounds produced by the fish. *After an illustration by Margaret Nelson, in Bass [66].*

In addition to these components, two other anatomically distinct elements ally themselves with the sonic motor nuclei. First, lying right next to each nucleus is a sheet of pacemaker neurons, which are believed to adjust the activity of the sonic motor neurons so that the frequency of the muscle contractions will yield a proper song. Second, in front of the pair of sonic motor nuclei are some special neurons that appear to connect the two nuclei, probably to coordinate the firing patterns coming from the left and the right nucleus, so that the muscle contractions and relaxations will be synchronous, the better to produce a humming sound [66].

Central Pattern Generators

The basic principles of motor control illustrated by crickets and midshipman fish are twofold. (1) Nervous systems are composed of clusters of nerve cells dedicated to the control of particular body parts and responses. (2) The vital timing and coordination of rhythmic locomotory activities are achieved with the help of cells that adjust and integrate the activity of motor neuron clusters.

To reinforce this point, look at the behavior of a *Tritonia* sea slug that comes into contact with chemicals from the tentacles of a predatory sea star. The slug quickly begins to "swim" in the ungainly fashion of sea slugs by bending its body up and down [1291]. The animal performs from two to twenty alternating bends, often continuing to swim long after it is out of range of the stimulus that triggered the reaction. The escape response of *Tritonia* is carried out through an alternating pattern of contractions of the animal's dorsal and ventral muscles (Figure 36). When the dorsal muscles contract, the slug's body is pulled into a U-shaped position; when the ventral muscles contract, the slug's body is pulled downward. These two sheets of muscles are regulated by two large motor cells, the dorsal flexion neuron (DFN) and the ventral flexion neuron (VFN). Electrical stimulation of the DFN causes the slug's body to bend upward; stimulation of the VFN generates the opposite response [1291]. The striking feature of the slug's escape behavior is the *alternating series* of upward and downward bends, which led A. O. D. Willows to suggest that the two flexion neurons are in contact with each other and with some other cell. Under his control model (Figure 37A), interactions between the two flexion neurons and a third general excitatory neuron (GEN) were responsible for producing an alternating cycle of activity in the DFN and VFN.

This hypothesis was tested by P. A. Getting, who ultimately rejected it because he found a cluster of central interneurons that control the two flexion neurons [433, 434]. One of these interneurons, labeled the C2 cell, has the capacity to excite certain other cells, the dorsal swim interneurons, that control the DFN, while inhibiting still other interneurons linked to the VFN. As a result, when the C2 cell fires, the DFN receives commands to fire, causing dorsal flexion. But the interaction between the C2 cell and the dorsal swim interneurons is such that after a period of excitation, the C2 cell begins to inhibit the DFN while exciting the interneurons that stimulate the VFN. As a result, the DFN ceases to fire when the VFN begins its burst of activity. The situation then reverses. This alternating activity in the interneurons regulating the DFN and VFN leads to alternating bouts of DFN and VFN firing, and thus alternation of dorsal and ventral bending.

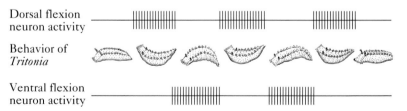

Dorsal flexion neuron activity

Behavior of *Tritonia*

Ventral flexion neuron activity

36 Escape behavior in *Tritonia* sea slugs. Alternating bouts of ventral and dorsal muscular contractions cause the slug to undulate away from a noxious stimulus, the scent of a predatory sea star. The dorsal and ventral muscles are controlled by two different groups of neurons whose activity is correlated with the alternating bouts of muscle contraction.

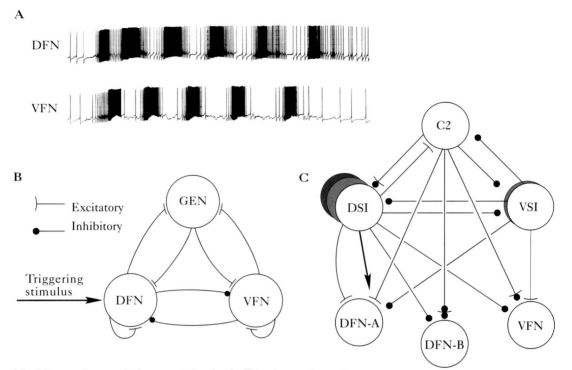

37   Neuronal control of escape behavior in *Tritonia:* two hypotheses.
(A) Recordings of dorsal flexion neuron (DFN) and ventral flexion neuron (VFN) activity, showing that activity in one cell somehow inhibits activity in the other. (B) One model put forth to explain this mutual inhibition proposed that the DFN and VFN are directly linked, so that signals from one cell directly inhibit the other. (Bars represent excitatory synapses; solid circles represent inhibitory synapses.) Both flexion neurons are also connected to a general excitatory neuron (GEN), which becomes active when the DFN begins to fire in response to a triggering stimulus. An active GEN excites the VFN, which begins to fire, thereby shutting down the DFN temporarily, and so on. (C) After rejection of this model, a new, more complex hypothesis was developed in which cerebral neuron 2 (C2), three dorsal swim interneurons (DSI), and two ventral swim interneurons (VSI), together constitute a central pattern generator that regulates the activity of the flexion neurons (DFN-A and -B and VFN), which do not communicate directly with each other. The DSIs modulate the strength of the connection from C2 to DFN-A (arrow). Several of the synapses in this model are multi-component synapses which are capable of sequential excitatory and inhibitory activity; for example, when active, the synapse between C2 and DFN-B is first inhibitory, then excitatory, then inhibitory again. *Sources: (A,B) Willows [1291]; (C) Katz and Frost [627].*

Still more recent work on this system has shown that activity in the dorsal swim interneurons affects the subsequent release of a neurotransmitter by the C2 cell, which increases the effect of signals from this cell on the other neurons with which it communicates [627]. Thus, a cluster of interneurons within *Tritonia*'s central nervous system interact with each other in a programmed manner to generate rhythmic output to specific motor neurons. These interneurons compose the **central pattern generator** that controls the sea slug's escape response (Figure 37B).

One key feature of this and other central pattern generators is that they do not require sensory feedback in order to produce their self-generated sequence of signals. For example, if one takes the brain stem and spinal cord from a lamprey (a primitive fish), the living tissue can still play out the rhythmic pattern of signals that would cause swimming motions in an intact lamprey [474]. This result shows that an independent central pattern generator within the brain stem or spinal cord regulates swimming behavior in this species (although in an intact animal, these cells would receive and use sensory feedback to adjust the basic swimming pattern in relation to particular conditions encountered by the lamprey [473].)

Central pattern generators that regulate locomotion are distributed throughout the animal kingdom, and have been especially well studied in certain invertebrates [173]. For example, when the grasshopper *Locusta migratoria* flies, it will remain airborne only if its wings move up and down in a precisely timed fashion. The wing muscles of these insects are controlled by separate motor neurons that either elevate or depress them. An elevator neuron begins to fire soon after air is blown onto the head of an immobilized, but living, grasshopper [1003]. The activity of this neuron begins an alternating cycle of elevator and depressor neuron signaling highly reminiscent of that just described for the dorsal and ventral flexion neurons of *Tritonia* sea slugs (Figure 38).

If this alternating pattern of motor neuron activity stemmed from reciprocal inhibitory arrangements between the elevator and depressor neurons, then we

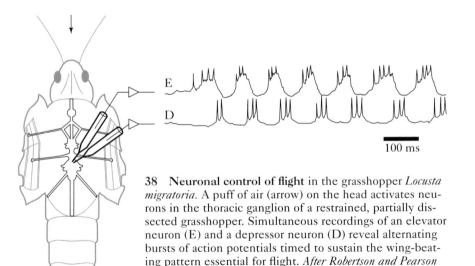

100 ms

38 Neuronal control of flight in the grasshopper *Locusta migratoria*. A puff of air (arrow) on the head activates neurons in the thoracic ganglion of a restrained, partially dissected grasshopper. Simultaneous recordings of an elevator neuron (E) and a depressor neuron (D) reveal alternating bursts of action potentials timed to sustain the wing-beating pattern essential for flight. *After Robertson and Pearson [1003, 1004].*

would predict the existence of direct connections between these two types of cells. But just as is true for *Tritonia* sea slugs, direct links do not exist, suggesting that the alternating pattern stems from a cell or cells elsewhere in the nervous system. Candidates for inclusion in a central pattern generator are some large interneurons in the thoracic ganglia of the grasshopper. These interneurons fire rhythmically at a frequency that matches the frequency with which the wings beat up and down in a flying grasshopper. Furthermore, if one experimentally interrupts the activity of one of these interneurons, flight muscle activity is correspondingly delayed. Finally, the pattern of activity in these interneurons persists even when sensory inputs to the central nervous system are largely eliminated by cutting the incoming sensory nerves. It appears, therefore, that these cells can generate a distinctive pattern of signal activity, a pattern that imposes fundamental order on the motor commands that the flight muscles must receive if the insect is to fly properly [1004].

SUMMARY

1. Neural mechanisms constitute a proximate cause of behavior. Nerve cells acquire sensory information, process these data, and order particular motor responses to key stimuli and events in the environment. Different species have different neural mechanisms and therefore perform these tasks differently, providing proximate reasons why species differ in their behavior.

2. The classic ethological approach to nervous systems focused on the hypothesis that animals possess neural elements designed by natural selection to detect key stimuli and to order appropriate species-specific responses to these biologically relevant cues. Modern neurophysiologists have confirmed that nervous systems contain specialized sensory receptors whose design facilitates the acquisition of critical information from the environment. By knowing what obstacles to reproductive success exist in an animal's natural environment, researchers can predict, search for, and find, specific neural "solutions" to these problems.

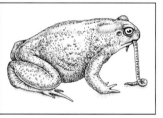

3. Stimulus filtering, the filtering out of irrelevant information, occurs at higher levels of the nervous system as well, and is a universal attribute underlying perception. It arises from the action of specialized decoder cells, typically within the central nervous system. Likewise, the reactions of animals to perceived events or objects are controlled by specialized motor elements of nervous systems, most evident in the central pattern generators of some species. Once activated, these cell aggregates can play out a long series of messages to selected muscles, generating complete and adaptive behavioral responses to events in the animal's environment.

4. Many unanswered questions remain in the study of the neural basis of behavior, as illustrated by the continuing debate over what role olfactory perception plays in the navigation of homing pigeons. Debates of this sort arise because of the complexity of the underlying neural mechanisms, which are especially difficult for us to understand because they are entirely absent in humans.

SUGGESTED READING

Kenneth Roeder's *Nerve Cells and Insect Behavior* [1008] is a wonderful book, an understandable, entertaining, and exciting account of how to conduct research on the physiology of behavior of moths, bats, and other animals. Nubuo Suga has reviewed bat sonar research [1156].

Books by Jeffrey Camhi [173], Jörg-Peter Ewert [373], and David Young [1319] explain how nerve cells promote adaptive behavior. The connection between proximate and ultimate aspects of neurobiology is reviewed by Richard Francis [402] and Walter Heiligenberg [525]. *Mechanisms of Animal Behavior* by Peter Marler and W. J. Hamilton [771] provides many examples of the relation between physiological mechanisms and animal ecology. Good general reviews of animal orientation include those by Stephen Emlen [348], William Keeton [629], and Rüdiger Wehner [1247], with a host of authors contributing to a recent collection of articles on the subject [1249]. I recommend two particularly readable articles on cricket flight control by Mike May [781] and song control in the midshipman fish by Andrew Bass [66].

DISCUSSION QUESTIONS

1. Kenneth Roeder discovered that the A1 receptor in certain noctuid moths does not discriminate between sounds of different frequencies. He made this discovery by formulating a hypothesis, developing predictions, performing tests of the predictions, and coming to a conclusion. Formally reconstruct this process.

2. When cockroaches are attacked by a toad, they turn and run away. A roach has wind sensors that react to puffs of air pushed ahead of approaching predators. These sensors are concentrated on its *cerci*, two thin projecting appendages at the end of its abdomen. One cercus points slightly to the right, the other to the left. Use what you know about moth orientation to bat cries to suggest how this simple system might provide the information the roach needs to turns away from the toad, rather than toward it. How might you test your hypothesis experimentally? See [173] after answering.

3. Sheep live in herds in which social status is affected by the size of the horns an individual possesses. Males threaten each other by frontal display of their horns. Given this knowledge and the information that the temporal lobe of the sheep brain is involved in visual perception, you hypothesize that specialized neurons exist in the temporal cortex that enable individuals to assess the size opponents' horns and the occurrence of threats from other sheep. What sort of activity would you expect to observe in these cells in response to different stimuli? See [634] after answering the question.

4. Produce a diagram or table in which you outline the different ways in which genetic information is involved in the growth and alteration of a nervous system that makes it possible for a white-crowned sparrow to learn the song of its species by listening to adult males sing that song.

5. Among plainfin midshipman fish, two types of males exist: large, humming, egg-guarding males and small individuals that do not sing and do not build or defend nests. What neuronal hypotheses can you develop to account for this difference between the two types of males, and how could you test them?

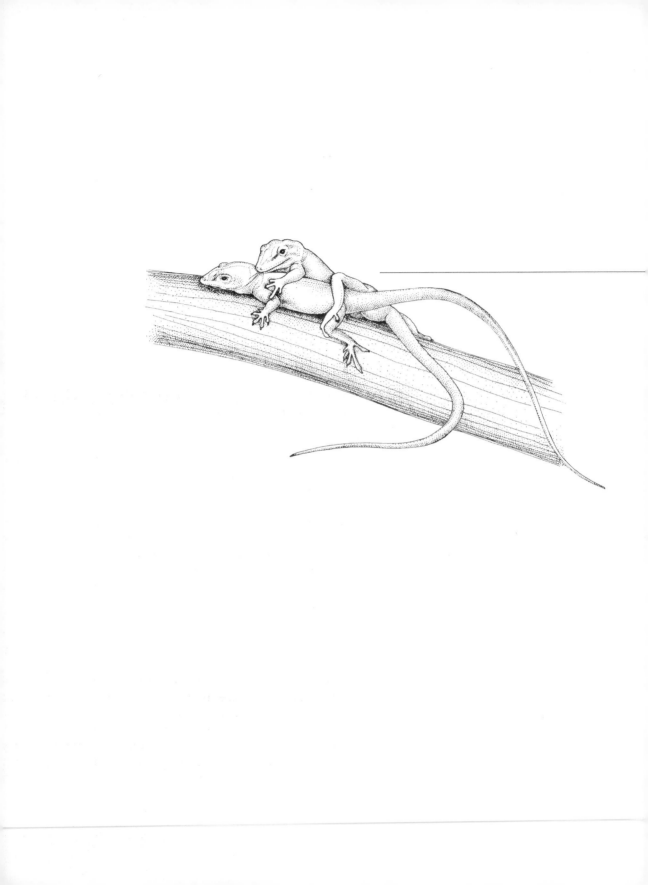

CHAPTER 6

The Control of Behavior: Organizing Mechanisms

W

E HAVE TALKED ABOUT ANIMALS as if they were machines endowed with neuronal computers that detect key stimuli, discriminate among patterns of inputs, and order adaptive responses to these "important" things and events. A flying moth or cricket has sensors that relay messages about acoustical stimuli produced by bats; these messages elicit behavior that (sometimes) guides the insect to safety. The capacity of neural mechanisms to filter out irrelevant information, to perceive some things very readily, and to order effective responses to perceived events makes adaptive sense. But that is not the end of the story. Neuronal computers help animals do more than merely activate response X in the presence of stimulus Y. Imagine a flying male moth hot on the scent trail of a female perched in a distant tree. If the male's nervous system operated simply by turning on flight behavior whenever female scent was present in the air, the scent-tracking moth would be unable to dive out of the sky away from an attacking bat. But the moth's nervous system does not operate in such a simpleminded fashion. Instead, it integrates inputs from many sources, making decisions about which of several possible behavioral options to activate. The ability of nervous and endocrine systems to make choices and deal with potential conflicts is the central topic of this chapter. The fundamental problem that we shall examine is how proximate mechanisms organize an individual's behavior—from moment to moment, over the course of a day, over weeks or a breeding season, and even a whole year. We shall examine three classes of mechanisms that carry out these functions: (1) neuronal command centers that "talk" to each other, (2) clocks that schedule the activity of these command centers, and (3) hormonal systems that track gradually changing environments and adjust the operating rules of nervous systems.

Organizing Behavior in the Short Term: Command Centers

Because animals usually have the capacity to do many different things in response to many different stimuli, at any given moment they face the problem of which of their responses to activate. One rarely observes an animal attempting to do two things at the same time. At an ultimate level, it is easy to understand why, given that simultaneous performance of two activities (e.g., searching for a mate and avoiding an onrushing predator) will almost never be adaptive. But at a proximate level, how are animal nervous systems organized so that maladaptive conflicts do not occur?

A major hypothesis that has guided the research of neurophysiologists interested in this question has been that animal nervous systems are endowed with "command centers," including the innate releasing mechanisms, central pattern generators, song control systems, and the like that we have met in preceding chapters. According to this theory, each command center—a unit within the nervous system—is responsible for activating a particular response, but the various centers are in contact with one another, and can inhibit, or block, one another. As a result, the animal can avoid incapacitating conflicts when carrying out a given behavior. (Let me note here that a "command center" need not be a single bundle of nerve cells found in a particular part of the brain; rather, it may consist of several interconnected structures distributed among several different brain regions that are capable of acting in a unified fashion.)

Kenneth Roeder used command center theory to examine decision making in the praying mantis [1008, 1010]. A mantis is capable of doing many things: searching for mates, sunbathing, copulating, flying, diving away from bats, and so on. Most of the time, however, the typical mantis remains motionless until an unsuspecting prey wanders within striking distance. When this occurs, the insect makes very rapid, accurate, and powerful grasping movements with its front pair of legs.

Roeder proposed that the mantis's nervous system sorts out its options thanks to inhibitory relationships between an assortment of command centers within its neuronal network. For example, the design of the mantis's nervous system (Figure 1) suggested that the command of muscles in each of the insect's segments was the responsibility of the **ganglion,** a dense cluster of neurons, in that segment. Roeder tested this possibility by isolating one segmental ganglion by surgically cutting its connections with the rest of the nervous system. Not surprisingly, the muscles within the neurally isolated segment subsequently failed to react when the mantis's nervous system became active elsewhere. However, if the isolated ganglion was stimulated electrically, the muscles and any limbs in that segment made vigorous, complete movements. These results confirmed the hypothesis that cells within each segmental ganglion control the motor output of that segment.

If the segmental ganglia are responsible for telling individual muscles to contract or relax to carry out a given movement, what is the mantis's brain doing? Roeder suspected that certain brain cells were responsible for inhibiting neural activity in the segmental ganglia, keeping them from doing anything until specifically ordered into action by a command center in the brain. If so, cutting the connection between these inhibitory cells and the segmental ganglia should have the effect of removing this inhibition and inducing inappropriate, conflicting respons-

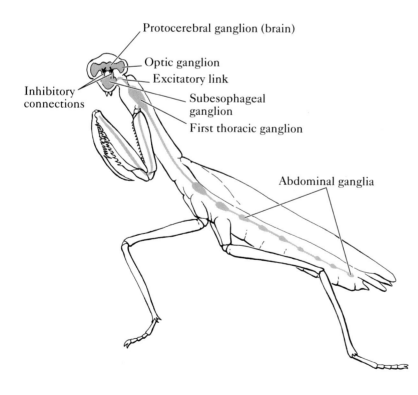

Protocerebral ganglion (brain)

Optic ganglion

Excitatory link

Inhibitory connections

Subesophageal ganglion

First thoracic ganglion

Abdominal ganglia

1   Nervous system of a praying mantis. If the connections between the protocerebral ganglion and the subesophageal ganglion are cut, the subesophageal ganglion sends a stream of excitatory messages to the thoracic and abdominal ganglia; the mantis then attempts to do several competing activities simultaneously.

es. When Roeder severed the connections between the protocerebral ganglion (the mantis's brain) and the rest of its nervous system, he produced an insect that walked and grasped simultaneously, something that would be disastrous in nature. The protocerebral lobes apparently make certain that a mantis either walks *or* grasps, but does *not* do both things at once.

When Roeder removed the entire head, however—a procedure that eliminates the subesophageal ganglion as well as the protocerebral ganglion—the mantis became immobile. Roeder could induce single, irrelevant movements by poking the creature sharply, but that was the extent of its behavior. These results suggest that the protocerebral ganglion of an intact mantis typically sends out a stream of inhibitory messages to the subesophageal ganglion, preventing these cells from communicating with the other ganglia. When certain sensory signals are received, however, specific cells in the subesophageal ganglion become unblocked as certain protocerebral ganglion cells stop inhibiting them. Freed from suppression, these subesophageal cells send excitatory messages to various segmental ganglia, where new signals are generated that order muscles to take specific actions. Depending on what sections of the subesophageal ganglion are no longer inhibited, the mantis walks forward, or strikes out with its forelegs, or flies, or does something else.

There is an interesting exception to the rule that the complete removal of a mantis's head results in a behaviorless animal. If a mature male's head is cut off, the animal, instead of losing its ability to behave, performs a series of rotary movements that swing its body sideways in a circle. While this is happening, the mantis's abdomen is twisting around and down. These actions are normally blocked by signals coming from the protocerebral ganglion, but are released if the male's head is removed. In nature, a male sometimes literally loses his head over a female, who captures and consumes him, head first (Figure 2). Nevertheless, the male can still copulate under these difficult circumstances, thanks to the nature of the control system regulating mating behavior. Headless, his legs carry what is left of him in a circular path until his body touches the female's, at which point he climbs onto her back and copulates with her.

Neural Inhibition among Command Centers

Whether male or female, adult or immature, the mantis, like many other animals, has a nervous system that appears to be functionally organized as a cluster of "centers," each with specific responsibilities. Some centers produce their own output, inhibiting the activities of other centers, which makes it possible for the mantis to do just one thing at a time.

The importance of inhibitory relationships within nervous systems is evident in Vincent Dethier's classic studies of blowfly feeding behavior [288]. These insects drink various exudates from plants, juices of liquefying animal corpses, and other savory fluids rich in sugars and proteins. During the night, the nutrients collected in the day's meals are metabolized to provide energy for the insect. By morn-

2 **Losing his head** does not extinguish the mating behavior of this male African praying mantis. The female has reached back to consume the head of her partner while they copulate. *Photograph by E. S. Ross.*

ing, the blowfly flies off to seek additional food, which it locates in part by olfaction and in part by tasting substances with its feet when it happens to step into a fluid. The appropriate sensory inputs activate the feeding commands: the fly extends its proboscis, spreads its labellum, and imbibes.

Dethier showed experimentally that the speed with which a fluid is sucked up and the duration of feeding are proportional to the concentration of sugar in the fluid. If the liquid is not very sugary, the oral receptors cease firing quickly, and sucking stops. If sugar concentrations are high, the oral receptors may keep firing for 90 seconds or thereabouts. In both cases, the sensory cells eventually recover and will respond again to stimulation. This causes reextension of the proboscis and a new bout of sucking.

Sooner or later, however, the fly stops drinking altogether, even when standing in the richest of sugary liquids. Dissection of a sated fly reveals that its crop, a storage sac off the digestive tract, is filled to overflowing, forcing fluid back into the foregut. Dethier hypothesized that distension of the foregut could be detected by

3 Nervous system and digestive system of a blowfly. Severing the recurrent nerve eliminates feedback to the brain on the degree to which the crop is filled with fluid, and eliminates the signals that eventually block feeding in a full blowfly.

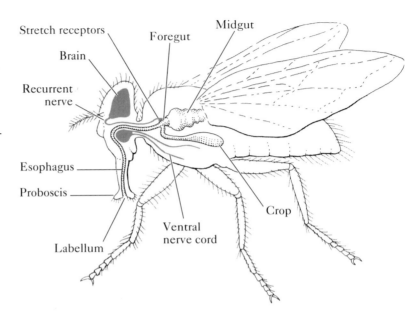

stretch receptors attached to this part of the digestive tract. He hypothesized that those receptors might send messages to the brain, stimulating cells there to inhibit the feeding response.

As predicted from this hypothesis, receptors similar in design to stretch receptors in other organisms are found in the fly's foregut. As predicted, these receptors feed their sensory input into a prominent nerve that runs from the foregut to the brain (Figure 3). As predicted, if one experimentally severs this pathway, the recurrent nerve, the fly cannot stop feeding. The insect continues on with bout after bout of drinking until its body bursts from overconsumption, in a scene reminiscent of a particularly gross episode in *The Meaning of Life*, a Monty Python movie from a bygone era. Intact flies do not explode, however, because their feeding behavior can be blocked, thanks to inhibitory arrangements between elements of their nervous systems. Instead, normal blowflies feed in sensible cycles as they respond to changes in their needs [288].

Mechanisms for Timing Behavior Appropriately

The neuronal command centers of blowflies and other animals may not only communicate with each other, but also receive inputs from a biological clock or pacemaker that imposes another kind of organizational structure on behavior, one that helps animals change their behavioral priorities in relation to the time of day, month, or year. The most familiar examples of changing behavioral priorities are those that occur over the course of a 24-hour day, since the transition from day to night is biologically significant for most species. For example, female crickets usually hide in burrows or under litter during the day and move about only after dusk, when it is relatively safe to search for mates [734]. In response, male *Teleogryllus* crickets start calling to attract mates in the evening of each day (Figure 4) [733, 735]. If the command center model applies, the inhibitory relationships between the calling center in a male cricket's brain and other neural elements responsible for different behavioral responses must change cyclically over 24-hour periods. What mechanisms might regulate these changes?

Students of behavior interested in this question have had two major competing theories to consider. The first is that animals change their priorities in response to a timing mechanism with its own built-in cycling schedule that is independent of cues from the animal's surroundings. That such environment-independent mechanisms might exist should be plausible to anyone who has flown across several time zones and then tried immediately to adjust his or her activity schedule in keeping with local time in the new location.

The second theory is that animals adjust the relationships between control elements in their nervous systems strictly on the basis of information gathered by mechanisms that monitor a changing environment. Such devices would enable individuals to shift their behavioral tendencies in response to particular changes in the world around them.

4   A calling male cricket at its burrow. *Photograph by E. S. Ross.*

Let's consider these two possibilities in the context of the calling cycle of male *Teleogryllus* crickets. Each day's bout of calling could get under way because the crickets possess an internal timer that measures how long it has been since the last bout began; they could use this environment-independent system to activate the onset of a new round of chirping. Alternatively, the insect's neural mechanisms might be designed to detect declining light intensity, or some other environmental cue, and to activate singing only when the critical cue appears.

If the environmental cue hypothesis is correct, then crickets held under constant environmental conditions in a laboratory should show no cyclical pattern of calling. But in fact, laboratory crickets continue to call regularly for a limited block of time each day, even when held in rooms in which the temperature stays the same and the lights are on (or off) all day long. Under conditions of constant

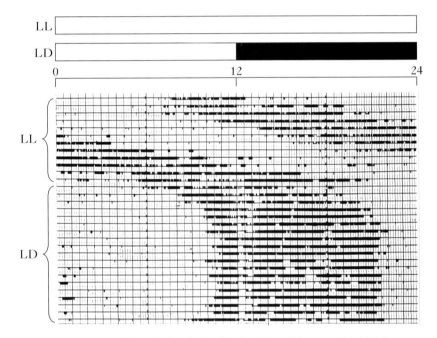

5 Circadian rhythms in cricket singing behavior. In this activity record (which is typical of those used in research on biological rhythms), each horizontal line on the grid represents one day; each vertical line represents one half-hour on a 24-hour time scale. Dark marks indicate periods of activity, in this case, singing. The bars at the top of the figure represent the lighting conditions; thus, for the first 12 days of this experiment, male crickets were kept in constant light (LL), and for the remainder were subjected to 12-hour cycles of light and dark (LD). Male crickets held under constant light exhibit a daily cycle of singing and nonsinging, but the singing starts later each day. The onset of "nightfall" on day 13 acts as a cue that entrains the calling rhythm, which now stops shifting and eventually begins an hour or two before the lights are turned off each day. *Source: Loher [733].*

light, calling starts about 25 to 26 hours later than it did the previous day (Figure 5). The deviation of this **free-running cycle** (a cycle of activity that occurs under constant environmental conditions) from a strict 24-hour period [901] offers evidence that the pattern is caused by an internal **circadian rhythm** (*circadian* means "about a day") that is independent of environmental cues. The earth rotates on its axis once every 24 hours, creating many cyclical changes in the environment, but all with periods of 24 hours, not 25 or 26 hours.

Now let's place our crickets in a regime of 12 hours of light and 12 hours of darkness. The switch from light to dark offers an external environmental cue that the crickets can use to *entrain* (i.e., set the starting point for) a circadian pattern of calling. In a few days, the males will all start to call about 2 hours before the lights go off, accurately anticipating nightfall, and they will continue until about 2.5 hours before the lights go on again in the "morning" (see Figure 5). This cycle of calling matches the natural one, which is synchronized with dusk; unlike the free-running cycle, it does not drift out of phase with the 24-hour day, but will continue to be reset each day so that it begins at the same time in relation to lights-out [733]. From these results we can conclude that the complete control system for cricket calling has both environment-independent and environment-dependent components: an environment-independent timer, or *biological clock*, set on a cycle that is not exactly 24 hours long, and an environment-activated entrainment device for synchronizing the clock with local conditions.

Exploring Circadian Mechanisms

To study the circadian mechanisms of crickets and many other animals, investigators have employed two main tactics. One is to infer something about the properties of the system by examining how it reacts to various environmental manipulations, usually involving changes in light and dark regimes, as just illustrated with *Teleogryllus* crickets. The other tactic is similar to that used by Roeder in his studies of mantis nervous systems—namely, to disconnect various parts of the nervous system surgically.

If one cuts the nerve carrying sensory information from the eyes of a male cricket to the optic lobes of his brain, depriving him of his vision, he enters a free-running cycle. Visual signals are evidently needed to entrain the daily rhythm, but a rhythm persists in the absence of this information. If, however, one separates both optic lobes from the rest of the brain, the calling cycle breaks down completely; the cricket will now call with equal probability at any time of the day (Figure 6). These results are consistent with the hypothesis that a master clock mechanism (Figure 7) resides within the optic lobes, sending messages to other regions of the nervous system [614, 898].

James Truman and Lynn Riddiford's study of adult emergence in two species of silkmoths involved a particularly ingenious use of surgery to investigate biological clocks [1203]. Adults of one of these silkmoth species usually emerge from their pupal cocoons at dawn; the other species enters the world as adults in the middle of the night. The removal of the brain from a silkmoth pupa of either species does

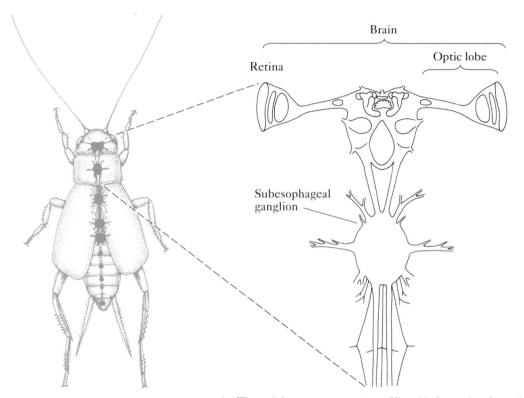

6 The cricket nervous system. Visual information from the eyes is relayed to the optic lobes of the cricket's brain. If the optic lobes are surgically disconnected from the rest of the brain, the cricket loses its capacity to maintain a circadian rhythm. *Based on diagrams by F. Huber and W. F. Shurmann.*

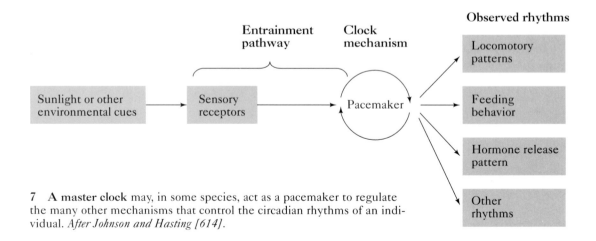

7 A master clock may, in some species, act as a pacemaker to regulate the many other mechanisms that control the circadian rhythms of an individual. *After Johnson and Hasting [614].*

not kill the creature or prevent metamorphosis, but it does destroy the emergence pattern. If the brain is transplanted from the head of the pupa to its abdomen, however, the animal is likely to emerge at the customary time for its species, suggesting that the clock mechanism is contained within the silkmoth's brain.

If this hypothesis is true, then transplanting brain tissue from species A to the abdomen of species B would be expected to impose the circadian rhythm of species A on a brainless member of species B, and vice versa. When this transplant experiment was performed, the subjects did adopt the emergence pattern of the other species, clinching the argument that the brain is the site of a biological clock that controls emergence behavior in the two silkmoths.

The search for the location and operating rules of the master clock in vertebrates has revealed a pattern of cyclical control roughly similar to that found in many invertebrates. The role of the hypothalamus in regulating many basic activities in mammals makes it a logical place to look for a mechanism involved in circadian rhythms. One candidate for this mechanism is the suprachiasmatic nucleus (SCN), a pair of cell clusters in the hypothalamus that receive inputs from nerve fibers originating in the retina. The SCN is therefore the kind of structure that could conceivably secure information about day and night length, information that could be used to entrain a master biological clock.

If the SCN contains a pacemaker critical for maintaining circadian rhythms, then damage to the SCN should cause individuals to lose those rhythms. Such an experiment has been done by selectively destroying SCN cells in the brains of hamsters and Norway rats, which subsequently exhibit arrhythmic patterns of hormone secretion, locomotion, and feeding [1333]. If these arrhythmic animals receive transplants of SCN tissue from fetal hamsters, they regain their circadian rhythms about 40 percent of the time. If the lesioned animals receive tissue transplants from other parts of the fetal hamster brain, they invariably remain arrhythmic. These results support the hypothesis that the SCN contains a master clock mechanism in these animals [284].

If the SCN contains a clock, then it should possess special cells that respond biochemically to the type of stimulation from optic nerve fibers that is thought to entrain circadian rhythms. Using techniques designed to detect the presence of specific proteins, Benjamin Rusak and his co-workers have detected precisely this response. They exposed laboratory hamsters and rats to pulses of light, some at times during the night when such pulses are known to alter the animals' circadian rhythms. Then, some hours later, they examined the animals' SCNs, and found large quantities of the protein product of a specific gene only in individuals exposed to the clock-shifting stimulus [1027].

You may recall that a single gene, the *period* or *per* gene, regulates the circadian rhythms and the pattern of wing-waving by courting male fruit flies (see Chapter 3). Some evidence suggests that in mammals, too, the activity of a single gene is essential for normal circadian rhythms. Laboratory mice with one copy of a mutant allele called *Clock* exhibit a free-running period of 25 hours, rather than the normal period of slightly less than 24 hours, when held in complete darkness.

Individuals that are homozygous for *Clock* (i.e., have two copies of this allele) and are kept in darkness display a period of 28 hours for a time, after which their circadian rhythm breaks down completely (Figure 8).

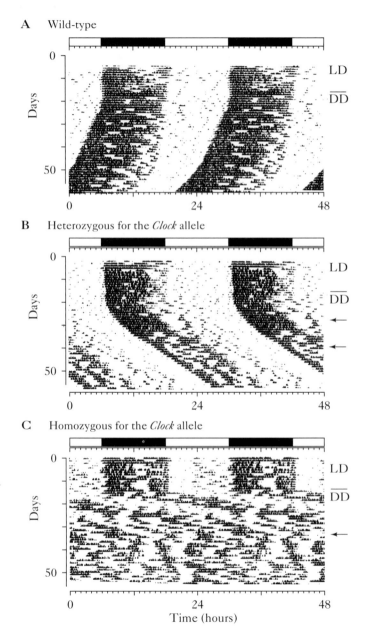

8 A single gene affects circadian rhythms in laboratory mice. Mice were exposed first to 10 days of 12 hours of light and 12 hours of darkness (LD), then to a series of days of total darkness (DD). (A) When exposed to constant darkness, wild-type mice exhibit a free-running cycle lasting somewhat less than 24 hours. (B) Heterozygous mice with one copy of the mutant allele *Clock* maintain a circadian rhythm based on a 25-hour period. (C) Mice homozygous for the *Clock* allele exhibit a 28-hour rhythm for the first few days after the lights go out completely, then their circadian rhythm simply disappears. *Source: Vitaterna et al. [1223].*

Just how this gene operates within SCN cells has not been established. However, one working hypothesis is that mammals are like fruit flies when it comes to circadian control. In fruit flies, the *per* gene has an innate schedule of activity; in the late afternoon, it turns itself on and begins a process that leads to the production of the Per protein in the cytoplasm of the pacemaker cells. As this protein is made, it travels into the nuclei of the cells, where it turns on other genes, while at the same time regulating the *per* gene itself. In some way, the buildup of Per protein eventually shuts the *per* gene down, which means that no more Per protein, with its gene-stimulating effects, is produced. As time passes, the Per protein that is inhibiting the *per* gene is gradually broken down, so that by the next afternoon the gene enters a new round of activity [1169].

If something similar happens in SCN cells, the result would be circadian patterns of gene activity, with resulting changes in the output of various proteins having multiple effects on other organs. For example, chemical activity in the SCN cells of mice and rats translates into signals to the pineal gland, a small structure located between the two cerebral hemispheres. In response to a circadian cycle of SCN messages, the pineal secretes a hormone, melatonin, in a rhythmic fashion, imposing a corresponding rhythm on many elements of the animal's physiology and behavior.

Remarkable progress has been made in recent years in understanding how specific genes in pineal cells participate in the control of the gland's activity [401, 1141]. In brief, chemical signals from SCN neurons activate a specific gene (CREM) in pineal cells, a gene whose protein product (ICER) is presumably involved in some way in melatonin manufacture. As the ICER protein builds up in pineal cells, however, it begins to inhibit activity in the very gene needed to make more ICER. (Note that this is another case of self-regulating negative feedback of the sort that may also be present in SCN cells.) However, the sensitivity of the CREM gene to feedback inhibition from its protein product varies depending on the length of the **photoperiod**—the hours of light in a 24-hour period—an animal has experienced in the previous few days. Animals that have been exposed to a series of "nights" that are 8 hours long show peak activity in the CREM gene about 6 hours after the lights go off; in contrast, animals that have been exposed to 12-hour nights for a couple of weeks reach peak CREM activity about 10 hours after the lights go off (Figure 9). This means that the pineal gland has a way to alter its biochemical output in relation to the recent history of photoperiods that the animal has experienced. This ability enables the pineal to adapt to seasonal changes in day length, thereby helping the animal adjust its daily schedule to these changes.

In conclusion, in some mammals, the SCN contains the central pacemaker that sends signals to the pineal gland. The pineal cyclically changes its production of melatonin with adjustments for shifts in photoperiod length, integrating the environment-independent and environment-dependent elements that regulate daily changes in behavior. The adaptive value of the environment-independent components of such mixed systems may be that they enable individuals to adjust the timing of their behavioral and physiological cycles without having to constantly

9 Gene activity varies with photoperiod. The CREM gene reaches peak activity longer after darkness falls as the number of hours of light during the day decreases. (A) The experimental light:dark (LD) regimes that white rats experienced. Samples of rats were killed at the times indicated by the arrows and (B) their pineal glands were analyzed to determine how much of the messenger RNA coded for by the CREM gene was present. (C) The peak production of this mRNA occurred from 3 to 12 hours after lights out, depending on the night length in the L:D regime that the animals had experienced. *Source: Foulkes et al. [401].*

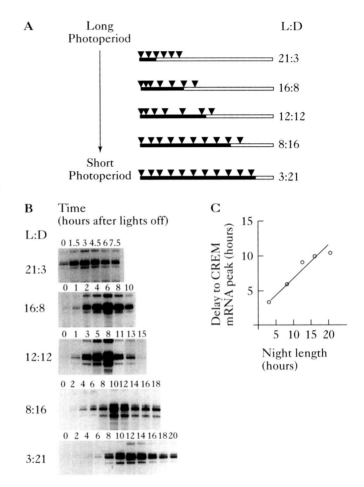

check the environment. At the same time, the presence of an environment-dependent element permits individuals to fine-tune their cycles in keeping with the subtle variations in their particular environment. As a result, a rat will become active at about the right time each night, but will accommodate to changes in day length as spring becomes summer, or summer becomes fall.

It would be intriguing to study the genetics and physiology of the SCN and pineal gland in an animal that is largely divorced from the day–night cycle. The naked mole-rat is such an animal. Naked mole-rats live in colonies that occupy elaborate underground tunnels. There they feed on roots and tubers, and almost never come to the surface. When they do open a burrow to the outside, it is usually just to throw out dirt from fresh tunnel excavations. They have no special dependence on what is going on above them during the day or the night. And, as

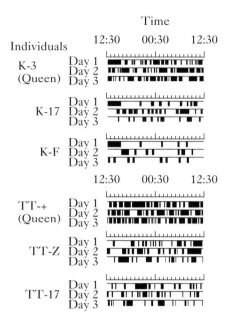

Time

10 Naked mole-rats show no circadian rhythms. Patterns of activity are shown for six individuals from two captive colonies held under constant low light. Dark bars indicate periods when the individual was awake and active. *Source: Davis-Walton and Sherman [276].*

you might predict, naked mole-rats exhibit no circadian rhythms of any sort (Figure 10). Instead, individuals scatter generally brief episodes of wakefulness among longer periods of sleep, with the pattern changing irregularly from day to day [276].

Long-Term Cycles of Behavior

Because of their very unusual lifestyle, naked mole-rats do not have to deal with cyclically changing environments, and they have apparently lost their circadian clocks as a result. But almost all other creatures confront not just daily changes in access to food or risk of predation, but changes that cover periods longer than 24 hours. Particularly common are cycles that last a year, with individuals adjusting their behavior to the annual cycles of seasonal changes that occur in many parts of the world. If circadian pacemakers enable animals to prepare physiologically and behaviorally for certain predictable daily changes in the environment, might not some animals possess a circannual clock that runs on an approximately 365-day cycle [486]? Such a mechanism might be similar to the circadian master clock. An environment-independent timer, but one that could generate a circannual rhythm, in conjunction with an entrainment device to set the annual clock to local conditions, would prime the animal to adopt a shifting set of priorities in tune with seasonal changes in the environment.

Testing the hypothesis that an animal has a circannual rhythm is technically difficult because individuals must be maintained under constant conditions for at least 2 years after their removal from the natural environment. One successful study of this sort involved the golden-mantled ground squirrel [916] of north-temperate North America, which in nature spends the late fall and winter hibernating in an underground chamber. Five members of this species were born in captivity, then blinded and held thereafter in constant darkness and at a constant temperature while supplied with an abundance of food. Year after year, these ground squirrels entered hibernation at about the same time as their fellows living in the wild (Figure 11).

In another similar investigation, several stonechats from equatorial Kenya were taken as nestlings to Germany. There the birds were reared in laboratory chambers in which the temperature and photoperiod were always the same. Needless to say, these birds, and their offspring, never had a chance to encounter the Kenyan rainy season, which usually begins in April and is associated with the appearance of many insects. Under natural conditions, the rainy season is when stonechats produce their babies, timing reproduction so that insect food will be available when their nestlings most need it. Under the nonseasonal conditions devised for them by their human captors, the transplanted stonechats nevertheless continued to exhibit a cycle of reproductive physiology and behavior, but one that was out of phase with the real year (Figure 12); for example, one male went through nine cycles of testicular growth and decline in the 7.5 years of the experiment. Cases of this sort offer evidence that circannual cycles are internally generated by an

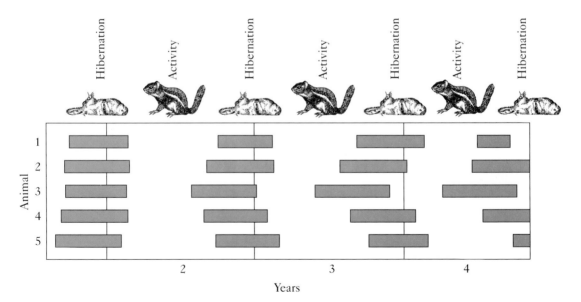

11 Circannual cycle of the golden-mantled ground squirrel. Animals held in constant darkness and at a constant temperature nevertheless entered hibernation (black bars) at certain times year after year. *Source: Pengelley and Asmundson [916].*

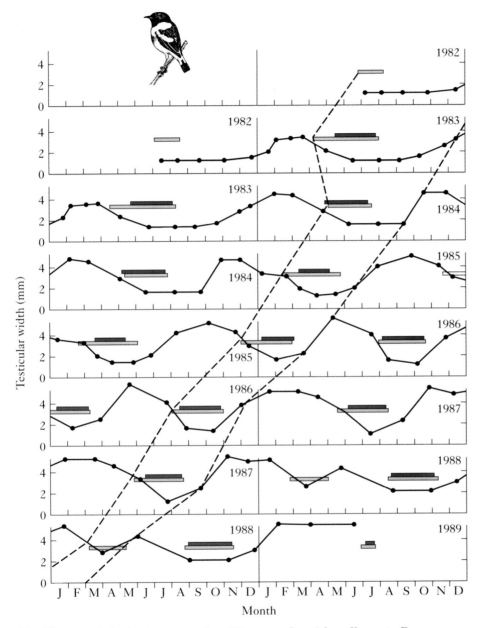

12 Circannual rhythm in a stonechat. When transferred from Kenya to Germany and held under constant conditions, this male stonechat still underwent a regular long-term cycle of testicular growth and decline (the line graph) as well as feather molts (bars). The cycle was not 12 months long, so that the period of testicular changes shifted over the years (see the dashed lines that angle upward from left to right). *Source: Gwinner and Dittami [484].*

environment-independent mechanism and are not the product of the birds' ability to detect subtle cues associated with the standard 12-month annual cycle [484].

Variation in the Physical Environment: Influences on Long-Term Cycles

In nature, environmental cues entrain circadian and circannual clocks so as to produce behavioral rhythms that match the particular features of the environment, such as the times of sunrise and sunset, or the onset of the rainy season in a given year, or the increasing day lengths associated with spring. This fine-tuning of behavioral cycles involves mechanisms of great diversity that respond to a full spectrum of environmental influences, which vary from species to species according to their special problems and demands. Let us first look at some examples of how different species use proximate cues from the physical environment to adjust to changing conditions.

The banner-tailed kangaroo rat is an animal whose readiness to forage for food is regulated by the lunar cycle [729, 730]. Robert Lockard and Donald Owings reached this conclusion by monitoring the activity of free-living kangaroo rats in a valley in southeastern Arizona. To measure kangaroo rat activity, Lockard invented an ingenious food dispenser/timer that released very small quantities of millet seed at hourly intervals. To retrieve seeds, an animal had to walk through the dispenser, depressing a treadle in the process. The moving treadle caused a pen to make a mark on a paper disk that turned slowly throughout the night, driven by a clock mechanism. When the paper disk was collected in the morning, it carried a temporal record of all nocturnal visits to the dispenser.

Data collection was sometimes frustrated by ants that perversely drank all the ink or by Arizonan steers that trampled the recorders. Nevertheless, Lockard's records showed that during the fall, when the kangaroo rats had accumulated a large cache of seeds, the animals were selective about foraging, usually coming out of their underground burrows only at night when the moon was not shining (Figure 13). Because the predators of kangaroo rats (coyotes and owls) can see their prey more easily in moonlight, banner-tails probably are safer when foraging in complete darkness. As a result, the kangaroo rats apparently possess a mechanism that enables them to shift their foraging schedule in keeping with nightly moonlight conditions.

Unlike the banner-tailed kangaroo rat, the ectothermic, or "cold-blooded," green anole of the southern United States relies heavily on temperature cues to regulate its activity patterns. These little lizards remain dormant under safe shelters for the entire winter from South Carolina to Florida. Males emerge first, in February or March, to feed and then claim territories, from which they repel other males with head-bobbing display threats and occasional physical combat. The winning males court females as they emerge, with the first round of reproduction occurring near the end of April.

The precise time at which an anole makes the transition from winter dormancy to reproductive activity is heavily influenced by its temperature receptors [713]. As temperatures rise in the spring, the pituitary gland releases gonadotropic hor-

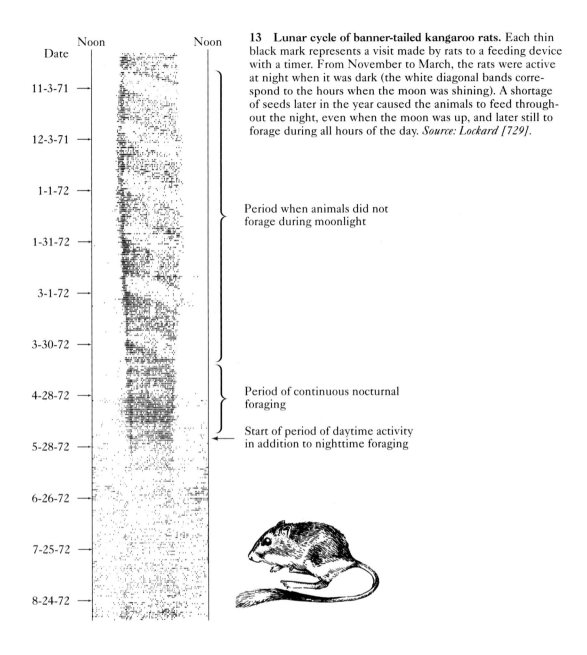

13   Lunar cycle of banner-tailed kangaroo rats. Each thin black mark represents a visit made by rats to a feeding device with a timer. From November to March, the rats were active at night when it was dark (the white diagonal bands correspond to the hours when the moon was shining). A shortage of seeds later in the year caused the animals to feed throughout the night, even when the moon was up, and later still to forage during all hours of the day. *Source: Lockard [729].*

Period when animals did not forage during moonlight

Period of continuous nocturnal foraging

Start of period of daytime activity in addition to nighttime foraging

mones that stimulate the growth of the ovaries in females and the testes in males. A series of cold days delays emergence and the start of reproduction; warmer days end dormancy sooner. As a result, green anoles reproduce at times when conditions enable them to pursue abundant insect prey.

White-crowned sparrows also undergo a remarkable shift in behavior once a year, when they enter their breeding mode after many months during which they exhibit no sexual behavior whatsoever. In the spring, males establish breeding territories, fight with rivals, and court sexually receptive females. In concert with these striking behavioral changes, the gonads of the birds grow with dramatic rapidity, regaining all the weight lost during the winter, when they fall to 1 percent of their breeding-season weight. In order to properly time the regrowth of their gonads and the onset of reproductive activities, the birds must somehow "anticipate" the spring breeding season. How can they manage this feat?

The sparrows can detect an increase in the length of the photoperiod, which grows longer as spring approaches in the temperate zone [380]. One hypothesis on how such a system might work proposes that the clock mechanism of white-crowns exhibits a daily cyclical change in sensitivity to light, a cycle that is reset each morning at dawn. During the initial 13 hours or so after the clock is reset, this mechanism is highly insensitive to light; this insensitivity then steadily gives way to increasing sensitivity, which reaches a peak from 16 to 20 hours after the starting point in the cycle. Sensitivity then fades very rapidly to a low point 24 hours later, at the start of a new day and a new cycle. Therefore, if the days are 12 or 13 hours long and the nights 12 or 11 hours long, the system will never become activated because no light is present during the light-sensitive phase of the cycle. However, if the days are 14 or 15 hours long, light will reach the bird's brain during the photosensitive phase, initiating a series of hormonal changes that lead to the development of its reproductive equipment.

If this model of the photoperiod-measuring system is correct, it should be possible to deceive the system. William Hamner, working with house finches [504], and Donald Farner, in similar studies with white-crowned sparrows [379], stimulated testicular growth by exposing captive birds to light during the hypothesized photosensitive phase of their circadian rhythms. In Farner's experiment, birds that had been on a regular schedule of 8 hours of light and 16 hours of darkness (8L:16D) were shifted to a 8L:28D schedule. Because the light periods were now out of phase with a 24-hour cycle, these birds sometimes received light during the time when their brains were thought to be highly photosensitive. The male birds' testes grew under these conditions, even though there was a lower ratio of light to dark hours than under the 8L:16D cycle, which did not stimulate testicular growth (Figure 14) [379].

In contrast to white-crowned sparrows, crossbills exhibit a much less constrained pattern of seasonal breeding. In fact, these songbirds were long thought to be able to breed at any time of the year, provided that pine seeds, their food specialty, were abundant. Craig Benkman showed that food intake, not photoperiod length, appears to be the primary, if not the sole, determinant of breeding in both the white-winged and the red crossbill [77]. Thus, these birds will breed in most months of the year, if they can secure enough pine kernels to sustain themselves and a brood of offspring.

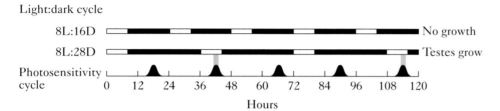

14 A cycle of photosensitivity. A test of the hypothesis that white-crowned sparrows possess proximate mechanisms that are especially sensitive to light between hours 17 and 19 of each day. The lower line represents these hypothetical bouts of photosensitivity. The open and solid sections of the two upper horizontal bars show the light and dark periods of two different light–dark regimens. Only sparrows under the 8L:28D experimental regimen were exposed to light during the supposed photosensitive phase of the cycle, and only they responded with testicular growth. *Source: Farner [379].*

Thomas Hahn, however, noticed a break in breeding in December and January (Figure 15). He proposed that even though the crossbill is more flexible and opportunistic in breeding than the average songbird, perhaps it too has an underlying reproductive cycle dependent upon photoperiod. When Hahn held cross-

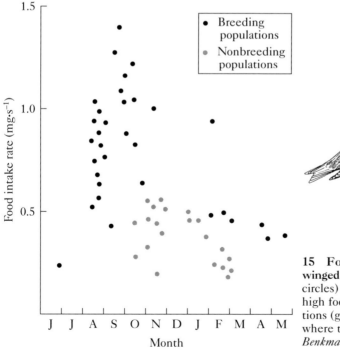

15 Food intake and nesting in the white-winged crossbill. Breeding populations (black circles) usually occur in areas with relatively high food availability. Nonbreeding populations (grey circles) generally occur in areas where the birds have low food intake. *Source: Benkman [77].*

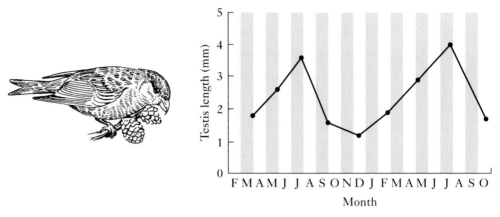

16 Photoperiod alters testis length in the red crossbill. Six captive birds were held under natural photoperiods, which changed over the seasons, but the temperature and food supply were held constant. The curve represents the average testis length among these birds at different times during the year. *Source: Hahn [493].*

bills in captivity with unlimited access to their favorite food, with temperatures held constant while the photoperiod changed naturally, he found that male testis length fluctuated in a cyclical fashion (Figure 16), becoming much reduced during October through December, even when the birds had all the pine seeds a crossbill could hope for. Therefore, the reproductive opportunism of the bird is not absolute, but rather is superimposed on the proximate photoperiod-driven mechanism characteristic of temperate-zone songbirds [493].

Changing Priorities in a Changing Social Environment

The studies we have reviewed so far show that many features of the physical environment, such as cycles of moonlight, changes in photoperiod or temperature, or the availability of key foods, are detected and used by certain animals to regulate their behavioral priorities. In addition, however, some species possess mechanisms that permit them to adjust behaviorally to the special social circumstances they happen to encounter. Thus, for example, when Hahn and several co-workers performed another experiment in which some captive male crossbills were caged with their mates, while others were forced into bachelorhood but were kept within sight and sound of the paired crossbills in a neighboring aviary, the bachelor males experienced a slower return to reproductive condition after the fall break than did the paired males [494]. Thus, social stimulation provided by the opposite sex also contributes to a return to reproductive condition in this species.

The behavioral influence of social interactions can be profound, as demonstrated by male house mice. When a male mounts a female and ejaculates, he immediately becomes highly aggressive toward mouse pups, which he will kill should he find them. He remains prone to commit infanticide for about 3 weeks, but then gradually switches into a paternal mode. He will then protect and care for young pups attentively until 50 days have passed since ejaculation, at which time he becomes infanticidal once again [918].

This remarkable cycle has clear adaptive value. After a male transfers sperm to a partner, 3 weeks pass before she gives birth. Attacks on pups during these 3 weeks will invariably be directed against a rival male's offspring, with all the benefits attendant upon their elimination (see Chapter 1). After 3 weeks, a male that switches to the paternal mode will almost always care for his own neonatal offspring. After 50 days, his weaned pups will have dispersed, so that once again infanticide can be practiced advantageously.

At the proximate level, what kind of mechanism could enable a male to delay the switch from infanticidal to paternal behavior for 3 weeks following mating? One possible explanation involves the social activation of an internal timing device, one that could record the number of days passed since the male copulated. If such a mechanism exists, an experimental manipulation that either increases or decreases the length of a "day," as perceived by the mouse, ought to have an effect on the absolute amount of time that passes before the he makes the transition from infanticidal killer to paternal caregiver.

Glenn Perrigo and his co-workers manipulated day length by placing groups of mice under two laboratory conditions, one with "fast days," in which 11 hours of light were followed by 11 hours of darkness (11L:11D) for a 22-hour "day," and another with "slow days" (13.5L:13.5D) that lasted for 27 hours. As predicted, it was the number of light–dark cycles that affected the infanticidal tendencies of males, not the absolute number of 24-hour periods that passed (Figure 17). Thus mice in the fast-day group experienced 22 light–dark cycles in 20 real (24-hour) days. Only a small minority of these males exhibited infanticidal behavior when exposed to neonatal mice at 20 real days (= 22 fast days) postejaculation. In contrast, males in the slow-day group had experienced only 18 light–dark cycles by the time 20 real days had passed. More than 50 percent of these mice attacked neonates at 20 real days, showing that these males' timing devices had not yet "registered" a sufficient number of "days" to inhibit the infanticidal response fully [918].

As we have seen, the annual cycle of green anoles is influenced by their ability to detect temperature changes (and to a lesser extent, photoperiodic changes) in their environment. But social factors also affect the onset of the yearly period of reproduction, as David Crews showed in the following experiment. He removed dormant females from their winter hiding places, brought them to a laboratory, and placed them in cages in one of four groups: (1) with no males, (2) with castrated males (which will not court females), (3) with males whose dewlaps had been surgically removed, or (4) with intact males, one of whom had established territorial dominance over the other males present [236]. Females in the first three groups did not have the opportunity to observe the normal courtship display of a territorial male, where-

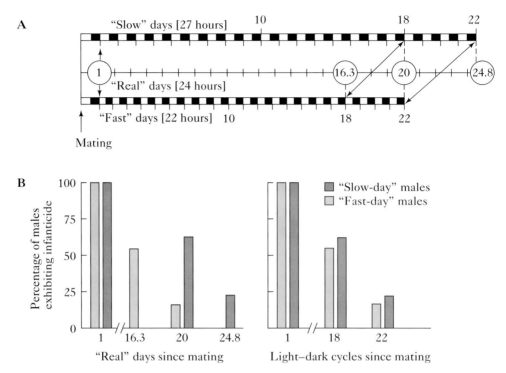

17 Regulation of infanticide by male house mice. Male mice held under artificial "slow day" and "fast day" experimental conditions (A) differ in how many absolute days must pass after a copulation before the male ceases to commit infanticide when he encounters mouse pups. (B) Males held under fast days (open bars) have largely stopped being infanticidal by 20 real days (= 22 fast days); males experiencing slow days (solid bars) do not show the same decline until nearly 25 real days have passed. *Source: Perrigo et al. [918].*

as females in the fourth group regularly saw territorial males showing off their colorful dewlaps (Color Plate 6). The ovaries of the fourth group developed much more rapidly than those of the females kept with castrated or "dewlap-deprived" males.

Thus, visual stimulation by displaying males prepares recently emerged females to become sexually receptive. Females that are frequently courted secrete more pituitary gonadotropins, which speed ovarian development and the consequent production of estrogen by the mature ovaries [236]. Estrogen travels via the bloodstream to the head of the lizard, affecting various endocrine and neural target cells there. The female's brain eventually becomes primed by specific hormones to activate precopulatory neck-arching behavior in response to a territorial male's courtship signals. Should such a female encounter an adult territorial male that flags her with his extended dewlap as he bobs his head up and down, she is likely to copulate with him.

If the amount of estrogen in the bloodstream is a key proximate factor that lowers the threshold for sexual behavior, then removal of an adult female anole's estrogen-producing ovaries should abolish sexual receptivity—which it does—and implantation of an estrogen pellet in an ovariectomized, nonreceptive female should restore her sexual receptivity—which it does. But once a female has mated, her receptivity drops precipitously. Within 5 to 7 minutes after copulating, she ceases to be sexually receptive and will ignore, run from, or even attack any male that dares court her. She will remain unwilling to mate for 10 to 14 days, until she has a mature egg once more (Figure 18). The transition from receptivity to non-receptivity is therefore extremely abrupt, and it occurs at a time when the female still has high levels of estrogen in her blood [236, 798].

This dramatically rapid change in the behavioral priorities of mated females might be related to stimuli associated with copulation. For example, the experience of being mounted and bitten on the neck by a copulating male might provide sensory stimulation that serves to "reset" the sexual receptivity command center, assuming that one exists in female anoles. If this hypothesis is correct, then if a female is courted and then mounted by a male anole whose double penis has been surgically removed, she should become unreceptive even though a functional mating has not occurred. However, such females retain their willingness to copulate after the penis-deprived male dismounts.

An alternative hypothesis is that females react to mechanical stimulation of the genital tract during copulation by producing prostaglandin (PG). PG concentra-

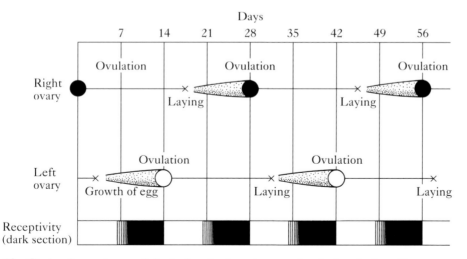

18 Cycle of sexual receptivity in female American anoles during the breeding season. As an egg matures, the female gradually becomes receptive and will mate before ovulation; copulation causes an abrupt cessation of receptivity. *Source: Crews [236].*

tions in a female's bloodstream shoot up immediately following a mating. Perhaps PG somehow acts as an internal cue that activates a change in sexual receptivity. To test this possibility, Richard Tokarz and David Crews injected minute amounts of PG into females whose receptivity had been established by their willingness to permit a mature male lizard to grip them by the neck—after which the male was removed [1195]. Shortly after the PG injection, these females reacted to courting males by dashing away or by attacking their consorts (Figure 19).

Just as species vary in what elements of the physical environment have an effect on their hormones and reproductive cycles, different kinds of social stimulation affect reproductive timing in different species. Visual displays are important for gonadal development in female anoles, but Barbara Brockway found that the gonads of budgerigars responded primarily to acoustical stimuli provided by their companions. These small, sociable parrots occupy the harsh, arid center of Australia, where flocks rarely encounter suitable conditions for breeding. When they do, however, males quickly begin to defend territories and court females. Brockway showed in laboratory studies that testicular growth in males was promoted if they heard tapes of the "loud warble," the territorial call of their species. Unpaired females, however, that heard only this aggressive call experienced *reduced* ovarian growth. Ovarian development required the male's courtship signals, particularly the "soft warble," which males produce when they are perched beak to beak with a female [129, 130].

Although female red deer are completely unrelated to budgerigars, they too are subjected to acoustical stimulation by potential partners. Male red deer bellow monotonously at the top of their lungs, producing roars at the rate of one to three per minute for hours on end during their breeding season. Karen McComb wondered if male roars might influence the timing of female sexual receptivity [787]. To find out, she worked with captive animals in New Zealand, where the deer are farmed for their meat. She divided a herd of females into three groups: one group

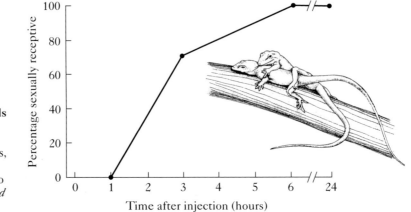

19 Prostaglandin controls receptivity in female American anoles. When injected into seven females, prostaglandin temporarily abolished their readiness to copulate. *Source: Tokarz and Crews [1195].*

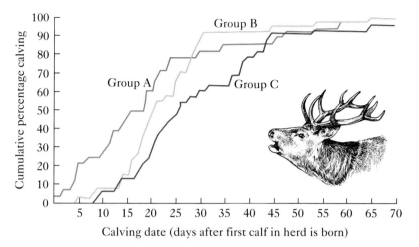

20 Roaring by stags affects the onset of estrus in female red deer. The cumulative percentage of female deer giving birth by each date is shown for three groups: group A females were first held with a vasectomized male (which roared regularly) before being placed with a fertile stag; group B females were exposed to tapes of roaring before gaining access to a fertile stag; group C females were held with no male and heard no tapes of roaring prior to being placed with a fertile stag. Note that females exposed to roaring, even if only on tape, generally gave birth—and therefore must have come into estrus—sooner than females without this stimulation. *Source: McComb [787].*

was penned with a vasectomized male that roared and moved among the females, but could not inseminate them; a second group was held away from all males, but was treated to long-playing tapes of male roars; and the third group was kept in isolation from both males and their calls. Subsequently all the females were given access to a number of nonvasectomized males, which impregnated them as soon as they were receptive. The date on which a female gave birth indicates when she must have come into estrus. As predicted from the social effect hypothesis, females that had had the company of a sterile but vocal male, as well as those that listened to tapes of male roars, gave birth sooner on average than females held in isolation from males and their roars (Figure 20).

Hormones as Mediators of Behavioral Changes

We have reviewed only a few examples of the great diversity of elements in an animal's physical and social environment that can influence its behavioral priorities, particularly with respect to the annual cycle of reproduction. These environmental cues, whatever they may be, often appear to exert their effects by changing the hormonal state of the individual. In such cases, hormones act as intermediaries between the environment and behavioral command centers, fine-tuning those centers to set new priorities under different conditions. Thus, for example, key sensory inputs alter hormonal releases from the brains of white-crowned sparrows; these hormones affect the development of the gonads, which in turn release their own hormones, which affect the bird's brain and behavior. Hormones underlie the integrated shifts in physiology and behavior that promote reproduction at times when environmental, social, and internal physiological conditions are most favorable.

The widespread occurrence of hormonal control of reproduction, particularly in vertebrates, has led most biologists to accept the view that hormones are the key arbiters of sexual behavior, serving as a means of communication between an animal's external environment and its various internal organs of reproduction (Figure 21A). According to this theory, certain hormones provide the causal basis for reproduction, leading to an *associated reproductive pattern* in which gamete production and sexual activity are linked by or associated with increases in particular hormones (Figure 21B). The male green anole offers a classic example of an associated reproductive pattern, with testosterone levels in the blood rising when males make mature sperm, defend territories, and mate with females (see also [834]).

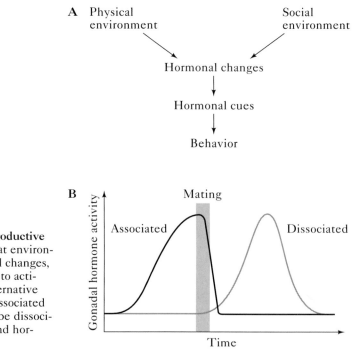

21 Associated and dissociated reproductive patterns. (A) The traditional view that environmental cues trigger internal hormonal changes, and that these changes are necessary to activate behavioral responses. (B) An alternative view in which mating can either be associated with a surge of gonadal hormones or be dissociated in time in relation to gonadal (and hormonal) activity. *Source: Crews [237].*

The same pattern applies to red deer, which breed during September and October. Stags that have been living peacefully with one another all summer become much more aggressive during the rut, and begin to court females. At this time, their testes generate sperm and testosterone, whose behavioral importance has been demonstrated by castrating adult males prior to the rut. The castrated individuals show little aggression, and do not try to mate with sexually receptive females. If the behavioral differences between castrated and intact males stem from an absence of circulating testosterone in the castrated stags, then testosterone implants should restore aggressive and sexual behavior during the mating season. The implants do have these effects, showing that aggressive and sexual behavior are at least partly under the control of testosterone, an association that promotes a seasonally adaptive pattern of physiology and behavior [722].

Although most species studied to date appear to possess associated reproductive patterns, the theory that hormonal control underlies reproductive behavior continues to be tested, as it should be [243]. For example, if testosterone is essential for reproductive activity in male white-crowned sparrows, then sexual behavior should not occur in individuals that have been castrated as young males, because removal of the testes eliminates a major testosterone source. But even without his testes, a male white-crown will mount females that solicit copulations, provided that he has been exposed to long photoperiods [835]. Thus, white-crowns differ from most birds, for which castration ends male sexual behavior [890], apparently because no testes means no testosterone to be converted to estrogen in the brain cells that control copulation [3]. Provide a castrated male Japanese quail with a testosterone implant, or with estrogen, and he is back in business. But castrated male white-crowned sparrows do not require hormonal therapy of this sort. Perhaps some other organ besides the testes produces testosterone in white-crowned sparrows, providing a hormonal explanation for the ability of castrated males to copulate. If so, one would expect to find testosterone in the blood of sexually active males, whether castrated or not. However, wild male white-crowns that copulate with a partner to produce a second, or third, clutch of fertile eggs in a breeding season do so at times when they have relatively low testosterone concentrations in their blood (Figure 22). It is also possible, of course, that in white-crowns, some other hormone altogether is responsible for the activation of copulation.

On the basis of these observations, John Wingfield and Michael Moore hypothesized that the primary function of testosterone for white-crown males is to enhance aggressiveness, rather than sexual drive. At an ultimate level, aggression is useful primarily when males defend their territories and guard their mates from other males [1301]. Later in a breeding season and in a well-established territory, high testosterone concentrations and increased aggressiveness might yield few fitness benefits, while perhaps interfering with a male's paternal behavior. Testosterone supplements have damaging effects on paternal behavior in some birds by causing males to spend more time singing and defending territories than feeding their nestlings [645, 1107].

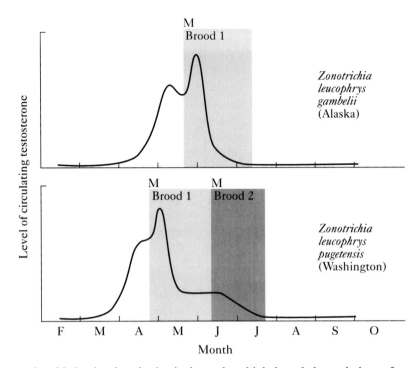

22 Hormonal and behavioral cycles in single- and multiple-brooded populations of white-crowned sparrows. Increased testosterone levels occur shortly before the time when males mate with females (M) in their first breeding cycle of the season, but in populations that breed twice in one season, copulation also occurs during a second interval when testosterone levels are declining in males. *Source: Wingfield and Moore [1301].*

In addition to testosterone's interference with parental behavior, the chemical can have a variety of other negative effects on fitness (see Chapter 11). During the breeding season, male impalas have higher testosterone levels and greater tick loads than females, in part because males are so involved in mating competition that they do not take the time to nip the ticks off their bodies [836]. High levels of testosterone may also suppress the immune system, making male mammals more vulnerable to ticks, nematode worms, and a host of other parasites and pathogens [392]. If so, selection may favor individuals that do not produce the hormone in large quantities except when circumstances enable the hormone's positive effects on fitness to outweigh its costly ones.

If testosterone has negative side effects, it is not surprising that male white-crowned sparrows, which are no longer territorial by late summer or earlier, have no circulating testosterone when migrating in the fall. But what about the song sparrow, a relative of the white-crown, whose males do not migrate, but rather defend a territory for much of the year, particularly in the fall? If testosterone promotes territoriality, then we would expect male song sparrows to pay the piper and maintain relatively high levels of the hormone into the fall. They do not [1300], nor do migratory male stonechats, which are highly aggressive in defense of their winter feeding territories, yet have almost no detectable circulating testosterone at this time [485]. These findings reinforce the point that the hormonal mechanisms of behavioral control are evolutionarily very diverse, with no one standard pattern applicable to all species of sparrows, let alone all birds, or all vertebrates.

The red-sided garter snake offers another powerful demonstration of this point. This snake lives as far north as southern Canada; as a cold-blooded reptile, it spends much of the year dormant in sheltered underground hibernacula, which may house thousands of snakes. On warm days in the late spring, the snakes begin to stir, and soon they emerge en masse (Figure 23). Before going their separate ways, they engage in an orgy of sexual activity, with males slithering after females and attempting to copulate with them. Although males compete for females by trying to contact receptive partners before their fellow males do, they do not fight with one another for the privilege of mating.

Examination of the sex hormone concentrations in their blood reveals that these nonaggressive snakes have almost no circulating testosterone, or any equivalent substance. Yet they have no trouble mating, so they, like the white-crowned sparrow, are animals with a *dissociated reproductive pattern* (see Figure 21B) [242]. Var-

23   Spring mating aggregation of red-sided garter snakes. Male snakes copulate avidly despite having almost no circulating testosterone. *Photograph by David Crews.*

ious hormonal manipulations have been performed on adult male garter snakes without effect on their sexual behavior, which is, however, dependent on the pineal gland, which integrates information about temperature changes in this species [243]. Removal of the pineal gland prior to hibernation produces male snakes that almost always fail to court the following spring. Thus, the garter snake possesses a critical mechanism for detecting temperature increases following a period of hibernation, and this mechanism suffices to activate sexual behavior independently of circulating levels of testosterone.

This does not mean that testosterone has no role to play in the sexual cycle of the snake. High levels of testosterone are present in males in the fall and contribute to the production of sperm, which are stored internally over the winter in anticipation of the spring mating frenzy. Furthermore, although temperature increases may be the activational cue for sexual activity, testosterone may play an organizational role in the development of the underlying mechanisms of reproductive behavior in the red-sided garter snake, as it does in so many other vertebrates (see Chapter 4).

Evidence on this point comes from experiments in which adult male snakes have been castrated. Without their testes, these individuals cannot produce testosterone, but they still exhibit courtship after a period of hibernation under laboratory conditions. If, however, the castrated snakes are tested again after a second bout of hibernation, their sexual activity falls sharply. These results suggest that the surge of testosterone that occurs prior to hibernation primes the neural systems that are responsible for stimulating sexual behavior the following spring (Fig-

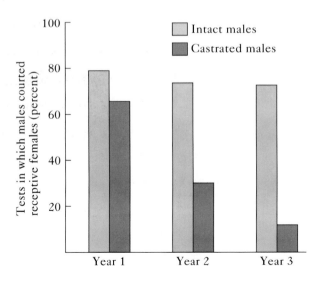

24   Testosterone and the long-term maintenance of garter snake mating behavior. Males whose testes are removed shortly before the breeding season (solid bars) in year 1 remain sexually active during that breeding season, despite the absence of testosterone. But in years 2 and 3, these males become less and less likely to court receptive females, compared with males that still possess their testes (open bars). *Source: Crews [239].*

ure 24). This hypothesis is further strengthened by the finding that sexual activity can be experimentally induced in 1-year-old males, which are normally sexually immature, by giving them testosterone implants in the summer preceding their first hibernation [239].

Therefore, testosterone production in male red-sided garter snakes, which normally begins in the animal's second or third year, appears necessary for the full development and maintenance of those mechanisms that control sexual behavior [239]. However, high levels of testosterone are completely unnecessary in order for copulation to occur in the spring, a reflection of the diversity of mechanisms regulating animal sexual activity [3, 240].

SUMMARY

1. Because the environment provides stimuli that could trigger contradictory responses, and because an animal's physical and social environment often change over time, individuals gain by having mechanisms that set priorities for the different elements in their behavioral repertoires.

2. Animals do not try to do two things at once, perhaps because they have behavioral command centers that are interconnected in a hierarchy of inhibitory relationships. If neural activity in one command center inhibits activity in others, the possibility of competing responses falls.

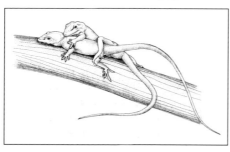

3. As the environment changes, the nature of the inhibitory relationships between neural centers may also change in a compensatory manner. Devices to achieve this end include the various pacemaker or clock mechanisms that regulate nervous system functioning and hormonal output in cycles that typically last 24 hours or 365 days. Circadian and circannual clocks have environment-independent components, but they can also adjust these cycles by accepting information from the environment about local conditions, such as time of sunrise or sunset.

4. Integrative hormones offer a third class of mechanisms for the adaptive organization of behavior. In many animals, changes in the physical environment (such as seasonal changes in photoperiod) and in the social environment (such as the presence of courting males) are detected by neural mechanisms and translated into hormonal messages. The consequent changes in circulating testosterone or estrogen can help set in motion a cascading series of physiological and behavioral changes that underlie seasonal reproductive activity.

5. The precise roles played by hormones in affecting behavioral change vary from species to species. For example, male sexual behavior may or may not be dependent on high testosterone levels in the blood, an illustration of the differences among species in the proximate mechanisms controlling behavior.

SUGGESTED READING

Kenneth Roeder's *Nerve Cells and Insect Behavior* [1008] and Vincent Dethier's *The Hungry Fly* [288] discuss how some animals avoid conflict and structure their behavior over the short haul. Randy Nelson has written a textbook that covers all the topics in this chapter in much more detail [857]. Terry Page has written a comprehensive review of biological clocks and circadian rhythms in insects [898], while two articles in the journal *American Scientist* also clearly explain how circadian clocks work [614, 1169]. David Crews and his colleagues are responsible for a beautifully detailed picture of the organization of anole behavior [236]. Crews and others also review the evolutionary significance of diversity in the hormonal mechanisms controlling behavior [238, 240].

DISCUSSION QUESTIONS

1. In Chapter 4, we discussed seasonal changes in the spatial learning abilities and size of the hippocampus in male meadow voles. Develop two alternative hypotheses on the control mechanisms that might be responsible for these changes.

2. California quail live in environments that experience great fluctuations in rainfall and seed production. Some years California quail do not breed at all. What are some possible proximate factors that might cause the birds to suppress their reproductive cycle? How would you go about testing which of the potential factors is actually responsible for regulating reproduction in this species? See [708] after planning your tests.

3. Think of an ultimate hypothesis for why kangaroo rats might use a circadian rhythm to time their daily activity, rather than simply checking from time to time on whether it is dark outside the burrow.

4. In studying the hormonal control of behavior, it is common to remove an animal's ovaries or testes and then inject the creature with assorted hormones to see what behavioral effects they have. What advantage does this technique have over another approach, which is simply to measure the levels of specific hormones in the blood of animal subjects from time to time? The direct measurement approach would show, for example, whether mating usually occurred when circulating testosterone or estrogen levels were elevated.

CHAPTER 7

The Evolution of Communication: Historical Pathways

UNTIL NOW OUR ATTENTION HAS CENTERED LARGELY, though not exclusively, on the proximate mechanisms of behavior. With this chapter the focus shifts primarily to the ultimate, or evolutionary, basis of behavior. The internal mechanisms that control behavior have an evolutionary history, and knowledge of this history contributes to our understanding of all elements of behavior. We will demonstrate this principle by looking again at animal communication, a topic that we used in Chapter 2 to introduce the different kinds of proximate questions about behavior. You will recall that male white-crowned sparrows sing because they have a battery of genes, hormones, muscles, neural mechanisms, and learning abilities that contribute in various ways to the production of special patterns of sound. But even with a full knowledge of the proximate causes of white-crown singing behavior, we can still ask about the pattern of evolutionary changes that produced the current mechanisms that enable white-crowned sparrows to sing their songs. And we can also ask whether this pattern of evolutionary changes was produced by natural selection—in other words, did white-crown song-producing mechanisms contribute in the past to male reproductive success, and if so, how? Thus, two allied, but somewhat different, kinds of ultimate questions exist, one dealing with a current trait's origin and subsequent modification, and the other with the reproductive consequences for individuals that possess a particular trait. These topics are covered in this and the next chapter. We begin with a case that shows both how to reconstruct the events leading to a complex communication signal and how to test whether the current form of the signal has adaptive value.

Evolutionary Levels of Analysis

Spotted hyenas are highly social mammals that live in permanent clans, often containing several dozen members, that hunt zebras and other large game [682]. Clan members regularly interact with one another, and one of the things they do together is erect their penises, which they present to their companions for nuzzles and sniffs (Figure 1). You may be surprised to learn that female spotted hyenas participate fully in this activity, for they too have something that looks exactly like a penis! In fact, human observers have a hard time telling male hyenas from females unless they can inspect the animals' genitalia at very close range—which most hyenas emphatically resist.

The observation that female hyenas have a mimetic penis and engage in elaborate "penis"-sniffing rituals raises two fundamental evolutionary questions: (1) What is the history behind this behavior? and (2) Did the pseudopenis evolve because of its adaptive value for females? We will deal with the historical question first.

The History of a Signal

Comparisons among related species sometimes provide a window into the past. If some close relatives of the spotted hyena had female pseudopenises, then we could reasonably claim that the common ancestor of these species probably also had this trait, which the descendant species had retained. (The other possibility, namely, that each of these related species had independently evolved pseudopenises, would require that one inherently improbable event occurred repeatedly, which strikes most evolutionary biologists as unlikely, a point that we will discuss below.) As it turns out, there are four species in the family Hyaenidae, which belongs to the mammalian order Carnivora, as do the Canidae (dogs) and Felidae (cats). The only species in the entire order whose females have a pseudopenis is the spotted hyena, so comparisons among living species do not help much in this case. However, males and females of many carnivores inspect the anogenital regions of their companions, a trait exhibited by the family dog as well as by various species of

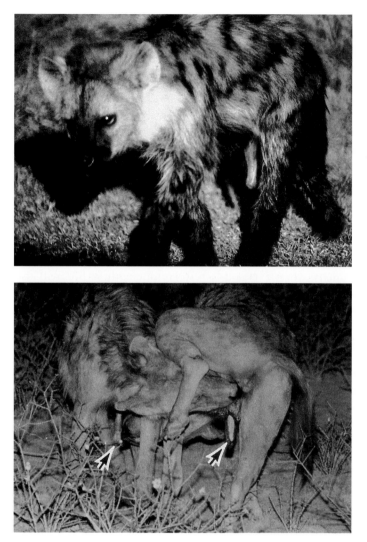

1 Pseudopenis of the female spotted hyena. (Top) A female with an erect pseudopenis walks toward another individual. (Bottom) The greeting ceremony of spotted hyenas. Here one female has pushed her head under the leg of the other to inspect the pseudopenis (arrow) of her companion. *Photographs by (top) Laurence Frank and (bottom) M. G. L. Mills.*

hyenas [683, 889]. Because of the widespread distribution of this behavior in related species, we can conclude that the ancestor of the spotted hyena probably also engaged in anogenital sniffing [682].

The novel feature of anogenital inspection by spotted hyenas is the use of an erect pseudopenis by females. How did they come to be endowed with this bit of anatomy? Studies of hyena genitalia reveal that the pseudopenis is a greatly enlarged clitoris. Thus, the historical question becomes, How did the spotted hyena clitoris come to look like a penis?

The hormone testosterone has key effects on genital development in mammalian embryos (see Chapter 4). If the reproductive organs of mammalian fetuses are exposed to testosterone, they develop into testes and penis. If, however, testosterone is absent, the fetuses develop the secondary sexual characteristics of females. The same tissues that develop into a penis under the influence of testosterone become a clitoris in the absence of the hormone.

In 1973, R. F. Ewer hypothesized that female spotted hyenas differ hormonally from typical mammals [371]. More recent studies have indeed shown that spotted hyena ovaries produce "male" hormones [446]. When a female hyena is pregnant, she produces an ovarian androgen that is converted to testosterone in the placenta and passed on to the developing fetuses, male and female alike. The hormone activates the basic mammalian system for developing male traits, thereby masculinizing young females, endowing them with an enlarged and penislike clitoris [453] (although puzzles remain about the hormonal regulation of this trait [405]).

We can test the hypothesis that the origins of the female spotted hyena's enlarged clitoris lie in the masculinizing effects of testosterone by predicting that females of other mammals, when exposed to testosterone early in life, will also exhibit a penislike clitoris. Supporting examples exist even in humans. A few pregnant women have received medical treatment that happened to expose their embryonic female offspring to high testosterone levels, and these children did indeed possess a greatly enlarged, penislike clitoris at birth [828].

It is conceivable, although unlikely, that a female spotted hyena living long ago happened to have ovaries that produced precisely the correct amount of masculinizing androgen to produce a near-perfect pseudopenis in her daughters. If so, the pseudopenis could have originated from a single mutation that yielded a developmental result identical to the current trait.

On the other hand, the transition from the ancestral state (a relatively small clitoris) to the current state (a large pseudopenis) may have required many generations in order for modifying mutations to occur and to spread through the species, gradually leading to a more perfect mimetic penis. The remarkable resemblance of today's pseudopenis to a "real" penis and the ability of modern spotted hyenas to erect the pseudopenis suggest that many changes have occurred since the original clitoris-enlarging mutation. The main point, however, is that the trait probably originated long ago in a mutant female hyena who happened to make novel use of the basic mammalian rules of genital development [453]. This is a statement about the possible evolutionary history of the pseudopenis.

An Adaptive Signal?

The other element of evolutionary studies deals with the processes that caused change to occur over time. The pattern of historical change in spotted hyenas suggests that at least one mutant allele that altered the hormonal environment of female embryos spread through populations of spotted hyenas in the past. All of today's spotted hyena females possess this allele, or more recently modified versions of it, and probably other later-appearing mutant genes as well. Together, these genes contribute to the development of the secondary sexual characteristics of female hyenas. As a result, all of today's spotted hyena females possess a pseudopenis and the ability to use it to communicate with others.

Why did this change or changes take place? Perhaps the mutant allele or alleles affecting hyena secondary sexual characteristics promoted the reproductive success of the individuals that possessed it. As discussed in Chapter 1, a mutant allele that did not increase fitness would disappear, doomed by the failure of its bearers to propagate it effectively [1289]. But why would the first spotted hyena female with masculinized external genitalia have left more descendants than the "normal" or typical females of her era? Surely this mutant female would have been harmed in some ways by the disruption of long-tested patterns of sexual development. Human females accidentally exposed to androgens as embryos become sterile adults, illustrating just how damaging the developmental effects of hormonal changes can be. And since even in modern populations of spotted hyenas, 10–20 percent of all females die as a consequence of having to give birth through the clitoris (Figure 2),

2 Cost of the pseudopenis for female spotted hyenas. The birth canal of this species extends through the pseudopenis, which greatly constricts the canal and creates considerable difficulties, especially for females giving birth for the first time. Protracted labor associated with blockage at the pseudopenis kills many mothers. *Drawing courtesy of Christine Drea, from Frank et al. [408].*

we can be confident that the spread of masculinized genitalia had to overcome major reproductive disadvantages [405, 408].

Obviously, some masculinized spotted hyena females in the past were not sterile and survived giving birth to their offspring. The fitness benefits these pioneers gained from the mutant allele(s) responsible for the development of the pseudopenis must have outweighed any problems associated with the gene. Perhaps the hormone-induced pseudopenis was used right from the start as a beneficial communication device. An alternative hypothesis is that the pseudopenis was initially a costly side effect of a mutant allele that had some other consequences, some so beneficial that they swamped any disadvantageous effects of the gene.

Several competing hypotheses exist on how females could benefit from a gene that exposed their fetuses to increased androgen levels. Marion East and his co-workers suggest that the gene for androgen production by ovarian cells originally spread not because females with the gene produced female offspring with pseudopenises, but because females that exposed their embryonic pups to androgen caused them to be intensely aggressive at a very early age [322]. Hyenas bear twin pups, which are born with their eyes open and their canine teeth fully erupted; they usually begin fighting with each other almost immediately, and one pup may kill the other. Surprisingly, siblicide can sometimes provide fitness benefits for parents (see Chapter 14). At the proximate level, intense aggression occurs between hyena siblings because the embryos receive placental fluids rich in androgen.

An alternative, but somewhat similar, hypothesis is that the mutant gene spread because it made adult females, rather than infants, larger and more aggressive. The alpha female of a clan can win the ferocious competition for food that occurs at carcasses of dead zebras and the like (Color Plate 7), and her offspring get to eat when the young of subordinate females are starving to death. Alpha females achieve their top hyena status by virtue of their large size and intense aggressiveness, traits that are enhanced by early exposure to androgen [405, 407].

Thus, the masculinization of the external genitalia of the daughters of the first mutant female hyenas could well have been purely a side effect of the unusual hormonal environment in which they developed, an environment that made them more aggressive. For reasons already discussed, the enlarged clitoris probably reduced, but did not eliminate, fitness gains due to the heightened aggressiveness of masculinized females or their neonates. Once it had originated and spread, however, the pseudopenis may then have undergone additional changes that enabled it to contribute to, rather than harm, female reproductive success. If we could demonstrate that the modern pseudopenis now has reproductive value in and of itself, then we could claim that the trait (and its genetic foundation) is currently being maintained by natural selection.

Today's female spotted hyenas apparently do use the pseudopenis to communicate something during their "greeting ceremonies." When communication occurs, one individual conveys information to another individual. If giving the signal *or* responding to it reduced individual fitness, selection would favor *non*signalers or *non*receivers. Two key questions, therefore, about any possible communication signal are, What benefits might the signaler gain by transferring information to others that compensate it for the costs of its action? and What benefits might a receiver gain by paying attention to this information?

Let's ask these questions of the spotted hyena in an effort to discriminate between two alternative hypotheses about its pseudopenis: the selectionist possibility that the structure is currently used to communicate adaptively with others, and the nonselectionist hypothesis that it was, and still is, merely an incidental side effect of elevated androgen levels in females.

As noted above, spotted hyenas live in large clans, hunt big game, and compete fiercely with one another for meat from the animals they kill. Dominant females and their offspring gain more food than subordinate ones [404, 1148], and as a result have three times the reproductive success of their inferiors (Figure 3). The sons of alpha females "inherit" their mothers' high dominance status. When fully mature, young male hyenas emigrate to another clan. If the son of a dominant female becomes the top male in his new clan, he will enjoy exceptional reproductive success, because only the dominant male mates with the many females in a clan.

Thus, the social system of spotted hyenas heavily rewards dominant females [407]. In turn, the dominance interactions among females may currently be mediated by pseudopenile displays because the tendency of a female to present her erect pseudopenis to others in a greeting display is related to her social status. Subordinate females and youngsters are far more likely than dominant animals to

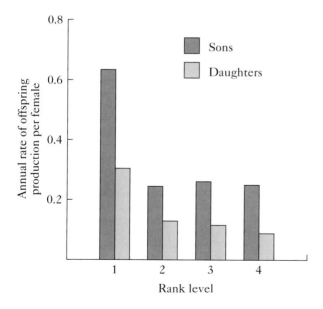

3   Dominant spotted hyena females enjoy great reproductive success. Alpha females produce about three times as many sons (dark bars) that reach sexual maturity as lower-ranking females do, and two to three times as many mature daughters (lighter bars) as well. *Source: Frank et al. [407].*

initiate these displays. This observation is consistent with the hypothesis that subordinates gain by transferring information of some sort to dominant individuals by way of the pseudopenis. Just what this information might be remains to be identified, although observers of spotted hyenas in the wild have all agreed that the greeting ceremony appears to promote cooperative behavior among the participants [404, 682].

Cooperation between dominant and subordinates might be advanced if the dominant could check the physiological (e.g., hormonal) state of a subordinate by inspecting its erect, blood-engorged penis or clitoris. Submissive subordinates may signal that they lack the hormones needed to initiate a serious challenge to the dominant, which then can afford to tolerate the subordinates, permitting them to enjoy the benefits of social life in the clan, since they do not pose an immediate social threat.

In another highly social mammal, the naked mole-rat, the colony contains only one breeding female, the "queen," who is dominant to all the other females. In a captive group studied by Susan Margulis and her colleagues [765], the queen was the only individual in the group to inspect the anogenital regions of her female colonymates, and she was especially aggressive toward one female, the only one to have ovulated during her tenure. Upon the removal of the queen, the victim of her aggression became one of the major contestants to become the sole breeding female in the group. The hormonal condition of the contestants changed markedly during this time, suggesting that the old queen could have learned about the hormonal state or reproductive capability of her fellow females via anogenital inspection.

The possibility that something similar happens in spotted hyena clans is supported by the finding that adult male immigrants rarely participate in greeting ceremonies [322]. These males do not affect the status or reproductive success of the top female members of the clan, who control the dominance hierarchy of their fellow females.

In summary, comparisons with other mammals can help us identify a possible origin for the spotted hyena's pseudopenis. The spread of the trait could have occurred via natural selection, either (1) indirectly because of the adaptive effects of hormonal mechanisms that promote aggression among newborn hyenas or among adult females, or (2) directly, because possessing a pseudopenis permits a signaler to participate in a mutually beneficial communication system.

Reconstructing the History of a Complex Signal

The hyena story illustrates the separate, but interrelated, issues of the historical patterns and the causes of evolutionary change. For the rest of this chapter, our focus will be on reconstructing the history behind complex traits, determining the steps from origin to current form for a few selected examples. Just as comparisons among related species helped make the pattern of pseudopenis evolution clearer, the **comparative method** has illuminated the history of many other strange communication signals in the animal kingdom. For an example, we turn to the pelagic shag's "rapid-fluttering wing-waving display," a courtship signal in which the bird points its bill straight up while raising both wings and fluttering them rapidly.

This bizarre behavior depends on the integration of many components of the bird's nervous, muscular, and skeletal systems, to say nothing of its wing feathers. The probability that such a complex and superbly integrated system could have arisen as a result of a single mutation is vanishingly small. Remember that a mutation is a random change in a single gene, and that a gene codes for the production of a single protein. Thus, the typical effect of one mutation is to alter the sequence of the amino acids that make up the one protein coded by the mutant gene. It is extraordinarily unlikely—in fact utterly impossible—for a single random change in one gene to produce an entire system of adaptively interlocked structures. A pelagic shag's wing alone is a developmental outcome involving thousands of genes, if not tens of thousands, all of whose protein products "cooperate" in the construction of the device.

An Accumulation of Small Changes

So how did pelagic shag's wings and wing displays and all the other million and one complexities of living things evolve? Most evolutionary biologists believe that they arose gradually from simpler ancestral patterns via a long series of mutations with small adaptive effects. The incremental process of layering one modification on top of preceding ones has been labeled **cumulative selection**, and it has immense power to generate adaptive complexity over time, as Richard Dawkins illustrates with a brilliant analogy [280]. Let us say that a complex current trait is like an English sentence, for example, a line from Shakespeare's *Hamlet:* METHINKS IT IS LIKE A WEASEL. What is the chance of producing this unique combination of letters by chance alone? For each of the 28 letters and spaces in the phrase, we could put in any one of 26 letters or a space. So if we generated a random sequence, it might be this: SWAJS MEIURNZMMVASJDNA YPQZK. If we kept at it, or had a computer keep at it, the time needed to come up with METHINKS IT IS LIKE A WEASEL would be enormous because there are so many possible combinations of letters and spaces (27 possibilities for the first position, times 27 for the next position, times 27 for the third . . . times 27 for the twenty-eighth position—you get the idea). The number of possibilities is immense, and yet only *one* combination is METHINKS IT IS LIKE A WEASEL.

However, instead of trying to get the "right" sequence in one go, let's change the rules so that we start with the random letter set shown above and use the computer to copy it over and over, but with a small error rate built into the program. Occasionally, the computer copies the sequence incorrectly, randomly inserting a new letter into one of the positions. Then we ask the computer to look over its list and pick the sequence that is closest to METHINKS IT IS LIKE A WEASEL. Whatever "sentence" is closest is used for the next generation of copying, again with a few errors. The sentence in this group that is most similar to METHINKS . . . is selected for "breeding," and so on. When Dawkins did this with a number of different starting sequences, he found that it took only 40–70 generations to reach the target sentence, not millions upon millions upon millions of attempts—a few seconds of computer time, not years [280].

Cumulative natural selection has the same effect on living systems. Some mutations induce small random changes in the genetic "sentence" possessed by individuals. Any changed genetic message that happens by accident to confer higher reproductive success on individuals propagates itself throughout the population. After the new "sentence," with its new developmental outcome, is widespread, additional small changes that improve the reproductive output of individuals will be incorporated in the same way, one after another. The cumulative effect of this process is to rework the genetic information, and thus the phenotypic attributes, of a species. Given enough time, a species may barely resemble its ancestor, but without requiring that the differences arise in one giant change.

The logic of the theory of cumulative selection requires that a series of changes lie between an ancestral pattern and a modified modern trait, with no major leaps of any sort. To test this proposition for the wing-waving pelagic shag, we first need to identify the probable starting point for the evolution of its display. We can begin by noting that the shag's behavior makes it look a little bit like a bird that is about to take off. Many bird species point the bill up and lift the wings slightly before leaping into the air and flapping away (see Figure 4). These "flight intention movements" are widespread in living birds, so perhaps ancestors of the pelagic shag exhibited them too [635]. Moreover, in one of these ancestors, wing movements might have provided information to another bird that the individual was about to fly away. If so, the first step had been taken in the evolution of a communication system with a signal provider and a signal receiver.

But how would such a relatively simple system evolve into the peculiar and complicated display of the pelagic shag? One way of getting information on possible intermediate evolutionary stages is to examine clusters of species related to the species of interest, groups with which it shares a fairly recent common ancestor. If species X exhibits a trait that is similar in some respects to the trait of interest in related species Y, species X may tell us how an ancestor of Y behaved. In a way, what we are doing is looking for living "missing links," species whose traits may be closer to those of now extinct ancestral forms because they did not undergo the evolutionary modifications that species Y did. Retention of an "older" trait can happen if mutations that would have resulted in evolutionary change simply never materialized in species X's lineage. Mutations are random events, and therefore cannot be induced when "needed." Alternatively, the modifications that increased the fitness of species Y in its environment may not have done the same for species X in its different environment, even if they did occur at some point.

When G. F. van Tets surveyed the relatives of the pelagic shag in a search for displays similar to, but not quite the same as, the rapid-fluttering wing-waving display, he hit pay dirt [1213]. Some other cormorants engage in a similar courtship

A Behavioral cladogram

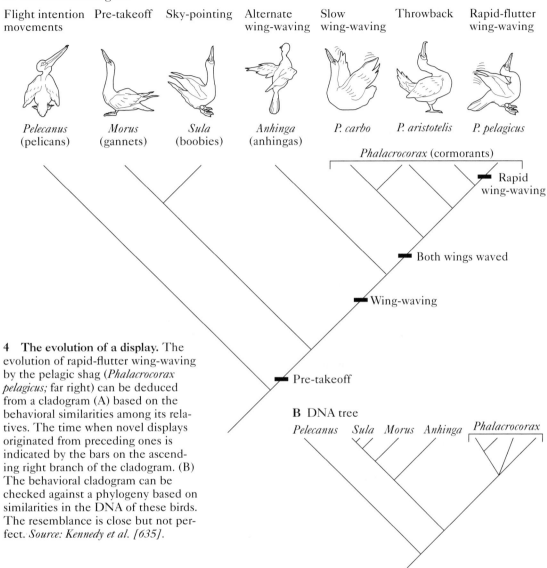

| Flight intention movements | Pre-takeoff | Sky-pointing | Alternate wing-waving | Slow wing-waving | Throwback | Rapid-flutter wing-waving |

Pelecanus (pelicans)   *Morus* (gannets)   *Sula* (boobies)   *Anhinga* (anhingas)   P. carbo   P. aristotelis   P. pelagicus

Phalacrocorax (cormorants)

Rapid wing-waving

Both wings waved

Wing-waving

Pre-takeoff

B DNA tree

Pelecanus   *Sula*   *Morus*   *Anhinga*   *Phalacrocorax*

4   The evolution of a display. The evolution of rapid-flutter wing-waving by the pelagic shag (*Phalacrocorax pelagicus;* far right) can be deduced from a cladogram (A) based on the behavioral similarities among its relatives. The time when novel displays originated from preceding ones is indicated by the bars on the ascending right branch of the cladogram. (B) The behavioral cladogram can be checked against a phylogeny based on similarities in the DNA of these birds. The resemblance is close but not perfect. *Source: Kennedy et al. [635].*

display, but they wave their wings slowly instead of rapidly. Anhingas, which are not cormorants but are thought to share a recent common ancestor with them, also have a wing-waving display, but they wave first one wing and then the other in alternation, rather than moving them together. And still other birds, the boobies, also related to cormorants, lift their head and wings in a highly exaggerated manner called sky-pointing, but do not wave their wings at all.

We can order these various displays into a sequence that minimizes the distance between the single steps in the series required to get from the presumed simpler ancestral state to the complex display of the pelagic shag. One plausible historical scenario goes from (1) flight intention movements to (2) sky-pointing to (3) slow wing-waving to (4) rapid-flutter wing-waving. The alternating wing-waving of the anhinga does not appear to fall into line, and so might be a separate offshoot from a species that used the sky-pointing display (Figure 4A).

What we have now is a hypothesis, the one presented by van Tets in 1965 [1213]. Let's try to test it. If the behavioral sequence that we have outlined is correct, the pelagic shag must have been preceded by a fairly recent ancestral species that engaged in slow wing-waving, while a more distant ancestor sky-pointed, and an even earlier ancestor used only flight intention movements.

In order to secure information about these hypothetical ancestral species, we can compare the living ones using characteristics other than the behavioral traits we have already used to produce our evolutionary scenario [635]. There are choices to make here. We could examine the skeletons and other structural features of the many species of cormorants, anhingas, and boobies to determine their relative similarity. Or we could use the DNA of the various species to do the same thing. The more similar the skeletal characteristics, or the DNA, of two species, the more likely it is that they had a recent common ancestor from which they have inherited the genes they possess, which influence the development of all their other characteristics. The longer the species have been separate, the more time there has been for genetic and structural changes to accumulate, and the more dissimilar the two species should be.

Figure 4B presents a diagram of the evolutionary relationships among the various cormorants and shags, anhingas, and boobies based on DNA similarities among them. You can see that this **cladogram** or **phylogenetic tree** matches the scenario of evolutionary changes that van Tets developed to explain the complex display of the pelagic shag. All the creatures called cormorants and shags are genetically more similar to one another than they are to the birds that ornithologists have named anhingas and boobies. Thus, the most recent common ancestor of the cormorants probably had a display in which the bird waved its wings. Judging from the DNA data, the ancestral species that eventually gave rise to the lineages of boobies, anhingas, and cormorants occurred much farther back in time. Its display behavior probably was similar to sky-pointing, a trait retained to this day in boobies.

To evaluate the reasonableness of this scenario, consider what it would mean to propose that the ancient ancestor of the pelagic shag and all the other cormorants, the anhingas, and the boobies engaged in rapid-flutter wing-waving. This would require a major transformation in the ancestral species, from simple flight intention movements to a highly complex display, in a single step (since none of the living relatives of the cormorants, anhingas, and boobies, such as the pelicans, do anything like rapid-flutter wing-waving). And the improbability of this scenario would be compounded by the requirement that this hypothetical rapid-flutter wing-waving ancestor then gave rise to a species ancestral to all living boobies that lost this trait and replaced it with the substantially different sky-pointing display. Van Tets's scenario does not involve these big jumps and major reversals, and so is more in keeping with what can be inferred about evolutionary probabilities [635].

Be aware, however, that every step of the procedures I have just described is a subject of debate, from determining what constitutes a "large evolutionary jump" to establishing what ancestral species preceded the one that gave rise to a cluster of related species of interest. Note that our reconstruction of the history behind the pelagic shag's display rests on a key assumption, one that will be used again in other historical analyses discussed later. This key assumption is that the *principle of parsimony* should be used in reconstructing the history of a trait—that is, that the simplest scheme, involving the fewest steps, should be preferred to more complex ones. If, however, evolutionary history actually follows a somewhat capricious course, forced one way or the other by odd accidents and improbable contingencies, then what seems most likely to a researcher may not be what actually occurred. Even infrequent violations of the assumption of parsimony would greatly complicate getting at the *real* history behind a trait, a point to keep in mind in the pages ahead [1032].

The Evolution of Flapping Wings

Although pelagic shags, and many other birds, use their wings in various displays, wings almost certainly did not evolve because they can be used to communicate with others. How did these structures originate, and for what purpose? Obviously, most birds now use their wings for powered flight, an immensely complex behavior. The initial phases of flight surely involved systems far simpler than the current elaborate and beautifully integrated mechanisms that make powered flapping flight possible.

Evolutionary biologists have proposed two different scenarios for the evolution of flight, a "ground-up" version and a "tree-down" alternative [888, 1026]. According to the ground-up view, birds evolved from a reptilian ancestor that developed feathered forelimbs for some reason other than flight. According to this hypothesis, the forerunner of birds was exactly that, a running animal that sprinted along the ground on its hind legs. At some point, its forelimbs began to serve an aerodynamic purpose, at which point they could be called wings. This proto-bird almost certainly used its wings not for flapping, powered flight, but perhaps to glide or to steer its body when it became airborne after leaping upward [174]. Such a creature could, through gradual modification, eventually yield a descendant species that might have gained additional flight range by flapping, setting the stage for an elaboration of the neuromuscular basis for powered flight.

The tree-down scenario differs from the rival ground-up proposal primarily in suggesting that the first bird lived in the trees, rather than on the ground. According to this view, an arboreal, rather than cursorial, proto-bird may have used its "wings" to glide from tree trunk to tree trunk in the manner of some current arboreal mammals, of which the flying squirrel is a familiar example (Figure 5) [1315].

5   Did the first birds use their wings to glide from tree to tree?

Descendants of this downward-gliding proto-bird may have added weak flapping flight to extend the glide, after which still more added muscle would have enabled individuals to use powered flapping flight.

How can we possibly test these two alternative hypotheses, which deal with events that took place millions of years ago? It would be helpful if we could examine the fossil feathers and bones of an ancestral bird close to the origins of the group, because the structure of the first bird could tell us something about how it behaved. Many persons have assumed that the famous crow-sized *Archaeopteryx*, which lived 150 million years ago, was one of the very first birds, since it had teeth and a skeleton remarkably similar to those of certain dinosaurs, which some believe to be the ancestors of the species that gave rise to the avian lineage [346, 1094]. According to this view, figuring out how *Archaeopteryx* flew would tell us how birds first used their wings.

Was *Archaeopteryx* cursorial, as required by the ground-up hypothesis, or arboreal, as required by the tree-down hypothesis? Did these ancient birds actually fly, and if so, how did their flight differ from that of modern birds? The history of the attempt to answer these questions offers a fascinating illustration of how scientific views evolve.

Fairly early on, some biologists concluded that *Archaeopteryx* had the structural features of an arboreal glider. They pointed to its feathered wings, which had thin, sharp-tipped claws (modified fingers of the forelimbs) of the sort that appear on the feet of some tree-climbing birds today. Moreover, the claws were positioned so that they could be used by an animal pulling itself up a tree trunk or onto a branch with its wings (see Figure 5) in a manner reminiscent of the climbing behavior of nestling hoatzins, a South American bird whose young have functional claws on their wings.

The wings of *Archaeopteryx* seem fully modern, with a shape like those of forest-dwelling grouse and individual flight feathers identical in aerodynamic design to those of today's flying birds (Figure 6) [383]. But the wings were evidently powered by a much reduced muscle mass compared with that of modern flying birds of the same size, since *Archaeopteryx* had a primitive-looking breastbone, or sternum, that lacked a prominent keel. Modern flying birds possess a large keel, which provides space for the attachment of the large breast muscles that power flapping flight—and which feed human consumers of turkeys and chickens.

6   A fossilized feather of *Archaeopteryx*. The curved central vein, or rachis, shows that the wings of this bird served an aerodynamic function, because all modern flying birds have similarly designed feathers. *Photograph courtesy of John Ostrom.*

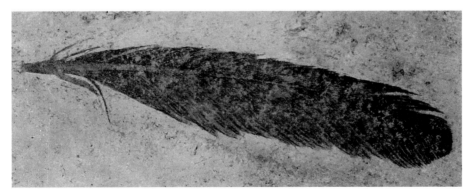

Although the anatomy of *Archaeopteryx* was such that it might have flapped its wings as it glided from tree to tree, it probably lacked sufficient muscle power to take off from the ground. If so, the tree-down hypothesis for the evolution of flight gains support, with a proto-bird gliding at first, then adding wing area for extended gliding flight, and then adding weak flapping for still greater range [866]. If in the early stages of gliding flight, the glider did not land with its feet on a branch as modern birds do, but instead used its clawed wings to grip the trunk as it swooped up at the end of the glide (see Figure 5), it could have landed without possessing the elaborate body control mechanisms required for branch-to-branch trips, mechanisms that could have evolved in steps over time.

In 1991, John Ruben noted that most persons had assumed that the flight musculature of *Archaeopteryx* was much the same as that of modern birds [1026]. He wondered, however, if the flight muscles of *Archaeopteryx* were reptilian in design, a plausible idea given the similarities between *Archaeopteryx* and the dinosaurs, which no one doubts were reptiles. It may surprise you, as it did me, to learn that the muscles of some modern reptiles can generate about twice the peak power of the muscles of modern birds and mammals because reptilian muscles have more contractile fibers per unit of mass. If *Archaeopteryx* was endowed with a reptilian set of shoulder muscles, this creature could have flapped its wings with twice the power of the corresponding muscles in a pigeon or other modern bird.

Ruben's calculations convinced him that *Archaeopteryx* had sufficient muscle mass to take off from the ground and fly for 20 meters or more, provided its muscles were indeed reptilian in design. Thus, *Archaeopteryx* could have been both cursorial and arboreal, spending much of its life walking and sprinting on the ground, but at home in shrubs and trees, where its clawed wings would have assisted it in moving about. In fact, any number of species of large living birds, such as wild turkeys, run rapidly on the ground and yet also fly up into trees for food and safety.

In 1993, however, Alan Feduccia reexamined the claws on the *feet* of *Archaeopteryx*. He argued that if this extinct bird did a lot of running on the ground, its claws should be similar in structural design to those of modern ground-dwelling birds. These species have claws that are relatively flat, whereas arboreal perching birds have highly curved claws, the better to grip a tree branch tightly. *Archaeopteryx* had a claw arc in the same range as modern perching birds, not cursorial birds (Figure 7). Furthermore, its long tail feathers were not frayed at the ends, judging from the beautifully preserved fossils of the bird, whereas modern cursorial birds with long tails, such as roadrunners, occasionally drag their tails on the ground, damaging their feathers [382].

So some persons believe that *Archaeopteryx* lived in the trees and flap-glided from limb to limb or trunk to trunk. If this species really was close to the root of bird evolution, then we could with some confidence accept that tree-down gliding was the very first stage of avian flight. But how close was *Archaeopteryx* to the ancestor of modern birds? Several smaller fossil birds have recently been discov-

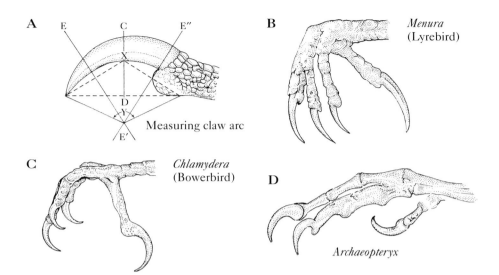

A   E   C   E″

X

D

Y

E′

Measuring claw arc

B   *Menura*
(Lyrebird)

C   *Chlamydera*
(Bowerbird)

D

Archaeopteryx

7   On the ground or in the trees? (A) Claw arc, represented here by the angle labeled "Y," can be measured in living birds known to run on the ground or perch in trees. (B) The lyrebird lives on the ground, and has a wide claw arc. (C) The bowerbird lives and perches in trees; it has a narrow claw arc, as does *Archaeopteryx*, whose fossilized feet have highly curved claws similar to those of perching birds (D). *Source: Feduccia [382].*

ered that are believed to be almost as old as *Archaeopteryx*. Among these species is one that had a broad sternum, modern wing-flapping anatomy, and a large opposable toe, features demonstrating that it flew and perched very much like a modern bird, despite living 135 million years ago (Figure 8) [1079]. Did it descend from an *Archaeopteryx*-like ancestor, or was it a member of a different lineage derived from a still older bird whose remains have yet to be discovered, as some argue [573]? Perhaps we need to find a species closer to the first bird than *Archaeopteryx* before determining the original function of wings.

Although not knowing precisely how flapping flight originated may be dissatisfying, remember that today's scientists do not know everything, and some of what they think they know is probably wrong. Many biological puzzles, some as interesting as the mystery of bird flight, remain to be solved, offering fascinating problems for scientists of the future, perhaps including some readers of this book.

But whether the original ancestor of the pelagic shag used its wings while sprinting across the ground or gliding from tree to tree, it probably did *not* have a courtship display like that of the pelagic shag. The display function of wings presumably followed their evolution for other purposes, with the gradual accumulation of many changes before the pelagic shag's recent ancestor happened to employ its wings in a novel, but adaptive, fashion to send a message to other members of its species (Table 1).

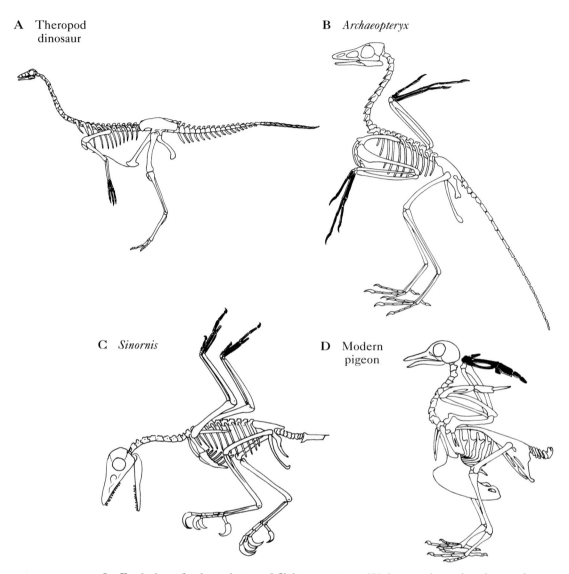

A Theropod
dinosaur

B *Archaeopteryx*

C *Sinornis*

D Modern
pigeon

8   Evolution of avian wings and flight apparatus. (A) A ground-running theropod
dinosaur may have been the ancestor of the birds. (B) Note the similarity between the
theropod skeleton and that of *Archaeopteryx*; the forelimbs of *Archaeopteryx* are only
slightly modified from the theropod pattern. (C) In contrast, the forelimbs of *Sinornis*,
which appeared only 15 million years after *Archaeopteryx,* are much more like those of
modern birds, such as (D) the pigeon. Note also that the tailbones of the pigeon and
Sinornis are greatly reduced compared with that of *Archaeopteryx*. The pigeon also possesses a large sternum, or breastbone, for the attachment of wing muscles. Neither of
the ancient birds has an equivalent skeletal feature. *Sources: (A,C) Sereno and Chenggang
[1079]; (B) Hou et al. [573]; (D) Colbert [212].*

Table 1 The historical pattern of changes leading to the rapid-fluttering wing-waving display of the pelagic shag

Structure	Function
1. Featherlike covering of forelimbs	1. Improved thermoregulation?
2. Feathered forelimbs become winglike	2. Weak gliding flight
3. Primary feathers become aerodynamic	3. Improved gliding/weak flapping flight
4. Wing muscles expand in size	4. Powered flapping flight
5. Neuronal mechanisms produce intention flight postures	5. Preparation for flapping takeoff
6. Neuronal mechanisms exaggerate takeoff postures	6. Posture has signal value
7. Neuronal mechanisms produce wing movements in exaggerated takeoff posture	7. Movements have signal value
8. Neuronal mechanisms produce rapid wing waving while in display posture	8. Rapid movements have signal value

The History of a Mechanism for Receiving Signals

Having just reviewed the possible evolutionary history of a visual signal transmitted by the pelagic shag, let's now take a look at the history of signal receiving. The species we will examine is the whistling moth, *Hecatesia exultans*, of Western Australia, which communicates with acoustical signals, a trait that occurs in few other moths [412]. Males of this small black, white, and orange moth are no bigger than your thumbnail, but they can generate loud sounds, mostly in the ultrasonic range, around 30 kHz (Color Plate 8). In order to listen in on these calls and track down the singers, humans need special equipment: a bat detector machine, which converts ultrasound to sounds of much lower frequency.

A male whistling moth calls from his perch by raising his wings and vibrating them so that unusual knobs of cuticle (called "castanets") on the edge of the forewings strike together very rapidly. The sound waves generated by this apparatus travel across the heath to other listening moths. One can demonstrate that other whistling moths can hear the ultrasonic signals of calling males by putting out a speaker in appropriate habitat and playing a tape of sounds recorded from a singing moth. On occasion, other males will fly to and even land upon the speaker (Figure 9) in their attempts to find and interact with the acoustical signaler [11]. Under natural conditions, males that are singing nearby sometimes come together for an aerial duel that involves much bumping into one another and ultrasonic buzzing before one stops singing and flees. Thus, the calls of males play a role in their attempts to defend or acquire a calling territory, where signaling males may be visited by receptive females willing to mate with territory owners.

9 **The ultrasonic signals of male whistling moths** are produced by striking the castanets (see arrow) on the forewings together (top). (Bottom) A male has been attracted to a recording of another male played back from a small speaker in his territory, confirming that this species uses acoustical communication. *Photographs by the author.*

Although only males have the sound-producing mechanism, which consists of the wing castanets and the neural apparatus for controlling the wingbeat pattern, both male and female whistling moths possess mechanisms for hearing their species' song [1161]. The sound-receiving system consists in part of an ear on either side of the thorax, near the point where the hindwing attaches to the thorax. As with the noctuid moths discussed in Chapter 5, the outer ear is composed of a small ellipse of thin cuticle, called the tympanic membrane, which covers an air sac. When the membrane vibrates in response to ultrasound, the air sac also moves, supplying mechanical energy to associated sensory receptors. These sensory cells then produce signals for relay to other parts of the nervous system. How might such a sophisticated device for detecting ultrasonic signals from other whistling moths have evolved?

Researchers interested in this question have operated with the same ground rules as those persons interested in the evolution of the pseudopenis in hyenas or flight in birds. They have assumed that in changing from the presumed ancestral state, which would have been an earless moth (even today, most moths lack ears), the whistling moth's ear probably did not arise via one or two mutations. The system is much too complex and composed of too many interrelated parts to have developed from a random change in a single gene, or even two or three. Instead, it is much more likely to have been assembled by degrees, with one small adaptive change layered on another, eventually producing the complex acoustical mechanism in its present form.

If this view is correct, the history of the whistling moth's ear should involve multiple modifications of a system that existed before the current ear. But where do we go to test this hypothesis? James Fullard and Jane Yack [413] decided to look for cues about the past in the bodies of moths living today. They examined species from the family Sphingidae, a group that lacks ears and therefore cannot hear any sounds. They paid special attention to those parts of the sphingid thorax where one can find the bat-detecting ears in noctuid moths (see Chapter 5), the family to which the whistling moth belongs. They found sensory cells attached to the sphingid thoracic cuticle that are remarkably similar in structure to the receptor cells in the ears of noctuid moths (Figure 10), suggesting that the ears of noctuids were not built from scratch, but rather involved modifications of sensory cells that performed some other function for the earless moths.

What do these "nonacoustical sensory cells" do for sphingids? They are attached to a part of the thorax that moves rhythmically as the wings move, particularly when the moth is "shivering," vibrating its wings rapidly to generate metabolic heat prior to takeoff. These mechanoreceptors translate mechanical energy from the motion of the cuticle into messages that are relayed elsewhere in the moth's nervous system, where information about the alignment of various body parts is integrated and used to adjust the moth's position in space. A similar kind of mechanoreceptor in an ancestral moth in the noctuid lineage could have conferred a weak ability to hear certain loud sounds, which also can make insect cuticle move [413].

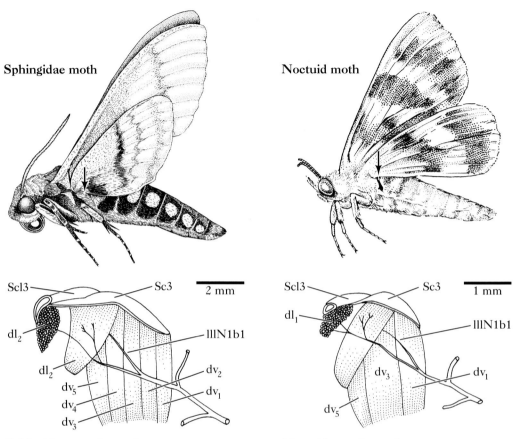

Sphingidae moth

Noctuid moth

Scl3 — Sc3   2 mm

Scl3 — Sc3   1 mm

dl_2

dl_1

lllN1b1

lllN1b1

dl_2

dv_3

dv_1

dv_5

dv_2

dv_5

dv_4

dv_1

dv_3

Sphingidae muscles and nerves

Noctuid muscles and nerves

10   Evolution of a sensory system. Sphingid moths are deaf, but noctuid moths can hear. The arrows in the upper panels point to the same location on the thorax of the two moths. Sphingids do not have an ear with a tympanic cavity, but noctuids do. The lower panel shows the external thoracic plates in the two families (labeled Scl3 and Sc3) with the attached muscles (all labeled with the letter "d") and a branching sensory nerve that innervates the region. Note the similarity between the nerves in the two moths. One branch of the nerve labeled IIIN1b1 carries information about the position of the hindwing in the sphingid; in the noctuid, this same branch relays acoustical information from the tympanic membrane to the moth's central nervous system. *Sources: Yack [1312] and Yack and Fullard [1313]; sphingid moth illustration by Diane Scott.*

Almost certainly, the first hearing noctuid moth had nowhere near the sophisticated sensory skills of the current descendants of that moth. But with the addition of other small changes, such as a slight thinning of the cuticle over the "acoustical" cells, an enlargement of the respiratory chamber behind this part of the thorax, and a sensory cell design that was somewhat more sensitive to sounds of particular frequencies, these mutant moths could have heard bats better and better. Receivers with superior ultrasonic hearing might have survived better, reproduced more, and spread the genetic basis for these variations, thus setting the stage for the next improvement to spread through the species via natural selection.

The Evolution of Insect Flight

Let us accept the conclusion that one stage in the evolution of hearing involved nonacoustical mechanoreceptors that monitored wing vibrations, rather than sound vibrations. Can we go back further still to determine how insect wings originated? If insect wings evolved, they must have done so by degrees, for the reasons we have already discussed in other contexts. The design and control of a noctuid moth's interlocked forewings and hindwings are extremely complex, and the production of such structures and their allied neural control systems cannot have occurred in one or two steps. Instead, we would expect that what came to be wings were once something else, with a structure and function different from those of any modern wing.

The precursors of insect wings might be the gill plates of aquatic immature forms of now-extinct insects, some of which became fossilized, luckily for us (Figure 11). In these extinct species, the gill plates probably moved water over the insect's gills, the better to move carbon dioxide out of and oxygen gas into

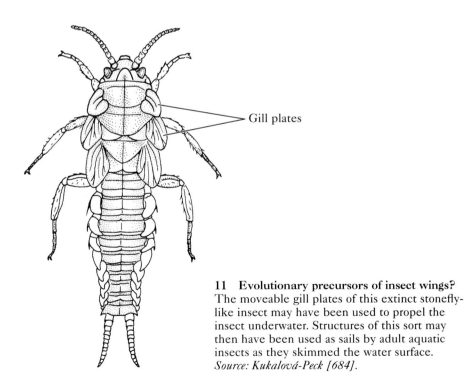

Gill plates

11 Evolutionary precursors of insect wings? The moveable gill plates of this extinct stonefly-like insect may have been used to propel the insect underwater. Structures of this sort may then have been used as sails by adult aquatic insects as they skimmed the water surface. *Source: Kukalová-Peck [684].*

the animal's body. Because these gill plates could move, judging from the way in which they were attached to the body via hinges, they might have been the ancient beginnings of wings, which are also movable and controlled by thoracic muscles.

If among the descendants of the gill-plated insects there were some that retained these structures when the immatures metamorphosed into adults, then they could have been used as sails to send the adult insects skimming across the water. Speedy movement in this environment may have been advantageous for creatures with predators eager to pick them off the surface. It is in fact the case that adult winter stoneflies, a living group of aquatic insects, use their wings as sails to catch the wind and propel themselves more quickly across the water surface [757]. Males of one stonefly species have ridiculously short wings and cannot fly, but on a windy day they can skitter rapidly over the water with all six feet in contact with the water surface (Figure 12). When James Marden and Melissa Kramer, an undergraduate at the time, clipped the wings of flying stonefly species so that they could no longer fly, these stubby-winged stoneflies could still go sailing. In fact, the experimentally flight-deprived adult stoneflies traveled faster than their immature forms could swim, another demonstration that the wings of ancestral nonflying insects might have had survival value, and so might have spread through populations in the past [759, 760].

12 A surface-skimming stonefly (top) raises its short, stubby wings (bottom) strictly as sails to catch the wind as it skitters over the water while keeping all six feet on the water's surface. *Photographs courtesy of James Marden.*

Once sailing wings had become standard for some species, mutant individuals able to flap their winglets even to a small degree would have been able to add muscle power to wind power, thereby promoting faster skimming and perhaps greater escape potential. These simple flappers would have set the stage for yet another modification, an increase in flight muscle capacity that could lift a powered skimmer off the water. Some modern stoneflies demonstrate how the transition between water skimming and full-fledged aerial flight might have come about. Adult stoneflies in the genus *Leuctra* beat their wings to achieve sufficient lift so that only the two hind legs remain in contact with the water (Figure 13). This position stabilizes the insect while permitting it to move its wings over a much wider arc than is possible for those stoneflies that keep all six feet on the surface of a stream. Six-leg skimmers travel about 40 percent slower than hind-leg skimmers, showing how mutant individuals of some ancestral insect species that happened to use the hind-leg technique might have enjoyed a selective advantage [673]. From hind-leg skimming it is a short additional step to full flight, which would have carried the first true fliers away from insect-eating fish into the relative safety of the air.

Evolution almost certainly did not cease at this point. A powered flier that happened to have a slightly more rigid, stronger wing structure would probably have

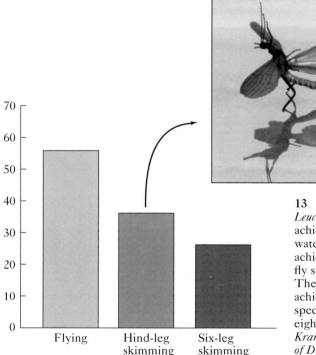

13 **Partial flight in a modern stonefly,** *Leuctra hippopus,* which beats its wings, achieving lift, but keeps its hindfeet on the water surface for flight stability. This species achieves flight velocities greater than stonefly species that keep all six feet on the water. The graph shows mean travel speeds achieved by samples taken from four flying species, two hind-leg skimmer species, and eight six-leg skimmer species. *Source: Kramer and Marden [673]; inset image courtesy of D. Hilfert.*

been a better flier than one with the more flexible wings of a water surface skimmer. We can check the prediction that ancestral fliers would have had less well supported wings than modern species by finding the appropriate fossils and examining the number, thickness, and connections of a wing's internal veins. A beautifully fossilized 260-million-year-old stonefly wing is not as elaborately cross-veined as those of modern species [757].

With the advent of powered flight on strengthened wings, mechanoreceptors in the thorax that once provided sensory information on the position of sailing wings relative to the thorax (useful information for controlling the wings so as to catch as much of the breeze as possible) could have evolved into sensors that monitored wing-flapping rate. When the ancestor of moths appeared on the scene, it presumably possessed wing position sensors. Over time, these cells gradually evolved into the ultrasound detectors that modern bat-detecting noctuid moths use to listen to their predators.

A moth ear that can sense the ultrasound of bats can have clear survival benefits, as Roeder's work suggests. However, ears that evolved because of their anti-bat advantages can also detect ultrasound coming from many other sources, including members of one's own species. Our whistling moth, *Hecatesia exultans*, is active during the day, so it does not have to contend with bat predators. Freed from the function of bat avoidance, the ear could take on other tasks, namely, the detection of ultrasonic signals produced as a consequence of wing beating [1161] (Table 2).

Table 2 Possible evolutionary sequence leading to the ability of the whistling moth *Hecatesia exultans* to hear its fellow moths' ultrasonic whistles

Structure	Function
1. Movable gill plates with associated mechanoreceptors in thorax	1. Promotion of gas exchange by gills in aquatic immature insects
2. Gill plates retained in adult stage	2. Used as sails for skimming across water surface by adult insects
3. Thoracic muscles flap skimming wings; associated mechanoreceptors in thorax	3. Powered skimming across water surface
4. Increased thoracic musculature	4. Powered flapping flight through the air
5. Modified thoracic mechanoreceptors	5. Monitoring of rapid wing movements for improved control of wing position or wingbeat rate
6. Some mechanoreceptors become sensitive to airborne vibrations	6. Detection of approaching predators
7. Nervous system capable of motor signals that cause moth to fly toward some ultrasonic signals	7. Detection of ultrasound used to locate signaling competitors and mates

Sensory Exploitation and the Origins of Signals

Whistling moths provide an example of what may be a common phenomenon: sensory abilities that had one function in the distant past may acquire a new function during evolution. Thus, it is conceivable that male whistling moths now use ultrasound to communicate with other whistling moths because the ancestor of *Hecatesia exultans* was subjected to bat predators, which made perception of ultrasound reproductively advantageous. Once the ultrasonic detectors were in place, their existence could have affected the payoffs to mutant individuals that happened to generate sounds with potential signal value. Those that happened to produce "signals" in the ultrasonic range around 30 kHz may have elicited responses from a receiver more readily than those that produced sounds of lower frequencies.

This biasing effect of existing perceptual mechanisms on the origins of communication systems has been labeled **sensory exploitation** [63, 196, 1033]. In one of the best-studied cases of this sort, Heather Proctor argued that modern courtship by male water mites began when males happened to "exploit" the predatory behavior of females waiting to ambush small aquatic invertebrates called copepods [959]. While the predatory female is in her attack position, the male vibrates his foreleg in front of her. She in turn often grabs him, using the same response that she uses to capture passing copepods. However, she releases the male unharmed. He turns around and places spermatophores—packets of sperm—near the female, which she picks up in her genital opening if she is receptive (Figure 14).

The apparent use of a predatory grab by the female in response to the male's "trembling" display suggested to Proctor that males mimicked the stimuli produced by copepod prey. Perhaps the female's reaction identified her as a potential mate and showed the male where to position his spermatophores to best effect (water mites cannot see). To test this hypothesis, Proctor predicted that swimming copepods would generate wave vibrations in the range (10–23 cycles per second) produced by trembling males. They do, producing vibrations between 8 and 45 cycles per second.

Furthermore, if males trigger the prey detection response of females, then unfed, hungry female water mites held in captivity should be more responsive to male signals than well-fed females. They are, providing further support for the contention that once the first ancestral male happened to use a trembling signal, the behavior and its hereditary basis spread because it activated a preexisting prey detection mechanism in females and so was especially effective in communicating with them [959].

Among the water mites, trembling courtship occurs only in a few species, namely, those in which females adopt the prey ambush position, called the net stance, shown in Figure 15. By measuring a large number of characteristics in several species of water mites, Proctor produced a cladogram of the possible evolution-

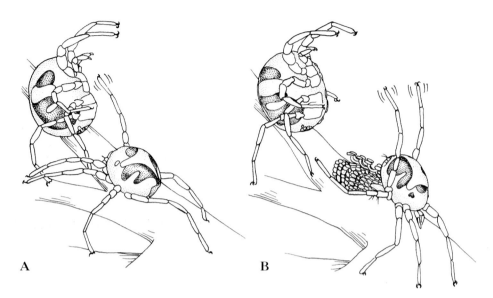

14 **Sensory exploitation** may have contributed to the evolution of a courtship signal in the water mite *Neumania papillator.* (A) The female, on the left, is in her prey-catching position (the net stance). The male approaches and waves a foreleg in front of her, setting up water vibrations similar to those a copepod might make. The female may respond by grabbing him, but releases him unharmed. (B) The male then deposits spermatophores on the aquatic vegetation in front of the female before leg waving over the spermatophores. *Source: Proctor [959].*

ary relationships among them. The cladogram indicates that the net stance of females originated in an ancestral water mite that eventually gave rise to eight descendant species. Within this lineage, male trembling subsequently appeared twice, once in the ancestor of the genus *Unionicola* and once in the ancestor that split into two species of *Neumania* (Figure 15). If this interpretation is correct, it would mean that male courtship trembling originated after females had adopted a copepod-ambushing lifestyle and had evolved a sensitivity to underwater vibrations of particular frequencies [960]. In this sense, courting males that first happened to mimic those frequencies were "exploiting" a preexisting female sensory mechanism and thereby improving their chances of mating.

Sensory Preferences May Precede the Appearance of a Preferred Signal

Sensory exploitation may be involved in the origins of many other communication systems. For example, the ability to detect waves in water may have originally assisted water striders in locating insect prey struggling on the water surface, but once in place, these same perceptual systems may have also been productively engaged in receiving signals from fellow water striders [1280]. Like-

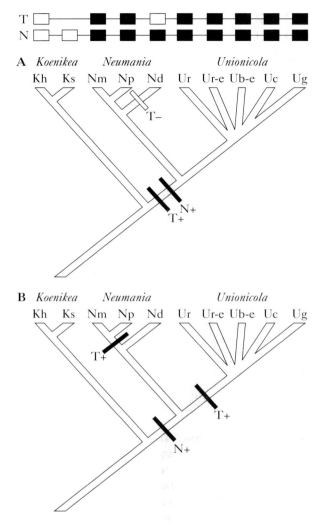

15 Two evolutionary scenarios for the evolution of male courtship trembling in water mites of the genera *Koenikea* (two species), *Neumania* (three species) and *Unionicola* (five species). (A) Male courtship trembling (T) may have evolved at the same time that predatory females adopted the net stance (N) in the ancestor of *Neumania* and *Unionicola* (N+, T+), with trembling being lost in the line leading to one species of *Neumania* (T–). (B) Another equally parsimonious scenario calls for the evolution of the net stance in the ancestor of the two genera *Neumania* and *Unionicola*, but instead of one origin for trembling, there might have been two origins, one in the ancestor of *Unionicola* and another in the ancestor of *Neumania papillator* and one close relative. *Source: Proctor [960].*

wise, many moths and other flower-visiting insects use scents to attract mates. In some instances, the sex pheromones of insects have a strong floral odor, suggesting that they may have had their origins as false flower scents given off by one sex to attract hungry foragers of the other sex [196]. This possibility is supported by the finding that the sex pheromones of male cabbage looper moths, which are designed to attract females, also incidentally attract more starved than unstarved *males* [697].

Sensory exploitation of another sort may have been involved in the evolutionary history of one of the most unusual of all communication systems, the pheromone system of male euglossine bees. These tropical bees collect volatile oils from orchid flowers, storing the plant-produced fragrances in their hind legs for use when displaying at mating territories far from the orchids. Klaus Lunau believes that this system may have evolved because some orchids happened to mimic the nests of orchid bees, which males of ancestral euglossines may have visited to search for emerging receptive females. By offering the same visual and chemical cues as bee nests (Figure 16), the orchids may have deceived males into landing upon the flowers, thereby participating in their pollination. If, however, odors from the orchid adhered to male bodies and made these individuals more attractive to receptive females, possibly by exploiting their floral odor receptors, the stage would have been set for a series of modifications that produced the elaborate current signal system of modern orchid bees [741].

Still other studies of sensory exploitation have been done with certain tropical frogs and fishes. Several researchers have asked whether a preference for a particular signal may have been present before the first individual to use the signal appeared on the scene. For example, males of the túngara frog (*Physalaemus pustulosus*) of Panama produce whining calls that may or may not be followed by one or more "chucks" (Figure 17). Females prefer males that add "chucks" to their calls, moving toward these males for mating while ignoring rivals that are only whining [1031].

Males of several close relatives of the túngara frog *never* produce chucks. Yet when females of *Physalaemus coloradorum*, one of the non-chucking species, are exposed to various artificial frog calls in laboratory arenas, they move toward taped

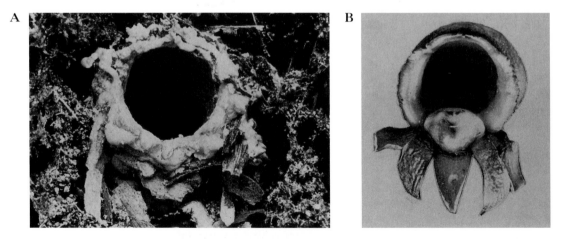

A   B

16 Sensory exploitation by an orchid? Males of the orchid bee *Eulaema longipennis* visit nests of their species (A) in search of females. Flowers of some members of the orchid genus *Catasetum* (B) look remarkably like bee nests, and may smell like them too. Males pollinate these flowers when they enter the "nests." *Photographs by Stefan Vogel, courtesy of K. Lunau.*

A

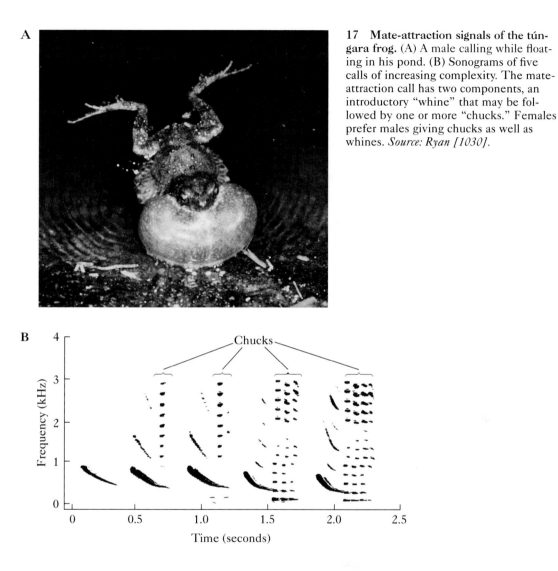

17   Mate-attraction signals of the túngara frog. (A) A male calling while floating in his pond. (B) Sonograms of five calls of increasing complexity. The mate-attraction call has two components, an introductory "whine" that may be followed by one or more "chucks." Females prefer males giving chucks as well as whines. *Source: Ryan [1030].*

B

calls of their own species that have a chuck added to them rather than taped calls without chucks. This result suggests that a preference for chucks arose in the common ancestor of *Physalaemus* frogs *before* chucks originated [652]. When the first mutant male of *P. pustulosus* added a chuck to his call, he apparently tapped into a preexisting sensory bias in females that had evolved for some other reason. Any of a wide range of acoustical stimuli might have had the same effect on females, judging from the discovery that female túngara frogs are as likely to approach a tape of a whine plus an unstructured "white noise" as one of a whine followed

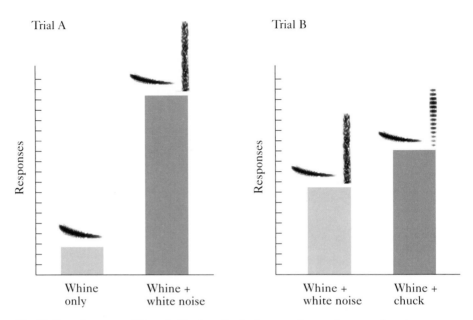

Trial A

Trial B

Responses

Responses

Whine
only

Whine +
white noise

Whine +
white noise

Whine +
chuck

18 Call preferences of female túngara frogs. In experimental arenas, females are more likely to approach speakers producing calls with two components than with one (trial A). It does not matter, however, whether the second component is a "chuck" (which male frogs produce) or a burst of white noise (which male frogs do not produce; trial B). *Source: Rand et al. [976].*

by a chuck (Figure 18) [976]. However, the original female preference for the whine-and-chuck signal yielded reproductive gains that helped propel the mutant gene for this kind of call through the species.

Michael Ryan and Stanley Rand conclude that females in this cluster of frog species acquired a preference for certain kinds of sounds, in this case low-frequency stimuli, prior to the appearance of signals that incorporated low-frequency sounds. They argue that the preexisting design features of a sensory system can affect the evolution of communication signals [1035]. This principle finds further support from studies of the fish genus *Xiphophorus*, in which males of some species have long tails while others do not (Figure 19) [64].

Females of the swordtail species *X. helleri* prefer males with relatively long tails, but so do females of the platyfish species *X. maculatus*, whose males do not have an extended caudal fin, as Alexandra Basolo demonstrated by surgically attaching yellow plastic "swordtails" to the tails of male platyfish. Females of *X. maculatus* consistently spent more time close to an aquarium compartment where they could see a male with an artificial yellow swordtail than near another compartment that contained a male whose tail had been amended with a piece of clear plastic [63].

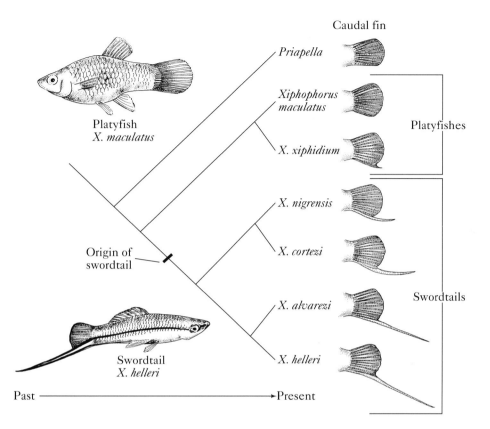

Caudal fin

Priapella

Xiphophorus
maculatus

Platyfishes

X. xiphidium

X. nigrensis

X. cortezi

Swordtails

X. alvarezi

X. helleri

Platyfish
X. maculatus

Origin of
swordtail

Swordtail
X. helleri

Past ——————————————→ Present

19 Sensory exploitation and swordtail phylogeny. The genus *Xiphophorus* includes the swordtails, which have elongated caudal fins, and the platyfishes, a group without tail ornaments. Because the closest relatives of the platyfishes and swordtails belong to a genus *(Priapella)* in which males lack long tails, the ancestor of *Xiphophorus* probably also lacked the elaborate tail. The long tail apparently originated in the evolutionary lineage that had diverged from the platyfish line. Even so, females of the platyfish *X. maculatus* find males of their species with experimentally lengthened tails more attractive, suggesting that they possess a sensory bias in favor of these kinds of tails. *Source: Basolo [64].*

According to the cladogram shown in Figure 19, the ancestor of the platyfishes preceded the species that gave rise to the swordtails. If the cladogram correctly reflects evolutionary relationships (and remember that a cladogram is only as good as the assumptions that underlie it), the ancestor of both the swordless platyfishes and the swordtails must have lacked a long tail, but possessed a female preference for this trait. Modern female platyfishes, according to this view, have retained the ancestral preference, even though males of these species did not evolve long tails. But when the first "swordtail" happened to have an enlarged tail, this "sworded" male enjoyed a reproductive advantage because females already preferred this phenotype.

But what if the cladogram is wrong? What if the platyfishes are actually derived from a swordtail ancestor, but platyfish males have lost the elongated tail that their ancestor once possessed? If so, then the preference of female platyfishes for males with long tails could have arisen in the ancestral species at a time when their mates had this trait; their platyfish descendants could have simply held onto the female preference for long tails that they inherited from their long-extinct ancestors.

Since some researchers believe that the platyfishes did indeed arise from an ancestor with a swordtail, Basolo decided to check a close relative of the genus *Xiphophorus*, namely, the fish *Priapella olmecae*, in which males have short caudal fins only. Here, too, Basolo found that females spent more time close to males with an artificial yellow sword; furthermore, the longer the sword, the greater the preference for the "sworded" male. From these results (Figure 20), Basolo concluded that a female preference for long tails evolved in an ancestor of *Priapella* before a long tail evolved in a more recent fish species, one that gave rise to the *Xiphophorus* lineage [65].

Why female fish whose males lack swords have a mating preference for elongated tails remains mysterious. However, Kern Reeve and Paul Sherman have suggested that the preference may not be for swords per se, but for any stimulus associated with large and healthy males [982]. An attached colored sword makes a male look bigger, and a receptive female in nature may gain by preferring large partners (see Chapter 12). If this scenario is correct, then in those lineages in which swords evolved, the original mutant males were not "exploiting" an *arbitrary* sensory oddity, but rather were tapping into an adaptive component of a sensory system that could provide females with useful information about mate size,

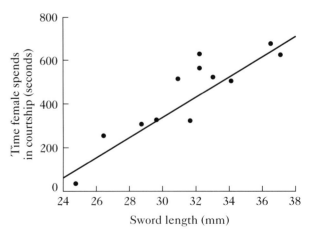

20 Female preference for males with swordtails is proportional to the length of the tail in experiments with *Priapella*, a swordless relative of the platyfish. *Source: Basolo [65].*

and thus mate quality. Note also that if it had been reproductively disadvantageous for female swordtails to respond positively to males with elongated caudal fins, those females that happened to have sensory equipment that could not be "exploited" in this manner would have had higher fitness, and the genes for less exploitable mate preferences would have spread through the population as a result of natural selection.

The sensory exploitation hypothesis represents a special case of the historical argument that what has already evolved has much to say about what additional changes are possible and which are not. If natural selection were responsible for the design of a jet plane, the first jet would have been a highly modified propeller-driven plane that had changed piece by piece, flying better after each change [279]. This is how evolutionary transitions must occur in nature. Stephen J. Gould labels this jury-rigged nature of evolutionary change "the panda principle," in honor of the panda's thumb, which is not a "real" finger at all, but instead is a highly modified wrist bone [455]. Pandas evolved from carnivorous ancestors whose first digit had become an integral part of a foot used in running. Thus, the first digit was not available for use as a thumb in stripping leaves from bamboo shoots when pandas became herbivorous bamboo eaters. Instead, selection acted on variation in the wrist bone, which is now exceedingly thumblike in both structure and function.

The panda principle can be seen in dozens of cases (see Figure 21 and [312]). Consider the persistence of sexual behavior in parthenogenetic whiptail lizards. These species are composed entirely of females, and yet if a female is courted and mounted by another female (and females do engage in such pseudomale sexual behavior, for reasons that are not fully understood), she is much more likely to produce a clutch of eggs than if she does not receive sexual stimulation from a part-

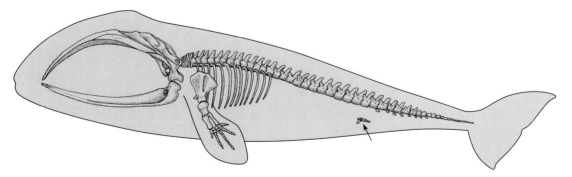

21 The "panda principle." The characteristics of a species' ancestors are reflected in its current characters. The skeletons of some whales include a vestigial pelvis (arrow). The hindlimbs have been lost altogether, but the forelimb bones in the flipper of the whale clearly reveal the five "fingers" of its terrestrial ancestor. Thus, whales had terrestrial ancestors that walked on four legs.

22 Sexual behavior in an asexual whiptail lizard. On the left, a male of a sexual species engages in courtship and copulatory behavior with a female. On the right, two females of a closely related parthenogenetic species engage in very similar behavior. *Source: Crews [238].*

ner (Figure 22) [244]. The relationship between courtship and female fecundity in unisexual lizards obviously exists because these reptiles had sexual ancestors. The parthenogenetic females retain characteristics, such as a need for courtship, that their nonparthenogenetic ancestors possessed, characteristics that a biological engineer would probably eliminate if he or she could play God in designing a new all-female species.

Natural selection cannot act like an omnipotent designer because it is a blind process: it has no goal in mind, and it has no means to get to any predetermined endpoint. Natural selection causes change when individuals differ in their hereditary attributes and some phenotypes secure higher reproductive success than others. Nothing else is needed for the spread of certain traits and the disappearance of others, which create a history of change for the species in which these events take place.

SUMMARY

1. We can ask two fundamental ultimate questions about behavior: What sequence of events took place over time that led to the behavior that we observe today? Why did this set of changes occur? The first question concerns the history or pattern of evolution, while the second deals with the causal processes that make changes happen.

2. The history of a behavioral trait can sometimes be reconstructed using the comparative method. If a cluster of related species exists, one can sometimes trace a pathway leading to a specific behavior in one of these species by looking for plausible transitional forms of the trait retained in the other living species. Hypotheses about behavioral history derived in this manner can be tested against cladograms developed independently of the behavioral data.

3. The probability that a complex, integrated trait could arise in one or a few large mutational changes is low. Therefore, the evolution of complex modern traits probably required many steps, with small changes gradually accumulating over time. The effect of cumulative selection over long periods of time is to produce major alterations in the original form of the evolving trait. When describing the evolution of a modern trait from an ancestral one, the pattern that minimizes the number and size of the transitions between the intermediate stages is usually considered most probable.

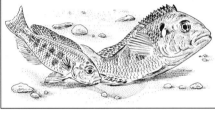

4. As traits change substantially in form over long periods of time, their functions often change as well. A trait that evolved for one purpose often becomes co-opted for another when a change serendipitously promotes individual reproductive success for a new reason. The wings of birds and insects, for example, had their origins as locomotory devices, but have more recently become involved in the communication systems of certain species.

5. Evolutionary changes must be compatible with what has already evolved. In this sense, the traits already in place constrain or bias the pattern of evolution. Thus, the sensory systems that have already evolved in a species make some signals much more likely to be detected by receivers than others. Therefore, mutant signalers that tap into preexisting sensory biases are more likely to cause evolutionary change in the communication system of their species than those whose signals are not readily perceived by other individuals.

SUGGESTED READING

The weird and wonderful spotted hyena communication system has been studied by two teams of researchers [322, 405]. Two papers on *Archaeopteryx*, one by John Ruben [1026] and the other by Alan Feduccia [383], show how hypotheses on the flight pattern of this ancient bird have been tested. The whole story is reviewed by Pat Shipman [1094]. James Marden tells the exciting evolutionary story of insect flight [757]. Stephen J. Gould discusses the panda principle in [455], and Richard Dawkins makes the power of cumulative selection clear in his great book, *The Blind Watchmaker* [280]. Technical approaches to historical methodology appear in [133, 775].

DISCUSSION QUESTIONS

1. Humans are unusual primates in moving about on two legs, rather than using all four of their limbs. What would you need to know in order to describe the sequence of changes that may have taken place over evolution as quadrupedal locomotion evolved into our bipedal system? What would you need to know in order to explain why some changes occurred and not others? You may wish at some point to refer to [1210, 1266, 1267] for views on these two levels of ultimate analysis.

2. Creationists frequently point to the complexity and apparent perfection of current traits, such as the human eye, arguing that existing attributes cannot have evolved by natural selection, given that it is so improbable that such beautifully designed traits could arise by chance. What is your response to claims of this sort?

3. In studying the courtship behavior of the empid fly *Hilara sartor*, E. L. Kessel was amazed to find a species in which males gathered together to hover in swarms, carrying empty silken balloons [644]. Females flew to the swarm, approached a male, and received a balloon, which they held while mating occurred. In the overwhelming majority of fly species, including some other empid flies, courtship does not involve the transfer of any object from male to female. But in addition to (1) the empty balloon gifts of *H. sartor* and (2) the "no courtship gifts" of some empids, males of other species in the group courted by offering (3) gifts of an edible food item (a freshly killed dead insect), (4) gifts of a dried insect fragment wrapped in a silken covering, or (5) gifts of a food item wrapped in silk. Construct a behavioral cladogram that minimizes the transitions from a possible ancestral pattern to the empty balloon gift-giving behavior. How would you test your hypothesis? (A recent comparative discussion of empid courtship behavior appears in [249].)

4. Females of an African cichlid fish pick up their large orange eggs almost as quickly as they lay them in depressions in the lake bottom made by males [1272]. The females brood the eggs in their mouths. As a female lays her eggs, the male who made the "nest" may move in front of her and spread his anal fin (see illustration on page 249 and chapter-opening page). The fin has a line of large orange spots. The female moves toward him and attempts to pick up the objects on the fin. As she does, the male releases his sperm, which are taken up in the female's mouth, where they fertilize her eggs. Use the theory of sensory exploitation to explain the evolutionary origins of the male's behavior.

5. If the cladogram of the Ensifera (crickets, katydids, and seven other related families) shown below is correct [489], how many times has acoustical signaling evolved in this group?

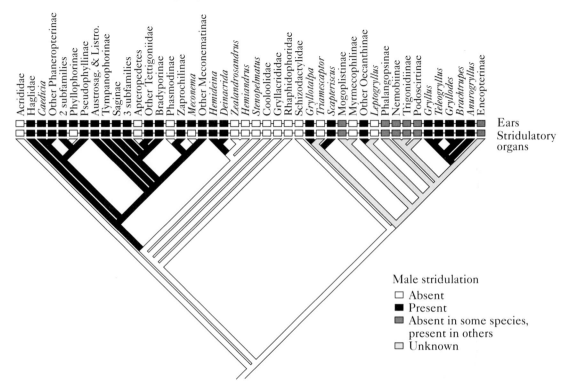

Ears
Stridulatory organs

Male stridulation
☐ Absent
■ Present
■ Absent in some species, present in others
☐ Unknown

CHAPTER 8

The Evolution of Communication: Adaptation in Signalers and Receivers

EVOLUTIONARY BIOLOGISTS DEAL WITH two complementary questions: (1) How did a current trait originate and change over time, and (2) What *processes* caused the trait to change the way it did? This chapter focuses on the second question, especially on what is arguably the single most important process underlying evolution—namely, natural selection. When natural selection occurs, a modified trait that confers higher fitness replaces an older version that now yields lower fitness. Understanding this point enables biologists to ask a key question about *any* trait: if the characteristic evolved, or is evolving, by natural selection, why did it spread through the species in the past, or why is it being maintained in the population today? What is the trait "good for"? Why did the first individuals that happened to have this attribute transmit more copies of their genes than individuals with alternative characteristics? Questions of this sort differ from those about the historical pattern of changes during evolution. Instead of trying to describe history, students of adaptation explore the fitness benefits—and costs—of a particular trait. Developing plausible reasons why a behavior pattern might have benefits greater than its costs is the first challenge. Even more challenging and entertaining is the task of testing the alternatives and winnowing out wrong explanations in a rigorous and convincing manner. This chapter presents some examples of research on animal communication that accomplishes these goals. We will start with some key features of bird communication and ask, Are these aspects of bird song the adaptive products of natural selection?

Questions about Adaptation and Signal Givers

Earlier, we asked three questions about bird song: (1) Why do species differ in their songs? (2) Why is it that males of some species sing while females do not, or do so only rarely? (3) Why do members of the same species sometimes sing different dialects? In Chapter 2, each of these questions received a proximate answer in terms of the genetic, hormonal, neuronal, and experiential differences among individuals. Proximate answers, however, do not cover the whole story, as we shall now demonstrate by examining the same three questions in ultimate terms.

Why Do Different Bird Species Sing Different Songs?

We begin by analyzing why bird species sing different songs, a fact that enables bird-watchers to identify most species of birds by their calls alone. To deal with this topic, we must first review a widely accepted theory on how new species arise. The theory of *allopatric speciation* argues that new species form when populations become separated from each other and then undergo genetic divergence in geographic isolation. This divergence could be a result of (1) various purely random events, including the occurrence of different mutations in the two populations and the accidental loss of genetic variation, which is especially likely in the small groups that may found new isolated populations, and (2) nonrandom natural selection for features suited to the particular environment in which each population finds itself. Given sufficient time in isolation from each other, large numbers of unselected and selected genetic changes could accumulate in the two populations. If the two groups then expanded their ranges and came into contact with each other, they might not interbreed because of the genetic changes that occurred while they were apart; some of these changes might have affected their mate recognition systems so much that interbreeding would not occur [783]. Since species are typically defined as populations of interbreeding individuals, these populations would now be classified as two distinct species.

However, when members of two such populations come together again, they may not be totally reproductively isolated. If "hybridization" occurred, and if the offspring of the hybrids were as fit on average as any other offspring, any genetic and behavioral differences between what were at the outset somewhat genetically differentiated populations would break down, and the two populations would merge into one species.

If, however, hybrids experienced reduced fitness in the zone of contact, then natural selection could act to reinforce any mechanisms that produced incipient reproductive isolation between the two populations. In this case, natural selection acting on sexual interactions across populations could lead to changes in the signals that males and females used to make mating decisions, favoring those individuals that avoided members of the other population. If new barriers to gene exchange spread through the two populations, they could become two distinct species, in part because of direct selection for reproductive isolating mechanisms.

Birds may indeed recognize members of their own species, because receptive females react very differently to songs of their own species than to songs of other species (Figure 1) [1071]. The question is, How did the songs of, say, yellow war-

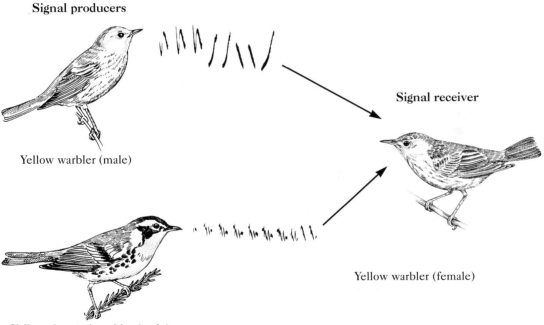

Signal producers

Yellow warbler (male)

Signal receiver

Yellow-throated warbler (male)

Yellow warbler (female)

1 Bird songs are species-specific. The songs of the yellow warbler and the yellow-throated warbler are different, and only the song of the male yellow warbler attracts sexually receptive female yellow warblers. But did the differences between the songs of yellow warblers and yellow-throated warblers evolve to help female listeners identify the species membership of the singer?

blers and yellow-throated warblers get to be different? On the one hand, it is conceivable that natural selection had nothing to do with the accumulation of changes that occurred randomly as one lineage split into the two that eventually gave rise to these two modern species. On the other hand, natural selection may have been directly involved in shaping the songs of the two species. During the time the populations were geographically separated, selection on male song may have differed in the two regions because (1) females differed in their mate preferences or (2) some types of songs transmitted information better than others in particular acoustical environments [431, 837]. The song frequencies used by great tits, for example, vary across Europe, North Africa, and the Middle East in a manner correlated with differences in forest density over the bird's range [584]. Variation in call type in response to open versus forested habitats has been found in frogs as well [1037], suggesting that whenever the acoustically communicating ancestors of once isolated populations first come into contact again, they may already be singing quite different songs. Indeed, many closely related species that live far apart sing highly distinctive songs, convincing evidence that geographic overlap among species is *not* necessary to produce species-specific songs. Thus, for example, yellow-throated warblers and Grace's warblers live many hundreds of miles apart, and yet these two members of the genus *Dendroica* sing highly distinctive songs.

But even though divergent song types can apparently evolve entirely in geographic isolation, natural selection might also act directly against hybridizing individuals in zones of contact between closely related populations by favoring females that preferred males of their own group with the most distinctive songs. If this has happened, then the greater the risk of maladaptive hybridization, the more distinctive the songs of the current species should be. So, when two warbler species have partially overlapping ranges, the songs of the species should diverge more in the zone of overlap than in areas in which only one of the two species occurs.

This prediction cannot be tested with respect to yellow-throated and yellow warblers, whose ranges overlap completely. However, blue-winged and golden-winged warblers coexist in only part of their collective range, and they do sometimes hybridize in these regions. Frank Gill and Bertram Murray report that in the area of overlap, the songs of the two species become less variable, and thus more distinctive [439]. As a result, the potential for confusion between the species is presumably somewhat reduced in the very places where hybridization can occur. On the other hand, the songs of the two warblers are highly distinctive everywhere, so it is hard to imagine how golden-winged females could ever have mated with blue-winged males because they could not tell the difference between their songs and those of golden-winged warblers.

Moreover, two other closely related species, the indigo bunting and the lazuli bunting, do *not* sing especially differently in their zone of overlap in Nebraska. Indeed, where they live together, a male indigo bunting may learn and incorporate song elements from a lazuli bunting neighbor, and vice versa [356]. And yet hybrids between the two species occur only rarely, suggesting that females have no difficulty telling them apart. Because these two buntings have different plumage, females could rely on visual cues to avoid hybridization mistakes.

Most of the few other bird studies on vocalization and reproductive isolation also fail to support the hypothesis that song differences are the sole, or even primary, basis for the reproductive isolation of species [60, 187]. Thus, there is good reason to doubt that selection against hybrids has played an important role in the evolution of species-specific bird songs. It seems much more likely that the differences among the songs of closely related species arose largely or entirely during the periods when an evolving species was geographically isolated from other populations of shared recent ancestry. But determining whether these differences were due to chance events or selection for various adaptive properties of the songs will require more work. And here we will leave this topic, having demonstrated that it is possible to propose, test, and tentatively reject a hypothesis on the possible adaptive value of the difference between bird songs. Remember that rejecting a hypothesis is every bit as scientifically valuable as confirming a possible explanation.

Why Do Only Males Sing?

A second question about bird song is why males, but not females, sing in many bird species. Here again, one can imagine ultimate hypotheses that explain the differences between the sexes in terms of nonadaptive processes of evolution *or* in terms of their possible adaptive value. For example, it could be that females fail to sing as a nonadaptive by-product of the way their sexual development occurs. The hormones that regulate the development of sex differences could be so important to the functional development of critical female attributes, such as the female reproductive tract, that certain side effects of female hormones, such as the absence of an elaborate song system in the brain, are tolerated. According to this hypothesis, hormonal mutations in the past that conferred singing ability on females would have so disrupted normal ovarian development, or egg production, or maternal care that they were selected against, despite the possible benefits that females might have gained from being able to sing.

This hypothesis is something of a straw man, if you will forgive the pun, because I knew it to be incorrect before I presented it. If the development of a song system really is incompatible with the development of female reproductive capability, then there should be no bird species in which females sing. And yet females of many species, especially those living in the tropics, sing elaborate and complex songs [56, 218], suggesting that there are no insuperable developmental constraints on the production of song systems in female bird brains. I present this nonadaptationist hypothesis here merely to illustrate what is meant by an ultimate explanation of this sort.

Moving on to adaptationist possibilities, perhaps the fitness benefits from singing go primarily to males, not to females, which is why genes that specifically promote male singing ability have spread through certain species. Because male songbirds regularly crank out several thousand full-throated songs a day [823], a costly activity in terms of time expenditure alone, males must derive great benefits from their acoustical signaling. One such benefit could be the attraction of females, which require information (a) about the species membership of a singer or (b) about the quality of a potential mate, or his holdings, relative to other males in the population. However, singing males may also gain by communicating with rival males, in which case their songs could contain information about their fighting ability, warning other males to stay away from (a) an occupied territory or (b) a fertile mate.

Although the various alternatives have been tested surprisingly few times [678], they produce testable predictions. For example, the mate attraction hypothesis generates the prediction that males of monogamous species should sing loudly and often before acquiring a mate, but stop afterward, in contrast to polygynous species, in which males should try to attract several mates in sequence [332]. Males of the monogamous California towhee conform to this expectation by ceasing to sing once they have paired off with a female [774, 969], and males of the polygynous house wren that have attracted one mate stop singing only temporarily, and soon resume their noisy efforts to draw in another partner (Figure 2).

The mate attraction hypothesis also predicts that females will approach taped song. Playback studies have shown that females of house wrens [616] and some other birds [365, 843] are more likely to visit nest boxes from which recorded songs of their species are being broadcast than "silent" nest boxes (although stronger evidence would have been provided if a group of nest boxes had been advertised by tapes of another species' song). Moreover, statistical analysis demonstrates that the only significant factor influencing how rapidly male starlings pair with females is the size of the male's song repertoire; that is, the number of different phrase types or distinctive song elements that he incorporates into his songs (Figure 3). Since other attributes of males and their territories were not correlated with speed of pairing, James Mountjoy and

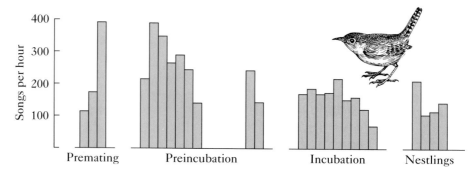

2 Bird song as a signal to females. Males of the polygynous house wren sing loudly to attract mates. Once they have succeeded in acquiring a partner, they cease conspicuous singing—until that female is busy incubating, after which they resume singing in an attempt to attract a second mate. *Source: Johnson and Kermott [615].*

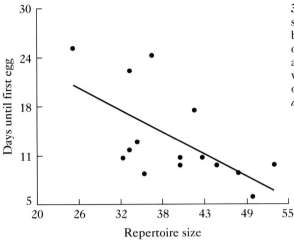

3 Female starlings prefer complex songs. The speed with which a female pairs with a male and begins to lay eggs is proportional to the number of song phrases in the male's repertoire. The y-axis shows the number of days from the date on which the male claimed a nest box to the date on which the first egg was laid. *Source: Mountjoy and Lemon [844].*

Robert Lemon concluded that song quality is what attracts females of this species [844], providing more support for the mate attraction hypothesis.

The rival repulsion hypothesis could explain why some monogamous males continue to sing steadily even after attracting their only mate [823]. If these males are singing to repel other males, then playing tapes of the songs of species X should keep males of species X away from the loudspeaker. In experiments in which male white-throated sparrows were removed from their territories and replaced by speakers broadcasting their songs, new males were slower to come into these territories than into others from which males were removed but not replaced with taped song (Figure 4) [376]. The neighbors of an established territorial male probably need only hear him in order to avoid him [1149]. Thus, male song could function to help established territory owners maintain their territories by keeping their

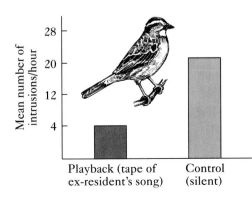

4 Does male song repel territorial intruders in the white-throated sparrow? Territories from which resident males were experimentally removed attracted fewer intruders when the taped song of the removed males was broadcast from the core of the vacant territories. *Source: Falls [376].*

neighbors at a distance, and perhaps also deterring other intruders, even competitors of other species [55].

Additional evidence in favor of this conclusion comes from a study of the ochre-bellied flycatcher. David Westcott used a simple surgical treatment to mute six territory-holding males [1258]. Having measured the frequency with which these males had repelled intruders before the operation, he compared it with the intrusion rate after they had been silenced. Territorial invasions shot up on the muted males' territories (Figure 5), and five of the six males eventually lost their territories to singing rivals.

Although singing may help males of some bird species to establish and maintain their territories, it may also have another allied function—namely, to keep rival males away from the territory owner's female(s) during the time when they are fertile [747, 823]. On these days, intruders could potentially "steal" egg fertilizations from the resident male by mating with the females, reducing the resident's reproductive success. But why would intruders stay away if there was a chance to fertilize another male's partner?

Anders Møller has argued that the frequency or quality of a resident male's song could convey information to other males about his capacity to guard his mate(s) effectively, giving receivers reason to avoid effective guarders, and their mates. The mate guarding hypothesis generates the predictions that (1) song rate will peak when females are *most* fertile, rather than during territory establishment and initial pair formation; (2) the better the physical condition of the male, the more songs he will produce; and (3) visits by intruders will be a function of the frequency (or quality) of the song of the resident male. Tests of these three predictions have produced the following results:

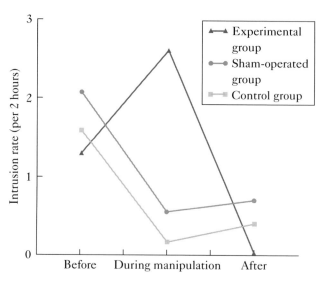

5 Male song deters rivals in the ochre-bellied flycatcher. Mean rates of intrusion on territories during observation periods before, during, and after an operation was performed on some territory holders that made them unable to sing. The graph also shows intrusion rates for the territories of males that received a sham operation, but were not muted, and control males. Some of the muted males regained the ability to sing, at which time they regained the ability to deter intruders. *Source: Westcott [1258].*

1. In some bird species, but not all [910, 1006], males do sing most frequently when their mates are most fertile (Figure 6), with singing declining thereafter [747].

2. As predicted, male European blackbirds that received a "CARE package" of "plums, mealworms, bread and pastries" on some evenings started singing sooner the following morning, and kept it up longer, than on days when they were not treated to a free meal [252].

3. The third prediction from the mate guarding hypothesis, however, is not consistent with another of Mountjoy and Lemon's discoveries—that tapes of starling songs played at nest boxes actually *attracted* male starlings to the loudspeaker, presumably to check out an unfamiliar intruder or to evaluate the resources that the "male" was defending [843]. But Mountjoy and Lemon also found that more males visited boxes where "simple" starling songs were played than boxes where a tape player belted out "complex" songs with many phrase types. If, as they suspect, only older, superior competitors can produce complex songs regularly, then these data provide some support for the hypothesis that singing by mate-guarding males affects the probability of direct challenges from opponents.

In summary, persons using natural selection theory have been able to identify possible benefits for singing by male songbirds, and they have tested some of their ideas. The resultant research has produced evidence that singing actually does help males of some species attract mates, while in some other species singing keeps fellow males out of the songster's territory or away from his mate. The current fitness benefits of male singing may help maintain the trait in some species, and perhaps the same benefits helped it spread through those species in the past.

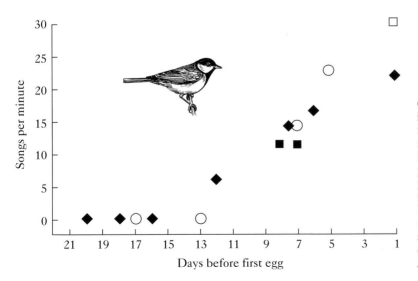

6 Song and mate-guarding in the great tit. The frequency with which males sing increases during the days when their partners are most fertile, perhaps as a warning to rival males that they will be attacked if they come near their mates. (Each symbol represents a different pair of birds.) *Source: Mace [747].*

Are Dialects Adaptive?

Our final question about bird song is, why do male white-crowned sparrows, among other species, sing different versions, or dialects, of their species' song? At the prox- imate level, dialects arise in part because males learn their songs by listening to other singers of their species (see Chapter 2). If a young male sparrow living around Sunset Beach, California, hears a different variant of white-crowned sparrow song than a male that happens to live in Berkeley, then their learning mechanisms will be exposed to different information, and the young males may sing different songs when they become adults.

But what about the evolutionary basis of dialects? As is true of almost every con- ceivable biological phenomenon, both nonselectionist and selectionist hypothe- ses are available. On the one hand, perhaps *differences* in dialects are not adaptive in and of themselves, but emerge as an incidental effect of song learning mecha- nisms that evolved for some other reason. Once learning plays a role in song acqui- sition, then a special kind of nongenetic historical accident can have a lot to say about what birds in a particular area will sing [123]. Imagine a region that is colo- nized by one or a few males, perhaps juveniles with incompletely formed full songs [1175]. If the population grows, young males born in the area will tend to imitate the idiosyncratic song of the older established males closest to them, forming an "island" of males singing the same distinctive song type. Alternatively, as the first youngsters to be born in a new area disperse to unoccupied habitat, they will lack acoustical models to imitate, and so will produce their own somewhat atypi- cal song types. Individuals of this sort will eventually serve as models for others to copy, producing dialects through a form of cultural transmission based on histori- cal accident. According to this nonadaptationist hypothesis, although an interest- ing history lies behind the divergence of song types sung in different areas by white-crowned sparrows, natural selection did not directly favor birds that hap- pened to sing distinctive dialects [27, 608].

A key prediction follows from the hypothesis that differences in dialects have no adaptive value: if males in area B were to sing dialect A, they should suffer no reproductive disadvantage. We can test this prediction by looking at the song pref- erences of female cowbirds from two different subspecies living in different parts of North America. Females of subspecies A find their males' song type more sex- ually stimulating than the songs of subspecies B, judging from the frequency with which they adopt the precopulatory position in response to tapes of these songs [649, 1255] (see Chapter 2). Because females prefer their own subspecies' dialect, males living with them gain by singing that dialect, evidence against the hypoth- esis that selection has not been involved in dialect evolution.

If dialects are also adaptive for white-crowned sparrows, then male white-crowns able to learn a dialect ought to gain higher reproductive success on average than males unable to acquire a particular dialect. Ornithologists have come up with sev- eral reasons why singing a specific dialect could be selectively advantageous for males. One hypothesis goes like this: In a species divided into stable subpopulations, males in population X have genes that have survived natural selection in an area occupied for generations by those males' ancestors. By learning to sing the dialect associated with their place of birth, males announce their possession of traits (and underlying genes) well adapted for area X. Females born in area X gain by having a preference for males that sing the local dialect, because they will endow their offspring with genetic information that promotes the development of locally adapted characteristics.

Some evidence in support of this hypothesis has been presented [42, 869], but one of its key predictions has not been confirmed. If females attempt to select mates born in their natal area, then females should prefer males with the dialect that they heard while nestlings, namely, their father's dialect. Glen Chilton and his colleagues found no such preference in females at a Canadian site [193]. Furthermore, young male white-crowns are not locked into their natal dialects, but can change them [292] (see Chapter 2); therefore, females could not rely on a male's dialect to identify his birthplace.

Thus, we will move on to a very different adaptationist hypothesis about song dialects, namely, that they help males communicate more effectively with fellow males, rather than with females. This hypothesis yields the prediction that young males should be able to change their dialects to match those of their neighbors, and as just noted, they can, in the case of the white-crowned sparrow [57] and some other birds [912, 913].

More generally, the male–male communication hypothesis yields the prediction that when territorial males hear a fellow male singing, they will reply to that male with a song that promotes communication with that individual. This prediction has been tested with the song sparrow, a bird that has a repertoire of many different, distinctive song types. (White-crowned sparrows sing just one song type, with different populations having their own dialects. Song sparrows sing many song types, with less sharply defined dialects in different regions.) Michael Beecher and his colleagues found through playback experiments that when a male heard a tape of a neighbor's song, he tended to reply to that tape by singing a song from his own repertoire that matched one in the repertoire of that particular neighbor. In fact, if the subjects heard a neighbor's song that was *not* in their personal repertoire, so they could not simply imitate it, they still picked a song from their repertoire that matched one in that neighbor's song bank (Figure 7). Thus, song sparrows recognize their neighbors and know what songs they sing, and they use this information to shape their replies [73].

The ability to sing a song that matches a neighbor's dialect or belongs in his repertoire is one way in which a singing male can send a special message to his rival. The young white-crown that modifies his song dialect to resemble another male's song is communicating an awareness of that male as an established neighbor. If it is adaptive to discriminate between neighbors and strangers because strangers are more likely to mount a challenge for territory control or interfere with a mate [1171, 1317], then territorial males should react more strongly to playback

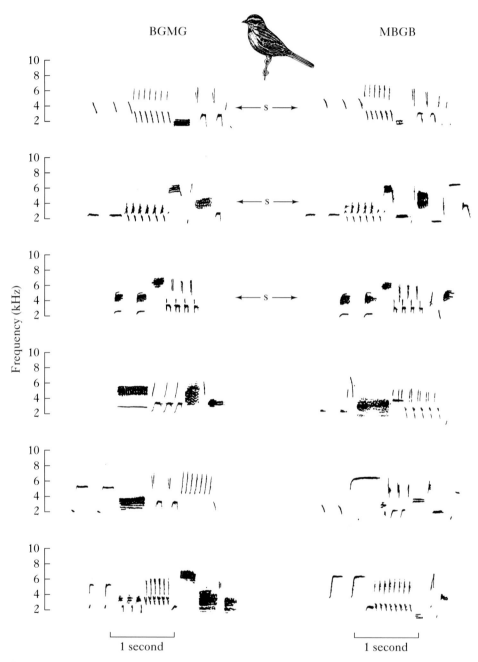

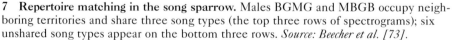

7 Repertoire matching in the song sparrow. Males BGMG and MBGB occupy neighboring territories and share three song types (the top three rows of spectrograms); six unshared song types appear on the bottom three rows. *Source: Beecher et al. [73].*

of a stranger's song than to that of a neighbor. Males of many bird species, including white-crowned sparrows [43], exhibit the predicted response, as do males of unrelated animals that are also territorial (Figure 8) [978]. Thus, we have good reason to conclude that signal differences among males of some species can have adaptive value.

A final demonstration of this point comes from an examination of how male chestnut-sided warblers acquire their songs. In this species, males sing two distinctively different song types, one of which seems designed to communicate with females, and the other, with males [165]. According to Donald Kroodsma, the "for-females" song is acquired in the first year of the male's life as a result of listening to social tutors in the youngster's neighborhood. This learning takes place quickly and results in a song that is not modified much thereafter, nor are there large differences among males with respect to this song type. In contrast, the "for-males" song takes longer to acquire, can be modified substantially in later years, and varies considerably from location to location. The differences in the proximate foundations of these two song types presumably reflect their different ultimate functions. In communicating with females, males have little to gain by providing different messages to different individuals. The goal instead is to provide the most attractive possible message in order to secure at least one breeding partner from a huge pool of potential female listeners. In communicating with other males, however, males can gain by directing their messages to a much smaller audience of immediate neighbors, an audience that can change from year to year. Signaling to this group selects for a different kind of song, and a different kind of proximate learning mechanism, than singing to attract a mate.

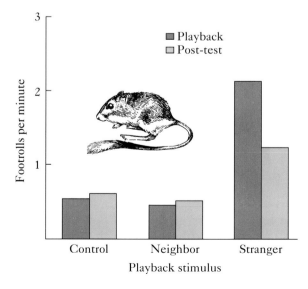

8 Response to foot-drumming signals of neighbors and strangers by banner-tailed kangaroo rats. The experimenter played recordings of the foot drummings of neighbors and strangers to residents in their burrows and noted the residents' responses. The control trial measured responses to a hammer striking the ground near the burrow. Post-test measurements were made during the 15 minute period immediately following a 5 minute playback test. *Source: Randall [978].*

The Meaning of Adaptation

Our review of studies on the ultimate causes of bird song has been founded on the assumption that all traits have a history, but that natural selection may—or may not—play a direct role in that history by favoring one particular trait over another. Even when differences in reproductive success among individuals in the past have been directly involved in the "selection" of a particular trait, we still have to figure out why the characteristic confers a reproductive advantage. Has bird song evolved to help individual males attract mates or to keep rivals away from their mates? As we have seen, each possibility requires testing, because any one could be right—or wrong.

Let us accept, for the moment, that male white-crowned sparrows currently sing to keep rivals at bay in a manner that promotes their reproductive success. The positive fitness effects of singing today do not necessarily demonstrate that the trait had the same effect in the past. Nor does the current function of a characteristic reveal for certain what it did for individuals, if anything, when it originated. The preceding chapter offered several examples of characteristics, such as insect wings and ears, whose original adaptive value was probably quite different from the current reproductive benefits that they provide. The general rule is that complex traits evolve by degrees, with the trait changing gradually over time in form, and often in function as well.

Many evolutionary biologists take a trait's history into account before calling it an **adaptation** [514, 701]. According to one popular definition, an adaptation is a characteristic that provides "current utility to the organism and [has] been generated historically through the action of natural selection for its current biological role" [67]. To apply this definition to a trait, we obviously need information about its origins and subsequent history. As Kern Reeve and Paul Sherman point out, however, this knowledge is missing for many interesting behavioral characteristics [982]. But we can still study the adaptive value of these traits by focusing strictly on their *current* fitness effects. If a trait is a current adaptation, as defined by Reeve and Sherman, it is spreading or being maintained by natural selection because it confers higher fitness than alternative forms of the trait. The process of selection occurring now does not differ from that in the past, depending as it does on hereditary differences in phenotypes that affect their relative reproductive success. If a trait that was spread or maintained by *past* natural selection is to be called an adaptation, why not give the same label to a trait that is *currently* being spread or maintained by natural selection?

No matter whether one insists on a historically based definition or uses one that considers only current effects, the trick for an **adaptationist** is to figure out why a particular attribute was, or is, selectively advantageous. Adaptationists have not been shy about coming up with ideas on the adaptive value of particular traits. You should know, however, that some evolutionary biologists, notably Stephen J. Gould and Richard Lewontin, believe that this approach is fatally flawed by what they ridicule as the Panglossian philosophy of adaptationists, who supposedly believe that all components of an organism are perfectly adapted products of natural selection [456, 459]. In contrast, Gould and Lewontin argue that most traits are imperfect for several reasons (Table 1), but especially because they are the constrained product of the process of evolutionary change, which requires that modifications be layered on what previously existed.

The preceding chapter presented Gould's panda principle, which helps us understand why some traits look as if they were invented by Rube Goldberg. Think of the parthenogenetic female lizards that require "sexual" interactions with other females in order to reproduce most effectively. Gould argues that historical constraints, and the constraints imposed by conflicting selection pressures, must often prevent a particular trait from achieving "perfection," in the special sense outlined above. If so, then adaptationists are wasting their time in analyzing every trait as if it were perfectly adaptive and directly produced by natural selection. To drive his point home, he claims that these researchers invent fables as absurd as the fictional "just-so stories" of Kipling, which contained inventive myths about the causes of the leopard's spots and the camel's hump.

Gould's objections have stimulated some sharp rebuttals [e.g., 145, 361, 784, 971], with much of the fuss revolving around the definition of adaptation. Gould claims that the adaptationists' research is founded on their notion that an "adaptation" is a characteristic made "perfect" by natural selection. But adaptationists have replied that since the 1966 publication of *Adaptation and Natural Selection* by G. C. Williams [1287], they have known that "better" traits, not "perfect" ones, are spread via natural selection. If an existing trait is being maintained by current natural selection, then it need only be superior to other possible alternatives, not the best of all possible imaginable traits.

Persons who accept this view have a powerful tool with which to identify puzzles worth solving. Although many characteristics appear patently adaptive, oth-

Table 1 Why maladaptive traits exist and why most evolved traits are "imperfect"

1. The trait evolved under conditions that no longer exist. Under current circumstances, the trait has a maladaptive effect; insufficient time or the absence of appropriate mutations has prevented the replacement of the maladaptive trait with a superior alternative [e.g., 491].

2. The trait develops as a maladaptive side effect of an otherwise adaptive proximate mechanism, i.e., one that generally causes an adaptive outcome [e.g., 256].

3. Gene flow from populations subject to different selection pressures prevents members of a local population from evolving the optimal trait for local conditions [e.g., 996].

4. The trait cannot be perfectly designed for one particular task, since it or its underlying structures are involved in more than one activity, requiring compromises in its suitability for any one task [e.g., 926].

5. The trait is the constrained product of previous selection, in which the already existing attributes of the organism limited what changes could occur in the past [e.g., 1247].

ers do not. What intrigues the adaptationist are cases in which individuals seem to lose, rather than gain, fitness as a result of one of their traits. These cases challenge the biologist to come up with a plausible hypothesis, or better still, several such hypotheses.

However, no adaptationist I know—and I know quite a few—can get away with simply producing a hypothesis, the "just-so story" so scorned by Gould. Scientific papers do not stop with the presentation of possible explanations, but continue on to show how these explanations were tested. Researchers who are able to test their ideas convincingly and reach defensible conclusions gain the attention and respect of their peers. And scientists like receiving attention and respect, with the result that peer pressure leads to scientific progress.

The Adaptationist Approach

Let's examine the utility of the adaptationist approach in the context of another study of animal communication. Bernd Heinrich has enjoyed watching ravens ever since he was a boy in Germany. When he moved to Maine, he continued to observe and listen to these big, intelligent birds, which have an amazing repertoire of sounds that includes quorks, quarks, queeks, yodels, warbles, pops, drummings, and thunks. They also occasionally yell, producing a song that Heinrich likens to the sound a dog makes when its tail has been caught in a slammed door.

One fall day in 1984, Heinrich found a mob of yelling ravens feeding together on a moose that a poacher had killed and hidden in the deep forest (Figure 9). He knew at once that he had an exciting evolutionary problem on his hands [529, 532].

9 **A yelling raven** calling still more juveniles to the carcass of a winter-killed moose that was once defended by a resident pair of adult ravens.

Ravens are uncommon birds in Maine, and yet somehow fifteen of them had assembled at a hidden carcass, almost certainly because some of the first birds to find the moose had yelled to other ravens. The behavior did not make sense to him. A yelling raven can be heard a mile away, but why attract competitors to a food bonanza? Why not stay silent and eat moose meat all winter long instead of sharing the bounty with a whole gang of ravenous ravens?

Had Heinrich found a behavior designed to help the species as a whole? He didn't think so, given the logical problems with group selection for species-benefiting traits (see Chapter 1). So, he asked himself whether the aggregated birds might be one big family, with parents helping offspring. No, a pair of ravens produces a maximum of six offspring per breeding season. Furthermore, DNA fingerprinting studies eventually confirmed that these flocks were composed of unrelated individuals [908].

Then just how could a yelling raven gain fitness from its actions? Heinrich wondered whether ravens yelled because the signals would get the attention of a bear or coyote, which could open up a tough-skinned moose. According to this hypothesis, a raven yells to attract large carnivores in order to be able to feed on the leavings after these animals have opened up a carcass.

To test his hypothesis, Heinrich hauled a dead, 150-pound goat through the Maine woods, taking it out to various sites during the day and storing the none-too-sweet-smelling carcass in his cabin at night to prevent its loss to a nocturnal scavenging bear. Ravens occasionally approached the decaying goat, though only after making Heinrich wait for hours on bitterly cold winter days. But counter to his prediction, the birds never yelled when they found the bait. Moreover, Heinrich sometimes observed ravens yelling at carcasses that had already been ripped open. These findings caused him to abandon the "attract-a-carcass-opener" hypothesis.

Instead, he switched his attention to an alternative explanation, the "dilution-of-risk-of-predation" hypothesis, an idea stimulated by seeing how cautiously some ravens approached carcasses when they first found them. Perhaps, he argued, a carcass discoverer yells to draw in others so that if a predator lurks nearby, the other birds would provide possible targets, reducing the yeller's risk of being taken by a hiding coyote or fox. The incoming birds would be attracted because they would gain a wealth of food in return for taking a small risk of feeding a predator.

However, this hypothesis generates the prediction that once a group has assembled and feeding has begun, the birds should shut up to avoid attracting more ravens, which would be unwanted and unnecessary for safety purposes. Heinrich continued to haul baits out into the woods over the winter, and on most occasions when he saw yelling birds at a carcass, the bait had already acquired a retinue of actively scavenging birds. This result convinced him to discard the "dilution-of-risk-of-predation" hypothesis.

As Heinrich carried on with his work, he came to realize that whenever he saw a single bird or a pair at a bait, these ravens were quiet. Yelling happened only

10   Yelling is a recruitment signal. Carcass baits were exploited either by nonyelling territorial singletons and pairs or by large groups of ravens, many of them yelling, most of them nonterritorial subordinates. The graph shows the percentage of days on which baits were visited by various numbers of birds. *Source: Heinrich [528].*

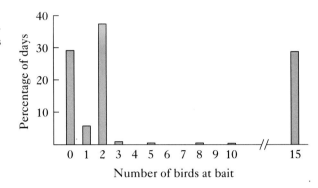

when three or more ravens were present, and it was then, and only then, that numbers of other ravens came to the site. Heinrich knew that older adult ravens form pairs that defend a territory year-round. Unmated young birds usually travel solo over great distances in search of food. If a singleton attempts to feed in a territory, the resident pair attacks. It occurred to Heinrich that yelling might be a signal given by nonterritorial intruders, a signal that attracts other unmated wanderers to a food bonanza that they can exploit if they can overwhelm the defenses of a resident pair.

This "gang-up-on-the-territorial-residents" hypothesis leads to a number of predictions: (1) resident territory owners should never yell, (2) *non*resident ravens should yell, (3) yelling should facilitate a mass assault on a carcass by nonresident ravens, (4) resident pairs should be unable to repel a communal assault on "their" resources, and (5) a food bonanza should be eaten either by a resident pair alone or by a mob of ravens. Heinrich collected data that supported all of his predictions, including prediction 5 (Figure 10) [528].

Thus, we can conclude that when a young raven yells, the consequences of providing information ("Food bonanza here") to other nonterritorial birds can include benefits (personal access to food for the yeller) that outweigh the energetic costs of yelling as well as the risk of attack by the resident ravens guarding the carcass [529]. Yelling is therefore an adaptation, as defined above, currently maintained by its fitness benefits to individuals who give the call under the appropriate circumstances.

Identifying Darwinian Puzzles

Heinrich's analysis of a raven call shows how an adaptationist approach helps to identify interesting problems and generates testable hypotheses on the adaptive value of a communication signal. He saw that yelling had clear disadvantages—obvious fitness costs to yellers in the energy involved in the signal's production as well as its attraction of competitors to a valuable food resource. If yelling had evolved by natural selection, then its net benefits to calling individuals (more food, and thus improved survival, and therefore more chances to reproduce eventually) must have been greater than the net benefits derived from alternative forms of communication that appeared in the population in the past. If natural selection shapes communication signals and systems, then individuals should transfer information to others only when it is in *their* own reproductive interest to do so. The logic of this argument informed Heinrich's attempt to understand why ravens yell. He came up with ideas and tested them, rejecting many perfectly plausible adaptationist hypotheses as the data came in. The approach worked in this case, as it has in many other evolutionary studies.

Let's use the adaptationist approach to identify another such **Darwinian puzzle** in signal production. As noted in Chapter 7, male túngara frogs call for mates at night from pools in the Panamanian jungle. Their signal has two components: a whine that may be followed by one or more chucks. Females tested in playback experiments moved toward speakers playing calls that contained both components while generally ignoring speakers broadcasting calls composed of whines alone (see Figure 18, Chapter 7). Given the female preference for whine–chucks, what do you suppose male túngara frogs do in nature? It may surprise you to learn that male frogs often give one-component calls—whines without chucks.

The failure of males to give their most attractive calls is exactly the sort of thing that catches the adaptationist's eye, or ear in this instance. If the frog's signaling behavior is the product of natural selection, males ought to maximize their mating chances, not reduce them by giving incomplete calls. To deal with the evolutionary puzzle posed by whining males, Michael Ryan and his associates asked whether the fitness *costs* of giving whine–chucks might not exceed the benefits under some circumstances [1036]. The researchers knew that the túngara frog has a special enemy, the predatory fringe-lipped bat, which sweeps down on calling males and hauls them out of the water (Figure 11). If males sometimes reduced the attractiveness of their calls to females, perhaps they did so to reduce their risk of becoming a bat's dinner, thereby improving their chances to live and reproduce on another night.

This hypothesis generates a number of predictions, which Ryan and his colleagues tested. First, fringe-lipped bats should be attracted to the signals produced by their prey, which they are; bat predators zero in on speakers broadcasting túngara frog calls while ignoring silent ones. Second, fringe-lipped bats should be more attracted to whine–chuck combination calls than to whine calls alone, which they are. The bats were more than twice as likely to inspect, even land upon, a speaker broadcasting a whine–chuck versus a whine alone [1031]. Third, the frogs should be more likely to give one-component whine calls when the risk of predation is higher, which they are. The chances of becoming a victim are higher for frogs calling alone or in small groups than for frogs in large choruses, primarily because a bat has fewer individuals to choose from in a small group, increasing the risk of death for any one member of that group. As expected, males calling alone or in small groups are much more likely to produce whine-only calls, which are less attractive to females but also harder for bats to locate. (An alternative, and not mutually exclu-

11 Predation pressure affects the evolution of communication signals. The fringe-lipped bat is an illegitimate receiver that tracks its prey, male túngara frogs, by listening for their calls.

sive, hypothesis is that in a small group, competition for mates is less intense, so the callers are under less pressure to produce signals that may be risky or energetically costly.) The frog team concluded that although the túngara frog's readiness to give "incomplete" calls may appear superficially to be maladaptive, the behavior actually is an adaptation that evolved in response to bat predation.

Illegitimate receivers are everywhere in nature, eavesdropping on signals intended for other individuals. For example, some parasitic phorid flies detect chemicals, including 4-methyl-3-heptanone, released from the mandibular glands by workers of a giant tropical ant. The ants produce these signals to recruit fellow workers to counter threats to the colony. The flies take advantage of the information to locate hosts on which to lay their eggs [384].

The risk of exploitation by an illegitimate receiver may be responsible for the evolution of differences between the mobbing call and the "seet" alarm call of the great tit (Figure 12) [769]. These small European songbirds sometimes approach a *perched* hawk or owl and give the mobbing call, a loud signal with most of its energy in the 4.5 kHz range. Mobbing signals may attract other birds to the site to join in harassing the predator, which is not dangerous while it remains perched.

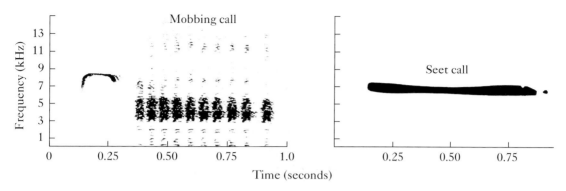

12 Great tit alarm calls. Sonograms of the mobbing alarm call and the "seet" alarm call. Note the lower sound frequencies in the mobbing signal. *Courtesy of (left) Peter Marler and (right) William Latimer.*

If, however, a great tit spots a *flying* hawk, it gives the "seet" alarm call, which appears to warn mates and offspring of possible danger. This is a much softer and much higher frequency call in the 7–8 kHz range. The properties of this signal are such that it attenuates (cannot be detected) after traveling a much shorter distance than the mobbing signal, thus compromising its effectiveness in reaching distant legitimate receivers. But the rapid attenuation of the "seet" call lowers the chance that a dangerous hunting predator can tell where the caller is. Moreover, the frequencies of the "seet" call lie outside the range that hawks can hear best, while falling within the range of peak sensitivity of the great tit (Figure 13). As a result,

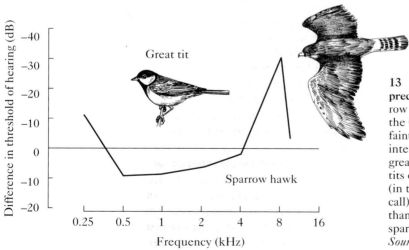

13 Hearing abilities of a predator and its prey. A sparrow hawk can hear sounds in the 0.5–4 kHz range that are fainter (5–10 dB lower in intensity) than those that great tits can hear. But great tits can detect an 8 kHz sound (in the range of the "seet" call) that is fully 30 dB lower than any 8 kHz sound that a sparrow hawk can detect. *Source: Klump et al. [656].*

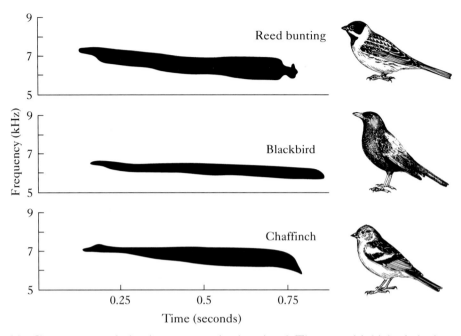

14   Convergent evolution in a communication signal. The great tit's high-pitched "seet" alarm call (see Figure 12) is very similar to the calls given by other unrelated songbirds when they spot an approaching hawk. *Source: Marler [769].*

a great tit can "seet" signal to a family member 40 meters away, while a sparrow hawk will not hear the call if it is a mere 10 meters distant [656]. Even when a hawk hears a "seet," the predator apparently finds it difficult to locate this somewhat ventriloqual sound [140]. Captive hawks and owls hearing a taped "seet" look around, but often peer in the wrong direction, whereas they have less trouble pinpointing the location of a mobbing signal of equivalent intensity.

If the "seet" call of the great tit has evolved in ways that reduce the risk of detection by its enemies, then unrelated species should produce alarm signals with similar properties for use under similar circumstances. The logic of this kind of comparative prediction is that similar selection pressures acting independently on different species should lead to the evolution of similar solutions. (We will have much more to say about this phenomenon, called *convergent evolution*, and about use of the comparative method in testing adaptationist hypotheses in Chapter 9.) The remarkable similarity in the "seet" calls of many unrelated European songbirds suggests that selection by bird-eating hawks has favored the evolution of alarm calls that are hard for hawks to hear, even if they might be somewhat less effective in reaching the intended receivers (Figure 14).

An Adaptationist Approach to Signal Receivers

We have been using an adaptationist approach to solve some cases in which signalers appear, at first glance, to be lowering their reproductive chances. Let's now examine some instances in which *receivers* of signals seem to lose fitness by reacting to certain messages from others.

Why do so many animals resolve their disputes with highly ritualized, symbolic displays? Rather than bludgeoning a rival, males of many birds settle their conflicts over a territory or a mate by singing and feather fluffing displays that do not involve the slightest contact between the two rivals, let alone an outright battle. Even when fighting does occur in the animal kingdom, it often has the appearance of comic opera. After a body slam or two, a subordinate elephant seal generally lumbers off as fast as it can lumber, inchworming its blubbery body across the beach to the water, while the victor bellows in noisy, but generally harmless, pursuit.

Prior to the recognition that "for-the-good-of-the-species" hypotheses have serious logical problems (see Chapter 1), many persons proposed that animal conflicts were resolved by threat displays and sham fights for the benefit of the species as a whole. For example, use of noncontact threats was said to prevent injury to the "superior" individuals who were needed to improve the genetic stock of the population. Species-benefit hypotheses imply that gentlemanly losers hold back for the welfare of all, even if it means they do not get to reproduce. However, the special genes of any truly self-sacrificing individuals would be selected against while others of their species propagated their genes "selfishly," without regard to the long-term survival of the species.

Therefore, adaptationists have suggested that even losers must gain fitness by resolving conflicts with little or no fighting [281]. Consider the European toad, *Bufo bufo*, whose males compete for receptive females. When a male finds another male mounted on a female, he may try to pull him from her back. The mounted male croaks as soon as he is touched, and often the other male immediately concedes defeat and goes away, leaving his croaking rival to fertilize the eggs of the female. How can it be adaptive for the signal receiver in this case to give up a chance to leave descendants simply on the basis of hearing a croak?

As it turns out, European toad males come in different sizes, and body size influences the pitch of the croak produced by a male. Therefore, Nicholas Davies and Timothy Halliday proposed that males can judge the size of a rival by his croak, thereby evaluating their chances of displacing him from a female. Because small males have little chance of ousting a larger opponent, it pays a smaller male to give up at once, using the time and energy that would be wasted in a fight to hunt for better mating opportunities elsewhere.

If this hypothesis is correct, deep-pitched croaks (made by larger males) should deter attackers more effectively than high-pitched ones (made by smaller males) [272]. To test this prediction, the two researchers placed mating pairs of toads in tanks with a single male for 30 minutes. The paired male, which might be large or small, had been silenced by looping a rubber band under his arms and through his mouth. Whenever the second male touched the pair, a tape recorder supplied a 5-second call of either low or high pitch. Small paired males were much less frequently attacked if the interfering male heard a deep-pitched call (Figure 15). Thus, deep croaks do deter rivals to some extent, although tactile cues also play a role in determining the frequency and persistence of an attack, as one can see from the higher overall attack rate on smaller toads.

Honest Signals

So why don't small males "pretend" to be large by giving low-pitched calls? Perhaps they would, if they could, but they can't. A small male toad apparently cannot produce a deep croak, given that body mass and the unbendable rules of physics determine the pitch of the signal that a male can generate. Thus, toads have evolved a defensive signal that accurately announces their body size. By attending to this honest signal—one that conveys accurate information—a male can determine something about the size of his rival, and thus, his probability of winning an all-out fight with him.

When smaller rivals withdraw upon hearing an honest signal from a larger male, both parties gain; small males do not waste time and energy in a battle they are unlikely to win, and large males save time and energy that they would otherwise have to spend struggling with annoying smaller toads. Imagine two kinds of aggres-

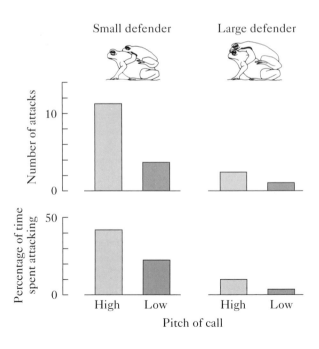

15 Deep croaks deter rivals. Single male European toads make fewer contacts and interact less with silenced mating rivals when a tape of a low-frequency call is played than when a higher-frequency call is played. *Source: Davies and Halliday [272].*

sive individuals in a population, one that fought with each opponent until physical-ly defeated or victorious, the other that checked the rival's fighting potential and then withdrew as quickly as possible unless victory seemed likely. The "fight no matter what" types surely would eventually encounter an opponent that would thrash them soundly. The "fight only when the odds are good" types would be far less like-ly to suffer an injurious defeat at the hands of an overwhelming opponent [1261].

Further, imagine two kinds of superior fighters in a population, one that gen-erated signals other lesser males could not produce, and another whose threat dis-plays could be mimicked by smaller males. As mimics became more common in the population, natural selection would favor receivers that ignored the easily faked signals, reducing the value of producing them. This in turn would lead logically to the spread of the genetic basis for an honest signal that could not be devalued by deceitful signalers.

If this argument is correct, we can predict that the threat displays of many unrelated species will exhibit similar properties (Figure 16; Color Plate 9; cover photograph). They should be expensive to produce and difficult for small or weak

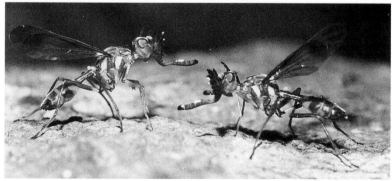

16 Convergent assessment displays. (Top) Male red deer clash aggressively, locking their horns and pushing in ways that enable opponents to judge the relative size and strength of their rivals. (Bottom) Males of this New Guinean antlered fly (*Phytalmia alcicornis*) con-front each other head to head, permitting each fly to assess their own size rela-tive to the other's size. *Photographs courtesy of (top) Timothy Clutton-Brock and (bottom) Gary Dodson.*

males to imitate, and they should reliably inform rivals about the size and fighting capacity of the displayer [905], given that body size, muscle mass, and strength are intercorrelated in many animals [303, 703, 1269]. The bizarre "antlers" and allied displays of certain flies meet these criteria beautifully. Males competing for access to mates confront one another, head to head, permitting rivals to compare the size of their large, and presumably expensive to produce, head projections. Antler span is tightly correlated with body size (Figure 17), so that individuals can quickly and accurately assess their relative sizes by matching up antlers [1285]. Smaller males generally abandon the field, avoiding energy-consuming contests they are likely to lose.

17 Antler span in two New Guinean flies provides accurate information about body size, permitting males to make accurate decisions about an opponent's fighting ability. *Source: Wilkinson and Dodson [1285].*

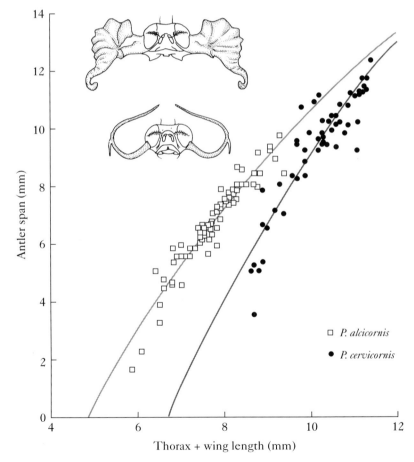

18   **An honest signal.** Only male red deer in top condition can sustain roaring contests for long periods. *Photograph by Timothy Clutton-Brock.*

Likewise, in the red deer, violent antler-clashing and shoving matches occur only rarely in the rutting season, even though this is the time when males compete seriously for mates. Instead, two males usually confront each other at a considerable distance and roar loudly (Figure 18) [206]. Only males in top condition can sustain roaring at a high rate for many minutes [205] because the activity is so costly in terms of energy expended, as is true of many other acoustical displays [e.g., 1067, 1251; but see 571]. Thus, just as in antlered flies and European toads, potential combatants can gain accurate information about the fighting ability of a rival by assessing his displays at a distance.

The Darwinian Puzzle of Deception

Although honest signalers provide useful information to each other for mutual benefit, even when the signals are threats [460, 1323], deceptive signalers also occur in nature. Recall the case of the orchids whose female wasp decoys lure male wasps into attempting to copulate with the flower (see Chapter 4). Just as signal givers can lose fitness by providing information to exploitative receivers, receivers can lose fitness by responding to signals generated by **illegitimate signalers.** A famous example of this phenomenon involves the firefly "femmes fatales" studied by James Lloyd [726]. The females of some predatory fireflies in the genus *Photuris* answer the flashes given by males of certain other species in the genus *Photinus*. Each species of *Photinus* has its own code, with a female answering her male's distinctive flash pattern by giving a flash of her own after a precise time interval. Some *Photuris* females can respond at the correct interval to male signals of three *Photinus* species [727]. If a *Photuris* female succeeds in luring a male *Photinus* close enough, she will grab, kill, and eat him (Figure 19).

19 **A firefly *femme fatale*.** Females of predatory *Photuris* fireflies attract male *Photinus* fireflies by imitating the flashes of *Photinus*
females. *Photograph by James E. Lloyd.*

Similar behavior has evolved in an Australian spider that specializes in hunting
females of other species that build nests in rolled-up leaves. Males of the victimized species court nest occupants by dropping onto the leaf and making it rock in
a particular way. Receptive females pop out of their nests to copulate. But when
females react to the signals of the predatory spider in the same way, they are killed
and eaten [595].

Deception of this sort poses a real puzzle for the adaptationist, not from the perspective of the predatory signaler, which gets a meal when its deception works,
but from the standpoint of the male firefly or female spider that pays attention to
the wrong "come hither" signal and gets killed. When an action is clearly disadvantageous to individuals, behavioral biologists generally propose one of the two
following causes (see also Table 1):

1. *The novel environment theory:* The maladaptive response is caused by a proximate mechanism that once was adaptive, but is not now. The current maladaptation occurs because modern conditions are very different from those
   that shaped the mechanism in the past, and because there has not been sufficient time for selection "to fix the problem."

2. *The exploitation theory:* The maladaptive response is caused by a proximate
   mechanism that is still adaptive in sum, but is being exploited by some individuals. The risk of being exploited is a maladaptive side effect that reduces,
   but does not eliminate, the net fitness gained by responding to a cue in a particular way [196].

Novel environments seem unlikely to account for the nature of interactions
between *Photinus* males and their predatory relatives. This theory is generally

20 A maladaptive response to an evolutionarily novel stimulus. A male Australian buprestid beetle attempts to copulate with a beer bottle. *Photograph by David Rentz and Darryl Gwynne.*

reserved for cases in which very recent human modifications of the environment appear responsible for eliciting maladaptive behavior, such as the attempts by certain Australian beetles to copulate with beer bottles left lying on the desert by Aussie litterbugs (Figure 20) [491]. The beer bottles happen to mimic the releasers of copulatory behavior provided by female beetles. These pseudofemale containers have been in the beetle's environment for only a few decades, too short a period for selection to have altered the response of males to them.

Instead, the exploitation theory seems more likely to account for male *Photinus* behavior. The argument here is that *on average*, the reaction of male *Photinus* to certain light flashes increases fitness, even though one of the costs of responding is the chance that the male will be devoured by an exploitative *Photuris* signaler. Males that avoided these deceptive signals might live longer, but they would probably ignore females of their own species as well, and would leave few or no descendants to carry on their cautious behavior.

This hypothesis highlights the definition of adaptation employed by most behavioral biologists. As noted above, an adaptation need not be perfect, but it must be superior in its fitness effects relative to possible alternatives. The male firefly that responds to the signals of a predatory female of another species possesses a mechanism of mate location that clearly is not perfect, but may be better than those alternatives that would improve a male's survival chances at the cost of decreasing his chances to mate.

If this adaptationist hypothesis is correct, deception by an illegitimate signaler should exploit a response that has clear adaptive value under most circumstances [196]. Thus, a host of predatory spiders successfully lure web-building spiders of other species to their deaths by mimicking the vibratory cues produced by prey trapped in a web. The deception works because these cues almost always signal the presence of food in the web, in which case a quick approach is highly appropriate [594]. Likewise, the anglerfish deceives its prey with a bizarre projection from the front of its head, a thin rod from which is suspended a bit of tissue with an uncan-

21 A deceptive signaler. A lure-using anglerfish has a "minnow" appendage on the front of its head (top), which it waves (bottom) to draw its prey within striking range. *Photographs by David Grobecker.*

ny resemblance to a small fish (Figure 21). This bait is waved about seductively, luring small predatory fish close enough so that the anglerfish can engulf them in its massive mouth [938]. Although the deceived fish pay a heavy price for their interest in the lure, members of their species may benefit on average from investigating stimuli that are almost always associated with prey, not with a lethal predator.

Many other cases of deception occur because an organism is able to exploit an otherwise adaptive communication system (see Chapter 4 on orchids that use mimetic sex pheromones to secure male wasp and bee pollinators [862]). The manufacture of deceptive sex pheromones has evolved in the red-sided garter snake, a species in which matings occur at the entrance to the hibernaculum where many snakes have spent the winter together (see Figure 23, Chapter 6). On warm days in spring, masses of males writhe about waiting for females to come to the surface and copulate with one of them. Some of the males produce the female sex pheromone, a chemical secreted on the skin, which causes other males to court them vigorously. These deceptive males may distract rivals to such an extent that they reduce the competition for the real thing when real females appear. Needless to say, the deceived males gain nothing from their attempts to mate with a fellow male, but a responsiveness to female sex pheromones seems clearly adaptive on average, given the alternative [778].

Or consider the nature of deceitful communication by the white-winged tanager-shrike of South America. This bird regularly travels through the forest with a variety of other species. In these mixed-species groups, the tanager-shrike pursues insects flushed out by the other species as they forage in dense foliage. Because the tanager-shrike scans its environment from a perch while waiting for insects to fly into view, it often sees predatory hawks, and when it does, it sounds the alarm, causing the other birds to dive for safety or freeze. The tanager-shrike, however, also sometimes gives this call even when no hawk is present, especially when the tanager-shrike and another bird are pursuing the same flying insect (Figure 22). The call generally causes the competitor to abort its chase, giving the tanager-shrike an uncontested shot at the prey. Why do the other birds permit this to happen? As Charles Munn points out, most of the time the tanager-shrike's companions gain by paying attention to its alarm calls, because the consequences of ignoring the signal may be lethal. Under these conditions, the benefits of treating each alarm call seriously outweigh the costs, permitting the tanager-shrike to take unfair advantage of the relationship from time to time [850].

If this hypothesis is correct, then deceptive alarm calls should have evolved in unrelated species in which the receivers of those signals would face serious consequences for failing to react to a nondeceptive alarm call. In animals as different as great tits [820] and vervet monkeys [191], some individuals occasionally produce an alarm call in the absence of a predator. Whether they do so "intentionally" is an interesting (and difficult to resolve) proximate question, but the ultimate consequence of these deceptive alarms is to give the false alarm caller access to rich food supplies or other resources abandoned by the individuals dashing for

22 **A bird that cries "wolf."** A white-winged tanager-shrike is far more likely to give its hawk alarm call when pursuing a flying insect that is also being chased by another bird than when it is chasing prey by itself. *Source: Munn [850].*

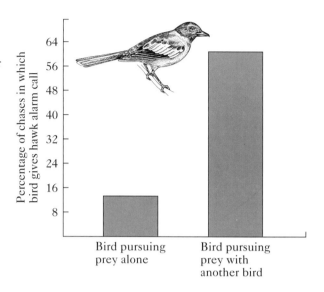

cover. Ignoring alarm calls is potentially so costly that receivers must react to them, just as male fireflies, wasps, and garter snakes cannot afford to ignore sexual signals associated with receptive females.

SUMMARY

1. Natural selection is a process that can lead to evolutionary change, provided there have been reproductive differences among individuals in the past caused by their genetically different traits. If these conditions apply, the current characteristics of living organisms ought to be adaptations; that is, they should promote greater reproductive success on average than alternative forms of those characteristics.

2. Not every current trait elevates individual fitness, however. Some characteristics may be the product of random processes of evolution, or they may be nonadaptive, or even maladaptive, by-products of genes that have been selected for their other adaptive effects. Therefore, in seeking ultimate explanations for a given trait, one must entertain both adaptationist and nonadaptationist hypotheses, while also remembering that there can be several competing alternative hypotheses under each general category.

3. The adaptationist approach has proved highly useful in identifying Darwinian puzzles and developing hypotheses to resolve them. A Darwinian puzzle is a trait that appears to reduce individual fitness rather than increase it. Seemingly self-sacrificing traits are a challenge to explain, whether they involve calling others to a scarce food resource or conceding resources to a rival when "threatened" by a harmless display.

4. Despite the challenges, persons employing the adaptationist approach have made great progress in resolving many Darwinian puzzles. Often behavioral traits that appear at first glance to reduce fitness have been shown either to raise reproductive success or to be a consequence of an underlying behavioral mechanism that has an overall positive effect on fitness. Thus, for example, costly responses to deceptive signals often occur because an illegitimate signaler is able to exploit a generally adaptive response to a signal that is usually produced by a legitimate signaler.

SUGGESTED READING

Two good recent books, one by Clive Catchpole and Peter Slater [187] and the other an edited volume by Donald Kroodsma and Edward Miller [680], offer up-to-date reviews of the evolutionary basis of bird song. Bernd Heinrich focuses exclusively on raven communication in *Ravens in Winter* [529]. Rosie Cooney and Andrew Cockburn's paper [218] on the adaptive value of singing by *female* fairy wrens shows nicely how adaptationists go about testing alternative hypotheses. Theoretical analyses of the evolution of communication, especially deception, appear in [281, 509].

DISCUSSION QUESTIONS

1. Randall Breitwisch and George Whitesides wanted to know what benefit male mockingbirds derived from singing their long and complex songs [120]. They considered two possibilities: singing helps males keep other males out of a territory, and singing helps a male attract a mate. Make predictions based on these two hypotheses about (1) the singing rates of mated versus unmated males, (2) the effect of experimental removal of a mate from a male that had acquired one, and (3) the direction in which a singing male will tend to project his songs before and after he has acquired a mate.

2. In male cricket frogs, the larger the male, the lower the dominant frequency in his mate-attraction calls. But male cricket frogs will sometimes lower the dominant frequency in their calls when they hear a larger rival calling nearby [1232]. This ability may surprise you in light of our discussion of European toads. Why? William Wagner found that when males lowered their dominant call frequency in this situation, they were significantly more likely to go on to attack the larger calling neighbor than when they did not adjust their call frequency. Does this finding help you evaluate the hypothesis that males who lower their calls are engaged in deceptive signaling?

3. Do the studies of communication covered in this chapter support A. J. Cain's claim (made in 1964) that "if we cannot see any adaptive or functional significance of some feature, this is far more likely to be due to our own abysmal ignorance than to the feature being truly nonadaptive, selectively neutral or functionless" [170]?

4. The pied flycatcher is usually monogamous, with one male pairing with one female on the male's territory, which contains a hollow tree or other nest site. After the eggs hatch, monogamous males assist in feeding the young. Some males, however, set up a second territory after attracting one mate, and call for another female. What evidence would you need to determine whether such a male is using a deceptive signal? If he is, how might you explain the female's maladaptive response to calls from a polyterritorial male? (See [435, 444, 1120].)

5. How many different techniques for testing predictions based on adaptationist hypotheses can you identify in the examples covered in this chapter?

6. Males of the scarlet-tufted malachite sunbird of Africa have tufts of red feathers on their chests that they puff up when engaged in aggressive interactions with rivals over possession of territories containing nectar-producing flowers. Researchers performed an experiment in which they clipped feathers from some birds, creating reduced-tuft males [369], then glued those feather pieces to the tufts of other males, creating enlarged-tuft males. What prediction does the honest signal hypothesis make about the increase or decrease in the number of flower stalks, or inflorescences, that males should be able to defend after their tufts have been altered? Based on the actual results (see the figure below), what is the appropriate conclusion?

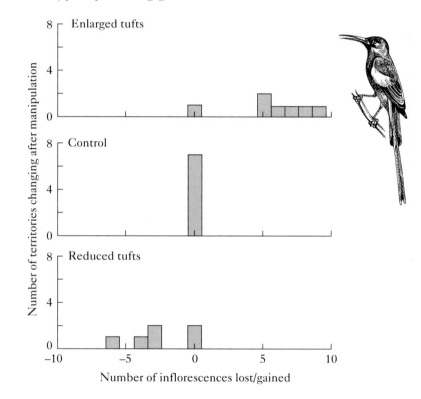

Source: Evans and Hatchwell [369].

CHAPTER 9

Adaptive Responses to Predators

THE PREVIOUS TWO CHAPTERS provide the foundation for the remainder of the book, in which we will examine either the history or the adaptive value of a full spectrum of behavioral traits, not just those involved in communication. This chapter examines characteristics that may help individuals cope with predators so that they or their offspring have a better chance of survival. An interaction between predator and prey can be divided into four stages, with the predator first detecting, then attacking, then capturing, and finally consuming its victim—in those encounters that have a satisfactory ending from the predator's perspective. An interaction with this conclusion, however, is rarely advantageous for the consumed animal. We would expect, therefore, to find that prey species have evolved antipredator adaptations designed to block attempts to detect, attack, capture, and consume them. You may recall a noctuid moth's response to an ultrasound-producing bat (see Chapter 5). The moth, by virtue of its great sensitivity to ultrasound, can sometimes avoid detection by turning away from a bat before the predator detects it. But even if the bat should sense the presence of the moth, the insect is not defenseless, thanks to the anti-interception power–dives and loops it performs in response to intense ultrasound, which make its capture a challenging task for an onrushing bat. If the moth's behavior works, the insect will postpone its inevitable demise, giving itself additional time to reproduce. We shall use an anti-detection, anti-attack, anti-capture, and anti-consumption scheme to organize this survey of the defensive behavior of prey. Our goal is to illustrate the full range of adaptationist hypotheses on antipredator behavior and the variety of methods that can be used to test these hypotheses.

Making Detection Less Likely

When a prey is overlooked by its predator, the prey is generally the better for it. Thus, both banner-tailed kangaroo rats in North America (see Chapter 6) and hairy-footed gerbils in Africa [582] forage more on moonless nights, and the same is true of some desert scorpions [1119]. Likewise, gull eggs that predators cannot see are more likely to hatch, producing offspring that may carry on their parents' genes and behavior patterns. One of these behavior patterns is the removal of conspicuous broken eggshells from the nest, thereby eliminating a visual cue that some predators could use to locate vulnerable but otherwise well-camouflaged chicks (see Chapter 1).

A parent may be able to conceal its offspring from predators in other ways as well. In the previous chapter, we discussed the mobbing call of the great tit, a little songbird that appears to have evolved a specific signal to attract others to a perched owl or hawk. As other songbirds arrive, they add their voices to the chorus and dart about the predator, flicking their wings and tails in a distinctive manner. At first glance, it is hard to imagine how **mobbing behavior** could help songbird parents improve their young's survival. Indeed, the behavior offers a particularly clear example of a Darwinian puzzle because the fitness costs of mobbing seem so obvious. The adaptationist Eberhard Curio has, however, generated a long list of potential benefits of mobbing by great tits (Table 1), including the possibility that mobbers distract enemies from their offspring [250]

The predator distraction hypothesis for mobbing does not seem to have been tested for the great tit. We shall therefore turn our attention back to gulls—in particular, the black-headed gull, a species whose mobbing behavior has been studied in some detail. These gulls form summer breeding colonies composed of dozens or hundreds of pairs nesting on the ground in open grassy areas, often on islands (Figure 1). Should a fox, crow, badger, hawk, or human appear in the colony, the breeding adults respond with a volley of loud cries. If the intruder continues

Table 1 Alternative hypotheses on the adaptive value of mobbing behavior by great tits

Hypothesis	Effect of mobbing	Beneficiary
1. Advertise uncatchability	Predator leaves mobbers alone	Mobbers themselves
2. Distract predator	Predator cannot find nestlings	Mobbers' offspring
3. Warn others	Offspring hide	Mobbers' offspring
4. Lure predator away	Predator pursues mobber	Mobbers' offspring
5. Educate offspring	Offspring learn to recognize predator	Mobbers' offspring
6. Injure predator	Predator harmed by mobbers' attacks	Mobbers and offspring
7. Attract larger predator	Predator attacked by larger predator	Mobbers and offspring

Source: Curio [250].

toward them, groups of gulls will fly toward it, calling raucously and defecating profusely (Figure 2).

The great ethologist Niko Tinbergen had many opportunities to ask himself about the adaptive value of mobbing, for he himself triggered the behavior during his frequent visits to gull colonies. As in the great tit, mobbing by black-

1 **A colony of gulls.** The small structures are blinds for close observation of nesting lesser black-backed and herring gulls. *Photograph by Niko Tinbergen and Hugh Falkus.*

2 Black-headed gulls mobbing a trespasser in their colony.

headed gulls occurs primarily during the breeding season, suggesting that the beneficiaries of the behavior could be the mobbers' young. Tinbergen believed that mobbing gulls confused and distracted predators intent on finding and eating their offspring [1190]. I too find the distraction hypothesis plausible, based on the personal experience of trying to get out of the way of dive-bombing gulls in a gull rookery. The roar of wind through their wings as they passed just overhead was unnerving, and the thwack of a gull's foot on the top of my head hurt.

Although the predator distraction hypothesis makes sense, we must test it before accepting it. We can begin by predicting that predators searching for black-headed gull eggs should be less successful at finding these prey while they are being mobbed by nesting gulls. We can test this prediction simply by observing what happens when mobbing black-headed gulls interact with nest-robbing predators, such as carrion crows and herring gulls [681]. These egg-eating birds can avoid swooping gulls, but only if they continually face their attackers. Thus, while being mobbed, they cannot search for nests and eggs, showing that mobbing gulls probably gain a fitness benefit from mobbing.

Another, more challenging prediction from the predator distraction hypothesis is that the degree of success experienced by gulls in protecting their eggs should be proportional to the degree to which predators are mobbed. Hans Kruuk used an experimental approach to test this prediction [681]. He placed ten hen eggs, one every 10 meters, along a line running from the outside to the inside of the gull nesting area. Eggs placed outside the colony, where mobbing pressure was low, were much more likely to be found and gobbled up by carrion crows and herring gulls than eggs inside the colony, where the predators were intensely mobbed (Figure 3).

Therefore, both observational and experimental tests support the hypothesis that mobbing of certain predators helps protect the eggs and young of black-headed gulls. Note that these tests did *not* involve measuring gull reproductive success

3 The effectiveness of mobbing in protecting eggs. Researchers placed hen eggs along a line from the outside to the inside of a gull nesting colony. The frequency of intense attacks (circles) on crows increased when these predators were within the borders of the colony. As mobbing increased, fewer hen eggs (triangles) were discovered and eaten. *After Kruuk [681].*

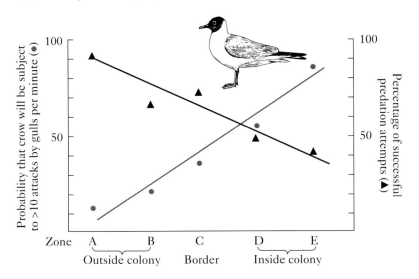

Table 2 Direct and indirect measures used by persons attempting to examine the fitness consequences of behavioral traits

---

DIRECT MEASURES OF REPRODUCTIVE SUCCESS
A researcher measures whether the trait in question improves

1. the production of gametes
2. the rate of copulations
3. the production of fertilized eggs
4. the production of newborn young
5. the production of independent offspring
6. the production of young during one breeding season that reach the age of reproduction
7. the production of young over the animal's lifetime that reach the age of reproduction

INDIRECT MEASURES OF REPRODUCTIVE SUCCESS
A researcher measures whether the trait in question improves the animal's

1. chance of survival
2. access to food
3. possession of living space
4. efficiency of locomotion or other activity

---

by counting the number of surviving offspring produced by individuals in their lifetimes. Kruuk looked instead at the number of hen eggs that were uneaten, on the reasonable assumption that had they been gull eggs in a gull nest, they would have had a chance to become surviving offspring for the gull parent.

Behavioral ecologists often have to settle for an indicator or correlate of reproductive success when they attempt to measure fitness (Table 2). In the chapters that follow, "reproductive success" is used interchangeably with such things as egg survival (Kruuk's measure), young that survive to fledging, number of mates inseminated, quantity of food ingested per unit of time, and so on. The reader should keep in mind, however, that the bottom line is the number of offspring produced by an individual that reach the age of reproduction, and that correlates of this ultimate measure of fitness will be accurate to varying degrees.

The Comparative Method of Testing Adaptationist Hypotheses

Experiments are highly valued in science, so much so that most persons believe that scientific research can be performed only in high-tech laboratories by researchers wearing white lab coats and goggles. As Kruuk's work shows, however, good experimental science can be done in the field. Moreover, the manipulative experiment is only one of several ways in which predictions from hypotheses can be tested. We just saw how it was possible to test the predator distraction hypothesis on mobbing by using nonmanipulative observations to find out whether certain predictions (e.g., hunting crows must stop searching for eggs while they are being mobbed) are correct.

In addition, biologists have another technique for testing adaptationist hypotheses: the **comparative method.** This approach involves testing predictions about the evolution of an interesting trait *in animals other than the species whose characteristics are under investigation* [207, 993]. Here is one such prediction: If ground-nesting black-headed gulls mob certain predators to protect their young, then closely related gull species that lack predators should lack mobbing behavior.

The rationale behind this prediction is that related species tend to share traits in common, having inherited the genetic basis for those attributes from their common ancestor (see Chapter 7). If, however, a species occupies an environment different from that of its relatives, where it encounters different selection pressures, then it is expected to undergo *divergent evolution* to evolve new and distinctive traits, provided mutations occur that confer reproductive advantages under the new conditions.

In fact, some gulls do not nest in groups on the ground where eggs are vulnerable to predators. For example, the kittiwake gull nests on nearly vertical coastal cliffs (Figure 4), where its eggs cannot be reached by mammalian predators and where even other gulls and hawks rarely venture because of dangerous swirling sea winds. The small, delicate kittiwakes have clawed feet, and they can land and nest on tiny ledges (Figure 5) where their eggs and young cannot be reached by predators. As a result, the predator pressures acting on kittiwakes differ markedly from those affecting their ground-nesting relatives. As predicted, nesting adult kittiwakes do not mob their predators, despite sharing ancestry with black-headed gulls. The kittiwake's behavior has diverged from that of their close relative, providing a case of divergent evolution in support of the hypothesis that mobbing evolves specifically in response to predator pressure [248].

Now here is another comparative prediction: Other ground-nesting gulls will exhibit the kind of mobbing behavior shown by black-headed gulls because they too face nest-robbing enemies that should select for this antipredator response. *This prediction represents an invalid use of the comparative method.* Why? Because similarities between different gull species could be caused by their shared ancestry. Related species have a relatively recent common ancestor from which they have inherited genetic information, which will tend to lead to the development of similar phenotypes (see Chapter 7). If we find that herring gulls, and indeed most ground-nesting colonial gulls, exhibit mobbing behavior like black-headed gulls, it could be because they all share similar selection pressures, *or* it could be because they have a common ancestor that transmitted the genetic basis of the mobbing trait to all its descendant species. In contrast, the discovery that the kittiwake *lacks* mobbing behavior cannot be attributed to the possession of a recent common ancestor, which makes the evolutionary cause of this species' behavior clear and unambiguous.

The other side of the comparative coin is that species from different evolutionary lineages are expected a priori to behave differently because they have different ancestors that endowed them with distinct genetic–developmental systems.

4   **Nesting habitat of kittiwake gulls** at Orkney Island, Great Britain. *Photograph by Arthur Gilpin.*

5   Adaptation for cliff nesting.
Kittiwake gulls have clawed feet that
help them cling to even the narrowest
of cliff ledges, such as this guano-
stained perch. *Photograph by the author.*

If we were to compare the black-headed gull with an *unrelated* species that nests
in relative freedom from predators, and if we were to find that the predator-free
species lacked mobbing behavior, we could not conclude that mobbing evolves in
response to predator pressure. Why not? Because the difference between the
two species in this case could as readily arise from their separate ancestries as from
the different selection pressures acting on them.

The only useful kinds of adaptationist tests involving unrelated
species are those in which the species have been subjected to *similar* selection
pressures. In this instance, two unrelated species can be predicted to have independently evolved similar behavioral traits through convergent evolution—*if* the
trait truly is an adaptation to a shared selection pressure that has caused species
with different genes to evolve functionally equivalent attributes (Figure 6).

The trick here is to identify truly *independent* cases of convergence in a trait.
To do so requires firm knowledge of both the evolutionary relationships and the
distribution of alternative traits among the species under examination. At times,
this can be a complex and tricky business [514]. With respect to mobbing, how-

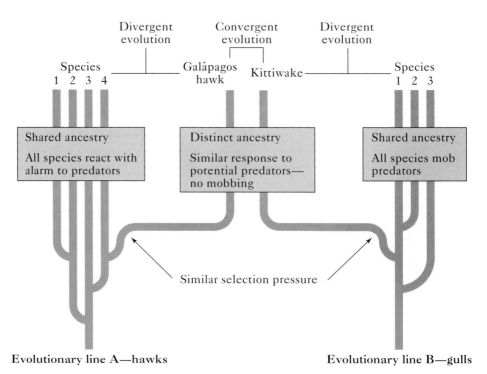

Evolutionary line A—hawks   Evolutionary line B—gulls

6 Divergent and convergent evolution. Members of the same evolutionary lineage (e.g., gulls of the subfamily Larinae) share common ancestry, and therefore often share the same traits, such as mobbing behavior. But the effects of shared ancestry can be overridden under different selection pressures, as in the evolution of the kittiwake's failure to mob potential predators. This gull species has diverged behaviorally from other gulls while converging in behavior with an completely unrelated species, the Galápagos hawk, that belongs to a different evolutionary lineage (hawks of the subfamily Buteoninae). Like the kittiwake, the hawk nests in a place (the Galápagos Islands) where predation pressure has been slight. It too shows little response to approaching potential predators, unlike most of the other easily alarmed members of its lineage.

ever, some animals clearly unrelated to black-headed gulls do exhibit the behavior. The colonial barn swallow and bank swallow both mob nest predators, including snakes and blue jays [569, 1093]. These swallows have been long separated from gulls, and yet they also harass their enemies. We can even find colonially nesting mammals that have convergently evolved the mobbing trait [538, 891]. For example, in the California ground squirrel, which lives in loose groups, the agile adults will approach a rattlesnake and kick sand in its face, thereby preventing the predator from exploring nest burrows for prey (Figure 7).

    Thus, mobbing behavior has evolved independently in several unrelated species whose adults can sometimes protect their vulnerable offspring by distracting preda-

7 Convergent evolution of mobbing behavior. (Top) A colony of bank swallows, a
social species that mobs predators in a manner similar to that of ground-nesting gulls.
(Bottom) Colonial California ground squirrels have also evolved mobbing behavior.
One squirrel kicks sand at a rattlesnake, while others give a variety of alarm calls.
Courtesy of (top) Michael D. Beecher and (bottom) R. G. Coss and D. F. Hennessy.

tors. But what if I have been presenting only supportive examples while ignoring many other colonial species in which parents do not mob the enemies of their vulnerable youngsters? If for every colonial species in which mobbing occurred under the expected conditions there were two in which it did not, you would be skeptical of the predator distraction hypothesis, and rightly so. For this reason, researchers increasingly require that the comparative method be used in a statistically rigorous fashion, a topic we will return to later [514, 993].

Cryptic Behavior

When baby gulls hear their parents' alarm cries and mobbing screams, they duck for cover or crouch low in the nest. Their camouflaged gray and brown plumage almost certainly makes them very hard for predators to find. Many other animals combine camouflage with hiding behavior. This phenomenon is especially common among the insects (Color Plate 10) [e.g., 531]. The ability of a moth to select a resting background that matches its particular color pattern has some fairly obvious benefits (an increase in survival gained by hiding from predators) and some less obvious costs as well (the time and energy expended by the moth to find an appropriate background) (Figure 8). If the benefits of selecting a matching background exceed the costs, then predators should have an especially hard time finding a cryptic moth perched on its preferred resting place.

In nature, the moth *Catocala relicta* rests head up with its whitish forewings over its body on white birch and other light-barked trees (Figure 9). When given a choice of resting sites, it selects birch bark over darker backgrounds [1048]. Alexandra Pietrewicz and Alan Kamil used captive blue jays, photographs of moths on different backgrounds, and operant conditioning techniques (Figure 10) to test whether the moth's behavior is adaptive [936]. As predicted, the jays evidently

8 Cryptic coloration and selection of a resting place. The degree to which this camouflaged moth is safe must depend on its ability to find appropriate dead leaf resting places during the day, when birds hunt for insects. *Photograph by the author.*

9 **Cryptic coloration and body orientation.** The body orientation of a resting *Catocala* moth determines whether the dark lines in its wing pattern match up with the dark lines in birch bark. *Photograph by H. J. Vermes, courtesy of Theodore Sargent [1048].*

saw the moth 10 to 20 percent less often when examining photos of *C. relicta* pinned to birch bark than when looking at moths on less appropriate dark bark backgrounds. Moreover, if the moth was oriented head-up on a birch trunk, the birds overlooked it more often than when it was in some other position. The moth's preference for white birch resting places and its typical perching orientation appear to be anti-detection adaptations against predators such as jays.

The possible importance of resting site selection for prey survival was earlier tested by H. B. D. Kettlewell in his classic study on industrial melanism in moths [646]. British lepidopterists knew that a completely black form of the moth *Biston*

10 **Does cryptic behavior work?** Images of moths on different backgrounds and in different resting positions are shown to a captive blue jay, which is rewarded for detecting moths when they are shown on the screen. *Photograph by Alan Kamil.*

11 Salt-and-pepper moths. One typical (salt-and pepper coloration) and one melanic form of the moth appear in this photograph of the m)ths posed on a silver birch tree. The pale form is well camouflaged on the pale birch bark (in the top third of the photograph); the dark form is hidden on the lower left limb scar. *Photograph by Bruce Grant.*

betularia became much more numerous in Britain after 1850, largely replacing the typical salt-and-pepper form of the moth, especially near big cities. This moth also occurs in North America, and here too the melanic type achieved frequencies of more than 90 percent in some areas by the mid-twentieth century [461]. Kettlewell proposed that the melanic form had spread because it was less conspicuous to bird predators in woodlands darkened by pollution from urban factories. He believed that the salt-and-pepper form remained abundant in unpolluted forests because in this environment whitish wings blended in nicely against pale lichen-covered tree trunks (Figure 11).

To test this hypothesis, Kettlewell placed samples of the two forms of moths on light or dark tree trunks and observed the reactions of insectivorous birds from a blind. As he expected, birds found whitish forms on dark tree trunks much more quickly than they found melanic individuals, but the dark moths quickly fell to predators when placed on a lichen-covered tree. In nature, however, the moths evidently do not perch on tree trunks. Instead, they may more often select the shaded patches just below the junction of a branch with the trunk. When R. J. Howlett and M. E. N. Majerus glued samples of frozen moths to open trunk areas and to the undersides of branch joints, they confirmed Kettlewell's original findings (Figure 12), while showing that birds were particularly likely to overlook moths on shaded limb joints [577].

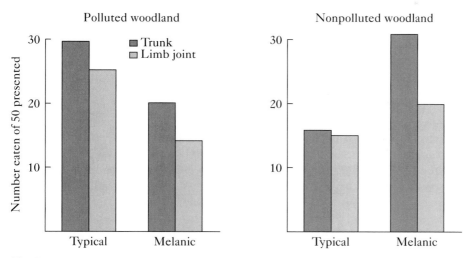

12 Predation risk and background selection by moths. Specimens of typical and melanic forms of the salt-and-pepper moth were pinned to tree trunks or limb joints. Melanic forms were discovered by birds less often in polluted (darkened) woods; typical forms "survived" better in nonpolluted (lichen-covered) woods. In every case, specimens on limb joints were less likely to be taken than those pinned directly onto trunks. *Source: Howlett and Majerus [577].*

Bruce Grant and his colleagues point out, however, that large shifts in the frequencies of the two forms of the moth have occurred in certain lichen-free woodlands of England as well as in some North American woodlands whose lichens have been unaffected by pollution [461]. Thus, there may well be more to this story than meets the eye.

The Costs and Benefits of Anti-Detection Behavior

In discussing cryptic behavior, we have talked mostly about the benefits that animals might gain by hiding from their enemies. But a moth sitting still for hours on end loses time that could otherwise be invested in feeding or finding mates. By considering these costs as well as the benefits of cryptic behavior, we can develop more sophisticated hypotheses about who should hide when, and for how long. The lower the cost, the greater the chance of a trait spreading through a population, all other things (the benefits) being equal.

The costs of cryptic behavior can vary for individuals within a species. For example, food-deprived animals might pay a higher price in terms of lost feeding time when hiding from predators than animals that have adequate energy reserves. If so, then animals that have lost weight should be less likely to hide than those that are plump and well-fed. This prediction has been tested in an experiment with Belding's ground squirrels. Juvenile females were captured each day for 6 days. Some were allowed to gorge on peanut butter, after which they were released. Others, however, were held in traps for most of each day and offered only low-calorie lettuce before being released in the late afternoon. The lettuce-fed squirrels lost weight over the experiment, and were subsequently more likely to continue foraging away from their burrows upon hearing a taped alarm call than their well-fed companions [39]. The food-deprived squirrels traded safety from predators in favor of the opportunity to regain lost weight.

In the case of Trinidadian guppies, the trade-offs for hiding are improved safety from predators at the cost of having less time to court females. These little fish live in clear forest streams where other fish predators target moving prey. Although immobile males are relatively safe, they are also unattractive to females. How can male guppies balance the benefits of reproducing against the costs of attracting predators?

One of the primary predators of guppies becomes active as light intensities increase, when it can see its prey more easily. Thus, the cost of conspicuous courtship increases as the sun moves overhead, fully illuminating the guppies' habitat by midday [362]. The prediction that male guppies should be less likely to give their courtship displays under bright light conditions has been confirmed (Figure 13A). Moreover, since the cost should be higher for larger, more conspicuous males than for smaller ones, we can predict that large males should be especially prone to abandon courtship under high light intensities, and they are (Figure 13B) [990].

The hypothesis we have just presented is an example of an explanation based on **optimality theory,** the assumption that traits with cost–benefit ratios better than those of alternative traits will spread through populations (Figure 13C). Thus,

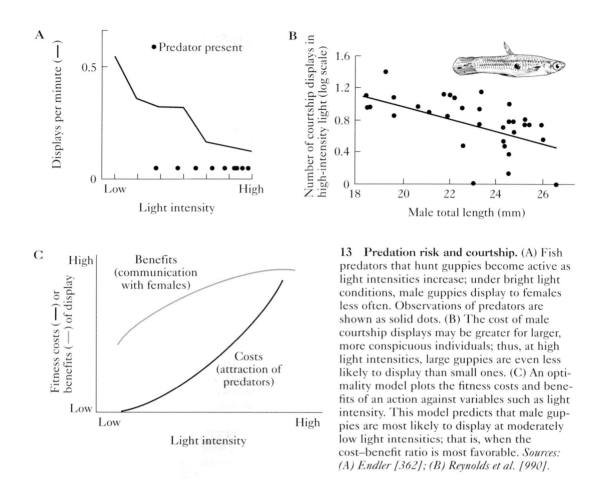

13 Predation risk and courtship. (A) Fish predators that hunt guppies become active as light intensities increase; under bright light conditions, male guppies display to females less often. Observations of predators are shown as solid dots. (B) The cost of male courtship displays may be greater for larger, more conspicuous individuals; thus, at high light intensities, large guppies are even less likely to display than small ones. (C) An optimality model plots the fitness costs and benefits of an action against variables such as light intensity. This model predicts that male guppies are most likely to display at moderately low light intensities; that is, when the cost–benefit ratio is most favorable. *Sources: (A) Endler [362]; (B) Reynolds et al. [990].*

if in the past some male guppies ignored light intensity and displayed with equal probability throughout the day, those males were presumably more likely to be detected and killed by predators, eliminating any genetic basis for their less-than-optimal behavior in populations with more cautious courters.

Making an Attack Less Likely

Even cryptic prey have some probability of being detected by a perceptive predator. One way to avoid a mano-a-mano confrontation with a predator that has spotted you is to keep it away with noxious tissues, stinging hairs, foul exudates, boiling sprays, sticky secretions, disgusting excretions, painful injections, and repellent regurgitates. "Better things for better living through chemistry," as the old DuPont Chemical Company slogan had it.

Consider the behavior of workers at nests of a tropical Asian relative of the honeybee, *Apis florea*. These little bees coat the branch on which their honeycomb is supported with rings of a sticky material (Figure 14A). Do the bees' secretions protect the nest's larvae and honey from ants? If so, we should see ants approach the ring of stickum and either retreat or get stuck, should they insist on trying to get to the nest. Observers have seen both outcomes, collecting direct observations to test this adaptationist hypothesis [1075]. We can also use the comparative method to test the ant repellent hypothesis by predicting that similar chemical repellents will have evolved in species unrelated to *Apis florea*, provided they share the same selection pressure: abundant ants hunting in the neighborhood of a vulnerable nest or cluster of young. Ant guards and ant-off substances have indeed evolved convergently in many other species (Figure 14B, C).

If related species differ in their risk of ant attack, they might also be expected to differ in their reliance on ant guards. In one paper wasp species in which solitary females must leave their nests unguarded while foraging, the females apply an ant repellent to the stalk of the nest (Figure 14B). In contrast, wasps of a closely related species live in colonies with large numbers of adults in attendance, and

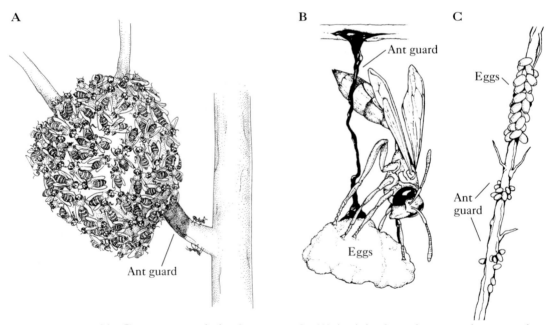

14 Convergent evolution in ant guards. (A) An Asian honeybee coats the approaches to its exposed nest with a sticky ant trap. (B) Paper wasps nesting alone rub the pedicel of the nest with a chemical ant repellent. (C) Owlfly females ring the grass stalk on which their eggs are glued with egglike objects that guard the eggs from ants. *After (A) Seeley et al. [1075]; (B) Jeanne [606]; (C) Henry [539].*

are therefore less vulnerable to ants because the nest is never unguarded. This species does not use anti-ant chemicals to protect the nest, a case of divergent evolution that supports the hypothesis that chemical ant guards evolve only when their benefits exceed their costs [607, 666].

Warning Coloration and Batesian Mimicry

Many species that rely on chemical defenses to deter vertebrate enemies (as opposed to ants) tend to be warningly colored in eye-grabbing reds, blacks, oranges, and yellows. Moreover, warningly colored species often behave in ways that enhance their already conspicuous appearance (Figure 15), perhaps to help remind predators of past unpleasant experiences with others of their species. After having tried to eat a warningly colored toxic prey, some predators quickly learn to back off whenever they see the same color pattern or warning display [134, 135]. A host

15 Warning predators not to attack. (Left) A tropical wasp that combines bright warning coloration with social defense. (Right) Red and black ladybird beetles are familiar examples of brightly colored, noxious animals that often cluster together, thereby enlarging the effect of their conspicuous appearance. *Photographs by (left) W. D. Hamilton and (right) the author.*

of edible prey have evolved a superficial resemblance to truly bad-tasting species, taking advantage of this ability of predators to learn particular color patterns on sight [1272]. These deceptive species are called *Batesian mimics*, after Henry Bates, an English naturalist who discovered their existence in Brazil during the 1800s.

Batesian mimicry can also work if animals have an innate fear of certain cues or color patterns. For example, consider the large, 2- to 6-inch larvae of several moths that immediately expand the anterior body when touched, forming a triangular "head" complete with "snake eyes" [1272]. Most predatory lizards are probably reluctant to press the attack on such a transformed caterpillar, which lunges forward as if it were a vine snake, a prime lizard killer (Figure 16).

Although no one has tested the Batesian mimicry hypothesis for snake caterpillars, it has been ingeniously tested for a tephritid fly thought to mimic a jumping spider [469, 780] (Figure 17). Jumping spiders defend their territories with leg-waving displays. The fly possesses a leglike pattern on its wings. When it waves its wings, something it does habitually, the "legs" appear (to human observers) as

16 Mimicry of a predator. A snake-mimicking caterpillar (left) resembles the Mexican vine snake (right) when the caterpillar lowers and expands the anterior part of its body to create a triangular snake head, complete with realistic eyes. *Photographs by (left) Lincoln P. Brower and (right) James D. Jenkins.*

17 Deterring jumping spiders by mimicry. A tephritid fly (left) waves its banded wings in ways that deceitfully mimic the aggressive signals of predatory jumping spiders (right). *From Mather and Roitberg [780].*

if they were the waving legs of a displaying jumping spider, which might persuade any nearby fly-eating spiders to retreat.

Two teams of researchers have established that jumping spiders are indeed reluctant to approach intact, wing-waving spider mimics. One team of fly surgeons armed with scissors and Elmer's glue performed wing transplants with clear-winged houseflies and pattern-winged tephritid flies. After the operation, the tephritid flies behaved normally, waving their now plain wings and even flying about their enclosures. But these modified tephritids with their housefly wings were almost totally ineffective in convincing approaching jumping spiders to back off. In contrast, tephritids whose own wings had been removed and then glued back on repelled their enemies in 16 of 20 cases. Houseflies with tephritid wings gained no protection, showing that it is the combination of leglike color pattern *and* wing movement that enables the tephritid fly to deceive its predators into thinking that it is a dangerous opponent rather than their next meal [469, 781].

Batesian mimicry may include acoustical, as well as visual, deception. Once I closely approached a snake I knew perfectly well to be a gopher snake, but when it made a sound like an aroused rattlesnake, I leapt back in fear. No one has carefully examined the possibility that gopher snakes startle their actual predators by imitating a rattler's tail-shaking rattle. But one well-studied case of acoustical Batesian mimicry involves burrowing owls (Figure 18), which give a call while in their nest tunnels that also sounds very much like a rattlesnake's rattle [1022]. Since rattlesnakes often spend the day in underground burrows, it is plausible that an owl that sounded like a rattler might persuade an unwanted intruder to go elsewhere.

As a comparative test of this hypothesis, Matthew Rowe and his co-workers pointed out that the burrowing owl is the only member of its family that nests underground (i.e., in rattlesnake habitat), and is the only owl in its family that possesses a rattling call. To test the acoustical mimicry hypothesis experimentally, the researchers examined whether Douglas ground squirrels discriminated between rattlesnake rattles and other burrowing owl vocalizations. Douglas ground

18 **An acoustical Batesian mimic:** the ground-nesting burrowing owl, whose hissing call resembles a rattlesnake's rattle. *Photograph by Stephan Schoech.*

squirrels are burrowing rodents that live in the same areas as burrowing owls and could destroy the nests and young of owls. If burrowing owls are producing a "rattlesnake" signal, then ground squirrels should treat the mimetic signal with the same respect they show the real McCoy. When captive ground squirrels were given an opportunity to enter an artificial burrow after having seen a rattler, they were equally reluctant to enter when they heard tape recordings of rattlesnake rattles and the burrowing owl equivalent, but much less hesitant if they heard an owl's scream–chatter or white noise coming from the burrow (Figure 19). These results support the hypothesis that burrowing owls engage in acoustical Batesian mimicry.

Associating with a Protected Species

Most Batesian mimics reduce the probability of an attack by looking or sounding like a dangerous, poisonous, or noxious species. Other animals use the defenses of well-protected species to deter assaults in a different way. For example, a variety of tropical birds, including rufous-naped wrens, often nest close to colonies of certain ants, bees, and wasps. Might they be taking advantage of their stinging neighbors to keep their enemies at a distance? If they are, then individuals that

A

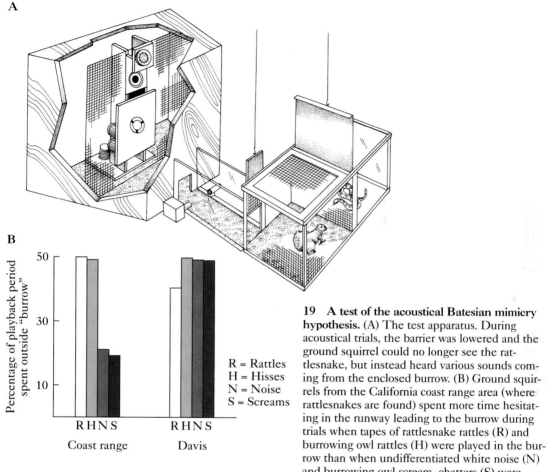

B

R = Rattles
H = Hisses
N = Noise
S = Screams

19 **A test of the acoustical Batesian mimicry hypothesis.** (A) The test apparatus. During acoustical trials, the barrier was lowered and the ground squirrel could no longer see the rattlesnake, but instead heard various sounds coming from the enclosed burrow. (B) Ground squirrels from the California coast range area (where rattlesnakes are found) spent more time hesitating in the runway leading to the burrow during trials when tapes of rattlesnake rattles (R) and burrowing owl rattles (H) were played in the burrow than when undifferentiated white noise (N) and burrowing owl scream–chatters (S) were played. Ground squirrels from Davis, where rattlesnakes do not live, did not discriminate among the various sounds. *Source: Rowe et al. [1022].*

nest near colonies of stinging social insects should be more likely to fledge their young successfully than those that nest without dangerous insects nearby.

Frank Joyce tested this prediction by moving paper nests of the wasp *Polybia rejecta* close to some nests of the rufous-naped wren. Since these wasps are notoriously quick to assault vertebrates of any sort, Joyce had to work at night while enclosed in a modified beekeeper's suit. But he eventually succeeded in gingerly transferring wasp nests to within a few feet of 28 randomly selected wren nests. The wasps soon settled down in their new homes, and the wrens benefited from their presence. About half of the wasp-protected wren nests ultimately produced fledglings, whereas fledging occurred in only 10 percent of the control nests, which were not experimentally blessed with wasp neighbors and could be approached with impunity by white-faced monkeys. These predators were smart enough to leave well enough alone when they found bird nests with wasps waiting nearby [619].

Some invertebrates also associate with powerfully protected social hymenopterans. For example, the caterpillars of a lycaenid butterfly are generally attended by a retinue of ants, which feed from "honeydew" glands on the larva's back. Do the caterpillars feed the ants to enlist their aid in repelling the many small wasps and flies that try to lay their eggs in or on butterfly caterpillars [934]? To find out whether ants do protect the caterpillars, Naomi Pierce and Paul Mead prevented ants from tending some larvae while permitting other caterpillars to retain their keepers. In order to create a group of untended larvae, Pierce and Mead placed sticky Tanglefoot around the base of a plant to keep ants away from the caterpillars on the leaves overhead. After establishing ant-free and ant-tended caterpillars, they inspected the larvae daily until they were ready to pupate. The caterpillars were then held until they either pupated successfully or were killed by parasitoids in their bodies, an outcome that occurred more than twice as frequently when the larvae lacked ant companions as when they grew up with a protective force of ants [934].

Advertising Unprofitability to Deter Pursuit

Warning coloration and behavior advertise the potential noxiousness or dangerousness of a prey, thereby making it advantageous for a predator to accept the information and not attack—except when the signal is a false one, as it is in Batesian mimicry. But even perfectly palatable prey may be unprofitable victims if, for example, they cannot easily be captured. When a prey communicates its physical fitness, a predator can avoid a futile attack while the prey can avoid a long run to prove the point.

One possible advertisement of this sort occurs when Thomson's gazelles have spotted a predator, such as a cheetah. The gazelles start to run away, but they may slow up a bit in order to stot—that is, to jump about a half meter off the ground with all four legs held stiff and straight and with the white rump patch fully everted (Figure 20). Stotting might be a means of communicating to the predator that it has been seen by a gazelle that is ready and able to flee. Predators may choose to let alert gazelles go because they are so hard to catch.

The unprofitability advertisement hypothesis is, however, not the only possible explanation for stotting. Perhaps a stotting gazelle sacrifices speed in escaping from one detected predator in order to scan ahead for other as yet unseen enemies lying in ambush, as lions often do [941]. The anti-ambush hypothesis predicts that stotting will *not* occur on short-grass savanna, but will instead be reserved for tall-grass or mixed grass and shrub habitats, where predator detection could be improved by jumping into the air. But gazelles feeding in short-grass habitats do stot regularly, so we can reject the anti-ambush hypothesis and turn to some others proposed by Tim Caro [180, 181]:

20   An advertisement of unprofitability? Stotting behavior by a Thomson's gazelle that has spotted a cheetah.

Alarm signal hypothesis: Stotting might warn conspecifics, particularly offspring, that a predator is dangerously near. This could increase the survival of off-spring and relatives, thereby improving the representation of the signaler's genes in subsequent generations (see Chapter 15).

Social cohesion hypothesis: Stotting might enable gazelles to form groups and flee in a coordinated manner, making it harder for a predator to cut any one of them out of the herd.

Confusion effect hypothesis: By stotting, individuals in a fleeing herd might confuse and distract a following predator, keeping it from focusing on one animal.

Table 3 lists the predictions that are consistent with the hypotheses under review. Because the same prediction sometimes follows from two different hypotheses, we must consider multiple predictions from each hypothesis if we are to discrim-inate among them.

Caro learned that a solitary gazelle will sometimes stot when a cheetah approach-es, an observation that helps eliminate the alarm signal hypothesis (if the idea is to communicate with other gazelles, single gazelles should not stot) and the con-fusion effect hypothesis (because the confusion effect can occur only when a group of animals can flee together). We cannot rule out the social cohesion hypothesis on the grounds that solitary gazelles stot, because there is the possibility that solitary individuals try to attract distant gazelles to join them. But if the goal of

Table 3 Predictions derived from four alternative hypotheses on the adaptive value of stotting in Thomson's gazelle

| | Alternative hypotheses | | | |
Prediction	Alarm signal	Social cohesion	Confusion effect	Signal of unprofitability
Solitary gazelle stots	No	Yes	No	Yes
Grouped gazelles stot	Yes	No	Yes	No
Stotters direct white rump toward predator	No	No	Yes	Yes
Stotters direct white rump toward other gazelles	Yes	Yes	No	No

stotting is to communicate with fellow gazelles, then stotting individuals, solitary or grouped, should direct their white rump patch toward other gazelles. Stotting gazelles, however, orient their rumps toward the predator.

The only hypothesis still standing states that gazelles stot to signal that they have seen the predator and will be hard to capture. Caro found that cheetahs were significantly more likely to abandon hunts when they saw a gazelle stot than when the potential victim did not perform the display (Figure 21) [181].

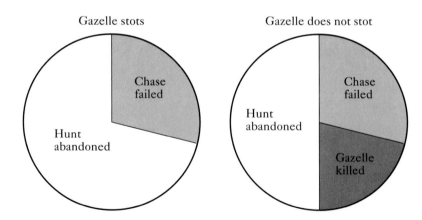

21 Cheetahs abandon hunts more often when gazelles stot than when they do not, supporting the hypothesis that these predators treat stotting as a signal that the gazelle will be hard to capture. *Source: Caro [181].*

Making Capture Less Likely

We have reviewed a sampler of traits that either increase the probability that a predator will overlook a prey or choose not to carry out a full attack. But what can an animal do if a serious attack is already under way? One possibility is to try to startle the predator momentarily, giving itself a few extra seconds to escape, as suggested for the wing-flipping, false-eye-exposing behavior of *Automeris* moths (see Chapter 1). Theodore Sargent and Debra Schlenoff have examined this hypothesis in detail while studying *Catocala* underwing moths. When a resting moth is grabbed by a blue jay, the moth struggles, suddenly exposing its hindwings, which typically have orange, red, or yellow bands on a dark background. The sudden flash of color appears to surprise the jay, which sometimes gapes involuntarily, allowing the moth to escape [1048].

To explore this response further, Schlenoff constructed artificial "moths" consisting of plastic wings with a pinyon nut seed attached underneath them, and trained captive blue jays to pick up the moths and remove the reward [1054]. Blue jays like pinyon nuts, and they soon learned to grab the "moths" without hesitation, after habituating to the plastic "hindwings" that popped into view whenever a "moth" was removed from its presentation board. Blue jays with experience grabbing pseudo-moths with gray hindwings were definitely startled when they picked up a distinctive moth with brightly colored hindwings like those of *Catocala* moths. When confronted with such a novel individual they often jumped back, dropping the "prey" and crying out in alarm.

Vigilance and Sociality

Another weapon in the anti-capture arsenal of some species is vigilance—staying alert so as to detect a rapidly approaching enemy in time to take effective action. Vertebrate prey generally keep an eye scanning, or nose sniffing, or ear listening for danger, no matter what else they are doing. The principle that many eyes, noses, or ears are better than a few could contribute to the widespread occurrence of flocks, herds, and other social groups. For example, a bird in a foraging flock may be able to escape from a predator that it has not personally detected by reacting to the behavior of its companions as they try to save their skins [720].

One way to test the many-eyes hypothesis would be to examine whether scanning rates fall as group size increases, as one would predict if group members can afford to rely on the reactions of others to detect an approaching predator. In many animals, ranging from shorebirds [235] to zebras [1052], the larger the group, the less often any one individual looks up to scan for predators.

And the more scanners, the quicker the response to danger, as G. V. N. Powell showed by releasing an artificial hawk that traveled along a line over an aviary containing either a single starling or a group of ten of these despised but adaptable birds [955]. The average reaction time of the solo bird was significantly longer than

Table 4   Effects of group size on the response of caged starlings to a simulated predator

Behavioral effect	Number of starlings in cage	
	One	Ten
Mean number of times per minute that bird stops foraging to look up	23.4	11.4
Percentage of foraging time spent in surveillance	47	12
Mean takeoff time (seconds) after hawk model is released	4.1	3.2

Source: Powell [955].

that of a bird in a flock, despite the fact that single birds spent about four times as much time looking around for danger as birds in a flock (Table 4).

R. E. Kenward took this test one step closer to reality by releasing a hungry, trained goshawk at a standard distance from wild flocks of wood pigeons [642]. As predicted, the larger the flock, the sooner the prey took flight, lowering the probability that the goshawk would make a kill (Figure 22).

Although these studies provide support for the hypothesis that social animals gain vigilance benefits, clumped prey may also pay a price for living together via increased competition for food. Considering this cost, Mark Elgar predicted that house sparrows would not form groups when the risk of predation was low, and as expected, house sparrows did not produce the "chirrup" calls that attract others to them when foraging at feeders close to safe cover and far from an observer (a potential predator) (Figure 23) [342].

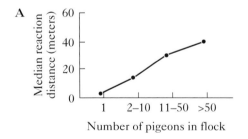

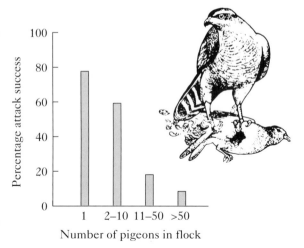

22   Vigilance and group size. (A) The larger the group of wood pigeons, the sooner the flock detected and flew away from an approaching goshawk, reducing the predator's chance of making a kill (B). *Source: Kenward [642].*

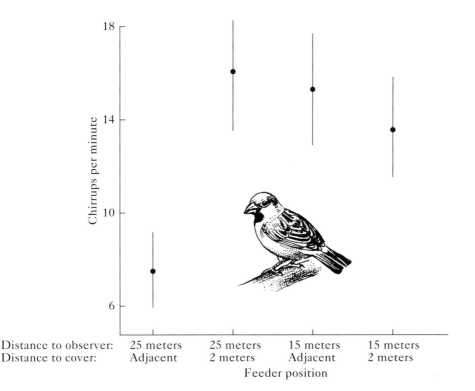

| Distance to observer: | 25 meters | 25 meters | 15 meters | 15 meters |
| Distance to cover: | Adjacent | 2 meters | Adjacent | 2 meters |

Feeder position

23 Social costs and defensive benefits. House sparrows were less likely to call for companions to join them at a food source when the risk of predation was low. The number of chirrup calls given per minute was lowest when the feeder was distant from an observer and close to cover. *Source: Elgar [342].*

Alarm Signals

Many animals that form flocks give special calls when they see an approaching predator, a costly response if the predator turns its attention to the signal giver. But perhaps alarm calls enable the members of a group to coordinate their escape, offering mutual benefits for all as the prey try to confuse a predator by fleeing en masse. Belding's ground squirrels, which live at high densities in mountain meadows, offer a probable case of this sort. When one of these small mammals sees a hawk or falcon sailing in for the kill, it gives a specific call, a high-pitched whistle, as it dashes for cover. The species has another alarm call for terrestrial predators such as coyotes (see Chapter 15). The ability to identify different enemies with different calls has evolved in some other animals as well [192]. In fact, Constantine Slobodchikoff and his associates have collected evidence that Gunnison's prairie dogs can learn to recognize specific individual predators and will give unique alarm calls for each individual, perhaps enabling the prey to take collective action appropriate for a particular enemy of known abilities [1122].

In any event, when Belding's ground squirrels hear the hawk alarm call, they all sprint for the nearest burrow, creating a brief moment of pandemonium. Whistling squirrels are almost never captured, a fact that enabled Paul Sherman to dismiss the hypothesis that callers warn others at risk to their own safety. In fact, noncalling animals attempting to escape were more than ten times as likely to be killed as alarm givers. If callers are safe, why do they call? And if noncallers are at risk, why do they run when they hear the alarm whistle? Sherman argues reasonably that if the first animal to see an incoming hawk failed to call as it ran conspicuously for safety, it might be targeted for attack. By calling, it can get lost in the chaos created as large numbers of squirrels dash in all directions. Noncallers are under selection to run in response to an aerial predator signal because if they do not, they will be sitting out in the open with a goshawk or prairie falcon headed their way, not a happy prospect for a ground squirrel. If so, the noncallers are safer if they react to the alarm than if they ignore it, producing benefits for all participating in the signal-giving and signal-responding interaction [1084].

The Selfish Herd

Belding's ground squirrels cooperate to foil hawks and falcons. But cooperation is not necessary to account for the formation of some "defensive" groups, as W. D. Hamilton emphasized by coining the phrase **selfish herd** [501]. What is a selfish herd? Imagine a population of antelope grazing on an African plain in which all individuals stay well apart, reducing their conspicuousness to their predators. Now imagine that a mutant individual arises in this species, one that approaches another animal and positions itself so as to use its companion as a living shield for protection against attacking predators. The mutant that employs this tactic will incur some costs; for example, two animals may be more conspicuous to predators than one, and so attract more attacks than scattered individuals. But if these costs are consistently outweighed by the fitness benefit of reduced capture, the mutation can spread through the population. If so, eventually all the members of the species will be aggregated, jockeying for the safest position in their groups. The result will be a clumping of individuals, a selfish herd whose members would actually be safer if they all could agree to spread out and not try to take advantage of one another. But since populations of such noninteractive individuals would be vulnerable to invasion by an exploitative mutant that takes fitness from its companions, the exploitative tactic can spread through the species, a clear illustration of why an adaptation is defined in terms of its contribution to individual fitness *relative* to other traits.

The selfish herd is a hypothesis based on **game theory,** the theory that the fitness payoffs for an individual as it attempts (consciously or unconsciously) to maximize its reproductive success depend on the actions of the other animals around it, particularly the actions of other members of its species. Thus, instead of focusing solely on the interaction between a prey and its predators, Hamilton asked how outcomes of prey–predator interactions might be influenced by what the other prey "decide" to do. Note that this is simply a form of the optimality approach in

24 Antipredator behavior in an Adélie penguin group, which is bursting out of the water in all directions in response to a leopard seal attack, presumably to confuse their attacker. *Photograph by Gordon Court.*

which an animal's social environment is considered when calculating fitness costs and benefits.

When Adélie penguins leave their breeding grounds to go out to sea to feed, they often gather in groups by the water's edge and then jump into the water together to swim out to the foraging grounds. The potential value of this social behavior becomes clearer when one realizes that a leopard seal may be lurking in the water near the jumping-off point (Color Plate 11) [224]. The seal can capture and kill only a certain small number of Adélies in a short time, and by swimming in a group through the danger zone, many penguins will escape while the seal is engaged in dispatching one or two unfortunate ones. If you had to run a leopard seal gauntlet, you would probably do your best to be neither the first nor the last into the water. If by going out in the middle of a wave of paddling penguins, an individual uses its companions as living shields to divert the predator, penguin flocks would qualify as selfish herds, although Adélies employ other group antipredator tactics as well (Figure 24). The selfish herd hypothesis generates testable predictions. If members of a group use one another for their own protection, we should observe competition for the safest positions within the "herd." Unfortunately, no one has systematically tested the prediction that departures into the water will be highly clumped, with many penguins waiting until the first few individuals take the plunge.

Evidence on the prediction that individuals should compete for the safest position has been gathered for some animals, including bluegill sunfish [480]. These fish breed in colonies, with males defending nesting territories against rivals. Males compete intensely for central territories, and larger, generally older, males are more likely to win. Visiting females prefer to lay their eggs in these territories, which are far safer from predators than eggs in peripheral nests. It is the peripheral male who first confronts a foraging bullhead or cannibalistic bluegill, and as a result, his eggs are twice as likely to be eaten by a predator as those of a male with a central nest.

If peripheral bluegills are at special risk, why do they accept their inferior position? If a younger or weaker individual has no reasonable probability of forcing a more powerful rival to yield his superior position, then the subordinate animal has two options: to nest on the outskirts of the group, or to nest alone. In the bluegill, solitary nesters (which do sometimes occur in this species) experience higher snail infestation rates, and must chase predators more often, than peripheral colonial males (Table 5). Note also that although mobbing of a predator is more likely to occur in the center of the colony, peripheral males also sometimes practice communal defense of their nests, an option unavailable to the solitary nester. Therefore, a subordinate fish may benefit slightly by nesting in a group even though more dominant males will use him as protection against their enemies [480].

The Dilution Effect

Perhaps the simplest antipredator advantage of living in a group is overwhelming the consumption capacity of the local predators, as Adélie penguins may do in leopard seal country. If five predators are hunting in an area, and each will kill two prey per day, a potential victim is ten times safer in a group of 1000 than in a group of 100. Will Cresswell wondered whether the dilution of risk accounted for the formation of flocks of redshanks, a sandpiper that winters on mudflats and marshes in Scotland [235]. Cresswell had determined that vigilance benefits could not explain why redshanks often form groups larger than 30 birds. In these large flocks, individuals scan for dangerous hunting hawks very much less often than birds in smaller groups, so much so that the probability that at least one member of a flock

Table 5 Predation pressure on bluegill nests in relation to nest position

Position of nest	Mean number of snails per nest	Mean number of times fish chases egg predator per hour	Percentage of chases in which two males pursue predator together
Center of colony	6.9	1.5	50
Edge of colony	13.7	8.7	8
Away from colony	29.7	10.4	0

Source: Gross and MacMillan [480]

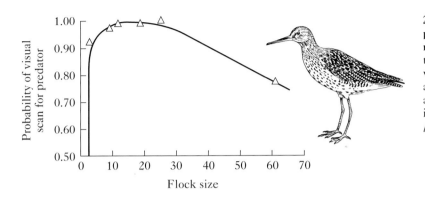

25 **Group size and predator detection in the redshank.** The probability that at least one redshank will have its head up in any 2-second interval rises and then falls as flock size increases. *Source: Cresswell [235].*

would have its head up during any given 2 seconds *fell* when flock size exceeded 30 (Figure 25). In 2 seconds, an attacking sparrow hawk can easily cover 25 meters.

Since members of large groups of redshanks do not detect predators sooner than members of intermediate-sized flocks, what do they gain from their choice of companions? Cresswell discovered that even though large flocks were more likely to attract a predator, the overall risk to an individual of being singled out by a rocketing peregrine falcon or sparrow hawk steadily declined as flock size grew, evidence for a strong **dilution effect** (Figure 26).

The dilution effect may also help explain why predator-prone túngara frogs call for mates in large groups [1036] (see Chapter 8), as well as why some mayflies make the transition to adulthood at the same time as many others of their species [1166]. Mayflies are apparently tasty insects, much enjoyed by various predators at all stages of life, but especially when they change from aquatic nymphs into winged airborne adults. This metamorphosis is one that entire populations may undertake over a few hours on a few days each year. Is synchrony in emergence from the water an anti-capture adaptation of these mayflies? To answer this question,

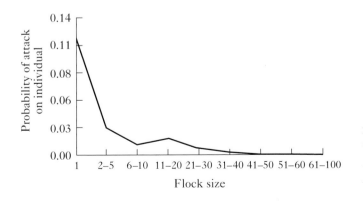

26 **Safety in numbers in the redshank.** The probability that any given individual will be targeted and attacked by a predatory hawk declines steadily with flock size. *Source: Cresswell [235].*

27   The dilution effect in mayflies. The more female mayflies emerging on a June evening, the less likely any individual mayfly is to be eaten by a predator. *Source: Sweeney and Vannote [1166].*

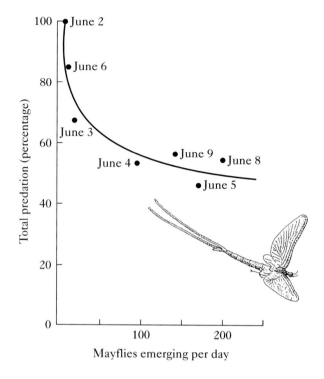

Bernard Sweeney and Robin Vannote first had to get a count of how many individuals emerged on a given evening from a segment of stream. To this end, they placed nets in streams to catch the cast-off skins of mayflies, which molt on the water's surface as they change into adults, leaving the discarded cuticle to drift downstream on the current. Counting these molted cuticles gave the observers the number of adults emerging on a particular evening. The nets also caught the bodies of females that laid their eggs and then died a natural death; a female's life ends immediately after she drops her clutch of eggs into the water, provided a nighthawk or a whirligig beetle does not consume her first.

Sweeney and Vannote measured the difference between the number of cast skins of emerging females and the number of intact corpses of spent adult females that washed into their nets on different days. The greater the number of females emerging together on a given day, the stronger the dilution effect, and the better the chance of each mayfly living long enough to lay her eggs before expiring (Figure 27) [1166].

Rapid Escape Flight

Many animals, social or solitary, will take evasive action if it appears that their number is up. The rapid, darting flight of certain butterflies may help these insects avoid being captured by birds, a possibility made more plausible by the discov-

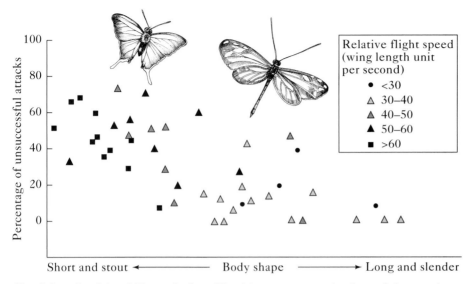

28 A benefit of the ability to fly fast. Chunkier, more muscular, faster-flying species of butterflies are less likely to be captured by a insect-eating jacamar than are thinner, less muscular, slower-flying species. *Source: Chai and Srygley [189].*

ery that some butterflies can accelerate in flight more rapidly than the birds that hunt them [758]. The hypothesis that rapid flight is an anti-capture adaptation produces the prediction that bird predators should miss rapidly flying butterflies more often than slow-flying species. To test this prediction, Peng Chai and Robert Srygley measured the flight speeds of a sample of Costa Rican butterflies using high-speed film [189]. They then released butterflies of various species into a large enclosure containing a captive jacamar, a bird noted for its consumption of fast-flying insects, especially butterflies. They found that speedy species were missed up to 80 percent of the time, whereas the slowest species were nailed on almost every attempt by the agile jacamar (Figure 28).

Why, then, are some tropical butterflies slow fliers? The ability to fly rapidly carries fitness costs as well as benefits. To beat one's wings with great force and power requires a large investment in thoracic muscles. In females a trade-off exists between muscle mass and ovarian tissue—between flight capacity and fecundity [758]. Fast-flying species invest as much as 40 percent of their body mass in flight muscles, leaving relatively little for ovaries. Slow-flying species use much less energy to build and maintain flight muscles, which leaves them with more to invest in a large abdomen filled with ovarian tissue (Figure 29).

Rapid flight has other costs in addition to the energetic costs of producing, maintaining, and operating the large muscles required for this activity. Srygley and Chai found that fast-flying species also apparently need to have high thoracic temper-

29   A cost of the ability to fly fast. The greater the investment that female butterflies make in flight muscles, the smaller their ovaries, creating a trade-off between flight speed and egg-producing capacity. *Source: Marden and Chai [758].*

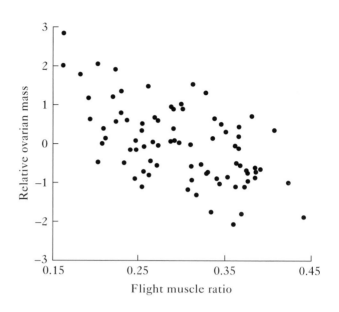

atures in order to initiate and maintain flight [1136]. This requirement means that fast fliers must remain inactive under cool conditions; when flying, they must stay in open, sunny patches. Slow fliers can fly during a greater part of each day, and are not restricted to areas that receive direct solar radiation.

Given the many costs of the mechanisms needed for rapid flight, Chai and Srygley predicted that only palatable species would fly fast, whereas unpalatable butterflies, which derive protection from predators primarily from their chemistry and coloration, would fly more slowly. Chai and Srygley determined the palatability of their Costa Rican butterflies by calculating the proportion of those individuals presented to caged jacamars that were actually eaten by the birds. As they expected, the rejected butterflies tended to be the weak-muscled, slow-flying species, whereas when a jacamar did capture a chunky, fast-flying butterfly, it almost always ate it with gusto.

Making Consumption Less Likely

A captured animal is not necessarily a consumed one, but may live another day if it can give its captor a reason to let go. Consider the notodontid moth caterpillar that I found feeding on a leaf. When I went over to take a closer look, the big larva suddenly dropped its head and abdomen from the leaf and formed an inverted U. Undeterred, I reached for the animal, at which point it everted a complex, red, antlerlike sac from the underside of its body and sprayed me with a strongly scented aerosol of formic acid (Figure 30). I withdrew my hand quickly.

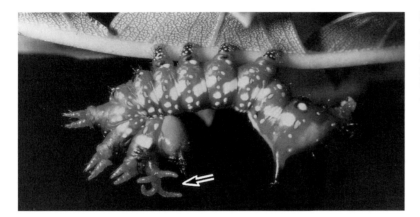

30 A chemical defense against consumption. This moth larva sprays formic acid from the osmeterium on the underside of its body to deter predators. *Photograph by the author.*

I do not know what natural predators the moth larva douses with formic acid, perhaps birds or ants. For a better understood case of chemical repulsion of a consumer, let's turn to the black widow spider. Black widows abound in my toolshed, and I once found a big female on her web under our kitchen table. I, like most Arizonans, treat these spiders with a mixture of fear and respect because of their toxic venom, but other animals are far less intimidated. It takes time for the spider to maneuver itself into position to use its fangs, and a small mammalian predator, such as a deer mouse, can immobilize a maneuvering spider with its own quicker bite. But about half the interactions between mouse and spider in a laboratory arena ended with the mouse repelled and the spider alive [1215]. In a large majority of these cases, the spider survived because of its rapid secretion and deployment of a strand of silk that it produced as soon as it was jostled (Figure 31). The silk strand, adorned with droplets of a profoundly adhesive substance, was applied by the spider to its attacker's face. The mouse typically recoiled and attempted to clean itself of the material, sometimes by rolling frantically on the ground. In nature, the spider could have used this reprieve to scramble from its web to a safe crevice.

Rick Vetter tested the value of the sticky silk strand for black widows by waxing over the spinnerets of some spiders. When attacked by mice, these individuals could not defend themselves with adhesives, and were three times as likely to be killed as unclogged spiders. Although the spider's glue is not toxic (consumption of the viscous substance did not affect the mice), it is sticky, forcing the mouse to remove it at once or suffer damage to its fur.

Chemical deterrents that come into play at the moment of capture are also used by some salamanders, which employ adhesives rather like those used by the black widow. When these salamanders are grasped by a garter snake, they writhe and thrash while releasing secretions from the tail and body. Because the secretions are sticky, the salamander's tail tends to adhere to the body of the snake. The goal of the salamander may be to frustrate attempts to eat it long enough to induce the

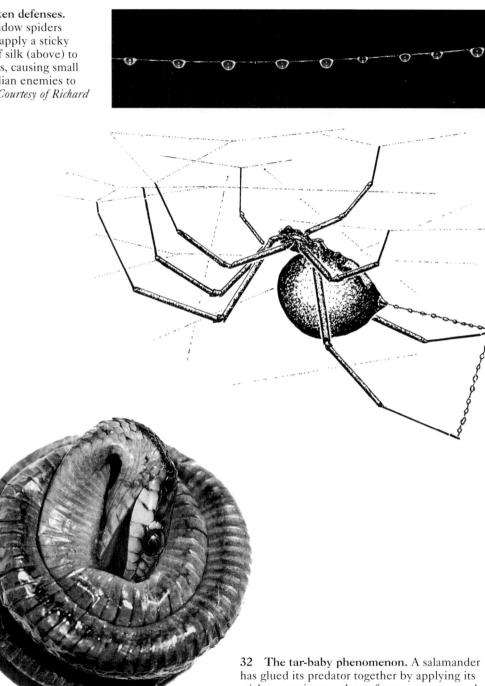

31 Silken defenses. Black widow spiders (below) apply a sticky strand of silk (above) to predators, causing small mammalian enemies to retreat. *Courtesy of Richard Vetter.*

32 The tar-baby phenomenon. A salamander has glued its predator together by applying its sticky secretions to the unfortunate garter snake. *Photograph by Stevan Arnold.*

snake to release it, somewhat the worse for wear, but still alive. Occasionally, a salamander may even succeed in coating its enemy so thoroughly with tar-baby glue
that the snake is rendered helpless (Figure 32), a most desirable outcome from the
salamander's perspective [33].

Misdirecting Consumers

Another way to thwart a consumer is to induce it to attack some body part other
than the head so as to give oneself time to make a last-second escape. Keeping
predators from striking the head is critically important because brain damage quickly immobilizes the prey and removes all chance of survival. As a result, animals
often hide their heads when under attack (Figure 33) [317].

A similar tactic is to induce predators to strike false heads on the posterior
portion of the prey's body, parts that can be sacrificed without incurring a mortal
wound [1272]. I sometimes find hairstreak butterflies alive and well with their
expendable false heads neatly snipped out of their hindwings (Figure 34), presumably by disappointed birds [1000].

Mark Wourms and Fred Wasserman investigated precisely how false heads work
by painting false heads of various sorts on the wings of living cabbage white butterflies [1308]. When the experimental butterflies and a set of unaltered controls
were released into an aviary inhabited by hungry blue jays, the birds set about adept-
ly capturing the insects. After grabbing a butterfly in its beak, a jay usually trans-

33 Protecting the brain.
A prairie rattlesnake hides
its vulnerable head from a
potential predator. *Photograph by Michael King, courtesy of David Duvall.*

34   Sacrificing a false head. This hairstreak butterfly has had one wing's false head (arrow) snipped out by a bird, but has escaped to tell the tale. *Photograph by the author.*

ferred the prey from bill to foot before killing it with blows to the head. At this stage, cabbage whites with false heads escaped almost three times as often as controls (Figure 35) because the jays mishandled the experimental butterflies by pecking at their false heads. This experiment supports the hypothesis that false heads induce life-saving mistakes, benefiting prey *after* they have been captured [1308].

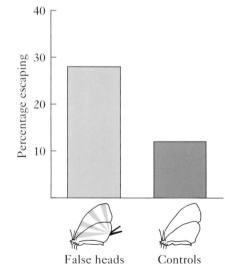

35   False heads misdirect attacks, increasing a butterfly's chances of escaping from a predator. In this experiment, blue jays mishandled captured cabbage butterflies with painted wing stripes that pointed to artificial false heads, permitting the butterflies to escape more often than unaltered controls. *Source: Wourms and Wasserman [1308].*

36 An expendable body part. (Left) This desert gecko raises and waves its tail, using it as an expendable lure when threatened by a potential predator. (Right) A desert gecko that has lived to regenerate its tail after losing it to a predator. *Photographs by (left) Justin Congdon and (right) the author.*

The use of an expendable, moving body part as a lure to deflect an *initial* strike occurs in some lizards, which twitch their tails when threatened, distracting their predators from their heads (Figure 36). Lizard tails may be brightly colored, as are the blue tails of young skinks. In one experiment, 9 of 19 blue-tailed skinks escaped when attacked by captive snakes, whereas only one of 15 skinks whose tails had been painted black had similar good fortune [220]. The blue-tailed youngsters escaped primarily because the snakes tended to attack the base of the conspicuous tail, and not the black body or head of the lizard. When a snake grabs a young skink's tail, it breaks off, thanks to the tail's fragile connection to the rest of the body. As a final touch, the automotized tail may thrash wildly on the ground. In another experiment, captive snakes were offered a skink's tail that was either still twitching or had been allowed to exhaust itself into immobility. The snakes took 40 percent longer to "subdue" and consume the thrashing tails than the nonmoving tails, which would give a freshly tailless lizard extra time to escape [295].

Attracting Competing Consumers

When a rabbit is in the clutches of a fox or hawk, it has no expendable tail to sacrifice. But it sometimes produces extraordinarily loud and piercing screams, despite the fact that it is normally the quietest of animals. Can it be adaptive to scream loudly in such a desperate situation? As described in a paper entitled "Adaptation unto Death," Goran Högstedt tested multiple hypotheses on the function of these vocalizations, including the possibility that a loud scream might so startle a predator that it would release its captured prey [838]. Högstedt noted that fear screams are noisy, as required by the startle hypothesis, but the calls are persistent rather than sudden and unexpected, which should reduce their potential to startle [555; but see 217].

Because fear screams contain both high and low frequencies of sound, features that enable others to locate the source of the sound easily, they might warn others of danger. Högstedt pointed out that if this second hypothesis were correct, one would expect conspecifics to respond to the cries by taking some action. Yet prey largely ignore the calls of captured neighbors, a reaction that makes a certain amount of sense, since a predator with a captured prey will be occupied with that victim for some time rather than hunting for uncaptured specimens nearby.

But perhaps fear screams are given by young animals to enlist the aid of their parents. This hypothesis produces the prediction that the calls should be given only by dependent young, and this is not the case. Adult birds captured in mist nets are just as likely to scream as juveniles when handled by a human being (Figure 37).

The evidence points to a fourth hypothesis, namely, that the about-to-die screams attract other predators to the scene, predators that may turn the tables on the prey's captor, or at least interfere with it, sometimes enabling the prey to escape in the confusion. This hypothesis requires that predators be attracted to fear screams, and they are, as Högstedt showed by broadcasting taped screams of a captured starling, which had hawks, foxes, and cats hurrying to the recorder.

Furthermore, the attract-competing-predators hypothesis produces the prediction that birds living in dense cover will be more prone to give fear screams than birds of open habitats. In areas where vision is blocked, a captured bird cannot rely on other nearby predators to see its mortal struggle. Therefore, as a final effort to avoid death, it attempts to call competing predators to the scene. Mist-netted birds of species that live in dense cover are in fact more likely to give fear screams when handled than are species that occupy open habitats [555].

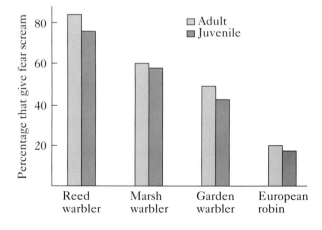

37   Fear screams and age in four European birds. In many species, juveniles are no more likely to give fear screams when handled by a human than are adults. This observation suggests that the function of the call cannot be to enlist parental assistance. *Source: Högstedt [555].*

The possibility of predator-attracting signals has also been explored in fathead minnows. Like many fish, fathead minnows that have been attacked release chemicals from specialized cells in the skin. Traditionally, these chemicals have been considered "alarm signals" designed to alert other members of the species to the presence of a predator, despite some evidence to the contrary [195, 753]. Douglas Chivers and his colleagues tested an alternative to the alarm signal hypothesis, namely, that injured minnows release these chemicals to attract other predators to the scene, where their interference might result in the release of the captured prey [194]. To perform their test, the researchers used divided aquaria with a juvenile pike in each compartment. After a fathead minnow had been placed in one compartment and was captured by the resident predator, the panel separating the two predators was removed. The second pike often approached its neighbor; other studies have shown that the "alarm odor" in and of itself can attract predators from a distance. Once near its neighbor, the intruding pike tried to steal the minnow by various means, with the effect that the time required for the first pike to swallow its victim increased threefold, compared with control trials in which the dividing panel was removed, but there was no second pike in the other compartment (Figure 38). Moreover, in 5 of 13 trials, the minnow escaped, at least temporarily, something that never occurred during the control trials. These findings confirm the utility of the "predator attraction signal" as a last-ditch survival aid for the signaler, rather than an alarm message for its fellow prey.

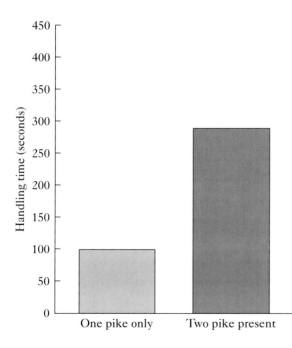

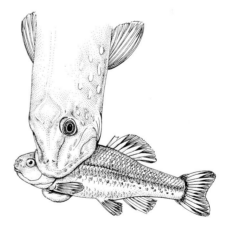

38 A predator attraction signal. By releasing chemicals from the skin when being bitten by a pike, fathead minnows can attract other pike to the scene. The presence of a second pike increased the time it took a pike to position a minnow properly for swallowing. *Source: Chivers et al. [194].*

Monarch Butterfly Defense Systems

Let's review the four components of antipredator behavior by looking at the inter-actions between the monarch butterfly and its enemies. This famous insect makes little or no attempt to avoid detection by its potential predators, but instead exhibits a brilliant orange and black color pattern. As might be expected from this classic warning coloration, monarchs are often poisonous, to the point of being fatal if con-sumed by a human [138]. In dealing with monarch butterflies, therefore, you would be wise not to follow the practice of the famous lepidopterist E. B. Ford, who wrote, "I personally have made a habit, which I recommend to other naturalists, of eat-ing specimens of every species which I study" [394].

Humans, however, are not the primary predators of monarchs. Instead, insect-eating birds constitute a major danger to these insects. But birds learn quickly to avoid toxic prey. A blue jay that eats a poisonous monarch becomes ill and vom-its 15 to 30 minutes after its meal (Figure 39). In the laboratory, a single such expe-rience suffices to educate the predator to avoid the butterfly thereafter on sight alone. Thus, the monarch's warning coloration apparently helps individuals avoid attack by educated predators.

Even if a poisonous monarch is captured, it may be released unharmed, espe-cially by an experienced predator that remembers what happens after consuming a creature whose wings taste like monarch butterfly wings. In a laboratory experiment, monarchs were usually dropped immediately after being attacked by quail [1279].

39 Effect of monarch butterfly toxins. The blue jay that eats a toxic monarch vomits a short while later. *Photographs by Lincoln P. Brower.*

In fact, from the monarch's perspective, the point of possessing toxic chemicals in one's tissues is not to cause a blue jay or some other predator to throw up. A vomited monarch does not fly off into the sunset. Instead, selection may have favored toxic individuals simply because they taste so bad that they may induce a prompt release. Vomiting and the learning it promotes may be protective adaptations of birds, which rid themselves of ingested toxins and avoid a second mistake.

But because predators can learn to avoid the monarch color pattern on sight, the door to deception opens. In this light, it is significant that not all monarch butterflies are poisonous. To understand why, it helps to know that these insects do not manufacture their own toxins, but instead make use of the poisons that may be present in their foods [138]. The same trick has independently evolved in a number of other insects [341] and even some vertebrates (Figure 40) [131]. The key recycled poisons for monarchs are cardiac glycosides contained in certain milk-

40 Recycled poisons. (Top) When threatened by a predator, Australian sawfly larvae regurgitate sticky aromatic oils sequestered from the eucalyptus leaves on which they feed. (Bottom) A hedgehog anoints itself with a foam containing toxins from a toad it has recently eaten. *Photographs by (top) the author and (bottom) E. D. Brodie, Jr.*

41 Monarch female lay-
ing an egg on a milkweed.
The concentration of car-
diac glycosides contained
in the plant will determine
the noxiousness of her offspring to predators.
*Photograph by Fred A.
Urquhart.*

weeds upon which monarch females lay their eggs (Figure 41). Monarch larvae
not only consume the poisonous milkweeds safely, they store and retain the plant's
poisons in their tissues when they undergo metamorphosis to adulthood.

Evidence that the butterfly acquires toxins from its foods came from Lincoln
Brower and his associates, who "persuaded" some monarch larvae (through artificial selection) to eat cabbage. When the adults reared on this harmless food plant
were fed to jays, the birds did not vomit. In contrast, jays that consumed butterflies reared on the milkweed *Asclepias curassavica* always vomited a short time later
[139]. Chemical analysis showed that the poisons in the milkweeds and in the butterflies reared on them were identical.

Thus, the monarch that feeds on a toxic food plant benefits from the evolved
defenses of the plant. But there are probable costs for this poison-eating individual as well. The larva must devote some of its metabolic energy to gathering the
toxins and storing them safely. If it did not have these expenses, it presumably
could grow faster or larger, and could reproduce more as a result.

In fact, monarch females in nature oviposit on a variety of food plants, ranging
from the very poisonous to the completely nontoxic [311]. The selection of a food
plant constitutes an evolutionary game between competing females. The best
option for a female will depend on what other females in her population are doing.
If many lay their eggs on toxic food plants, those females that lay eggs on nontoxic
food plants will produce edible offspring that deceptively advertise "unpalatability" to predators that will be familiar with other, toxic monarchs.

Edible monarchs do exist, but they have been discovered and exploited by some birds that have the capacity to counter deception by their prey. In Mexico, certain grosbeaks and orioles frequent huge winter aggregations of monarch butterflies (see Chapter 11). The grosbeaks can eat monarchs with impunity because they are insensitive to the butterfly's poisons. Black-backed orioles cannot simply gobble down one monarch after another, but they can find the relatively edible "automimics" by giving each captured monarch a taste test. The orioles throw away those butterflies that taste bad, and feast on those that are not poisonous, or only mildly so [389]. Between them, grosbeaks and orioles killed about 2 million monarchs in a single winter at one location [136].

The birds can perform their taste tests with ease because monarchs spending the winter in the Mexican mountains are chilled and often unable to fly. No grosbeak would bother pursuing a monarch in summer, when the insect's powerful flight capacity would make its capture highly unlikely. But in the winter aggregations, the dormant monarchs cannot flee from their attackers. However, they can still reduce their probability of being an unlucky one selected for consumption by gathering by the tens of thousands on a single tree. By doing so, they bring the dilution effect into play, swamping the feeding capacity of their predators. Brower and colleagues have shown that the risk of dying for any one monarch declines in relation to increases in the number of butterflies clustered within an aggregation site [172]. Birds feed heavily on the edges of colonies, so that even within a colony, monarchs in the more densely populated central core have a much higher chance of survival than those in the less densely occupied periphery [136].

Monarchs and their avian enemies offer an illustrative example of how prey species can counter attempts by predators to find, attack, capture, and consume them (Table 6). In turn, the antipredator attributes of prey may favor the evolution of predatory tactics to overcome prey defenses, a point that we will develop further in the next chapter.

Table 6 Evolutionary interactions between monarch butterflies and some predators

Abilities of avian predators	Responses of monarch butterflies
Ability to learn to avoid noxious prey on sight	**Avoiding attack** Incorporation of toxins from foodplants by larvae; warning coloration Automimicry of noxious individuals by edible ones
Ability to select profitable prey	**Avoiding capture** Powerful flight; migration to winter aggregations for dilution effect
Ability to sample prey	**Avoiding consumption** Concentration of toxins in wings

SUMMARY

1. Most of the antipredator hypotheses considered in this chapter focus on the fitness benefits derived by prey as they deal with predators. However, adaptationist hypotheses can take costs as well as benefits in account. A trait is considered optimal if it has a better cost–benefit ratio than alternative traits. Optimality models also include those based on game theory, which examine how the fitness benefits and costs of a behavioral trait are affected by the actions of other members of the species.

2. To test adaptationist hypotheses, a scientist must check the validity of predictions derived from these potential explanations. Evidence can be gathered through (1) field observations, (2) controlled manipulative experiments, or (3) "natural" experiments involving comparisons of living species.

3. The comparative method yields two key predictions: (1) convergent evolution of traits will occur among *unrelated* species that share the same selection pressures, despite their having different ancestors, and (2) divergent evolution will occur among *related* species that face different selection pressures, despite their having a common ancestor and thus a similar genetic heritage. Therefore, we cannot, for example, test the hypothesis that mobbing behavior by black-headed gulls is an evolved adaptation that thwarts predators by predicting that *other* gull species will also exhibit this trait, because all gulls may share certain traits as a result of their shared ancestry, and not necessarily because of shared selection pressures. Appropriate cases of convergent and divergent evolution are required to test adaptationist hypotheses via the comparative method.

4. Predators exert selection pressure on most animals. Encounters between predator and prey can proceed from detection to attack to capture to consumption. At each level, prey may invest in traits that serve to make detection, attack, capture, or consumption less likely.

5. Thwarting detection often involves cryptic behavior in which an animal enhances the effectiveness of its camouflage color pattern by remaining motionless on an appropriate background. Blocking an attack often involves either honest advertisement of unpalatability through warning coloration or deceitful Batesian mimicry. Some animals also communicate to a potential attacker that an assault has a low probability of success because the prey is alert and physically fit.

6. Prey may prevent their capture through (a) vigilant behavior and alarm calls, (b) startling behavior that makes a predator hesitate during an attack, (c) the formation of selfish herds whose members attempt to use one another as living shields, or (d) the formation of large groups that dilute the risk to any one individual of being captured. Even captured animals may have a "doomsday option," such as fear screams or misdirection tactics that reduce the probability of being consumed.

7. The monarch butterfly illustrates that one species can have several different kinds of antipredator adaptations that come into play at different phases of encounters with predators. Furthermore, monarch behavior demonstrates that each tactic has costs as well as benefits, that none works equally well against all enemies, and that many elements of the behavioral repertoire of a species have interlocking consequences (oviposition choices by females, for example, affect the palatability of their offspring).

SUGGESTED READING

Malcolm Edmunds's *Defense in Animals* [331] and Wolfgang Wickler's *Mimicry in Plants and Animals* [1272] contain many examples of amazing behavioral defenses. John Endler has provided a modern review of the interrelationships between predator adaptations and prey counteradaptations [363].

The cost–benefit approach to antipredator behavior has been described by Steven

Lima and Lawrence Dill [721]. Tim Caro's papers on gazelle stotting [180, 181] are excellent, as is his comparative review of the pursuit deterrence hypothesis and how it has been tested [182]. These are additional interesting studies not discussed in the text that provide tests of alternative hypotheses on the antipredator function of behavioral traits [198, 517, 540].

DISCUSSION QUESTIONS

1. Many North Americans have seen white-tailed deer lifting their tails to expose their conspicuous white underside. Construct a list of hypotheses on the possible antipredator function of tail flagging and a list of predictions that would enable you to discriminate among the various candidates. Consider what may happen to a predator following a tail-flagging deer when that deer abruptly drops its tail, hiding the white undersurface. After completing your answer, you may wish to check references [90, 183, 547].

2. Bombardier beetles earn their name by firing a boiling mixture of quinones and hydrogen peroxide in the face of ants and other predators that molest them [338]. Consider both the benefits and costs of this behavior [1316] and develop testable predictions on when the beetle would *not* fire upon being contacted by a potential predator. (Keep in mind that the beetle has only so much defensive material stored in its glands, and that it must metabolize the substance from foods it consumes.)

3. Consider the following situation (with thanks to Jack Bradbury): In a population of

prey animals, most individuals are solitary and stay well apart from others. But some mutant types arise that search out others and use them as living shields against predators. The mutants take fitness from the would-be solitary types by making them more conspicuous to their predators.

Set the fitness payoff for solitary living at P in a population composed only of solitary individuals. But when a solitary individual is found and used by a social type, the solitary animal loses some fitness (B) to the social type. There is a cost (C) to being social in terms of the time required to find a solitary individual to hide behind, as well as the increased conspicuousness to predators of groups composed of two individuals rather than one. When two social types interact, we will say that they each have one chance in two of being the one that happens to hide behind the other when a predator attacks.

A game theory diagram (see p. 338) summarizes these interactions. If B is greater than C, what behavioral type will come to predominate in the population over time? Now compare the average payoff for individuals in populations composed entirely of solitary versus social types. If the average fitness of indi-

	Opponent	
	Solitary	Social

Focal animal	Solitary	$P + B - C$	$P + B - C$
	Social	$P + B - C$	$P + \frac{B}{2} - \frac{B}{2} - C$ $= P - C$

viduals in a population composed only of social types is less than that of individuals in a population composed only of solitary types, can hiding behind others be an adaptation?

4. In my front yard, I sometimes find several dozen to several hundred small native bees clustering in the evening on a few bare plant stems. These bees, which spend the night "sleeping" high in shrubs (see the photograph at right), are all males. An assassin bug sometimes approaches the cluster and kills some bees as they are settling down for the night. Devise at least three alternative hypotheses on the possible anti–assassin bug value of these sleeping clusters, and list the predictions that follow from each hypothesis.

Photograph by the author.

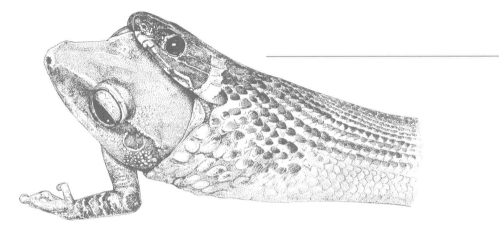

CHAPTER 10

Adaptive Feeding Behavior

A S THE PRECEDING CHAPTER DOCUMENTED, living prey have evolved a complete spectrum of responses to the risk of being detected, attacked, captured, and consumed. Thus, when a predator forages for food, it is faced with a whole series of obstacles that lie between it and a good meal. First, the forager has to find a food item, which may be a living prey skillfully hiding from its predators. Second, having found an item, the hunter may then have to decide whether an attack is likely to yield benefits in excess of its costs. Third, having launched an attack, the next trick is to capture the prey, which may very well have another outcome in mind. Finally, even after a prey is in hand, there may still be difficulties to overcome before the food (and its useful calories and nutrients) can be consumed.

This chapter examines each of these obstacles in turn, showing how behavioral ecologists have treated foraging decisions as cost–benefit problems that animals solve, enabling individuals to collect food as efficiently as possible. Food gatherers that do a better job of increasing the benefits of foraging and decreasing its costs should propagate their genes more effectively than others whose feeding activities yield lower net benefits. Optimality theory will play a particularly important role in this chapter as we explore its utility as a tool for understanding the adaptive value of feeding behavior.

Locating Food

Almost everything biological is eaten by one animal or another. Even the most extraordinarily elusive prey, or poisonous plant, or repellent organic substance (from a human perspective) contains exploitable calories and nutrients. These edible materials inadvertently provide stimuli that may be detected by a foraging animal. Thus, to take a humble example, dung contains volatile substances, and dung-eating beetles, of which there are many (Figure 1), take to the air when they smell far-off feces and quickly zigzag up the odor trail to what, for them, is a food bonanza [530].

In fact, urine and dung can be used by certain predators to locate the otherwise well-hidden living producers of these waste products. A small hawk, the kestrel, travels over the Finnish countryside looking for evidence that voles are present in numbers that warrant an intensive search. These small rodents deposit urine and feces on trails around their home shelters. The scents from their waste products signal occupation of the area to other voles. Vole urine and feces, however, not only release olfactory signals, but also incidentally reflect a certain amount of ultraviolet radiation, something that you and I cannot see, but kestrels can [1216]. Once a kestrel has found a sufficient density of ultraviolet-reflecting scent marks, it settles down to hunt, providing a fine example of an illegitimate receiver taking advan-

1   Dung-eating beetles. Even cow dung contains nutrients. This dung ball may nourish this male and female beetle, or the female may lay an egg on it (after the pair have buried it underground), and the resulting grub will feast at its leisure. *Photograph by the author.*

tage of another species' communication system (see also the case of the túngara frog in Chapter 8).

Thus, the kestrel, in its evolutionary "arms race" with its prey, has evolved a counteradaptation to the tendency of voles to hide from their predators. Likewise, visually hunting insect-eating birds may have evolved a hunting technique that enables them to overcome the remarkable cryptic coloration and behavior of certain prey. Luuk Tinbergen found that when a new species of camouflaged moth caterpillar started to appear in Dutch woodlands in late spring or early summer, nesting songbirds at first usually overlooked it. But after finding a few of these caterpillars, parent birds began to come back to the nest with more and more. Tinbergen suggested that insect-eating birds learned from their experience with a newly profitable but well-hidden prey by formulating a *search image* that emphasizes subtle visual characters associated with this food [1190]. By concentrating on these cues, the predator might improve its efficiency in detecting prey that blend into their environment.

Alexandra Pietrewicz and Alan Kamil used operant conditioning techniques to test Tinbergen's search image hypothesis. They trained captive blue jays to respond to slides of cryptically colored moths positioned on an appropriate background (see Figure 11 in Chapter 9). When a slide flashed on a screen, the jay had only a short time to react; if the jay detected the prey, it pecked at a key, received a food reward, and was quickly shown a new slide. But if the bird pecked when shown a slide with no cryptic moth present, it not only failed to secure a food reward, but had to wait a minute for the next chance to get some food [937].

In the course of these tests, the jays were shown a series of 16 slides, half of which contained an image of a moth. In one trial, the 8 slides were all of the same moth species, but in another 16-slide trial, two species of moths with different color patterns and resting positions appeared in the series in random order. When the birds were given repeated experience with just one species, they apparently formed a search image, because they made fewer errors as the series proceeded. But if they were shown a slide series with two prey species, they did not improve, suggesting that they either failed to form a search image at all or sometimes used the wrong search image on a given trial, thus reducing their prey detection accuracy under conditions when both prey were at equal frequency (Figure 2). An operant conditioning study of laboratory pigeons has yielded similar results [944].

In a field study, European blackbirds also became better at finding camouflaged food as they gained experience with it. In this case the "prey" was a flour-and-lard cylinder dyed green or brown and placed on trays covered with gravel that had been painted green or brown. The birds were familiar with this food because they had been given opportunities to forage for undyed versions of these pastry "worms." When hunting for the dyed prey on the background that made them inconspicuous (i.e., a green gravel background for green pastry cylinders), the birds found the last four items more quickly than the first four. This result is consistent with the notion that prey detection improves because the bird has learned a specific search image of a particular prey [702].

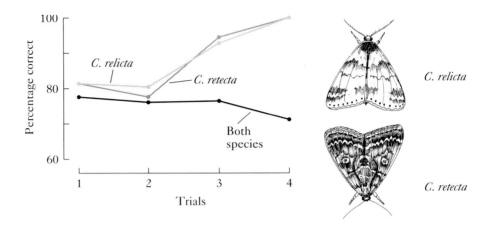

2 **Search image formation in blue jays.** When given the task of finding a moth image that appeared in 8 of 16 slides, blue jays did not improve if there were 4 slides of one species and 4 of another intermixed in the series. But if all 8 slides showed just one species, the jays did better at finding the moth over the course of the test, suggesting that they had learned to search for the key cues associated with that species. *Source: Pietrewicz and Kamil [937].*

The History of Prey Detection Mechanisms

Dung beetles detect the olfactory cues given off by their food, while birds locate their meals visually. Why do species often differ in the cues they use to find food? One historical answer to this question is that different animals inherit distinctive perceptual systems from their different ancestors. An example comes from an analysis of how various lizards secure their food. As it turns out, lizards fall into two camps: (1) active foragers that wander about, usually searching for insects or sometimes for edible plants, and (2) sit-and-wait ambushers that launch surprise attacks on passing prey. The active foragers invariably flick their tongues as they hunt, collecting chemical cues from the air that help them zero in on distant foods. In contrast, the ambush predators rely strictly on visual cues provided by moving prey.

William Cooper plotted the occurrence of the two types of foragers and their allied prey-detecting systems on a cladogram of the lizards [219]. The result (Figure 3) indicates three independent origins of active foraging with tongue flicking and one transition from this mode to ambush predation without tongue flicking. Thus, we learn that the shared attributes of certain families of lizards are derived largely, but not exclusively, from a common ancestor with a particular foraging characteristic. However, since tongue flicking has evolved three times in association with active foraging, we can also conclude that chemical detection of food is an adaptive component of this hunting technique.

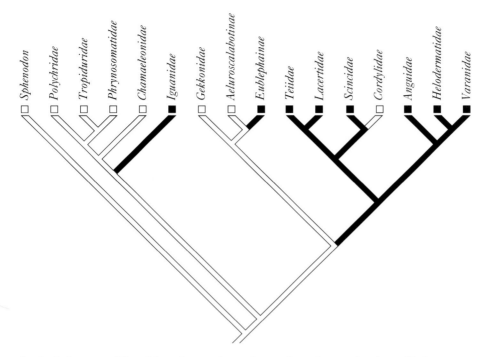

3 A cladogram of lizard foraging modes and prey detection mechanisms. Dark branches indicate that the members of this family of lizards actively search for food, flicking their tongues to detect prey odors. Open branches represent ambush predators, which employ vision rather than chemoreception to find their meals. *Source: Cooper [219].*

Getting Help from Companions

Foragers do not necessarily have to rely only on themselves to track down food, but may instead secure information about food location from their companions [816, 1236]. The social insects, including bees, wasps, ants, and termites [1293], and some social mammals [620] provide many examples of this phenomenon, the dances of honeybees being perhaps the most famous case.

The complex dances of worker honeybees are performed when a forager that has found pollen or nectar returns to the hive [1225, 1226]. These dancers attract other bees, which follow them around their circuits on the comb in the complete darkness of the hive. Humans watching dancing bees in special observation hives have learned that the dances contain a surprising amount of information about the location of a food source, such as a patch of flowers. If the bee executes a round dance (Figure 4), she has found food fairly close to the hive—say, within 50 meters of it. If, however, the worker performs a waggle dance (Figure 5), she has visited a nectar or pollen source more than 50 meters distant. By measuring (1) the number of times the bee runs through a complete waggle dance circuit in 15 seconds, or (2) the length of the straight-run portion of the waggle dance, or (3) the fre-

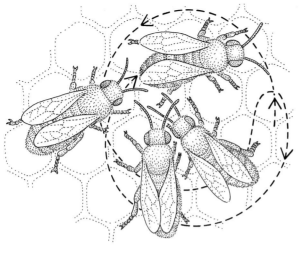

4 Round dance of honeybees. The dancer (the uppermost bee) is followed by three other workers, who may acquire information that a profitable food source is located within 50 meters of the hive. *Source: von Frisch [1225].*

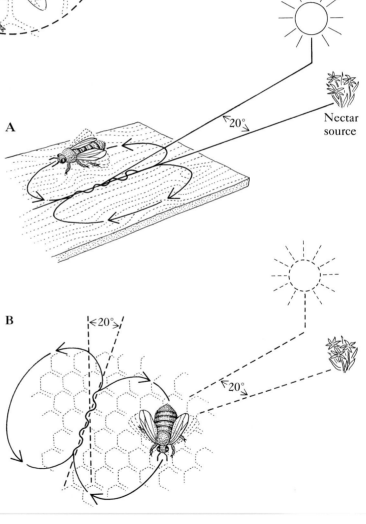

5 Waggle dance of honeybees. As the worker performs the straight run, her abdomen waggles; the number of waggles and the orientation of the straight run contain information about the distance and direction to a food source. In this illustration, workers attending to the dancer learn that food may be found by flying 20° to the right of the sun when they leave the hive. (A) The directional component of the dance is most obvious when it is performed outside the hive on a horizontal surface, in which case the bee runs right at the food source. (B) On the comb, inside the dark hive, the dance is oriented with respect to gravity so that the displacement of the straight run from the vertical equals the displacement of the location of the food source from a line between the hive and sun.

quency with which the bee produces bursts of sound, a human observer can tell approximately how far away the food source is. The fewer the number of dance circuits performed over 15 seconds, the more distant the site.

Moreover, by measuring the angle of the straight run with respect to the vertical, someone observing a waggle-dancing bee can also tell the *direction* to the food relative to the sun. Apparently, a foraging bee on the way to a relatively distant food source she discovered on a previous trip notes the angle between the food, hive, and sun. She transposes this information onto the vertical surface of the comb when she performs the straight-run portion of the waggle dance. If the bee walks waggling straight up the comb, the nectar or pollen will be found by flying directly toward the sun. If the bee waggles straight down the comb, the food is located directly away from the sun. A patch of nectar-producing flowers positioned 90° to the right of a line between the hive and the sun triggers waggle runs oriented at 90° to the right of the vertical on the comb. In other words, sun compass information is converted into a symbolic code based on gravity.

The conclusion that the dance displays of bees encode fairly specific information about the distance and direction to good foraging sites was reached by Karl von Frisch after 20 years of experimental work. His basic research protocol involved training bees (which he daubed with dots of paint for identification) to visit particular feeding stations, which von Frisch had stocked with rich sugar solutions or honey. By watching the dances of these trained bees, he saw that their behavior changed in highly predictable ways depending on the distance and direction to a food source. More importantly still, his dancing bees were able to direct others to a food source they had found (Figure 6).

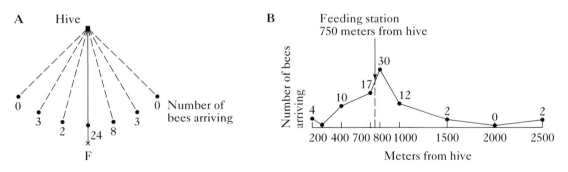

6 Distance and directional communication by honeybees. (A) A "fan" test to determine whether foragers bees can convey information about the direction to a food source they have found. After training recruiters to come to a feeding station at *F*, von Frisch collected all newcomers that arrived at seven feeding stations with equally attractive sugar water. Most new bees arrived at the site in line with *F*. (B) A test for distance communication. Recruiters were trained to come to a feeding station 750 meters from the hive. Thereafter, all newcomers were collected at feeding stations placed at various distances from the hive. In this experiment, 30 newcomers were collected at the site closest to 750 meters, far more than came to any other location in the same time. *Source: von Frisch [1225].*

Debate continues on whether von Frisch's experiments and more recent studies show unequivocally that recruited workers learn from the dances per se where to hunt for food [451, 723, 1253]. Some believe they might rely exclusively on the flower odors present on the dancing bee's body as a guide for their search [1253]. However, the finding that a mechanical or robot "bee" can apparently induce recruits to search for food in a particular direction offers evidence in favor of the proposition that workers derive useful information from the dances [806, 807]. Indeed, it would be most surprising if the dance movements were totally irrelevant, given that they carry energetic costs to the dancers and given that human observers can accurately decode the distance and directional information that they contain.

The History of Honeybee Dances

Before we analyze the adaptive value of honeybee dances, let us reconstruct their evolutionary history. Martin Lindauer used comparisons among living bee species to uncover the steps that may have led to this highly specialized behavior [723]. He found that all three other members of the genus *Apis* perform dance displays identical to those of the honeybee (*Apis mellifera*), except that *A. florea* dance on the horizontal surface of a comb built in the open over a limb (Figure 7). To indicate the direction to a food source, a worker simply orients her waggle run direct-

7 The nest of an Asian honeybee, *Apis florea*, has a flat upper surface. Workers use this surface when dancing and therefore point in the direction of a food source when performing the waggle component of the dance. *Photograph by Martin Lindauer.*

ly at it. Because this is a less sophisticated maneuver than the transposed pointing done in the dark on vertical surfaces, it is probably similar to an ancestral form of dance communication.

Lindauer then looked to other bees for recruitment behaviors that might give hints about the steps that might have preceded the first waggle dancing. Stingless bees of the tropics employ diverse communication systems, which Lindauer used to create the following evolutionary scenario:

The possible first stage: Workers of some species of *Trigona* stingless bees perform unstructured excited movements coupled with a high-pitched humming when they return to the nest with high-quality nectar. This behavior arouses hivemates, which request food samples from the "dancer" and detect the odor of the source on her body. With this information, they leave the nest and search for similar odors. The actions of the recruiter do not offer any specific cues about the direction and distance to the desirable food.

A possible intermediate stage: Workers of other species of *Trigona* convey more precise information about the location of a food source than those bees that recruit foragers with unstructured dances. In these species, a worker that makes a substantial find marks the area with a pheromone produced by her mandibular glands. As the bee returns to the hive, she continues to chew on grasses and rocks every few yards. At the hive entrance there may be a group of bees waiting to be recruited. The forager leads these individuals back along the trail she has marked (Figure 8).

A still more sophisticated pattern: A number of stingless bees in the genus *Melipona* separate distance and directional information, unlike *Trigona* bees. A dancing forager communicates information about the distance to a food source by producing pulses of sound; the longer the pulse of sound, the farther away the food is. In order to transmit directional information, the recruiter leaves the nest with a number of followers and performs a short zigzag flight that is oriented toward the source of nectar. The recruiter returns and repeats the flight a number of times before flying straight off to the nectar site, with the novice bees in close pursuit.

Lindauer used this comparative information to suggest that communication about the *distance* to a food source by an ancestor of the honeybee probably at first involved only agitated movements by a food-laden worker. Other workers that were stimulated by the movements of the returning forager would then have left the hive to find food. In some species, selection may have subsequently favored standardization of the sounds and movements made by an "excited" worker, as in *Melipona*. This trend culminates in the round and waggle dances of *Apis* bees that contain symbolic information about how far food is from the hive [723, 1293].

In contrast, communication about the *direction* to a food source appears to have originated with personal leading, with a worker guiding a group of recruits direct-

8 Communication by scent marking in a stingless bee. In this species, workers that found food on the opposite side of the pond could not recruit new foragers to the site until Martin Lindauer strung a rope across the pond. Then the recruiters placed scent marks on the vegetation hanging from the rope and quickly led others to their find. *Photograph by Martin Lindauer.*

ly to a nectar-rich area. Here the evolutionary sequence has involved less and less complete performance of the guiding movements as queens produced workers with a greater and greater tendency to perform incomplete leading. At first this may have taken the form of partial leading (as in *Melipona*) and later involved simply pointing in the proper direction with a waggle run on a horizontal surface (as in *A. florea*). The final step is the transposed pointing of *A. mellifera*, in which directional cues based on a sun compass are converted into cues based on gravity.

The Adaptive Value of Honeybee Dances

Having outlined the possible historical sequence leading to the complex honeybee dance display, we now ask, what is its adaptive value? Thomas Seeley and Kirk Visscher proposed three hypotheses on how the time and energy costs of dancing might be more than repaid by fitness benefits to the queen whose workers perform this behavior [1076]. Since worker honeybees are sterile, their activities cannot promote their own reproductive success, but their behavior can raise the fitness of family members that are able to reproduce (see Chapter 15).

The most obvious possibility is that a dancing recruiter helps her nestmates find food supplies more quickly than they would otherwise, reducing the costs of food location for workers and thus increasing the caloric cost–benefit ratio for the colony. This hypothesis predicts that a scout bee that hunts for new food supplies will need more time to find a food source than recruit bees that attend to her dances and then go out to find the flowers advertised by the successful scout.

In fact, however, although recruit bees from Seeley and Visscher's hives followed dancers before leaving the hive to forage, they rarely found the target site on their first flight out from the hive. As a result, recruits took a little over 2 hours on average before they appeared at the advertised food source, whereas scouts needed a little less than 90 minutes on average to find a new food source worth dancing about back at the hive. We must reject the time-saving hypothesis for recruitment by dancing bees.

A second possibility is that recruiters direct foragers to better sites than their colonymates could find on their own, increasing the calories gained per unit of time spent foraging by workers. Based on this hypothesis, we predict that when recruits do return with food from an advertised patch, they will bring back more pollen and nectar on average than scouts do on trips unguided by other bees. Seeley and Visscher found that after having been directed to a patch, recruited foragers tended to make several highly productive trips in a row to this site, whereas scouts often came back from their searching excursions with no food at all (Figure 9).

Finally, recruitment of a large workforce to a food patch could mean that more of its resources can be harvested before other bee colonies and competitors can get to it. If this is true, then the number of recruits that come to a food source after some scouts have begun advertising it should be greater than the number of workers that discover the site without guidance from colonymates.

Seeley and Visscher moved a number of flowerpots from place to place and measured the time it took scout bees to locate each patch. New food finders came by

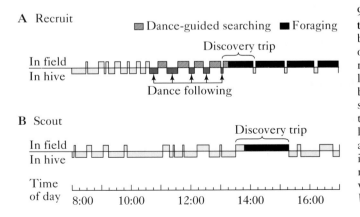

A Recruit

▪Dance-guided searching ■Foraging

Discovery trip

In field
In hive

Dance following

B Scout

Discovery trip

In field
In hive

Time
of day 8:00   10:00   12:00   14:00   16:00

9 Foraging behavior of a representative recruit and scout honeybee. Both bees made many short flights in search of food. (A) The recruit followed a number of dancers and eventually located a new food source advertised by a nestmate. Thereafter, she foraged steadily at this source. (B) In contrast, the scout, after making a series of fruitless searches, discovered a food source and collected food there before returning to the hive. However, she then resumed searching for food elsewhere without success. *Source: Seeley and Visscher [1076].*

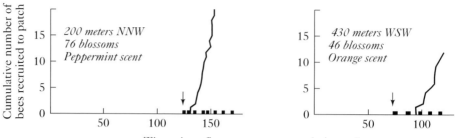

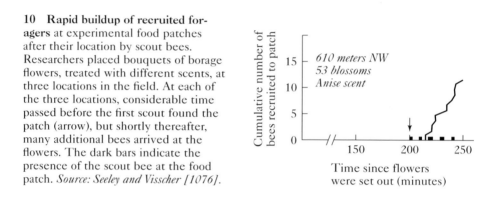

10   Rapid buildup of recruited foragers at experimental food patches after their location by scout bees. Researchers placed bouquets of borage flowers, treated with different scents, at three locations in the field. At each of the three locations, considerable time passed before the first scout found the patch (arrow), but shortly thereafter, many additional bees arrived at the flowers. The dark bars indicate the presence of the scout bee at the food patch. *Source: Seeley and Visscher [1076].*

at the rate of about one bee every 5 to 10 minutes. But once a scout returned to its hive and danced there, recruited bees quickly flooded the area. Even though any one recruit took an average of 2 hours to find a newly advertised food source, some of the many bees following dancers arrived at the advertised site quickly, resulting in a fairly rapid buildup of recruits at a given site (Figure 10). As a result, the pollen and nectar at the food patch wind up in the colony of the recruiters and recruits, not in the guts of other nectar and pollen foragers.

Roosts and Breeding Colonies: Information Centers?

Active recruitment of others to a food source is much less well developed in vertebrates than in the social insects, although ravens are something of an exception (see Chapter 8). However, animals need not signal others about a food supply in order to give away information about where to get a good meal. An observant bird might monitor the foraging success of returning neighbors and simply follow the successful hunters back to a rich food source. In this way, an animal that had been unsuccessful in its own attempts to detect food could find good foraging areas [203].

The idea that colonies and roosting aggregations of birds serve as "information centers" [1236] generates several predictions. One is that foraging birds should leave their nesting colonies in groups and fly off in the same direction. If clumping of departures does not occur, or if the birds head out in all directions, social transmission of information seems unlikely to have occurred. Some studies of colonial birds (but not all [957]) have documented a tendency of birds to leave the colony together, heading in the same direction [e.g., 449, 468]. Thus, for example, when one barn swallow flies away from its nest to gather flying insects, other birds are likely to follow, synchronizing departures from the nesting area [520]. Moreover, departing swallows tend to adopt similar bearings, suggesting that some individuals are following others.

A second prediction from the information center hypothesis is that when there are followers and leaders, the followers will have previously foraged without success and the leaders will be birds that have done well recently. This prediction failed its test for the population of barn swallows studied by Margaret Hebblethwaite and William Shields [520]. Birds that came back empty-beaked on a preceding trip were no more likely than successful birds to follow another bird away from the colony on the next foraging trip (Figure 11). Nor were successful foragers more likely to be followed than unsuccessful ones.

The same negative result was obtained in an experimental study of colonial black-headed gulls. Researchers dumped a load of dead fish on a raft some distance offshore, out of sight of a nesting colony. Wandering gulls soon found the food and gorged themselves. An observer near the raft radioed information about the identity of successful foragers back to a colleague at the colony, who could then record what happened upon the return of the color-marked birds. Each successful gull landed by its nest with a conspicuously distended crop and often with fish in its beak. Despite these cues, no neighbor flew behind a successful forager when it headed back to the raft. Moreover, when the next gull did leave the colony, there was absolutely no ten-

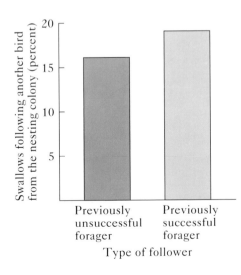

11 Information transfer in the colonial barn swallow? If barn swallows gain information about the location of food from successful colonymates, then foragers that have failed to find food on the previous trip should be especially likely to fly after a departing neighbor—but they are not. *Source: Hebblethwaite and Shields [520].*

dency for it to fly in the same direction as the successful fish finder [25]. These results suggest that black-headed gulls having trouble finding food probably do not learn where to go from birds that have found abundant food elsewhere.

On the other hand, Erick Greene found support for the information center hypothesis in the behavior of ospreys—large fish-eating hawks that form loose nesting colonies in some coastal areas. These birds do learn from their colonymates where to find certain kinds of fish—namely, those that appear in large schools but in unpredictable locations [468]. Not only are ospreys more likely to begin foraging shortly after a fellow osprey returns to the colony with a schooling alewife or smelt, but they also tend to fly out in the direction taken by the successful forager. In addition, birds that fly off after seeing a successful hunter with prey are able to capture the same prey much more quickly than ospreys that hunt without benefit of observing a successful forager (Figure 12).

Still more support for the hypothesis comes from John Marzluff and his colleagues, who watched young, nonterritorial ravens leaving communal roosts in groves of trees. On winter morning after morning, dozens of ravens left in the space of a few minutes, with most individuals headed in the same direction. To test experimentally whether naive followers could gain information about the location

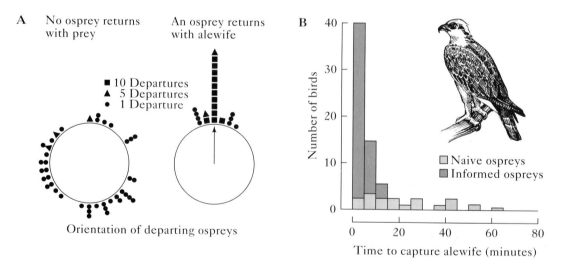

12   Information transfer in colonially nesting ospreys? (A) When ospreys leave their nests in the colony at a time when no colonymate has returned with prey in the preceding 10 minutes, they fly off in all directions. But if another osprey has come back with an alewife, a schooling fish, the departing birds tend to fly off in the same direction (represented by an arrow) as the successful hunter. (B) Naive ospreys (those that have not seen a prey-carrying colonymate prior to departure) take much longer to find alewife schools than informed ospreys (those that have just observed a returning neighbor carrying an alewife in its talons). *Source: Greene [468].*

of food, the raven research team captured some ravens and held them for a number of days in an aviary. They then released some of the birds near a roost of ravens that had just found a new food source— a dead moose provided by the researchers while the captured ravens cooled their heels in the aviary. All 14 birds released at roosts in the evening followed their roostmates out the next morning to the bait. In contrast, only 4 of 15 captured ravens that were released away from roost sites found the food on their own [776].

The colonial roosts of evening bats also appear to be places where foragers can secure help in finding food. Some of these bats follow others when they depart to hunt for night-flying insects [1284]. Here too, rich hunting patches are ephemeral, as clouds of nocturnal insects appear and disappear in various places. Bats often leave the roost when others head out, and those that follow previously successful foragers gain significantly more weight on that foraging excursion than on their prior trip.

Locating Prey by Deceit

Solitary predators do not have the chance to locate prey by watching others. However, as we saw in Chapter 8 with anglerfish, they do sometimes come up with ingenious alternatives by using deceptive signals to bring prey within attack range. Consider the bolas spider, which releases a scent identical to the sex pheromone of certain female moths. Males of these moths fly upwind in search of a mate, but may instead encounter a bolas spider waiting with a sticky globule on the end of a silken thread (Figure 13). If the spider swings the blob and hits the moth, the predator will feast on the captured insect. Because the spider makes its living throwing a

13 A bolas spider swings its lure, a ball impregnated with a scent identical to the sex pheromone of certain female moths. Male moths that approach the odor source are often captured by the sticky ball. *Photograph by William G. Eberhard.*

ball, William Eberhard, an entomologist who also happens to be a dedicated sports fan, gave the name *Mastophora dizzydeani* to one species of bolas spider [323].

Another deceptive spider, the golden orb-weaving spider, *Nephila clavipes*, gets its common name from the yellow orb webs that it constructs in sunny openings. Tropical stingless bees, a major prey of the golden orb weaver, find yellow webs attractive, so much so that even after having escaped one, they are likely to return

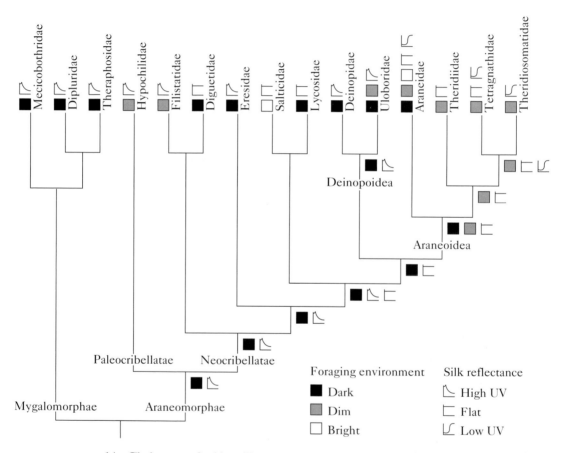

14 Cladogram of spider silks with different reflectance spectra. Three types of spider silk have evolved with different light-reflecting properties: (1) high UV-reflecting silk that strongly reflects ultraviolet light compared to light of other wavelengths, (2) flat reflecting silk that reflects light of all wavelengths equally and (3) low UV-reflecting silk that absorbs ultraviolet light more strongly than light of other wavelengths. The most recently evolved spiders (in the superfamily Araneoidea) are the only ones to produce low UV-reflecting silk, suggesting that this kind of silk is a relatively recent evolutionary innovation, which may have spread because it helps spiders catch prey during the day in relatively sunny environments. Multiple icons at the ancestral nodes of the diagram indicate uncertainty about the trait exhibited by the ancestral species. *Source: Craig et al. [230].*

and blunder into the same web again. In contrast, the bees readily learn to avoid otherwise identical webs constructed with white silk or silk artificially dyed blue or green [231]. Yellow flowers provide much of the pollen and nectar collected by stingless bees, making it advantageous for the bees to approach this color, and generally disadvantageous for them to learn to avoid yellow objects. Thus, the golden orb weaver is an illegitimate signaler that exploits a communication system whose legitimate signaler (yellow flowers in need of a pollinator) and legitimate receiver (bees in need of pollen and nectar) derive mutual benefits from their interaction (see Chapter 8).

Different spiders make different silks, which Catherine Craig has categorized into three groups based on their reflectance of light [228]. When these silk types are plotted on a cladogram of the spiders (Figure 14), we can see that ultraviolet-reflecting silk appears to be an ancestral characteristic that is no longer used by the recently evolved orb weavers (Araneidae) to make the main part of their web. Instead, modern orb weavers employ a kind of silk that *absorbs* ultraviolet light.

What is the adaptive significance of using ultraviolet-absorbing silk to make an orb web? Craig suggests that since many insects can see ultraviolet light well, a web that does not reflect these wavelengths would be less conspicuous to them. Because some orb weavers catch prey during the day in bright light, a less conspicuous web could mean that fewer insects see the trap in time to avoid it.

But if this is true, why throw in some highly conspicuous zigzags of ultraviolet-reflecting silk, as many orb-weaving spiders do (Figure 15)? These spiders have

15 Ornaments in the garden spider's web are composed of thick, conspicuous zigzagging lines of ultraviolet-reflecting silk woven into the web and radiating out from spider's central resting point. *Photograph by Catherine Craig.*

evidently retained the ancestral capacity to make this kind of silk. Perhaps these web decorations have nothing to do with foraging [326], but instead advertise the position of the web to large animals such as birds, which might otherwise destroy it inadvertently [340], or make the spider appear larger than it is, thus discouraging some relatively small spider consumers, such as lizards [1059].

The web advertisement and size enhancement hypotheses could be evaluated by testing the prediction that web destruction by blundering birds or mammals, as well as spider mortality, should occur more frequently in webs with fewer decorations (there may be one to four decorations per web). This prediction remains to be checked, but the size enhancement hypothesis has been supported by observations that medium-sized females of the garden spider, *Argiope argentata*, are the most likely to add decorations to their webs. These individuals may use their web decorations to boost their apparent body size past the upper limit favored by spider-eating lizards [1059].

What about the suggestion that the decorations of *A. argentata* lure prey into the web? Working with a sample of webs that contained only one line of "decorative" silk, Catherine Craig and Gary Bernard watched each web for an hour and compared the number of insects flying into the half of the web that contained the decoration with the number captured in the other, "decorationless" half [229]. More prey were entangled in the web section with the decoration than in the other half of the web (Figure 16). In work with a close relative of the garden spider, I-Min Tso [1207] also found that flying insects were more often trapped in webs with decorations than in webs without.

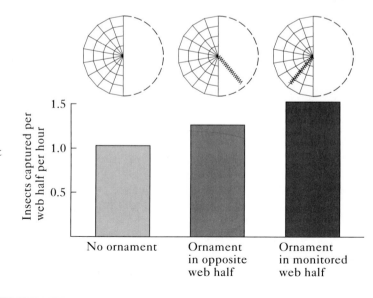

16 Do web ornaments lure prey? Garden spider webs without ultraviolet-reflecting ornaments attract fewer prey per hour than those containing ornaments. Furthermore, in webs with only one ornament, more flying insects are trapped in the half of the web containing the ornament than in the half lacking these structures. *Source: Craig and Bernard [229].*

These results indicate that at least one function of web decorations for some spiders is to increase prey capture. But why do insect victims find the ultraviolet-reflecting decorations so attractive? Spiders in the genus *Argiope* are called garden spiders because they so often build their webs in flower gardens, and flowers often contain ultraviolet-reflecting patches and patterns. Thus, the web decorations of these spiders may lure flower-visiting insects to their doom by mimicking flower signals. If this is true, then the primary prey of *Argiope argentata* should be pollinating insects, and one study found that over 60 percent of the spider's victims were indeed small flower-visiting bees [229].

Selecting What to Eat

Once potential foods have been attracted or otherwise located, foraging animals may still benefit by deciding which to consume and which to ignore. In the preceding chapter, we noted that toxic prey, such as milkweed-reared monarch butterflies, often advertise their noxiousness through warning coloration and behavior. Predators pay attention to these signals. The existence of potentially toxic foods has evidently shaped the foraging decisions of herbivores as well as meat-eaters. Plant-eating animals as different as tassel-eared squirrels [377] and leaf-cutter ants [575] select foods with low concentrations of toxic terpenoids, poisons that many plants incorporate into their tissues to repel consumers.

Selection of plant foods low in toxins is also important to the howler monkeys that Kenneth Glander studied in a riverbank forest in Costa Rica (Figure 17) [445]. Glander discovered the following "rules" of howler leaf selection: (1) The more common a tree species, the less likely the monkeys were to feed on its leaves. Instead, they spent considerable time searching out the scarcer species. (2) Even among the less common tree species, the howlers refused leaves from most individuals available to them. For example, the monkeys took plant material from only 12 of 149 specimens of one "acceptable" tree species growing in the forest. (3) The monkeys preferred the scarcer, smaller new leaves to the more abundant, larger mature leaves. (4) The monkeys often fed "wastefully," eating only the petiole and dropping the larger leaf blade.

All four choices raise the costs of foraging for the howlers. Glander hypothesized, however, that these seemingly perverse feeding choices actually overcame the poisons and other defenses characteristic of tropical forest trees. He found that (1) the most common tree species had relatively high alkaloid or tannin concentrations in their leaves. Alkaloids poison howlers; tannins bind plant proteins and make them difficult to digest. (2) Among the scarcer, preferred tree species, howlers sought out those individuals with especially low levels of alkaloids and tannins. (3) New young leaves contain more water and less nonnutritive fiber than mature leaves do. When the monkeys did eat mature leaves, they selected specimens that had higher (12.4 percent) protein levels than the mature leaves of the trees they rejected (which averaged only 9.4 percent protein). (4) "Wasteful" feeding occurred

17   Food selection by howler monkeys. Although surrounded by leaves, these monkeys forage very carefully, avoiding toxic leaves and leaves low in nutritional value. *Photographs by Kenneth Glander.*

because the monkeys were eating the leaf part, the petiole, that is lowest in toxins while discarding the more poisonous leaf blade. Howlers may be surrounded by leaves, but by feeding selectively, they ingest fewer toxins and more usable proteins.

Optimal Clam Selection by Northwestern Crows

But what about foods that differ only in size, not in toxicity? Howard Richardson and Nicholas Verbeek noticed that crows in the Pacific Northwest often leave littleneck clams uneaten after locating them [992]. The crows dig the clams from their burrows, but they often leave the smaller ones on the beach and only bother with the larger ones, which they drop on rocks and eat. Their acceptance rate increases with prey size: they open and eat only about half of the 29-millimeter-long clams they find, while consuming all clams in the 32–33 millimeter range.

The two Canadian researchers determined that the most profitable clams were the largest, not because they broke more easily, but because they contained many more calories than smaller clams. So why bother with 29- or 30-millimeter clams

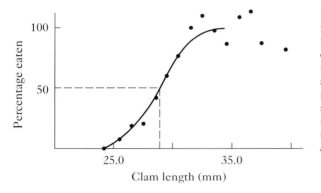

18 Optimality model of prey selection in relation to prey size. The curve represents the predicted percentages of small to large clams that crows should select for consumption after locating them, based on the assumption that the birds attempt to maximize the rate of energy gain per unit of time spent foraging for clams. The solid circles represent the actual observations, showing that the model's predictions were supported. *Source: Richardson and Verbeek [992].*

at all? Perhaps because there is a time cost to abandoning one clam in order to find a larger replacement. By considering the caloric benefits from clams of different sizes and the costs of searching for, digging up, opening, and feeding on clams, Richardson and Verbeek were able to construct a mathematical model—a hypothesis—based on the assumption that the crows would select an optimal diet, in this case one that maximized their caloric intake. The model predicted that clams about 28.5 millimeters in length would be opened and eaten about half the time, given the search costs required to find clams of different sizes; the crows' behavior shows that they agree with the researchers' math (Figure 18) [992].

Richardson and Verbeek's work is based on *optimal foraging theory*. If the behavioral mechanisms underlying diet selection by crows are indeed optimal, they should enable crows to choose clams that contribute the most to their reproductive success. Note that Richardson and Verbeek did not measure crow fitness directly. Instead, they measured net energy gained while crows were foraging for clams, on the assumption that by maximizing its caloric intake, a crow would gain more energy for the production of offspring or more time to spend on other fitness-promoting activities.

This assumption has been tested experimentally with captive zebra finches, which were given the same foods, but under different experimental feeding regimes that resulted in different daily energy gains for individuals. The more net energy acquired per day, the better the survival and reproductive success of the finches (Figure 19) [705, 706]. Likewise, the biomass of prey provided by a water pipit (a songbird) to its young in the field correlates with the number of fledglings produced [411], and well-fed orb-weaving spiders produce more egg sacs than poorly fed individuals [1081].

Optimal Mussel Selection by Oystercatchers

The oystercatcher is another seashore bird whose foraging decisions can be matched against predictions taken from optimality models. Two Belgian researchers, P. M. Meire and A. Ervynck, developed a calorie maximization hypoth-

19   Food intake affects fitness. Four captive populations of zebra finches were treated identically except for the amount of work they had to do to secure their food. Group A received 200 grams of seed in 800 grams of inedible chaff; Group B, 200 grams of seed in 600 grams of chaff; Group C, 200 grams of seed in 400 grams of chaff; and Group D, 200 grams of seed. The different treatments affected the birds' intake of seeds per minute, which in turn affected their production of offspring. *Source: Lemon [705].*

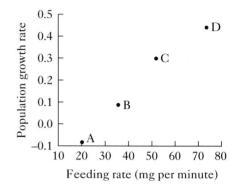

esis to apply to oystercatchers feeding on mussels [803]. They too calculated the profitability of different-sized prey, based on the calories contained in the mussels (a fitness benefit) and the time required to open them (a fitness cost). Even though mussels over 50 millimeters long require more time to hammer or stab open, they provide more calories per minute of opening time than smaller mussels. Therefore, the model predicts that oystercatchers should focus primarily on the largest mussels. But in real life, the birds do not prefer the really large ones (Figure 20). Why not?

Hypothesis 1: The profitability of very large mussels is reduced because some cannot be opened, reducing the average return from handling these prey.

In their initial calculations of prey profitability, the researchers had considered only those prey that the oystercatchers actually opened (Figure 21, model A). As it turns out, oystercatchers select some mussels that they find impossible to open, despite their best efforts. The larger the mussel, the more likely that it will have

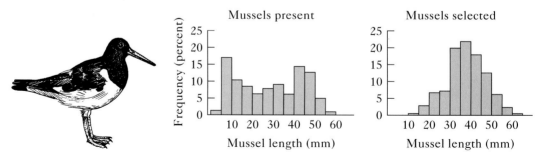

20   Available prey versus prey selected. Foraging oystercatchers choose mussels to attack that are larger than the average available mussel, but they do not concentrate on the very largest mussels. *Source: Meire and Ervynck [803].*

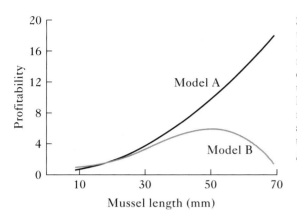

21 Two optimality models yield different predictions because they calculate prey profitability differently. Model A calculates the profitability of a mussel based solely on the energy available in opened mussels of different sizes divided by the time required to open these prey. Model B calculates profitability with one added consideration, namely that some very large mussels must be abandoned after being attacked because they are too difficult to open. *Source: Meire and Ervynck [803].*

to be abandoned, unopened and uneaten. The handling time wasted on large impregnable mussels reduces the average payoff for dealing with these mussels. When this factor is taken into account, a new optimality model results, yielding the new prediction that the oystercatchers should concentrate on mussels 50 millimeters in length, rather than the very largest size classes (Figure 21, model B). The oystercatchers, however, actually prefer mussels in the 30- to 45-millimeter range. Therefore, time wasted in handling large invulnerable mussels fails to explain the oystercatchers' food selection behavior.

Hypothesis 2: Many large mussels are not even worth attacking because they are covered with barnacles, which makes them impossible to open.

This additional explanation for the apparent reluctance of oystercatchers to feast on large mussels, despite their calorie richness, is supported by the observation that oystercatchers never touch barnacle-covered mussels. The larger the mussel, the more likely it is to have acquired a coat of tough barnacles, which interfere with an oystercatcher's attempt to break through the mussel shell. In effect, these barnacled mussels are not available as potential prey to oystercatchers, and so should not be considered when calculating the return rate expected for birds attempting to consume mussels of different sizes. According to a mathematical model that factors in (1) the effect of prey opening time, (2) time wasted in trying but failing to open a mussel, and (3) the actual range of realistically available (non-barnacle-covered) prey of different sizes, the birds should focus on 30- to 45-millimeter mussels—and they do.

Criticisms of Optimality Theory

By testing optimality models, researchers have concluded that northwestern crows and European oystercatchers choose prey that provide the maximum caloric benefit in relation to time spent foraging. But some persons criticize this approach on the grounds that animals do not always behave as efficiently as possible. This criticism is reminiscent of the complaint that adaptationists believe that all behavior is perfectly adaptive, and like that complaint, it misses the point. Optimality models are not constructed to make statements about perfection in evolution, but to make it possible to test whether one has correctly identified the conditions that shaped the evolution of an animal's behavior. As we have just seen, the factors included in an optimality model have a large effect on the predictions that follow from that model. If an oystercatcher is assumed to consider every mussel in a tidal flat as potential prey, then it is predicted to make different foraging decisions than if the modeler assumes that oystercatchers ignore all barnacle-covered mussels. If the predictions of a model fail to match observations, researchers can make progress, rejecting those explanations based on unrealistic conditions and developing alternative hypotheses that come closer to reality.

If factors other than caloric intake affect oystercatcher prey selection, then a caloric maximization model would fail its test, as it should. And for many foragers under some conditions, prey selection has consequences above and beyond the acquisition of calories. For example, foraging decisions can affect the survival chances of an animal. If you were to suspect that predators had shaped the evolution of an animal's foraging behavior, then the kind of optimality model you might choose to construct and test would not focus solely on calories gained versus calories lost. If foraging exposes an animal to predators, then we might predict that when the risk of attack is high, the animal will sacrifice short-term caloric intake for long-term survival.

For example, if whirligig beetles dedicated their lives strictly to maximizing caloric intake, they would wander widely in search of food. In reality, whirligig beetles severely restrict their movements. They rarely venture far from the dense clusters of beetles that form on the surface of the water. In fact, when beetles secure food outside the cluster, they soon move to the center of the circling mass of companions, sacrificing the chance to encounter more floating food in favor of occupying a central position where they are less likely to be the first in line to meet an approaching predator [1013].

Likewise, hoary marmots—relatives of the familiar groundhog—are usually reluctant to range far from their burrows in rocky talus slopes; they feed primarily on the heavily grazed margin of meadow closest to their retreats rather than going farther to greener pastures. As a result, they consume less food per unit of foraging time, but reduce their risk of being food for a golden eagle or coyote. Since juvenile marmots are smaller and less experienced than older ones, they are favored prey, which makes the costs of foraging away from cover even greater for juveniles than for adults. As you might predict, therefore, juvenile marmots compromise their foraging efficiency (defined strictly in caloric or nutritional terms) even more than adults do by keeping closer to their burrows and by spending more time looking around for danger [564].

Similarly, a parasitoid fly has been identified by Matthew Orr [885] as a key factor in reducing caloric foraging efficiency in a leaf-cutter ant. If the ants' goal

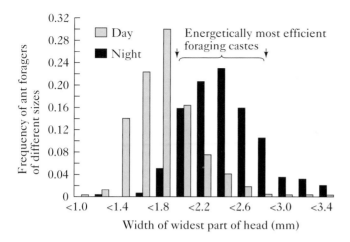

22 Compromising foraging efficiency when the risk of predation is high. During the day, but not at night, foraging leaf-cutter ants tend to be smaller than the optimal size for food-collecting efficiency. But during the day, large foraging worker ants are at risk from parasitoid phorid flies, which ignore ants with head widths of 1.8 mm or less. *Sources: Orr [885] and Wilson [1296, 1297].*

was to maximize the collection of leaf pieces to fuel the underground fungus gardens that maintain their colonies, then workers should forage day and night for leaves to cut, and the size of the workers that gather food should be the optimum in terms of net energy gained for the colony. In reality, workers of *Atta cephalotes* are primarily nocturnal, foraging only at a low level during the day, and the foragers at work during the day are smaller than the optimal size in terms of energy acquisition.

These constraints on energy maximization arise because small phorid flies lay eggs on leaf-cutter ants; the resulting fly larva burrows into its host's head, ultimately killing it. The flies, however, are active only during the day, and they refuse to lay eggs on small workers, presumably because these lack the necessary mass to support the full development of a larva (Figure 22). Now we can better understand why the leaf-cutter ant behaves as it does. Nocturnal foragers of all sizes are safe from the fly, as are small diurnal leaf collectors. The colony trades off short-term energy gain for a longer total foraging life for its valuable leaf-collecting workers.

Studies such as Orr's demonstrate the utility of the optimality approach in creating testable hypotheses that, if rejected, may turn a researcher's attention to factors not considered in the original model, thereby eventually leading to an improved understanding of a given case.

Alternative Tactics within Species

The hoary marmot study demonstrated that the age of an individual can affect its optimal foraging tactics. The same point has been established in a study of yellow-eyed juncos. When these sparrows have just fledged, they handle large prey clumsily, taking a long time to crush a big grub, for example, before swallowing it. But fledglings can pick up and swallow small items quickly without much handling. Kim Sullivan demonstrated that because of the long handling times associated with large mealworms, recently fledged juncos actually gain more energy per unit of feeding time by choosing small mealworms [1159]. Experienced older foragers, however, have learned to manipulate insect larvae skillfully, and can prepare a large mealworm in less than half the time it takes a novice junco. This means that adults do not pay a great time penalty for selecting large mealworms; in fact, the caloric payoff per second of foraging time for a mature junco is the same for large and small mealworms. Sullivan predicted that if one were to give fledgling and adult juncos free access to an ample supply of small and large mealworms mixed together, fledglings should go for the small prey, unlike adult juncos. Figure 23 shows the results of a test of this prediction.

Chapter 4 presented another quite different case of a species whose members forage for different foods: in the tiger salamander, specialist cannibal morphs feed upon the insect-eating forms of their own species in ponds where the two types coexist—uneasily. A somewhat similar phenomenon occurs in the New Mexico spadefoot toad, whose tadpoles also come in two forms, a rapidly developing type that specializes in eating small invertebrates called fairy shrimp, and a slowly developing type that feeds on odds and ends, but especially on bits of dead algae and other detritus on the bottom of the temporary desert ponds where the tadpoles live [928]. A high-protein fairy shrimp diet evidently enables the carnivore

23   Prey profitability and prey selection by yellow-eyed juncos. Young juveniles take so long to prepare large mealworms for consumption that they actually gain more energy per unit of time by feeding on small mealworms (lighter bars), and they prefer this size class of prey (darker bars). Older juveniles and adults can handle large prey efficiently, and these more mature juncos find large and small mealworms equally profitable. *Source: Sullivan [1159].*

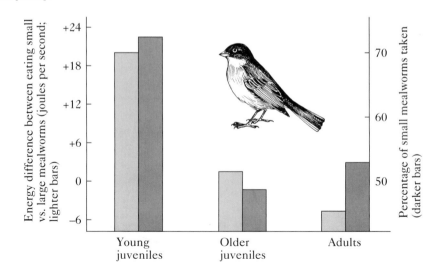

morphs to reach the toadlet stage faster than their detritus-eating companions. So why don't all tadpoles eat shrimp?

As it turns out, the difference between the two types is not caused by a genetic difference between them, but rather by dietary differences. A young tadpole that is finding fairy shrimp fairly regularly can become a carnivore that specializes on this prey, but if the shrimp become scarce, the tadpole can revert to the omnivorous detritus-eating type. Thus, spadefoot toads have evolved the capacity to assess the available food supply and adjust their feeding behavior accordingly. If this hypothesis is correct, then an experimental shift in the ratio of carnivore to omnivore tadpoles in a pond should cause some of those of the type made unnaturally common to switch to the other type, bringing the ratio of the two forms back to its original level. We have, then, a game theoretical hypothesis (see Chapter 9) in which the adaptive feeding mode depends on what other members of the population are eating.

David Pfennig tested this hypothesis by dividing five natural desert ponds into compartments of equal size and then changing the frequency of carnivores in some of the compartments, while leaving others untouched as controls. After the initial manipulation, the frequency of carnivores was sampled in the experimental and control compartments over a number of days. In compartments in which carnivore numbers were boosted experimentally, carnivores gradually declined relative to omnivores until the original proportion of fairy shrimp consumers had been reached. In sections in which carnivore numbers were depleted initially, carnivores became more common until reaching the same frequency as in the unmanipulated control compartments (Figure 24). These results strongly suggest that when competition for fairy shrimp is high relative to their availability in a pond, the benefit from being a shrimp-eating specialist declines, favoring individuals with the capacity to switch their selection of food accordingly [928].

Capturing Prey

Finding a food and selecting it for consumption are necessary steps, but they are not sufficient to guarantee that the food will be eaten. Often the selected meal resists being captured. Touch an *Eleodes* darkling beetle ambling across the desert, and the beetle stops to elevate its abdomen. Continue to molest the beetle, and it discharges a mix of irritating and poisonous chemicals. Although this may protect some beetles against some enemies some of the time, grasshopper mice counter this defensive maneuver by quickly grabbing the beetle before it fires and stuffing the tip of its abdomen into the sand. The mouse then crunches its way through the insect from the head down [337]. Skunks use a different tactic, rolling the beetle vigorously on the ground with their forepaws until its glands are completely emptied, after which the now defenseless tenebrionid can be eaten with impunity [1121].

Nonpoisonous prey may also require special capture techniques. For example, a tightly packed school of fish can suddenly break apart as its members rush

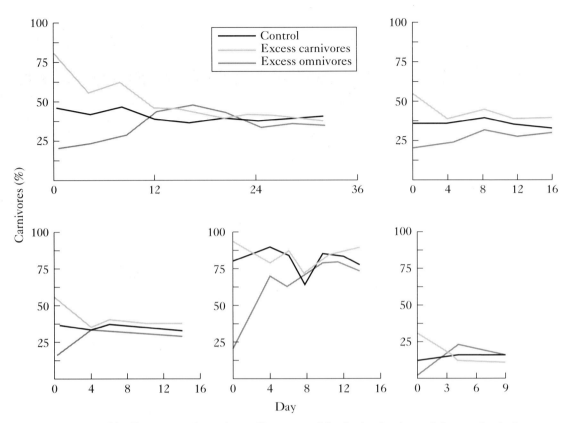

24 Frequency-dependent adjustment of food selection in spadefoot toad tadpoles. Five ponds were divided into compartments, and the ratio of carnivore (shrimp-eating) to omnivore tadpoles was experimentally manipulated in some compartments. Each graph represents one pond. The proportions of the two types quickly moved back toward the equilibrium level maintained in the control (nonmanipulated) compartments. *Source: Pfennig [928].*

off in all directions to avoid an attacking gull. Gulls often forage for fish in flocks, perhaps to counter the anti-capture tactics of their prey. This hypothesis has been nicely tested by a Swedish team of researchers, who "persuaded" some captive black-headed gulls to help them test the prediction that groups of gulls should capture food more rapidly than birds foraging alone [450].

The gulls were introduced into a large indoor aviary containing a pool in which fish had earlier been released. The gulls were permitted to forage alone or in groups of three or six. They caught more fish per unit of time *per individual* when they were part of a group of three or six than when they fished alone (Figure 25). They

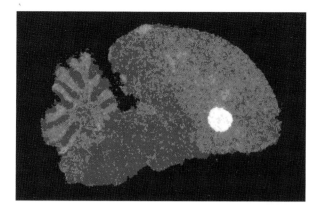

1 A section of the brain of a zebra finch that had just been singing, "probed" using in situ hybridization to visualize the location of the ZENK gene product. The image shows a high level of activity (white and yellow areas) in a circular area known as "Area X," which is part of the song control circuit; there are also a few purplish regions scattered about where a lower level of gene induction can be detected. Blue areas indicate regions with little or no gene activity. *Courtesy of David Clayton.*

2 A tiger salamander that specializes in cannibalism caught in the act of consuming a typical member of its own species, which feeds only on aquatic insects and other small invertebrates. What environmental factors could trigger the development of such different kinds of individuals in the same species? *Photograph by Tim Maret, courtesy of James Collins.*

3 Males of this cichlid fish species change their color (and the size of certain brain cells) in response to changes in their social status. The dull-colored male currently lacks a territory, whereas the more colorful male owns a site that may attract egg-laden females. *Photograph by Russell Fernald.*

4 Males of a thynnine wasp species attempt to mate with female decoys—highly modified flower petals produced by this Australian orchid. The plant secures pollinators by exploiting the "simple" way that the male wasp's nervous system operates; note the yellow pollinia attached to one of the males, which has been "deceived" by another orchid earlier in the day. *Photograph by the author.*

5 The star-nosed mole has an amazing nose and an equally astonishing nervous system. The animal uses both to find prey in the pitch-black underground tunnels where it hunts for earthworms. *Photograph by Kenneth C. Catania.*

6 Male anoles communicate to rival males and to females by expanding the dewlaps on their throats. *Photograph by Thomas Jenssen.*

7 The intense competition for food in the social spotted hyena, shown here devouring a wildebeest, might be indirectly responsible for the evolution of the strange pseudopenis possessed by females of this species. *Photograph by Jonathan Scott/Planet Earth Pictures.*

8 Males of the Australian whistling moth have evolved odd devices ("castanets") on their forewings which they use to produce ultrasonic calls on their territories. The evolutionary history behind this type of signaling behavior may involve the "exploitation" of moth hearing abilities that may have evolved originally as aids in the detection of bat predators. *Photograph by the author.*

9 A pair of antlered male flies (*Phytalmia mouldsi*) in ritual combat, illustrating a remarkable convergence between an insect and antlered mammals in the aggressive displays used by territorial competitors. *Photograph by Gary Dodson.*

10 Keen-eyed predators may be responsible for the behavior of this cryptically colored grasshopper, which not only looks like the ground on which it rests, but also hides itself further by settling into a trench it dug with its hindlegs and then drawing pebbles over its back with its midlegs. *Photograph by the author.*

11 In the Antarctic, leopard seals lurk near the places where Adélie penguins jump into the water to forage for food to bring back to their chicks in nearby nesting colonies. The risk of dying in the jaws of a seal may explain the tendency of the penguins to enter the water in groups, to better reduce each individual's chance that they will be unlucky enough to be captured and eaten before they can get through the danger zone. *Photograph by Gerald Kooyman/Hedgehog House.*

12 Clay-eating has evolved independently in a number of animals, including humans and red-and-green macaws, shown here ingesting clay from an exposed riverbank in tropical Peru. *Photograph by Charles Munn/Wildlife Conservation International.*

13 A male satin bowerbird at his bower using a yellow flower in his courtship display to a female that has come to inspect the bower and its builder before accepting or rejecting the male as a sexual partner. *Photograph by Bert and Babs Wells.*

14 Female katydids, like this thin, brown Australian species, receive a nuptial gift, a spermatophore, when they copulate. After the male transfers the spermatophore and leaves, the female eats it and gains valuable nutrients from the gelatinous mass attached to the underside of her abdomen. The female's two thin antennae and shorter, thicker ovipositor point downward. *Photograph by Darryl Gwynne.*

15 The evolution of monogamy in birds is intriguing, particularly in species like the willow ptarmigan in which males (one appears on the right) do not feed their offspring (four cryptically colored chicks follow their mother, on the left). *Photograph by Kathy Martin, courtesy of Susan Hannon.*

16 In animals that practice lek polygyny, males often use bizarre and elaborate displays to persuade females to mate. Here a female raggiana bird of paradise (left) inspects a territorial male displaying special ornament plumes. Just what females gain by making mate choices based on these displays remains uncertain. *Photograph by Bruce M. Beehler.*

17 Parental behavior is more often provided by females than males, particularly among the insects. Here a female fungus beetle guides her larval brood from one edible fungus patch to another. *Photograph by Edward S. Ross.*

18 An exception to the rule that males eschew parental care, males of the burying beetle, *Nicrophorus orbicollis*, sometimes remain with their mates after they have buried a ball of carrion in the ground. When the young hatch from eggs, males may guard them against infanticidal intruders as well as feeding them bits of food from the brood ball. *Photograph by Mark Moffett.*

19 Males of the very large water bug, *Lethocerus medius*, leave the water and climb up exposed perches to brood the eggs deposited there by their mates. How did this behavior evolve, and why? *Photograph by Robert L. Smith.*

20 Utterly ugly naked mole-rats have a beautifully strange and complex social system more like that of ants and termites than that of the overwhelming majority of their fellow mammals. *Photograph by Raymond Mendez.*

21 The accepted standards of human body ornamentation vary greatly from culture to culture. This Huli tribesman from New Guinea has a pierced nose and wears red and yellow ochers on his face and shoulders, elaborate headgear made from human hair and bird-of-paradise feathers, and necklaces made from boar's tusks, hornbill beaks, cassowary quills, and cowry shells. Are the differences between societies purely random artifacts of cultural evolution? *Photograph by Art Wolfe.*

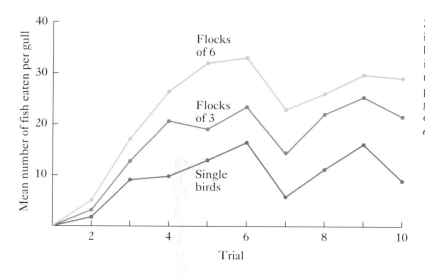

25 Social foraging improves prey capture by black-headed gulls. A gull in a flock of six consistently captured more prey per 3-minute trial than singletons or gulls in groups of three. *Source: Götmark et al. [450].*

did better in a group because fish fleeing from one hunter sometimes swam toward another gull, often enabling it to capture the fish by its head or body. Single birds had to try to catch fish by the tail as they swam away, a more challenging and less rewarding task.

Sociality and the Capture of Large Prey

To the extent that the gull experiment has relevance for natural populations, we can conclude that gulls flock together to catch more food. Cooperation occurs in other animals when the predators are attempting to subdue unusually large prey. For example, some social spiders capture grasshoppers much larger than they are because of the extra-large web they build together [1038]. In their collective web, large prey that bounce off one sheet often fall into another, and are eventually entangled. This "ricochet effect" provides the communal web builders with an energetic bonus denied spiders that build smaller, less extensive webs by themselves (Figure 26) [1211].

Cooperative hunting has also evolved in some members of three different families of terrestrial carnivores: the Felidae (e.g., lions), Canidae (e.g., wolves and African wild dogs), and Hyaenidae (e.g., the spotted hyena). These social carnivores can bring down prey that weigh from six to twelve times as much as any one adult hunter (Figure 27) [233, 799, 1051]. In contrast, solitary species of cats, dogs, and hyenas usually forage for much smaller victims. These patterns of convergence among non-relatives and divergence between closely related species support the hypothesis that group foraging by carnivores and social spiders is an adaptation for the capture of prey too large, or too dangerous, to be subdued by predators hunting on their own.

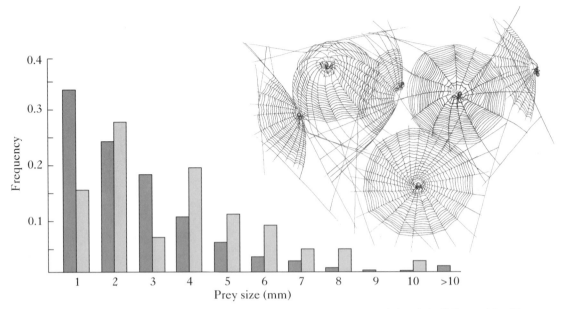

26 The ricochet effect provides a benefit to the colonial web-building spider *Metapeira incrassata*. These spiders construct their webs communally, so that prey bouncing off one web often falls into another web and is captured. Dark bars represent the size of prey captured in the first web they hit; lightly shaded bars represent the size of prey captured upon ricochet into other webs. *Source: Uetz [1211].*

27 The benefits of group hunting. Three lions have captured and partly eaten a large prey, a wildebeest. They are attempting to defend their kill from another social carnivore, a pack of spotted hyenas. *Photograph by Norman Myers.*

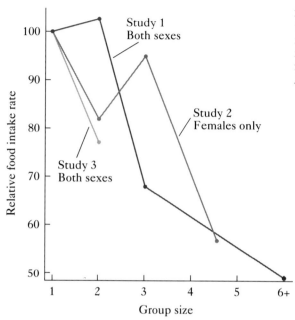

28 Group size and food intake in lions. In three studies, solitary lions generally consumed more meat than social lions. The relative food intake rate is expressed as a percentage of the food intake of a solitary lion in that study. *Source: Packer [894].*

This hypothesis has been subjected to optimality tests by researchers working with lions and African wild dogs. Let's assume that these carnivores hunt in groups that will return the best fitness benefit–cost ratio for each individual participant. Let's also assume that net fitness gain is positively correlated with the amount of food consumed by an individual. With these assumptions in place, we can measure the food eaten by individuals in groups of different sizes to test the prediction that members of groups of the preferred size will maximize their food intake.

When Thomas Caraco and Larry Wolf analyzed George Schaller's data on lions (Figure 28, Study 1), they found that food intake per lion per day was higher for lions hunting in pairs than for lions hunting alone. However, in Schaller's study and in those of some other researchers (Studies 2 and 3), groups of three or more lions actually did *less* well per lion than groups of two or singletons [176]. In fact, in more recent research, Craig Packer, David Scheel, and Anne Pusey found that single lions did as well as, or even better than, group-foraging lions in terms of kilograms of meat eaten per day (Figure 29) [897]. They also established that small prides of two to four females did not disband to hunt alone during times of food scarcity, despite a very low rate of foraging success at these times. If group foraging is not advantageous in terms of increasing net food intake per individual, then why stick together?

Scott and Martha Creel [234] point out that the optimality model tested above focuses only on the *benefit* of cooperative hunting as measured in terms of meat

29 Group size and feeding success in lions. When prey were scarce (shaded line), solitary female lions did much better than females hunting in groups of 2 to 4 in terms of kilograms of meat eaten per day. *Source: Packer et al. [897].*

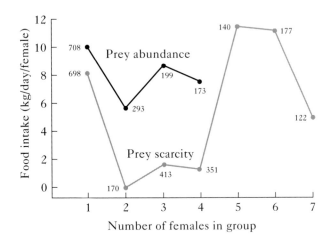

eaten per day. The model changes if one considers some of the *costs* of hunting for individuals in groups of different sizes. If the distance traveled to make a kill changes with group size, then the energy expended to reap the same benefit also changes, altering the benefit–cost ratio for hunters in different group sizes. What if cooperative hunters are really trying to maximize the difference between energy gained and energy expended in the hunt?

The Creels collected data on the benefits of social foraging in terms of kilograms of meat gained by wild dogs hunting in packs of 3 to 20. But they also measured the costs of hunting in terms of such things as the number of kilometers that the dogs ran while rapidly pursuing their prey. With these data, they calculated two different measures of foraging success: (1) the kilograms of meat gained per wild dog each day and (2) the kilograms of meat gained per dog divided by the number of kilometers of energetically expensive chases required to secure the meat that the hunters consumed. Plotting these values against the size of the hunting pack, they showed that the optimal group size for wild dogs would be 20 or higher if one considered only foraging benefits (measure 1), whereas it would be 14 if both foraging benefits and costs were taken into account (measure 2).

In reality, the mean size of wild dog hunting packs is about 10, which tells us two things. First, the more realistic model for wild dogs is the one that factors in the costs of chases as well as the caloric benefits of successful kills. Second, the mismatch between the predicted optimal group size in wild dogs and the actual mean value suggests that optimal foraging considerations, at least as calculated by measures 1 or 2, cannot be the whole story in setting group size.

How Many Prey per Trip?

Optimality theory can be applied to foraging decisions other than how large a hunting group a lion or wild dog should belong to. For example, when a solitary bird forages for its brood of young nestlings, it must leave its nest, find a suitable patch of food, and then "decide" how many prey to capture before returning to the nest to deliver the goods. As a bird collects food item after food item, its efficiency declines because it gets harder and harder to pick up new prey with a beak already full of food; furthermore, the density of prey in the patch declines as the bird feeds there. Given the law of diminishing returns, at some point it will pay the forager to stop trying to add one more prey to its collection. The optimal quitting time will result in the best benefit–cost ratio per foraging trip.

As just illustrated, however, the assumptions of an optimality model affect the nature of the hypothesis. For example, a parent starling gathering prey for its young might be attempting to maximize the overall rate of energy extraction from its foraging behavior. Or, alternatively, it might be attempting to maximize the energy gained by its family, a somewhat similar, but not identical, goal.

Let's look at how these two hypotheses differ. The *rate* of energy extraction, or yield from foraging, is a function of the number of prey collected (N) multiplied by their useful energy (V) divided by the time (T) the bird spends foraging and delivering food, which includes the round-trip flying time to and from the food patch, the time spent in the patch, and the time spent stuffing the collected food into the gullets of its nestlings: Yield = NV/T.

In contrast, the family gain hypothesis states that the individual forager is out to maximize the energy available for the growth of its nestlings. This energy will be the yield (NV/T) minus the energy (E_p) spent by the parent collecting prey (some have to be consumed by the parent to keep its energy budget in balance) and minus the energy (E_c) that all the chicks together use just to stay even at their current weight: Family gain = $NV/T - E_p - E_c$.

Alejandro Kacelnik figured out how to test these two hypotheses experimentally. He devised a mealworm dispenser and trained two wild, free-flying starlings with nestlings to come to it [622]. There the birds received mealworms one at a time, with the interval between prey deliveries increasing to simulate the diminishing return that a foraging starling would experience at a natural patch of food. The starlings had to decide when to give up and take their haul back to the nest.

Kacelnik realized that the optimal number of mealworms to carry back to the nest per foraging trip would be a function of the distance from the food dispenser (alias "food patch") to the nest. When a patch is far from the nest, travel time becomes longer, making an increase in search time a less dominant factor in the overall rate of energy gained from the patch. The mathematics of the family gain model predicted that the optimal number of prey to take back to the nest would be three when the dispenser was located right by the nest, four when the starling had 10 to 20 seconds of round-trip flight time, five when flight time was 30 to 60 seconds, and six when the dispenser was hundreds of meters away, requiring a round-trip flight time of over a minute. The energy yield model generated somewhat different predictions.

Kacelnik then observed his cooperative starlings when the dispenser was placed at different distances from the nest. He found that the number of prey actually taken back to the nest on 34 trips matched the number predicted by the family

30   The number of mealworm prey "captured" and taken back to the nest by starlings is a function of the travel time between the mealworm dispenser and the bird's nest. The different symbols represent data from individual birds. *Source: Cuthill and Kacelnik [251].*

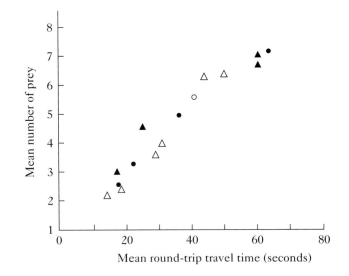

gain model 63 percent of the time. The predictions of the yield model were on target only half the time. Although additional factors must be involved, Kacelnik concluded that the primary ultimate goal of his mealworm-capturing starlings was to promote the growth of their nestlings [622].

In later studies, however, Kacelnik and Innes Cuthill discovered that starlings do not always appear to make sensible prey "capture" decisions [251]. Give a brood-provisioning starling access to one prey every 5 seconds, so that there is no decline in payoff over time, and you would expect the bird to pick up the maximum load of about seven or eight mealworms no matter where it was relative to the nest. Such a strategy would maximize the rate of prey delivery to the nest whether the foraging parent was near or relatively far from its nestlings. In reality, when the feeder is close to the nest, starlings grab just two or three mealworms before dashing home; only when travel time between nest and feeder is substantial do they load up fully (Figure 30).

Thus, theory and reality collide. To resolve the matter, Kacelnik and Cuthill tested the proposition that they had overlooked some costs of carrying large numbers of mealworms. They found that although a full load of mealworms adds only about 1 percent to an adult starling's weight, even this small amount can interfere with efficient flight. The larger the number of prey, the longer it took the starlings to return from the feeder to their nests (Figure 31), indicating that heavily loaded birds may pay a heavy metabolic price for their attempts to bring back large amounts of food on a single trip [251]. Once again, by checking optimality predictions against actual results, these researchers learned something about animal behavior.

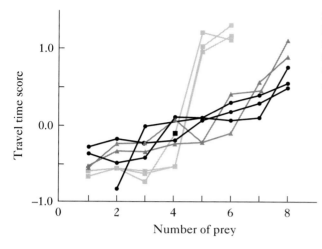

31 Time to travel back to the nest is increased by the number of mealworm prey the starling attempts to transport to its offspring. The different sets of plots represent different individuals. *Source: Cuthill and Kacelnik [251].*

Consuming Food

Once prey have been detected, attacked, and captured, the job of a forager is not necessarily over. Even after lions bring down a wildebeest or zebra, these predators often have to contend with would-be thieves from outside the pride before they can enjoy the fruits of their labor. On another front, predators that have captured toxic prey items may have to discard them, as black-backed orioles demonstrate in their dealings with monarch butterflies (see Chapter 9). An alternative to throwing away a chemically protected prey, such as a noxious grasshopper, is to impale it on thorns and let the toxins degrade for a few days. This tactic works for loggerhead shrikes; when these birds come back to retrieve their long-dead and well-aged victims, they swallow them without harm [1318].

Preparing Food

The general point that captured prey may need some additional preparation before they can be consumed is also illustrated by northwestern crows, which, as we have seen, must open captured shellfish if they are to feast upon them. A beachcombing crow spots a clam, snail, or whelk, picks it up, flies into the air, and then drops its victim. If the mollusk's shell shatters on the rocks, the bird plucks out and eats the exposed body. The adaptive significance of the bird's behavior seems straightforward. It cannot use its beak to crack open the extremely hard shells of certain mollusks. Therefore, it breaks its prey by dropping them on rocks. This seems adaptive. Case closed.

But we can be much more ambitious in our analysis of this component of the crow's foraging decisions. When a hungry crow decides to consume a whelk, it has many choices to make—which whelk to select, how high to fly before dropping the prey, and how many times to keep trying if the whelk does not break on the first try.

Reto Zach observed that (1) crows dropped only large whelks about 3.5–4.4 centimeters long, (2) they flew up about 5 meters to drop their selections, and (3) they kept trying until the whelk broke, even if many flights were required. Zach sought to explain the crows' behavior by applying the now-familiar hypothesis that the birds' decisions were optimal in terms of maximizing the whelk flesh available for consumption per unit of time spent foraging [1321]. This hypothesis yields the following predictions: (1) large whelks should be more likely to shatter after a drop of 5 meters than smaller ones, (2) drops of less than 5 meters should yield a much reduced breakage rate, whereas drops of much more than 5 meters should not greatly improve the chances of opening a mollusk, and (3) the probability that a whelk will break should be independent of the number of times it has already been dropped.

Zach tested each of these predictions in the following manner. He erected a 15-meter pole on a rocky beach and outfitted it with a platform whose height could be adjusted and from which whelks of various sizes could be dropped. He collected samples of small, medium, and large whelks and dropped them from different heights (Figure 32). He found, first, that large whelks required significantly fewer 5-meter drops before they broke than either medium-sized or small whelks. Second, the probability that a large whelk would break improved sharply as the height of the drop increased—up to about 5 meters. On still higher drops, the improvement in breakage rate was very small. Third, the probability that a

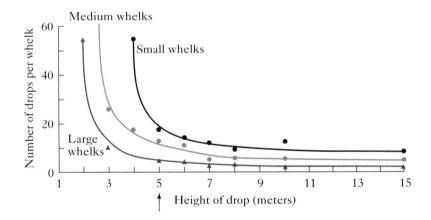

32   Optimal consumption decisions by northwestern crows when feeding on whelks. The curves show the number of drops at different heights needed to break whelks of different sizes. Northwestern crows drop only large whelks, and drop them from a height of about 5 meters, thereby minimizing the energy they expend in opening whelks. *Source: Zach [1321].*

large whelk would break was about one in four on any drop. Therefore, a crow that abandoned an unbroken whelk after a series of unsuccessful drops would not have a better chance of breaking a replacement whelk of the same size on its next attempt. Moreover, finding a new prey would take time and energy.

Zach went one step further by calculating the average number of calories required to open a large whelk (0.5 kilocalories), a figure he subtracted from the food energy present in a large whelk (2.0 kilocalories) for a net gain of 1.5 kilocalories. In contrast, medium-sized whelks, which require many more drops, would yield a net loss of 0.3 kilocalories; trying to open small whelks would have been even less profitable. Thus, the crows' rejection of all but large whelks was clearly adaptive, and their selection of dropping height and persistence in the face of failure enabled them to reap the greatest possible energy return for their foraging activities [1321].

Where to Consume Captured Food

As we saw earlier in this chapter, energetic considerations are not always the dominant factors affecting decisions about which prey to attack, and the same applies to decisions about what to consume. Some ingenious experimental studies confirm the point that individuals will opt for a less profitable, but safer, feeding behavior when the risk of predation is high [1194]. Chickadees, for example, are small birds with many enemies, and they regularly seek cover after capturing a food item, rather than consuming the food out in the open. To study the costs and benefits of chickadee shelter-seeking, Steven Lima placed a feeding tray with sunflower seed bits at 2, 10, or 18 meters from dense cover. If the birds' behavior was designed simply to maximize food intake, they should have stayed at the feeder, working through one food item after another, no matter how close they were to cover. In actual practice, they often picked up a seed and flew with it to the relative safety of pine foliage, where they could prepare and consume the seed out of view of predators. The tendency to seek shelter increased if the birds were close to it, which reduced the time and energetic costs of flying to a tree or shrub [719]

The willingness of chickadees to pay the price of flying to cover increased when they sensed the presence of a predator. Lima arranged to heighten the birds' awareness of apparent danger by sending a model hawk sailing overhead while a chickadee was at the feeder. Once exposed to the simulated predator, chickadees took a higher proportion of "captured" sunflower seeds to cover, even when this required a long and relatively costly flight (Figure 33). In other words, the birds were (sensibly) willing to sacrifice some foraging efficiency in order to improve their chances of avoiding a killer in the neighborhood.

When to Eat Dirt

That caloric considerations are not the only factors influencing what animals choose to eat becomes particularly clear when animals seek out indigestible substances to consume. Chimpanzees, for example, are believed to eat certain plant materi-

33   Predation risk and consumption decisions by chickadees. After a simulated hawk has passed near chickadees (grey circles), they are more likely to carry seeds from an exposed feeding site to cover before preparing and eating them than when no "predator" is present (black circles). *Source: Lima [719].*

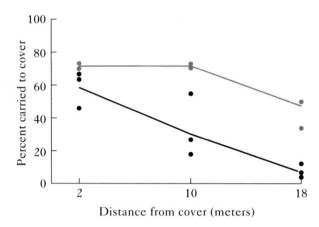

als for their antibiotic or other medicinal properties [200]. Or consider the human practice of eating dirt, which occurs in various societies around the world. Usually the earth-eaters invest time and energy in collecting and eating clays that have no caloric value whatsoever. Numerous hypotheses have been advanced to account for this habit, which violates the optimality prediction that individuals should consume only those things that maximize their caloric intake. Among the alternative ideas is the nonadaptationist pathology hypothesis (clay eating is an aberrant behavior of no functional significance) and the adaptationist detoxification hypothesis (clays are consumed to detoxify certain kinds of food, thereby improving their nutritional value) [613].

The pathology hypothesis predicts that only a relatively few, possibly mildly deranged, individuals will engage in clay eating, a prediction that does not stand given the observation that clay eating is, or was, standard practice in a diverse set of cultures, including the Aymara of Andean Bolivia, the Hopi of the American Southwest, and the natives of Sardinia, an island in the Mediterranean Sea (Figure 34). What the people in these cultures have in common is a dietary reliance on bitter acorns, with a high concentration of tannins, or varieties of bitter potatoes, with a load of toxic alkaloids. When potatoes are dipped in a clay slurry and then baked, or when acorns are baked into a clay-containing bread, the tannins and alkaloids in these foods are either bound to the clays or altered chemically by their components, rendering the foods more palatable and less toxic [613].

A comparative test of the detoxification hypothesis involves the prediction that animals unrelated to humans will also seek out certain clays when their diets feature foods high in tannins or alkaloids. Red-and-green macaws are such a species. These huge, spectacular parrots feed on seeds, unripe fruits, and leaves, many of which are apparently high in toxins. The parrots regularly gather on certain exposed riverbanks to eat clay, which probably helps them utilize the useful nutrients in their dangerous foods (Color Plate 12) [851].

34   Two Sardinian men collecting anti-toxin clays for use in cooking acorns.
Photograph by Timothy Johns, from Johns [613].

SUMMARY

1. Animal species possess a remarkable array of tactics to overcome the defenses of prey. Predators may locate prey by means of special perceptual systems, by relying on information provided by successful forag-ing companions, or by deceiving prey into approaching them. Foraging animals that have found potential foods may nevertheless be highly selective, taking only those items that provide maximal caloric payoffs. Overcoming the defenses of prey in order to capture them may be advanced by social cooperation in some cases. Even foods that have been located, attacked, and captured may still pose problems for the consumer, which may employ special devices to secure the nutrients in well-protected or toxic food items.

2. Optimality theory can be used to make predictions about what an optimal (fitness-maximizing) solution to a foraging problem should be. The optimality prediction that animals should maximize energy gained per unit of foraging time in order to maximize their reproductive success has been used extensively in the study of feeding behavior. Although some foraging animals do behave in ways that match this prediction, many other species compromise energy maximization to reduce the risk of predatory attack or to deal with some other constraint on acquiring calories.

3. Tests of optimality hypotheses are designed to help researchers identify the factors involved in the evolution of animal behavior, not to support claims that all behavior is perfectly adaptive.

SUGGESTED READING

A general review of predator decisions is provided in a book chapter by John Krebs and Alejandro Kacelnik [676] and in *Foraging Theory* by David Stephens and John Krebs [1144]. Krebs and Kacelnik also provide an introduction to mathematical models based on optimality theory, an approach that has often been applied to foraging behavior. The mathematics of modeling are covered in detail in books by Dennis Lendrem [707] and by Marc Mangel and Colin Clark [755]. Bernd Heinrich's *Bum-* *blebee Economics* [526] is a good companion to this chapter, as it deals clearly and simply with plant–pollinator interactions and optimality theory as applied to bumblebees. Reto Zach's article [1321] on optimal foraging in the northwestern crow is a model of clarity. For a critique of optimality models, see "Eight Reasons Why Optimal Foraging Theory is a Complete Waste of Time," by C. J. Pierce and J. C. Ollason [933], which can be contrasted with Krebs and Kacelnik's chapter [676].

DISCUSSION QUESTIONS

1. When sharks attack humans, they often bite the victim once and then swim away. Researcher X argues that sharks rarely eat these unfortunate persons because "sharks prefer the odor and taste of their customary prey, seals and sea lions." Researcher Y disagrees: "Sharks release humans because they are not fat enough, and therefore provide too few calories to be part of the sharks' optimal diet." Clear up the confusion here.

2. Black-headed gulls regularly rob lapwings and golden plovers of the worms they have captured in agricultural fields. Develop a simple optimality model that yields predictions about (a) which of the two species, lapwings or golden plovers, gulls should prefer to attack, and (b) the distance a gull should fly to launch an attack. Remember that the average energy gained per thievery attempt will equal the energy in a stolen worm times the probability that the attempt will succeed minus the energy expended in flight during the attack. See [1177] after you develop your model. How would a game theoretical perspective affect the kind of optimality model you build for this case?

3. In Canadian lakes containing two species of stickleback fish, one species typically forages on plankton in the open water, while the other specializes on insects taken from the lake bottom. One hypothesis argues that these differences evolved as a result of competition for limited resources, which favored individuals that diverged in their diets from members of other similar species. Provide testable predictions from this hypothesis with respect to (*a*) the diets of other pairs of closely related species that co-occur in various Canadian lakes, (*b*) the diet of a stickleback species when it is the sole occupant of a given lake, and (*c*) the survival and growth rates of individual sticklebacks that come from single-species lakes when they are experimentally placed in enclosures with members of a population of plankton-feeding specialists. (Note: Individuals of all stickleback species vary to some extent in their dietary specialization.)

Check your predictions against the data contained in [1055, 1056].

4. Many small birds flock together when foraging for seeds on the ground. The more birds present, the less often any one bird interrupts its foraging to look around for danger [e.g., 175]. (Why?) But the more birds present, the more competition there is for food, and the longer it takes an individual to find each food item. How might you incorporate these conflicting pressures into a model that predicts what the optimal flock size is? How would changing food abundance or predator pressure affect optimal flock size?

5. Various persons have proposed that when a horse switches from slower trotting to faster galloping, it changes gaits to minimize the energetic expense of locomotion, under the assumption that animals able to minimize the energetic costs of getting from A to B will enjoy greater reproductive success than individuals that squander their energy supplies in inefficient locomotion. The energy minimization hypothesis was tested by Claire Farley and Richard Taylor with the help of some cooperative horses willing to run on a treadmill while outfitted to provide data on their oxygen consumption, a factor directly related to energy use [378]. What scientific conclusion is justified on the basis of the figure below? Does this study support those who claim that optimality theory is not useful because the assumptions underlying particular hypotheses are often oversimplified and incorrect?

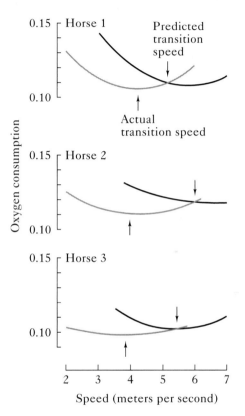

CHAPTER 11

Choosing Where to Live

THE DECISIONS THAT ANIMALS MAKE ABOUT WHERE TO LIVE usually have important consequences for their ability to avoid predators, find food, and reproduce successfully. This chapter looks at various "where to live" decisions, sometimes to trace the history behind a behavioral trait, but more often to inspect adaptationist puzzles—namely, those traits that appear to harm, rather than promote, an individual's fitness. For example, why should an animal pass up some potential places to live and keep searching for a patch of real estate somewhere else? When an animal has found a place to live, why abandon it to disperse to another spot when this requires travel across dangerous and unfamiliar terrain? Why do some migratory animals leave one area only to return to that very spot some months later? How can flying almost halfway around the world and back each year raise a migratory arctic tern's fitness? Finally, why do some animals aggressively defend living space against intruders, while others live together without risky and energetically expensive fights over territories? This chapter shows how testing of alternative hypotheses has provided some tentative answers to these questions about habitat preferences, dispersal, migration, and territorial behavior, all elements of the behavioral ecology of choosing a place to live.

Habitat Selection

Naturalists know that certain species are reliably found in certain habitats. When I was a teenager, each spring my father and I would set aside one Saturday in May to see how many different species of birds we could find within a few miles of our home in southeastern Pennsylvania. We knew that yellowthroats would be skulking in marshy spots, and that yellow warblers would be high in the sycamore trees by the creek, while we would add blue-winged warblers to the list when we visited abandoned farm fields.

The rule that certain species live in particular places applies to all groups of animals, perhaps because members of most species can choose where they will live. This decision, like any other, can be treated as an optimality problem to be solved by maximizing benefits and minimizing costs. Steve Fretwell and his colleagues were the first to use the cost–benefit approach to develop a general theory of habitat choice [410]. Using this **ideal free distribution** theory, they examined how animals ought to make choices among available habitat patches of varying resource quality when they could settle free from interference by aggressive rivals. The algebra suggested that animals would take the presence of others into account when making a habitat choice. As the density of resource consumers in prime habitat increased, there would come a point when an animal could gain higher fitness by leaving what had been the preferred habitat to settle in the next lower-ranked one, with fewer settlers and less competition.

We can apply this approach in a qualitative way to the habitat choices made by honeybee colonies in selecting a nest site. Habitat selection in honeybees occurs when a colony has grown sufficiently large to split in two, with the old queen and half her worker force flying off in a swarm, relinquishing the old hive and the remaining workers to a daughter. The departing swarm settles temporarily in a tree, where the workers hang from a limb in a mass around their queen (Figure 1). Over the next few days, scout workers fly out from the swarm in search of small openings that lead to chambers in the ground, in cliffs, or in hollow trees. Of the many such sites within range of the waiting swarm, only those with a volume of 30 to 60 liters cause a worker to perform a dance back at the swarm [1072], a dance that communicates information about the distance, direction, and quality of the potential new home. Other workers around a dancing scout may be sufficiently stimulated to fly out to the spot themselves. If it is attractive to them, they too will

1 Finding a new home. Honeybee scouts dance on the surface of the swarm, announcing the location of potential nest sites for the colony. *Photograph by E. S. Ross.*

dance and send still more workers to the area. Eventually most scouts will be advertising one location, at which time the swarm leaves its temporary perch and flies to the nest site that has received the most "votes."

What is especially interesting about this process from the perspective of ideal free distribution theory is that when some bees are experimentally offered two identical high-quality nest sites at different distances from the home hive—say, 50 and 200 meters away—the bees will choose the more distant of the two [723]. One might think that they would prefer the closer spot, since covering even a few hundred meters can exhaust their queen, a prodigious egg layer but a poor flier. However, by moving a substantial distance from the daughter's hive, a dispersing queen and her companions presumably reduce competition for food between the two colonies, increasing the survival chances of both groups.

We can test the reduction of competition hypothesis for honeybee habitat selection by predicting that a swarm's readiness to fly the extra distance should be correlated with the intensity of the potential competition between mother and daughter hives. As it turns out, the size of bee colonies varies geographically; for example, northern European colonies are much larger than those in southern Europe, a difference linked to the value of having a large number of bees as a living insulation blanket for the queen and her core workers in places with cold winters [604, 605]. In an area with large colonies, two hives are likely to compete more intensely for food, favoring dispersing queens that settle with their swarms far from the original hive. As expected, large honeybee colonies in cold Germany move farther when finding new hive sites than do the much smaller swarms of bees that live in warmer Italy [452].

Habitat Preferences in a Territorial Species

Honeybee colonies do not attempt to defend a foraging area as their exclusive feeding preserve, and therefore dispersing swarms can settle wherever they choose without direct interference from other colonies. For other animals, however, an ideal free distribution is not possible because some members of the species actively exclude others from superior sites. Thomas Whitham examined the effects of this kind of aggressive competition on habitat selection in an insect, the poplar aphid [1268–1270].

Each spring in Utah, vast numbers of aphid eggs hatch in crevices in the bark of cottonwood poplar trees. The tiny black females walk to leaf buds on cottonwood branches. Each female—and there may be tens of thousands per tree—actively selects a leaf, settles by its midrib, almost always near the base, and in some way induces the formation of a hollow ball of tissue—a gall—in which she will live with the offspring she bears parthenogenetically (Figure 2). When her daughters are mature, the gall splits, and the aphids disperse to new plants.

Whitham found that newly emerged females quickly attempted to occupy all the very large leaves on poplars, but that aphids who had settled on leaves previously were prepared to fight intruders for possession of their gall sites. Since there were about 20 aphids for every large leaf on the trees he studied, foundress females

2 Poplar aphid galls. A poplar aphid gall at the base of a leaf (left). A scanning electron micrograph of a female inside her gall (right). *Courtesy of Thomas Whitham.*

3 Territorial dispute between two poplar aphids. Females may spend hours kicking each other to determine who gets to occupy a preferred leaf or the superior location on a leaf. *Courtesy of Thomas Whitham.*

regularly encountered competitors. The aphids battled for up to 2 days (Figure 3), during which time the combatants sometimes died [1271]. Given the costly nature of this physical competition for control of the largest leaves, major benefits should come from their possession—and they do (Figure 4).

Defeated females or small ones incapable of effective fighting are forced to accept inferior options by winners of the competition for top sites. The optimality expectation for these individuals is that they will make the best of a bad situation, accepting the best of the lower-value habitats available to them. The options for these individuals include finding a smaller, unoccupied leaf or settling with

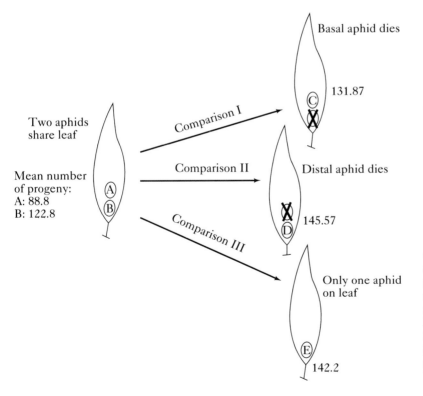

4 Territories and reproductive success. The average number of progeny produced by aphids that succeed in monopolizing a poplar leaf versus those that are forced to share a leaf of the same size with a rival. *Source: Whitham [1271].*

Table 1   Effect of leaf size and position of gall on the reproductive success of female poplar aphids

Number of galls per leaf	Mean number of progeny produced			
	Mean leaf size (cm)	Basal female	Second female	Third female
1	10.2	80	—	—
2	12.3	95	74	—
3	14.6	138	75	29

Source: Whitham [1270].

an established foundress on a large leaf. If a leaf already has a resident female, the latecomer will have to form her gall farther out on the midrib, where it will provide fewer nutrients than the gall at the prime location near the leaf petiole. At this stage, an ideal free distribution is possible, since poplar aphids usually do not face despotic rivals for these less valuable gall sites. These second- or third-rate sites yield fitness gains that depend on the size of the leaf and whether it already has a gall on it. The second colonist on a large leaf can do just as well as a single aphid on a medium-sized leaf, and substantially better than a lone colonist on an even smaller leaf (Table 1). As predicted, when aphid females do double up, they choose large leaves, despite their scarcity [1270].

Dispersing from One Place to Another

Once a poplar aphid finds a spot to live in, she will spend the rest of her life at that location. But for other species, such as the honeybee, dispersal from an established home base to another living area happens regularly. Moving exposes the disperser to predators and requires energy, as can be seen in the fat-laden bodies of those naked mole-rat males destined to leave their natal colony (see Chapter 15) in search of another group (Figure 5)[878].

For an example of a predation cost of dispersal, consider what happens to vervet monkeys that abruptly leave their familiar home ground to join another band in a new and unfamiliar area—something that happens when their original band has lost many members. The mortality rates for the immigrants are much higher over the first year in their new homes than for the vervets already resident in the area (Figure 6). Since vervet mortality is almost always caused by predators, monkeys unfamiliar with new terrain are apparently more likely to become leopard chow [592].

Since dispersing may mean death, why are animals so often willing to leave home? This question is particularly pertinent for species in which some individuals do *not* disperse, and so do not pay a price for leaving a familiar area. In colonies of cliff swallows, some young of the year return to breed in their natal colony, while others strike out for other places. Past parasite experience apparently determines which stay and which leave. The dispersers had suffered more from nasty fleas

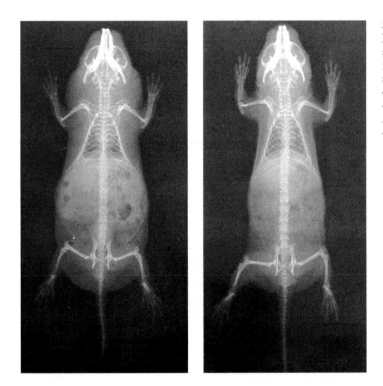

5 Dispersing requires energy. X-ray images of disperser (left) and non-disperser (right) naked mole-rats show that, while there are no significant skeletal differences between the two types, the disperser has much more body fat. *From O'Riain et al. [878], courtesy of J. O'Riain and R. Alexander.*

and blood-sucking bugs when they were nestlings than birds that remained faithful to their natal colony [142].

In Belding's ground squirrels, some individuals disperse much farther than others. These small mammals spend their first 4 weeks of life largely underground in their mother's burrow in the high meadows of the California Sierras. However,

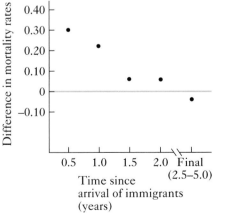

6 Dispersers are at risk. Immigrant vervet monkeys that join a new group suffer higher mortality during their first year, probably from predators, than do established residents of the band. *Source: Isbell et al. [592].*

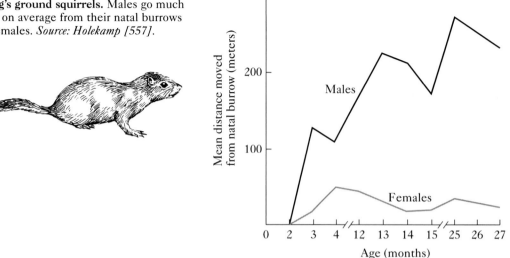

7 Distances dispersed by male and female Belding's ground squirrels. Males go much farther on average from their natal burrows than females. *Source: Holekamp [557].*

by the time males are just 9 or 10 weeks old, they head out, leaving the safety of the home burrow for the uncertainty of new terrain. Their sisters also eventually leave the natal burrow, but they usually settle down right next door, moving an average of only 50 meters from their mother's territory, whereas their brothers usually wind up more than 150 meters from home (Figure 7) [557].

The Inbreeding Avoidance Hypothesis

Why should young male Belding's ground squirrels go farther from home than their sisters? According to one argument, costly dispersal by juvenile animals of many species may arise because of the risk of **inbreeding depression** [967]. When closely related individuals mate, the offspring they produce are more likely to carry damaging recessive alleles in double doses than are the offspring produced by unrelated pairs (but see [1092]). The risk of genetic problems should in theory reduce the average fitness of inbred offspring, and high juvenile mortality actually does occur in inbred populations of a number of species [975]. For example, when inbred and non-inbred white-footed mice were experimentally released into a field from which their ancestors had been captured, the inbred mice survived only about half as well as the non-inbred ones [612]. And even if inbred mice manage to reach adulthood, they may be less likely to reproduce than outbred individuals (Figure 8) [764].

If avoidance of inbreeding is the point of dispersing, then one might expect as many female ground squirrels as males to leave the natal site. But they do not, perhaps because the costs and benefits of dispersal differ between the sexes. Females

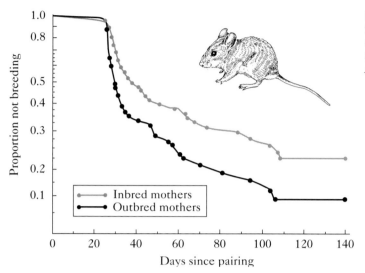

8   Inbreeding depression in old-field mice may result in the failure of inbred females to reproduce as soon as outbred females. *Source: Margulis and Altmann [764].*

bear most of the costs of inbreeding because mothers alone endure the energetic expenses of nurturing defective offspring. Thus, female squirrels would be expected to resist breeding with close male relatives, which might then disperse to gain access to as large a number of unrelated females as possible.

Furthermore, as Paul Greenwood has suggested for female mammals in general [470], female Belding's ground squirrels may remain at or near their natal territories because their reproductive success is especially dependent on possession of a territory in which to rear their young. Female ground squirrels, and some other mammals [839], that remain near their birthplace enjoy assistance from their mothers in territorial defense and protection of burrows against rival females. Thus, for females, the benefits of remaining on familiar ground are greater than for males, as are the costs of inbreeding, resulting in sex differences in dispersal [557].

The Mate Competition Hypothesis

Jim Moore and Rauf Ali offer an alternative hypothesis to account for male-biased dispersal in Belding's ground squirrels and other animals. They suggest that differences in the nature of sexual competition, not differences in the benefits of staying near one's birthplace or avoiding inbreeding, may often force males to move farther than females. Because males typically fight with one another for access to mates (see Chapter 12), losers may find it advantageous to move on. Females rarely fight one another for mates, and thus may generally be able to remain near home [833].

If the mate competition hypothesis applies to Belding's ground squirrels, males should compete violently for access to mates—and they do. But we would also expect male fighting and dispersal to go hand in hand, and this prediction fails. The male squirrels invariably disperse in their first year of life, *before* reaching sexual maturity and at a time of year when competition for mates is absent. Thus direct competition among males for mates cannot be the immediate cause of male dispersal.

Young immature males might, however, disperse in order to avoid intense sexual competition in the future with more experienced rivals. If this version of the sexual competition hypothesis is true, then the density of breeding males at sites chosen by dispersing squirrels should be lower than around their natal burrows. It is not; competitor densities are on average the same in both locations. The mate competition hypothesis also suggests that males should stay put when they have had a successful breeding season. Winning males, however, always leave one year's mates behind and move away in anticipation of the next breeding season, with the result that they avoid their sexually mature daughters.

Although we can probably rule out the mate competition hypothesis for male dispersal in Belding's ground squirrels, it deserves to be tested in other cases. Take lions, for example. These mammals live in groups, or prides, dominated by one or several adult males. As is true among mammals generally, young males disperse from their natal group, while females usually spend their entire lives right where they were born (Figure 9) [965]. The sedentary females benefit from their familiarity with good hunting grounds and safe breeding dens in their natal territory.

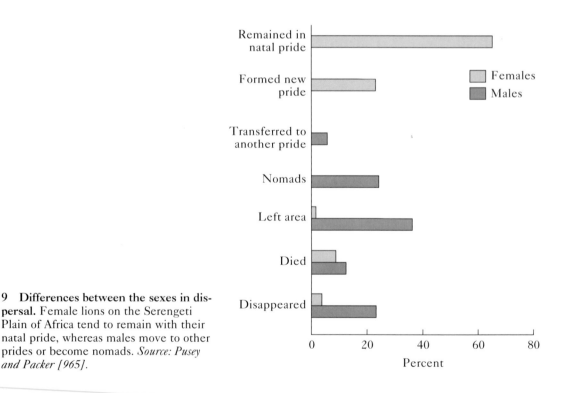

9 **Differences between the sexes in dispersal.** Female lions on the Serengeti Plain of Africa tend to remain with their natal pride, whereas males move to other prides or become nomads. *Source: Pusey and Packer [965].*

The proximate reason for the departure of many young male lions is the arrival of new mature males that violently displace the previous pride owners and chase off the subadult males in the pride as well. These observations support the mate competition hypothesis. However, if young males are not evicted after a pride takeover, they often leave anyway, without any coercion from adult males and without ever having attempted to mate with their female relatives. Moreover, mature males that have claimed a pride will sometimes disperse again (in the manner of Belding's ground squirrels), expanding their range to annex a second pride of females at the time when their daughters in pride number one are becoming sexually mature. Proximate inhibitions against inbreeding apparently exist in lions. Ultimately, dispersing males gain by mating with nonrelatives, even though the timing of their departure from their birthplace is not always under their control [505].

Changing Breeding Locations or Staying Put

It is not just mammals that sometimes leave one breeding location for a new home. In many birds, some adults that have reproduced in one place may shift to a new site in their next breeding season, while others return to the same spot for another breeding attempt. One hypothesis for differences in site fidelity within a species is that the dispersing birds experienced nest failure the preceding year, while the returning birds bred successfully. In one study, male bobolinks that returned to the same meadow or hayfield to breed again had in fact reproduced more successfully in the preceding year than males that did not come back to their old breeding grounds (Figure 10).

But did the nonreturning males actually go to new breeding sites, or did they die? It is possible that male bobolinks that have low reproductive success in one year are in relatively poor condition. Their failure to return may stem from their deaths, not their refusal to breed in the same location in a subsequent year. We could eliminate this hypothesis if we could find the nonreturning birds in a new location—a challenging requirement, but one that has been partly met for bobolinks [430] and a few other birds.

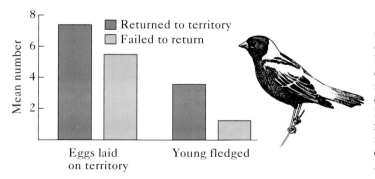

10 Site fidelity and reproductive success in bobolinks. Males that exhibited site fidelity by coming back to the same territory in successive years had enjoyed higher reproductive success in the preceding year (as measured in eggs laid per territory and young fledged) than males that failed to return to the same territory in successive years. *Source: Gavin and Bollinger [430].*

11 Dispersal and reproductive success in red-winged blackbirds. Males move when they have reared relatively few fledglings; after moving, their reproductive success is likely to improve in the next breeding season. *Source: Beletsky and Orians [75].*

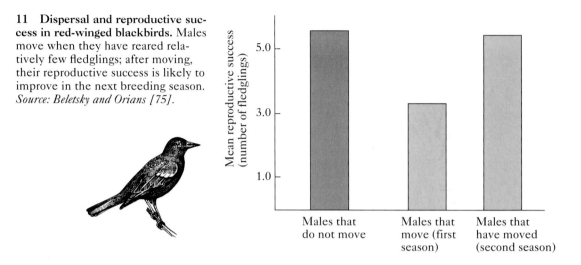

Birds that shift voluntarily from one breeding place to another, and survive the move, should do better in the new spot than in the one they left behind. This prediction has been tested for male red-winged blackbirds and female sparrow hawks, among others [e.g., 308]. Male redwings defend territories in marshes; females inspect potential nesting areas and choose a male on the basis of the resources he controls [370]. In one population, more than 70 percent of returning males reclaimed the territories they had held the year before. But, as predicted [75], those that moved (1) had had relatively poor reproductive success in the place where their mate(s) had nested the previous year, and (2) were more successful after making the move (Figure 11). Likewise, 72 percent of female sparrow hawks nested on the same territories they had used the previous year. Most of those that moved had not bred successfully in the site they abandoned, and many did better in their new nesting territories [860].

Migration

The capacity of bobolinks and red-winged blackbirds to return to the same nesting site in successive years becomes considerably more impressive when one realizes that for much of the year these birds may be hundreds or thousands of miles away. Nearly half of all the breeding birds of North America take off in the fall for a trip to Mexico or Central America or South America, only to return in the spring [227]. Tiny ruby-throated hummingbirds, the weight of a penny, fly nonstop 850 kilometers across the Gulf of Mexico twice a year. Arctic terns breeding in Canada may complete a roughly 40,000-kilometer migration each year (the equivalent of seven trips across the continental United States) (Figure 12), a journey that the average human would find intimidating even if he had a burning desire for frequent-flier mileage [227].

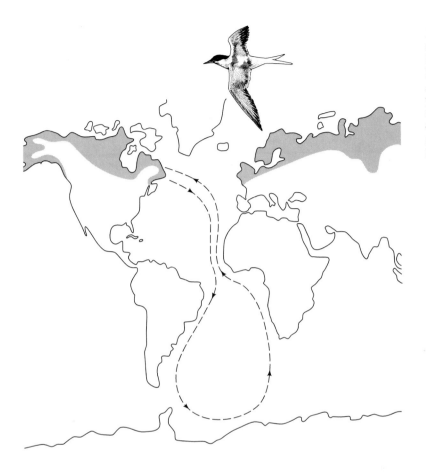

12   The migratory route of arctic terns. These birds fly from high in the Northern Hemisphere to Antarctica and back each year. Some young birds may spend two years circling Antarctica before returning to the northern breeding grounds.

Among the mammals, wildebeest, caribou, bison, seals, and whales also make round-trip journeys of thousands of miles each year. Ditto for some reptiles [958], including the green sea turtles that nest on Ascension Island, a tiny speck of land a mere 8 kilometers wide in the center of the Atlantic Ocean between Africa and Brazil. The adult female turtles visit the island only to deposit their eggs in beach sands. They then swim 1600 kilometers or so to warm, shallow water off Brazil, where they feed on marine vegetation for several years before returning, usually to the same beach, to lay another clutch of eggs [185].

The Historical Basis of Migration

Migration poses a major historical problem: How could the ability to make a monumental round-trip journey evolve in a species that once did not migrate, since sedentary species were probably ancestral to modern migratory ones? One of the lessons of history (see Chapter 7) is that gradual evolutionary change is more probable than an abrupt jump from one trait to another very different one. Can we imagine successful intermediate states between a sedentary life and a life involving two long-distance trips per annum? We can, especially since even today, many tropical bird species exhibit short-distance "migratory" movements of dozens to hundreds of miles, with individuals moving up and down mountainsides or from one region to another immediately adjacent to it. Douglas Levey and Gary Stiles point out that short-range migrants occur in nine families of songbirds believed to have originated in the tropics. Of these nine families, seven also include long-distance migrants that move thousands of miles from the tropics to temperate zones. The co-occurrence of short-range and long-distance migrants in these seven families supports the argument that short-range migration preceded long-distance travel [711].

But what benefits might an ancestral short-range migrant have gained? Levey and Stiles think the answer lies in the diets of these birds. A heavy reliance on fruit characterizes eight of the nine families of Neotropical birds with some current short-range migrants. To exploit fruit efficiently, these birds must search out widely scattered trees that happen to be producing heavily. If the same were true for an ancestral premigratory species, it may have evolved good navigational systems because these devices helped it move among widely separated fruit trees, setting the stage for future refinements needed for longer travels between more distant resource patches.

Today's long-distance songbird migrants in the Americas travel from warm wintering grounds to temperate regions notable for their production of insects in the spring and summer. These birds may be derived from an ancestral fruit-eating species that incorporated protein-rich insects into its diet, or that of its nestlings, to compensate for the lack of protein in its mainstay, carbohydrate-rich fruits [227, 711].

The Costs and Benefits of Migration

In tracing the history of songbird migration between tropical and temperate America, we have reviewed the possibility that the long-distance migratory abilities of today's birds evolved in stages from a sedentary ancestor that gave rise to a species capable of navigation across short distances, with the capacity for guided travel slowly increasing in a lineage that eventually produced species capable of flying thousands of miles twice a year. If natural selection was involved in producing this pattern, each change in the sequence must have been adaptive, with the costs of the trait outweighed by the benefits of the new migratory ability.

I have already alluded to the high energetic costs of migration, which must be especially great for those birds that spend several months each year on vast journeys across oceans and over major mountain ranges. (Lawrence Swan once saw a migrating hoopoe, a bird with the exquisite scientific name *Upupa epops*, reduced to hopping up a Himalayan pass at 20,000 feet [1165]!) In addition to the time and energy expended, migrants run a gauntlet of predators, including Eleonora's falcon, which literally makes its living off exhausted songbirds that have crossed the Mediterranean Sea in the fall [1235].

Many migrating individuals act to reduce the costs of the trip. It is common for migrants to travel in groups, perhaps to dilute the risk of being captured by a predator (see Chapter 9). In addition, a host of European songbirds travel to central Africa by way of Spain and Gibraltar in order to cross the Mediterranean at its narrowest point [86]. This lengthens the journey but reduces the overwater component of the trip, perhaps decreasing the risk of drowning at sea or encountering an Eleonora's falcon at the end of an exhausting journey.

In the fall, red-eyed vireos migrating from the eastern United States to the Amazon basin of South America also face the option of crossing a large body of water, the Gulf of Mexico, or staying close to land, moving in a southwesterly direction along the coast of Texas to Mexico and then south. The trans-Gulf flight is shorter, but birds that run out of energy before reaching Venezuela are dead ducks, so to speak. In light of this danger, Ronald Sandberg and Frank Moore predicted that red-eyed vireos with low fat reserves would be less likely than those with considerable body fat to risk the 1000-kilometer journey due south across the Gulf of Mexico. They captured migrating red-eyed vireos in the fall on the coast of Alabama, classified each individual as lean or fat, and placed the birds in orientation cages similar to those described in Chapter 3. Vireos with less than about 5 grams of body fat showed a mean orientation at sunset toward the west–northwest, whereas vireos that had been classified as having more fat tended to head south, as Sandberg and Moore had predicted (Figure 13) [1045].

Some songbirds even smaller than the red-eyed vireo attempt a still more impressive flight across water in the fall [1290, but see 853]. At first glance, a blackpoll warbler that elects to go from Canada to South America over the Atlantic

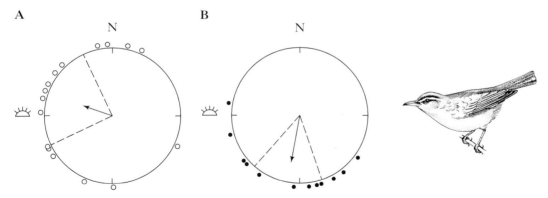

13   Body condition affects the migratory route of red-eyed vireos. (A) Birds with low fat reserves do not risk flights south across the Gulf of Mexico. They oriented toward the west at sunset (sunset symbol), the direction they would take for an overland flight. (B) Fat birds have the energy reserves needed for the long-distance flight south over water; they oriented due south. The central arrow shows the mean orientation for the birds tested in each group. *Source: Sandberg and Moore [1045].*

Ocean seems to have a death wish, since the trip requires a nonstop flight over water of more than 3000 kilometers (Figure 14). One would think it would be far safer to travel along the coast of the United States and down through Mexico and Central America. But migratory warblers do appear regularly on islands in the Atlantic and Caribbean, so some, perhaps most, must survive their ocean crossing.

Special fitness benefits accrue to the blindly courageous blackpoll warbler that manages the over-the-ocean trip. First, the sea route from Nova Scotia to Venezuela is about half as long as a land-based trek. Timothy and Janet Williams estimate that a blackpoll warbler flying under good weather conditions can go from Maine to South America in 50 to 90 hours of continuous flight [1290]. Second, there are very few predators lying in wait in mid-ocean or on the islands of the Greater Antilles that the blackpolls strive to reach. Third, the birds leave the Canadian coast only after a west-to-east-traveling cold front pushes them out over the Atlantic Ocean for the first leg of the journey, after which the birds pick up westerly breezes in the southern Atlantic that blow them to an island landfall. Likewise, without assistance

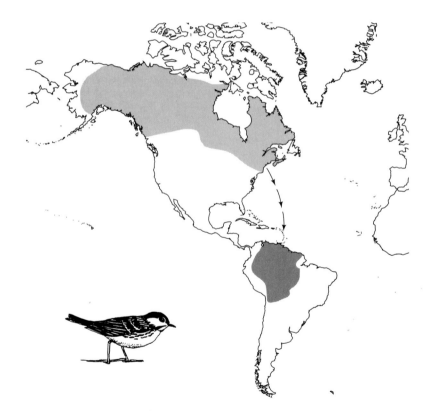

14 **Transatlantic migratory path of blackpoll warblers** (arrows) from southeastern Canada and New England to their South American wintering grounds (darkly shaded area). *Courtesy of Janet Williams.*

from prevailing tailwinds, the bar-tailed godwit, a large shorebird, probably could not fly 5000 kilometers nonstop from northwestern Australia to China [1208].

For all their navigational and meteorological skills, migrant birds still pay a steep price for their trips across oceanic waters, especially if they are forced down during storms. Although migrants may reduce the costs of their travels, they cannot eliminate them. What ecological conditions might elevate the benefits enough to outweigh the dramatic costs of migration?

As noted, many birds in the Americas follow a migratory route from a warm wintering area with ample supplies of fruit to a temperate region that experiences a spring and summer surge of insect production, which may support a large brood of young [144]. Changes in resource availability appear to be linked to fish migrations as well. In any number of unrelated fishes, individuals migrate from fresh water to the sea and back during their lifetimes, or from the sea to fresh water and back. In northern latitudes, most migratory species, including the familiar salmon, spend the bulk of their lives in the ocean (Figure 15). In the tropics, migrants are far more likely to move into fresh water from the sea. This pattern reflects the fact that food production in the northern oceans exceeds that in bodies of fresh water, while just the reverse is true in tropical latitudes [477].

Another major migratory pattern involves movement from a feeding area to a protected breeding or birthing location. When an Adélie penguin waddles onto the Antarctic mainland, it moves into an area a long walk from the nearest feed-

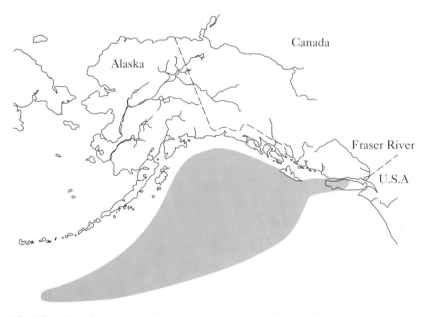

15 Migration by salmon. Salmon that are born in British Columbia's Fraser River move far out into the northeastern Pacific Ocean (shaded area) before returning to breed in the river in which they were hatched. *Source: Quinn and Dittman [974].*

16 Breeding grounds of Adélie penguins in coastal Antarctica. These birds use rookery sites near open water in the breeding season, with good landing beaches and bare ground for nests. *Photograph by Colin Monteath/Hedgehog House.*

ing grounds (Figure 16). The windy, rock-strewn slopes chosen by the penguins offer the best sites for rearing young on land in the Antarctic region [347]. Like-wise, marine mammals, fishes, and reptiles may abandon rich feeding grounds and swim without food for hundreds or even thousands of miles to spots suitable only for reproduction [868]. Seals that mate and bear their young on land, as well as marine turtles that lay their eggs in sand, tend to choose island coves that are isolated, well sheltered, and relatively free of predators [185].

The Migration of the Monarch Butterfly

Each fall millions of monarch butterflies join the many birds moving from southern Canada and the northeastern United States to central Mexico, with some individuals traveling as much as 2500 kilometers [1212], only to turn around and head north the next spring [137] (Figure 17). In the fall, the traveling monarchs all wind up in a very few high mountain patches of Oyamel fir forest in central Mexico, where

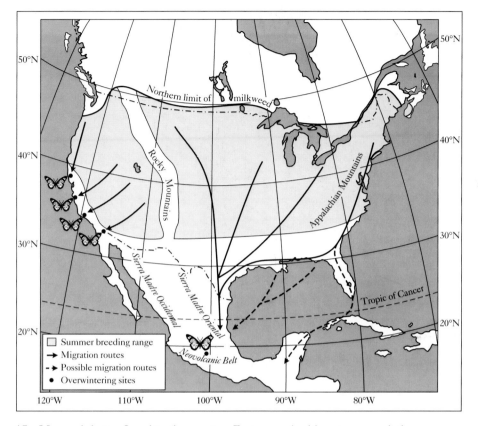

17 Monarch butterfly migration routes. Eastern and midwestern populations move
to a few small patches in the mountains of central Mexico. Western populations move
to the Pacific coast. *Source: Brower [137].*

they spend the winter in spectacularly dense aggregations [754] (Figure 18). The
seasonal fluctuation in availability of the plants needed by monarchs is linked to their
migration. During the winter, the milkweed plants that monarch larvae feed upon
disappear from the eastern United States and Canada, but will grow again in spring.

But why should some eastern monarchs fly so far to spend the winter in the cold
high mountains of Mexico? Surely there are more suitable places far closer to the
spring and summer breeding grounds. But perhaps not, since killing freezes occur
regularly at night throughout the eastern United States. In contrast, freezes are very
rare in the Mexican mountain refugia used by the monarchs. In these forests, at
about 3000 meters elevation, temperatures usually stay within 4–11°C during the
coolest winter months. Occasionally, however, snowstorms do strike the mountains,
and when this happens as many as 2 million monarchs can die in a single night of

18 Monarch butterflies spend the winter in huge clusters that form on fir trees in a few locations in Mexico. *Photograph by Lincoln P. Brower.*

subfreezing temperatures. The risk of freezing to death could be completely avoided in many lower-elevation locations in Mexico. But William Calvert and Lincoln Brower note that in warmer and drier areas, the monarchs would quickly use up their water and energy reserves. By remaining moist and cool—but not frozen— the butterflies conserve vital resources, which will come in handy when they begin to reproduce and start back north after their 3 months in the mountains [171].

The hypothesis that the stands of Oyamel fir used by the monarchs provide a uniquely favorable microclimate that promotes butterfly survival in the winter may be tested shortly in an unfortunate manner. Mexican lumber companies have already cut trees from areas bordering some winter aggregations, and there have been suggestions that the stands of firs where the butterflies themselves cluster could be safely thinned. Brower and his associates believe that timber removal would greatly increase butterfly mortality, even if some roosting trees were left intact. When the butterflies are rained upon or directly exposed to the night sky, they are more likely to freeze (Figure 19). Opening up the forest canopy will increase the chances that the butterflies will become wet and exposed. Thus, even partial forest cutting may destroy the conditions needed for the survival of monarch aggregations. If timber removal causes the local extinction of overwintering monarch populations, it will be a powerful demonstration of the value of a particular habitat for butterfly survival [21].

Another test of the hypothesis that monarchs migrate in the fall to reach areas with special survival advantages comes from examining the major overwintering sites in coastal California (see Figure 17). Here, too, immense numbers of butter-

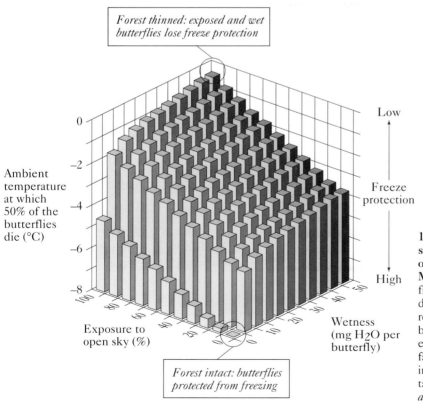

Forest thinned: exposed and wet butterflies lose freeze protection

Forest intact: butterflies protected from freezing

Ambient temperature at which 50% of the butterflies die (°C)

Exposure to open sky (%)

Wetness (mg H_2O per butterfly)

Freeze protection — Low / High

19 Habitat quality and survival of monarchs overwintering in central Mexico. Protection from freezing depends on a dense tree canopy that reduces wetting of the butterflies and their exposure to open sky, factors that greatly increase monarch mortality. *Source: Anderson and Brower [21].*

20 Cluster of monarch butterflies overwintering in a coastal Californian location. *Photograph by Edward S. Ross.*

flies gather from regions of western North America where their food plant is absent in winter (Figure 20). Because these locations are near the Pacific Ocean, wintering monarchs experience moderate temperatures and moist air, which enable them to wait out the nonproductive months with reduced risk of death from freezing, exhaustion, and desiccation. Apparently the locations that offer this combination of benefits are so limited as to favor a major investment in long-distance travel.

The Coexistence of Migrants and Nonmigrants in the Same Species

An especially interesting problem for adaptationists is posed by the existence of both migratory and resident individuals within the same species in the same area. The European blackbird is an example, since in portions of its range, some individuals migrate in the fall, leaving the area to others that overwinter there. If the differences between these two phenotypes were hereditary, you might think that one type would have completely replaced the other by now, since natural selection would produce this result if one type had even a slight reproductive advantage over the other. The long-term coexistence of two genetically distinct types demands that they have *exactly* the same average fitness over the long haul. It is hard to imagine how migrants and nonmigrants could secure exactly equal fitness, although frequency-dependent selection (see Chapter 3) provides one evolutionary route to this end.

If migratory and nonmigratory individuals differ because of genetic differences between them, they are, in the lingo of game theory, said to employ two different **strategies** [278, 479, 742]. Evolved "strategies" are not consciously adopted plans of the sort humans often make use of, but rather behavioral traits that differ among individuals because of their hereditary makeup. The two-strategies hypothesis for blackbird behavior generates the predictions that (1) the differences between migratory and nonmigratory individuals are caused by differences in their genetic constitution, (2) individuals will not change their behavior from year to year, and (3) the two types will achieve equal fitnesses on average.

However, European blackbirds that migrate in one year often switch their behavior to become residents in a subsequent year (Figure 21) [1060]. This finding makes it unlikely that the two-strategies hypothesis can apply in this instance. Instead, let's consider another game theoretical hypothesis, the **conditional strategy** alternative, in which the genetic makeup of individuals confers on them the ability to adopt either the migratory or the resident option, depending on the conditions they experience in their environment. In a conditional strategy, the differences between individuals are *not* caused by genetic differences; instead, possession of the strategy per-

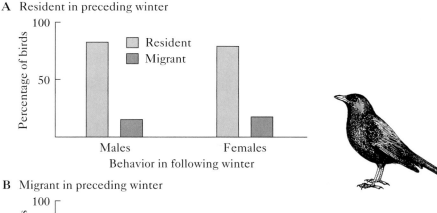

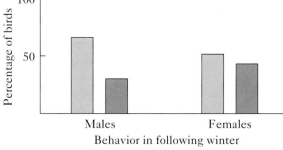

21 Migratory behavior: the preferred tactic for European blackbirds? (A) Birds that were resident in the preceding winter tended to be nonmigratory the next winter as well. (B) In contrast, birds that were migrants the preceding winter were quite likely to switch to the resident option the following winter. *Source: Schwabl [1060].*

mits an individual to engage in different **tactics** under different conditions. Thus, the ability of individual New Mexico spadefoot toad tadpoles to adopt either a shrimp-eating or a detritus-eating lifestyle, depending on the availability of the two foods in their desert ponds, represents a conditional strategy (see Chapter 10).

Applying a conditional strategy hypothesis to the blackbird case, we would expect individuals to adopt whatever tactic yields the higher fitness for them. Socially dominant individuals should be in a position to select the better of the two tactics, forcing subordinates to "make the best of a bad situation" [278] by adopting the option with a lower reproductive payoff (but better than what they could get from futile attempts to be dominant). For example, perhaps an area can support only a few permanent resident blackbirds, especially during winter. Under such circumstances, subordinates faced with a coterie of more powerful residents might do better to migrate away from the competition, returning in the spring to occupy territories made vacant by the deaths of some rivals over the winter.

Given the logic of this hypothesis, we can predict that (1) blackbirds will have the ability to switch between tactics, rather than being locked into a single behavioral response, (2) individuals using the two tactics need not achieve equal reproductive payoffs from the different options, and (3) when choosing *freely* between tactics, individuals will choose the option with the higher payoff (as already illustrated by the behavior of poplar aphids when choosing between occupied and unoccupied leaves of different sizes). In light of these predictions, it is significant that European blackbirds do switch from one tactic to another, and when they do, the typical pattern is for an older, more experienced bird to become a resident after having been migratory when it was younger [742]. We do not yet know, however, whether blackbirds tend to achieve higher reproductive success when they are permanent residents than in years when they migrate.

Territoriality

Thus far, we have examined why animals show preferences for some living areas over others, and why animals might move from one location to another. We now turn our attention to yet another decision that many animals face: having settled temporarily or permanently in one place, should the individual set up a **territory** that it defends against intruders, or should it coexist peacefully with others on an undefended **home range**? As we have noted, female poplar aphids and Belding's ground squirrels can be extraordinarily aggressive in competing for a living area, while honeybees and monarch butterflies tolerate or even ignore their fellows. Likewise, some species in the mammalian family Carnivora have territorial and nonterritorial members (Figure 22). It is revealing that the carnivores that live on home ranges typically occupy much larger areas than those that defend territories. The energetic costs of patrolling and defending a territory appear to put an upper limit on the size of the area that a territory owner can attempt to claim for itself.

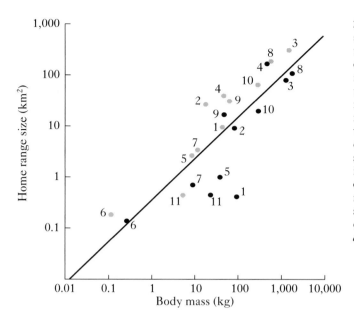

22 Territory size versus home range size. Territories tend to be smaller than undefended home ranges among carnivore species that exhibit both territorial defense and occupation of undefended home ranges. The circles represent the size of a territory as a multiple of the body mass of the individual (or group, in social species) that defends a site. Each number represents a different species of carnivore, including the golden jackal (1), the coyote (2), and so on. The grey circles represent undefended home range sizes; the black circles represent defended territory sizes. *Source: Grant et al. [462].*

Consider the energetic price paid by some Yarrow's spiny lizards, participants in an experimental demonstration of the costs of territoriality [766, 767]. The researchers, Catherine Marler and Michael Moore, inserted small capsules containing testosterone beneath the skin of some male lizards they captured in June and July, a time of the year when the lizards are only weakly territorial. These experimental animals were then released back into a rockpile high on Mt. Graham in southern Arizona. Marler monitored the behavior of the experimental lizards as well as that of a control group that had been captured and operated upon, but had received inert capsules. The testosterone-implanted males patrolled more, performed more push-up threat displays, and expended almost a third more energy than the controls [768]. As a result, the testosterone-implanted males had less time to feed, captured fewer insects, stored less energy in fat, and died at a faster rate than those with normal levels of the hormone (Figure 23) [767]. Similar results have been secured for the red grouse. Territorial males that received supplemental testosterone expanded their territory sizes greatly (Figure 24), but were less likely to survive to the following year than were males whose testosterone levels were not artificially elevated [842].

If a territory is worth defending, then some reproductive compensation must come from the defended resource. In nature, male Yarrow's spiny lizards become *highly* territorial only during September and October, when females are receptive; the same pattern of seasonal changes in the intensity of territoriality holds for red grouse as well.

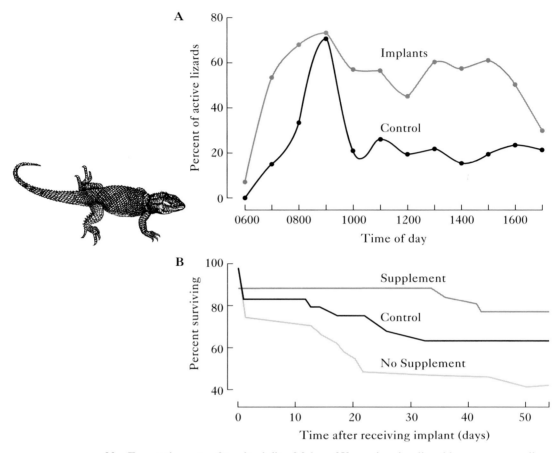

23 Energetic costs of territoriality. Males of Yarrow's spiny lizard became unusually territorial during the summer when they received an experimental testosterone implant. (A) The experimental males spent much more time moving about than did control males. (B) Testosterone-implanted males that did not receive a food supplement disappeared at a faster rate than did control males. Testosterone-implanted males that received a food supplement (mealworms) survived as well as or better than controls; thus the high mortality experienced by the unfed group probably stemmed from the high energetic costs of their induced territorial behavior. *Source: Marler and Moore [766, 767].*

Territoriality and Fitness

The general expectation that owning a resource-containing territory advances the owner's reproductive success has been examined in many animals. For example, most songbirds defend breeding territories, where they gather a large proportion of the food needed for their offspring, although territorial males may also benefit by

A 1992

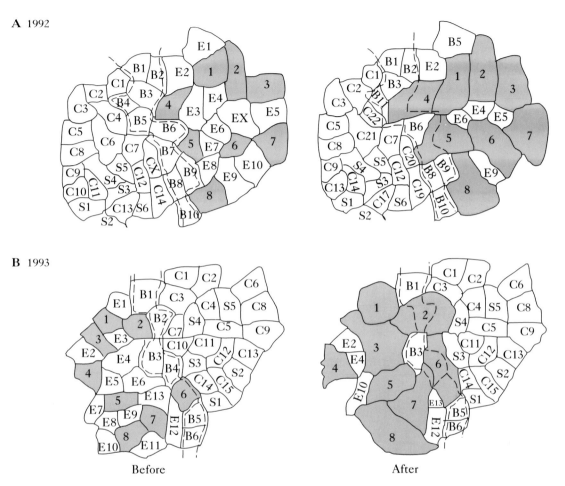

B 1993

Before   After

24 Testosterone and territoriality in the red grouse. (A) Eight male red grouse on eight territories (represented by numbered shaded areas) were selected in 1992 to receive testosterone implants, after which these males greatly expanded the areas they defended. (B) The experiment was repeated in different territories in 1993 with the same result. *Source: Moss et al. [842].*

keeping other males away from their mates [822] (see Chapter 12). If rich feeding territories really are in short supply, nonterritorial individuals should fail to breed.

As expected, nonterritorial rufous-collared sparrows do not reproduce, but instead move about secretively within the breeding territories of other birds. Susan Smith made this discovery by capturing birds in a mist net (a fine-meshed, black nylon net that can be strung between poles in areas traveled by birds). Each captured sparrow received a unique color band combination so that Smith could plot the locations of known individuals and record their territorial and breeding activity, if any. Although the nonterritorial birds remained inconspicuous, they could not breed, because the territorial birds attacked them if they tried. Therefore, members of this underworld population waited for an opportunity to acquire a territory when a resident of the appropriate sex disappeared. Immediately upon the demise of the male owner, the male "floater" in a territory would claim the site and the female in it. Female floaters did the same when the resident female died or moved away. Nonterritorial birds from elsewhere had little chance to secure the area in competition with an established floater, who, if able to make the transition, would then reproduce [1130].

Smith's study supports the prediction that territory owners will outreproduce individuals that cannot claim a territory. The same is true of the great tit, in which floaters sometimes pair up and attempt to breed on another pair's territory without ever claiming a territory of their own. These pairs essentially sneak into an established territory and build a nest as quickly and quietly as possible. If detected by the resident territorial male, they may be evicted, but they sometimes succeed in nesting successfully, especially if they enter a territory after the residents have finished establishing their territory and are incubating eggs.

If territoriality is an adaptive conditional tactic preferred by the great tit, then territorial pairs, which voluntarily pay the price of fighting, singing, and patrolling, should do better reproductively than sneaky intruders. The Belgian ornithologists André Dhondt and Jeannine Schillemans found that intruding birds laid clutches of eggs only slightly smaller than those of territorial birds, but more of their nests failed, largely because of interference from the territory owners. As a result, intruders averaged only about four young fledged, compared with eight for territorial pairs [294].

The Belgian researchers studied a woodland population of the great tit, but this species sometimes nests in hedgerow habitat as well. In an English population of this bird studied by John Krebs, males first occupied woodland habitat. Only after the woodland became filled with territorial individuals did some birds settle in hedgerows adjacent to the woodland, where they attempted to breed. Because the birds preferred woodland territories to hedgerow territories, Krebs predicted that great tits in woodland habitat would produce more fledglings than birds in hedgerows. This proved to be the case, as only 2 of 9 nests (22 percent) in hedgerows succeeded in producing fledglings, compared with 54 of 59 woodland nests (92 percent) [674].

To find out whether the hedgerow birds really were being excluded from superior breeding territories that they would have preferred to occupy, Krebs removed six pairs from the woodland. Within a few days, most of the vacant territories had been claimed by new defenders—primarily banded birds from the hedgerow area, which they had abandoned as soon as superior habitat became available (Figure 25).

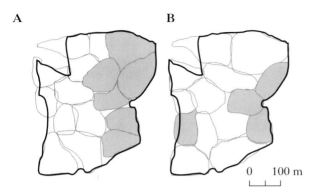

25 Territories of great tits before (A) and after (B) a removal experiment. When six pairs were removed (from the shaded territories in A), the vacated areas were soon occupied, some by neighbors that expanded their territories and some (the shaded territories in B) by replacement pairs that arrived within 3 days. *Source: Krebs [674].*

0   100 m

Territoriality and Calories

We have looked at some studies that directly tested whether possession of territories yielded more descendants for the territory owners. More often, researchers only indirectly test whether territorial possession enhances reproductive success by measuring traits that may be correlated with fitness. For example, in those species in which individuals compete for feeding territories, the winners presumably increase their reproductive chances by monopolizing the food in their territories. If so, then animals should at the very least gain more calories from their territories than they expend in defending their real estate.

For part of the year, African golden-winged sunbirds compete for possession of patches of flowering mint (Figure 26). Frank Gill and Larry Wolf conducted an economic study to see whether territorial birds derived a net caloric gain from territory defense [440, 442]. To measure calories gained, Gill and Wolf first counted the number of flowers in a patch of the sunbird's food plant. They next measured the rate of nectar production per flower by covering a number of flowers to prevent the removal of nectar. At the end of the day they inserted a glass micropipette into the flower to take up the nectar by capillary action. The micropipette was then sealed and taken to a laboratory to determine the nectar's sugar, and thus caloric, content. The quantity of nectar produced per flower per day times the number of flowers in a patch yields an estimate of the total daily nectar production there. If a sunbird could monopolize this nectar, then it alone would get the calories from this food, obtaining a benefit from its territorial defense of the patch.

But sunbirds pay a price for being territorial, including the calories lost when chasing intruders from the area. Because physiological studies have measured the caloric costs of perching, foraging, and chasing for a bird of the sunbird's size, one can compare the calories spent in defending a territory with the calories gained by having a private foraging preserve.

26 Golden-winged sunbird on a nectar-producing mint. These birds sometimes defend patches of mint.

As Table 2 shows, a sunbird that owns a territory with flowers producing nectar at the rate of 2 microliters per bloom per day can meet its daily energy requirement during the nonbreeding season in just 4 hours of foraging. If its other option were to feed in an undefended area where the flowers were giving 1 microliter of nectar per bloom per day, the sunbird would have to forage for 8 hours instead of 4, expending an extra 2400 calories while foraging instead of resting quietly on a perch. However, any such savings for territorial sunbirds must be weighed against the costs of defending a rich patch against intruders. Each hour of defense flight

Table 2 Energetic costs and benefits of territoriality by golden-winged sunbirds under different conditions

Activity	Cost in calories/hour			
Resting on perch	400			
Foraging for nectar	1000			
Chasing intruders from territory	3000			
Nectar production (microliters/blossom/day)	1	2	3	4
Hours of foraging to meet caloric need for one day	8	4	2.7	2

Nectar production		Hours of resting gained	Calories saved[a]
In territory	In undefended site		
2	1	8 – 4 = 4	2400
3	2	4 – 2.7 = 1.3	780
4	4	2 – 2 = 0	0

[a] For each hour spent resting instead of foraging, a sunbird spends 400 calories instead of 1000, saving 600. Total calories saved = 600 × hours of resting gained by not having to forage for nectar.
Source: Gill and Wolf [441].

burns up 2000 more calories than would be spent if the bird were foraging nonaggressively elsewhere. An hour's worth of defense would be worthwhile if the bird were defending a 2-microliter area while forcing other sunbirds to forage in 1-microliter patches (net caloric gain = 2400 – 2000 = +400 calories). But territoriality becomes disadvantageous if the bird spends an hour chasing intruders to protect flowers producing at the 3-microliter rate when undefended 2-microliter patches are available (net loss = 780 – 2000 = –1220 calories).

If sunbirds are adaptively territorial, they should be sensitive to the relative rates of nectar production in various patches and the time required to deal with intruders. As predicted, when nectar production is uniformly high throughout a large area, the birds are not territorial. But if nectar productivity in different patches begins to diverge, some individuals will almost immediately begin to defend the richer areas. As the density of birds increases, so does the cost of repelling intruders. In response to increased defense expenditures, territory owners tend first to contract their territories. But if the rate of intrusion continues to rise, the birds simply abandon their territories altogether. Thus, sunbirds have the conditional capacity to switch from foraging in a home range to defending a territory, making decisions that tend to maximize their caloric gain [441].

More Examples of Conditional Territoriality

Many animals in addition to sunbirds adjust their investment in feeding territories in relation to the energetic costs and benefits of their defense. For example,

John Craig and Murray Douglas found that New Zealand bellbirds (a species unrelated to African sunbirds) defend nectar-producing trees under some circumstances but feed pacifically in flocks under other conditions [232]. Craig and Douglas watched some bellbirds switch from nonterritorial foraging at one species of flowering tree to territorial defense of another tree species. If the birds switched for economic reasons, those that became territorial should have increased their food intake. The researchers established that the territorial bellbirds took in more than 125 kilojoules a day, up substantially from about 100 kilojoules on days when they were not territorial.

If the conditional territoriality hypothesis is correct, then whenever a territorial animal *voluntarily* switches to nonterritorial foraging, it should benefit. The white-fronted bee-eater, an African insect-eating bird, provides an opportunity to test this prediction [524]. A bee-eater typically travels out from its nesting colony in a clay bank to a specific foraging site, which it defends against intruders. But during the period when they are feeding nestlings, some bee-eaters give up their feeding territories to gather food in other places that they do not defend. A sample of nesting birds that had abandoned their territories collected about 100 milligrams of insects per hour on average in their new feeding ranges. In contrast, a group of nesting adults that continued to utilize their foraging territories averaged about 250 milligrams of insects per hour. Thus, some white-fronted bee-eaters *reduced* their insect-collecting success by abandoning their rich foraging territories precisely when they needed to gather as much food as possible to feed their hungry nestlings.

Needless to say, these results puzzled two Darwinian biologists, Robert Hegner and Stephen Emlen [524]. But perhaps the time and energy costs of commuting back and forth between a nest and a distant foraging territory led to the abandonment of some feeding territories. A bird without chicks can afford to stay on a foraging territory until it has collected all the food it needs for itself. In contrast, a bee-eater with chicks has to cart insects to its youngsters, one beakful at a time. Under these circumstances, a bird with a distant foraging territory might actually lose energy by continuing to forage so far from its nest. By switching to undefended locations closer to home, the bird may capture fewer prey *while on the foraging grounds*, but may still be able to deliver more prey per unit of time to its brood than it would otherwise.

The predictions that territory-abandoning adults should (1) forage close to home and (2) have relatively distant feeding territories compared with territory-maintaining adults both proved correct. Birds fortunate enough to have feeding territories close to the colony retained them while breeding; nesting bee-eaters whose feeding territories lay over 2 kilometers away quit defending them. Thus, bee-eaters that give up their territories make the best of a bad situation, delivering less food than others with nearby foraging territories, but probably more food than they would have if they had persisted in making long commutes between their distant foraging territories and their nests.

Long-Term Effects of Territoriality

Another examination of the foraging gains made by territorial birds involves the pied wagtail, a European songbird that defends feeding territories during the nonbreeding season [265]. In southern Britain, wagtail winter territories are 600-meter stretches of riverbank. Territory owners consume aquatic insects washed up by the river. Wagtails, like sunbirds and bee-eaters, have the ability to switch back and forth between territoriality and nonterritorial behavior. When the river delivers relatively few insects to a territory, the owner often temporarily joins flocks of nonterritorial birds that forage widely over the countryside. When the river edge is productive, a territorial bird will remain at its site and vigorously defend it against outsiders.

But two odd features of wagtail behavior seem to violate the assumption that individuals defend feeding territories in order to monopolize a rich food resource. First, a resident sometimes tolerates another bird as a "satellite" on its territory for a time, in effect letting the satellite remove calories from the site [274]. Second, the food-collecting rate for territorial wagtails never exceeded 20 items per minute during Nicholas Davies's study in the winter of 1974, whereas flocking birds collected 21–29 prey per minute during the same period [265]. Examination of the feces of the birds (one of the less glamorous activities of a behavioral ecologist) showed that the sizes of the insects consumed by territorial and nonterritorial birds were the same, and therefore that the territorial birds were not compensating by eating larger prey.

Why satellites? As it turns out, owners tolerate satellites on their territories only on days when food is abundant, so that the cost of having a satellite share one's territory is relatively low. Moreover, on those days, other intruders are especially likely to invade the territory. The satellite assumes 20 to 50 percent of the defense of the area against these outsiders, and as a result the resident actually increases its net energy gain by associating with a satellite under these conditions (Figure 27). End of puzzle number one.

But how can we account for the territorial birds' persistent return to areas that yield less prey than the undefended regions visited by flocking birds? Why bother to be territorial when the hunting is better elsewhere? Davies suggests that when snow covers the meadows, the river edge may be the only place with a predictable and constant supply of food. To maintain its body weight, a wagtail must feed 90 percent of the time and must collect about 7500 food items per day. A single day of starvation can kill it. Thus, territory ownership may provide long-term insurance against exceptionally bad weather, although the hypothesis that the *reliability* of food production in an area over the long haul influences a wagtail's decision to be territorial remains to be formally tested [265].

How Large Should a Territory Be?

Territory size varies enormously both among species and within species [1171], reflecting in part the size of the species (the minute poplar aphid defends a territory a few millimeters long, whereas the larger wagtail defends 600 meters of river-

27 Territorial pied wagtails tolerate satellites only when the presence of the other bird helps them drive off numerous intruders, resulting in a net gain in food eaten for the owner. *Source: Davies and Houston [274].*

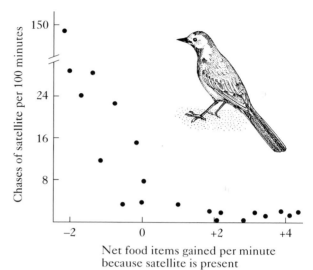

Chases of satellite per 100 minutes

150

24

16

8

−2   0   +2   +4

Net food items gained per minute because satellite is present

28 A small display territory. A male gerenuk stands on a termite mound, which he uses as a platform for his mate-attraction display. *Photograph by Walther Leuthold.*

bank) or the number of individuals jointly defending a site in social species [1206]. In addition, the size of a territory is correlated with its function. Males of many species defend a small space (Figure 28) that they use only for the performance of mate-attracting displays (see Chapter 13). All-purpose territories that supply food for the territory defender tend to be much larger than such display territories.

Even within a species that defends feeding territories, the size of the defended area may vary considerably. For example, the feeding territories of New Holland honeyeaters, an Australian nectar-consuming bird, may expand and contract greatly as the winter progresses (Figure 29). David Paton showed that when *Banksia* flowers were diffusely distributed, territories were generally larger, but when many inflorescences started to produce nectar in a small area, territory size decreased [911].

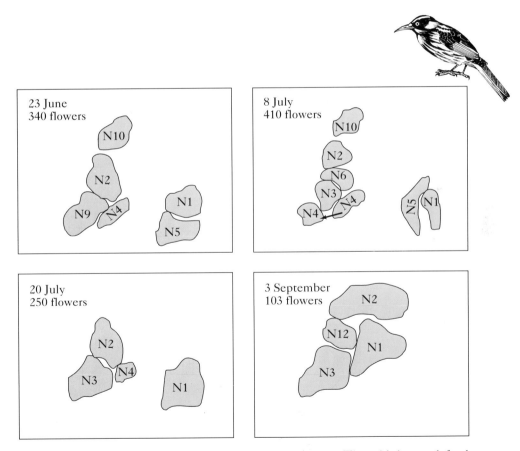

29 Changes in territory sizes of New Holland honeyeaters. These birds may defend patches containing nectar-producing *Banksia* flowers. Numbers within the territories refer to individual birds. *Source: Paton and Ford [911].*

Likewise, migrating rufous hummingbirds making their way through California to Mexico in the fall also vary in how large a patch of Indian paintbrush they defend. These birds make *daily* adjustments in the number of nectar-producing flowers in their territories (Figure 30), with the result that they gain more weight each day during their brief sojourn in Californian mountain meadows. Lynn Carpenter and her colleagues induced individual hummers to reveal their weights in an ingenious manner. Within each territorial patch, they placed a conspicuous perch on a very sensitive scale that could be read through binoculars from a distance. Territorial birds like to be where they can survey their domain easily, and some accepted the experimental perches, thereby weighing themselves at the convenience of the observers, who gathered the relevant data at the start and end of each day [184].

But what caused the Australian honeyeaters and the rufous hummingbirds to change the size of the areas they defended? Were they simply responding to

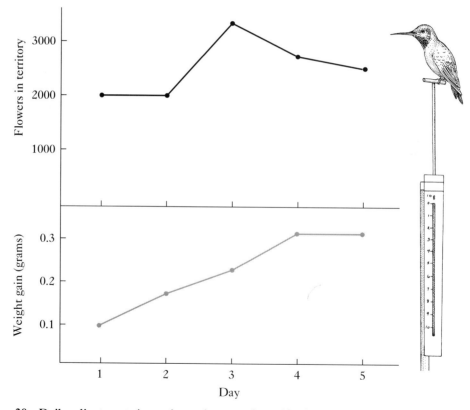

30 Daily adjustments in territory size, as reflected in the number of flowers defended, by one rufous hummingbird during the fall migration. The bird steadily improved its weight gain each day. *Source: Carpenter et al. [184].*

changes in the *benefits* (nectar) available to them in a given area (reducing a territory to the minimum size needed to gain a certain number of calories), or were they reacting to changes in the *costs* of defending real estate (high-nectar-producing sites might attract many intruders, favoring a decrease in territory size)?

Doug Armstrong investigated whether the territorial behavior of New Holland honeyeaters and white-cheeked honeyeaters in eastern Australia was affected by changes in the benefit end of the equation. At his study site near Sydney, the honeyeaters exhibited seasonal changes in their willingness to defend the area around a favored perch. During the winter, the birds attacked only a small proportion of the nectar-feeding intruders that came within 30 meters of their main perch. At this time, the local *Banksia* flowers were so common that the birds could secure all the food they needed in little time. In the spring and fall, however, *Banksia* flowers became much less abundant, and the time required to get sufficient nectar increased sharply. Under these conditions, the honeyeaters attacked a much higher proportion of the competitors that came near [29].

If these seasonal changes in aggressiveness were directly caused by changes in nectar availability, then providing sugar water supplements ought to have decreased the birds' territory sizes. Armstrong placed feeders some distance away from his subjects' main perches so that they would be able to supplement their food supply, but would not be inundated with intruders near the core of their territories. The honeyeaters did locate and use the feeders during the periods when they were available, but did not defend them against competitors, always returning instead to their main perches. However, the supplemental feedings did not alter the seasonal pattern of territorial aggression. In spring and fall, the honeyeaters were always much more aggressive than during the winter (Figure 31).

These results rule out the possibility that the honeyeaters monitor their food intake and adjust their investment in territorial defense accordingly. But as Armstrong notes, his findings do not eliminate the possibility that food scarcity provides an *ultimate* reason why certain honeyeaters are more aggressive in the spring and fall. His original prediction focused on food intake as the proximate factor reg-

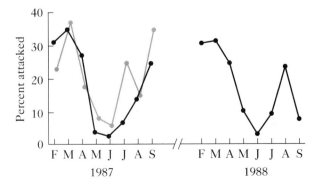

31 **Seasonal changes in aggression in territorial New Holland and white-cheeked honeyeaters** occurred whether extra food was available from sugar water feeders (shaded plots) or not (solid plots). *Source: Armstrong [29].*

ulating territorial aggression. However, other proximate mechanisms could also produce seasonal changes in territorial aggression that correlate with food availability. For example, if the relationship between *Banksia* nectar supplies and the seasons is highly predictable, selection may have favored honeyeaters that used photoperiod changes or visual stimulation from *Banksia* flowers, rather than the amount of nectar ingested, to adjust their territorial behavior. According to this untested hypothesis, individuals with "season sensors" become more aggressive in fall and spring, regardless of short-term fluctuations in food supply, because over evolutionary time, food shortages have occurred so reliably in these seasons [29].

Why Do Territory Holders Almost Always Win?

Most studies of territoriality have found that winners in the competition for territories gain substantial caloric, survival, or other benefits that ought to translate into heightened reproductive success. Given the value of owning a territory, it seems paradoxical that when a territory holder is challenged by a rival, the owner almost always wins the contest, usually within a matter of seconds. Why are intruders so quick to give up?

First, game theory analysts have shown algebraically that an **evolutionarily stable strategy**—that is, one that cannot be replaced by an alternative strategy—for resolving conflicts between residents and intruders would be to use the simple *arbitrary* rule, "The resident always wins." (Similar algebraic maneuvers demonstrate that the contrary convention, "The intruder always wins," is also a possible evolutionarily stable strategy.) If all the competitors for territories were to adopt the "resident wins" rule so that intruders always gave up, a mutant with a different behavioral strategy could not spread its special gene and thus its special response to territory holders. Thus, the "resident always wins" strategy would persist indefinitely.

In his study of the speckled wood butterfly, whose males defend small patches of sunlight on the forest floor, Nicholas Davies found that, as predicted from game theory, territorial males did in fact always defeat intruders, which invariably departed rather than engaging in intense, escalated conflicts [266]. The butterflies were capable of such fights, as Davies showed by capturing and holding territorial males in an insect net until a new male had arrived and claimed the "empty" sunspot territory. When the prior resident was released, he returned to "his" territory, only to find it taken by the new male. This male, having claimed the site, reacted as if he were the resident, with the result that the two males engaged in a fierce fight, at least by butterfly standards. The combatants spiraled upward, clashing wings, before diving back to the territory, only to repeat this spiral "fight" many times until the previous resident gave up and flew away (Figure 32). Davies saw these fights only when he manipulated the males to create two "residents" for the same territory.

Although Davies's results were consistent with the "resident always wins" hypothesis, other observers of the speckled wood butterfly subsequently noted takeovers of occupied territories in this species [1274]. Davies's study had been done at an unusually sunny time when sunspots were not in short supply. Under

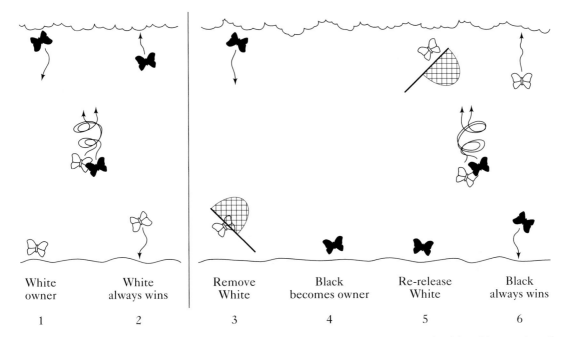

White owner	White always wins	Remove White	Black becomes owner	Re-release White	Black always wins
1	2	3	4	5	6

32 The resident always wins? An experimental test of the hypothesis that territorial resident males of the speckled wood butterfly always win conflicts with intruders. When one male ("White") is the resident, he always defeats intruders (1,2). But when the resident is temporarily removed (3), permitting a new male ("Black") to settle on his sunspot territory (4), then "Black" will defeat "White" upon his return after release from captivity (5,6). *Source: Davies [266].*

these conditions, territory-seeking males had little or nothing to gain by fighting intensely with rivals for access to an abundant resource. Instead, searching males simply kept moving, and soon found an unoccupied spot. Under cooler, partly cloudy conditions, however, sunspots become scarcer and more valuable, and males will fight for their possession. Even then, resident butterflies rarely lose to rivals, but the fact that they sometimes do helps eliminate the possibility that males operate with a "resident always wins" strategy, a strategy that most researchers now doubt applies to any species.

One alternative to the arbitrary rule hypothesis is that territorial residents *usually* win because they have nonarbitrary attributes that really make them superior territory holders. If, for example, they are larger and physically more powerful than intruders, then they should usually defeat their rivals. If an asymmetry in resource-holding power is responsible for the ability of residents to avoid takeovers, then (1) residents should be larger, stronger, faster, or more agile than the intruders they defeat, and (2) intruders that actually oust residents should be superior in whatever physical traits are critical to territory maintenance.

Many studies demonstrate that body size in particular is related to success in holding a territory. For example, males of the beewolf wasp *Philanthus basilaris* compete for small mating territories that consist of patches of sand or harvester ant mounds. Kevin O'Neill tested whether large size contributed to success in holding territories by removing beewolf territory owners and then capturing the replacements that occupied their vacant sites [876]. As predicted, the replacements were consistently smaller than the original territory holders, which presumably had the strength to prevent their smaller rivals from staging takeovers (Figure 33).

Being large is also of paramount importance in the territorial contests of the tropical pseudoscorpion *Cordylochernes scorpioides*. The life cycle of this arthropod involves periods of colonization of recently dead or dying trees, followed in a few generations by dispersal from the now older dead trees to fresh sites. Dispersing pseudoscorpions make the trip under the wing covers of a harlequin beetle, disembarking when the huge beetle touches down on an appropriate tree. Matings occur both on trees and on the backs of beetles, but male pseudoscorpions attempt to be territorial only when they occupy the limited, defensible space on a beetle's back. Under these circumstances, the largest males in the population have much higher reproductive success than smaller rivals. On tree trunks, however, small males do better than large ones, in part because they develop more rapidly and need not fight for mates there (Figure 34). These changes in the value of territory ownership resulting from the life cycle of the pseudoscorpion lead to strong oscillations in selection for or against large size in this species [1325].

Even in species that are consistently territorial, being larger than average may not always be critical to a male's ability to dominate rivals and expel them from a territory. In the black-winged damselfly, males fight for mating territories along streams visited by sexually receptive, egg-laden females. Territory owners typi-

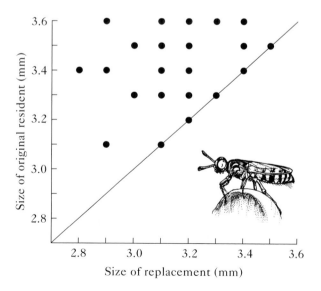

33 Resource-holding power and the resident advantage in a beewolf wasp. The graph plots the size of the original resident (as measured by head width) against the size of the replacement male that occupied his territory upon his removal. Points that fall above the ascending line represent cases in which the original resident was larger than the replacement. *Source: O'Neill [876].*

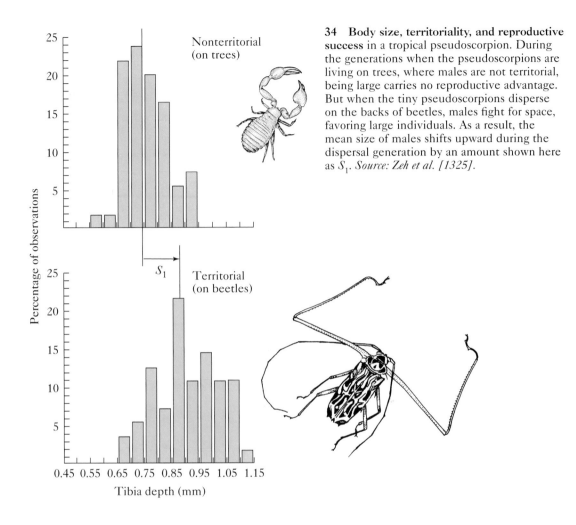

34 Body size, territoriality, and reproductive success in a tropical pseudoscorpion. During the generations when the pseudoscorpions are living on trees, where males are not territorial, being large carries no reproductive advantage. But when the tiny pseudoscorpions disperse on the backs of beetles, males fight for space, favoring large individuals. As a result, the mean size of males shifts upward during the dispersal generation by an amount shown here as S_1. *Source: Zeh et al. [1325].*

cally chase intruders away quickly, although rarely, the owner and intruder become involved in a lengthy back-and-forth flight contest, which may result in the displacement of a territory holder by a newcomer. James Marden and Jonathan Waage showed that the winners of these escalated contests were no larger than the losers, nor were they relatively more muscular [762]. They did, however, have a higher fat content than the males they defeated in 21 of 24 cases (Figure 35). Thus, the energy reserves available to males of this species apparently influence their chances of winning a fight. However, the males do not simply fight until one combatant can continue no longer. Instead, they seem to assess each other's fat reserves during their aerial contests; the individual with the lower capacity for rapid and continuous flight eventually leaves the field before it is totally exhausted [761].

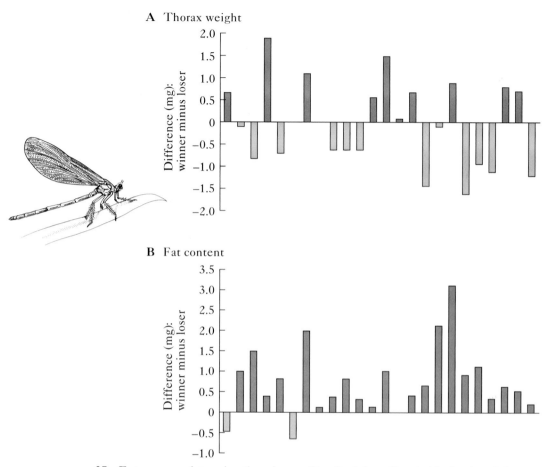

A Thorax weight

B Fat content

35 Fat reserves determine the winner of territorial conflicts in black-winged dam-selflies. In this species, males may engage in prolonged aerial chases that determine ownership of mating territories on streams. (A) Larger males, as measured by dry thorax weight, do not enjoy a consistent advantage in territorial takeovers. (B) Males with the greater fat content, however, almost always win. *Source: Marden and Waage [762].*

The hypothesis that an asymmetry in resource-holding power determines the winner of territorial conflicts generates the following prediction: Territory owners should be larger, or in better physiological condition, or in possession of greater fat reserves than nonterritorial intruders or floaters. But this is not always the case. For example, in one study of red-winged blackbirds, no differences could be found in the physical traits of territorial and nonterritorial males [1100], and even in the northern elephant seal, in which fierce fighting among males occurs, smaller resident males are often able to defeat larger intruders [495].

Therefore, let us consider a third hypothesis for resident success in territorial defense—namely, that the fitness payoff from holding a territory increases over time, motivating the owner to fight more intensely than an intruder. The value of a territory might grow for an established resident after he has settled disputes with his neighbors over their mutual boundaries in the early phases of their relationship. Once these boundaries have been agreed upon, everyone calms down for the mutual benefit of all (see Chapter 8) [321]. If a resident is ousted, however, he will have to go through these battles with new neighbors at a new site. He therefore has more to gain by holding onto his present territory than the new intruder has to gain by acquiring it, since any new boy on the block will have to expend a great deal of energy in dealing with the neighbors before they get their boundaries straight.

The payoff asymmetry hypothesis predicts that when a territory holder is experimentally removed from his territory and later released, the probability that he will be able to reclaim his territory will be a function of how long the replacement male has occupied the site. This experiment has been done with birds, such as the redwinged blackbird, as well as with insects, such as the speckled wood butterfly. In redwings, when captive ex-territory holders are released from an aviary and allowed to return to their home marsh to compete again for their old territories, they are more likely to fail if the new males have been on the experimentally vacated territories for some time [76].

The payoff asymmetry hypothesis also predicts that contests between an ex-resident and his replacement will become progressively more intense and lengthy as the territorial tenure of the replacement increases. John Krebs tested and confirmed this prediction with the great tit (Figure 36) [675]. But perhaps the birds in these experiments had been stressed by their captivity and their resource-holding power

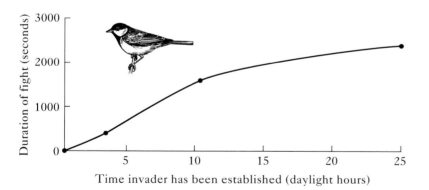

36 A test of the payoff asymmetry hypothesis. In great tits, the more time a new resident has been on a territory, the longer the fights between that individual and the original resident (which was temporarily removed from his territory by the experimenter). *Source: Krebs [675].*

thereby reduced. Or perhaps the replacements gained more resource-holding power the longer they were in residence on a good territory. We cannot, therefore, conclude with certainty that the payoff asymmetry hypothesis has been supported in these experiments. More refined tests of the alternative hypotheses are required before we can claim to understand why territorial residents so often defeat intruders.

SUMMARY

1. This chapter introduces the concept of an evolved strategy, an inherited behavioral mechanism that enables an individual to respond adaptively to problems it faces in its environment. An evolutionarily stable strategy is a behavioral mechanism that is superior to alternative strategies that have occurred in the species. A conditional strategy is a decision-making mechanism that provides its owner with the ability to select among several tactics in response to different patterns of resource distribution or social competition.

2. In choosing where to live, many animals actively select certain places over others. Individuals able to occupy preferred habitats should have higher fitness, a prediction that has been tested with affirmative results in a number of species.

3. The selection of living space often occurs in the context of leaving one spot for another, as when juvenile animals abandon the place where they were born to go in search of a new home. Cost–benefit analyses have helped to explain why male mammals typically disperse farther than females of their species, probably because the costs of dispersal and inbreeding are greater for females than for males, thus making it advantageous for males to leave their female relatives in search of receptive mates elsewhere.

4. Migration is a special form of dispersal in which the migrant eventually returns to the place it left. The costs to individuals of migratory journeys appear to be high; counterbalancing benefits may arise if migrants can exploit seasonal bursts of food productivity during their reproductive season or if migrants can secure especially safe sites for breeding or giving birth.

5. When choosing living space, some animals invest additional time and energy in defense of the site. Territorial behavior is strongly correlated with the occurrence of valuable resources in small, economically defensible patches. Owners of defended breeding sites typically reproduce more successfully than individuals that fail to secure a territory; when some breeding territories are preferred to others, owners of the preferred sites enjoy greater reproductive success. When feeding territories are established, owners usually gain caloric benefits in excess of the caloric expenses incurred during defense of the site, or else they are able to secure their minimum daily caloric requirements more quickly, saving time that can be spent resting rather than foraging.

6. There are several competing hypotheses on why territory owners usually defeat challengers for control of their territories. In some, but not all, cases, residents are physically stronger, more capable defenders than their would-be replacements.

SUGGESTED READING

The concept of evolutionary strategy is reviewed by Richard Dawkins [277, 278] and Mart Gross [479]. Papers by Thomas Whitham [1268, 1270, 1271] and by Lincoln Brower and his colleagues [21, 171] offer useful analyses of habitat selection. For competing views on the evolution of dispersal, see [470, 1237]. Dispersal in Belding's ground squirrels is analyzed at the proximate and ultimate levels by Kay Holekamp and Paul Sherman [558]. A classic exposition of the cost–benefit approach to territoriality is given by Jerram Brown and Gordon Orians [151]. The papers by Nicholas Davies and his colleagues on wagtail territoriality are especially well done [265, 274].

DISCUSSION QUESTIONS

1. Some animals defend territories against members of other species. How can this behavior be adaptive? Develop at least one cost–benefit hypothesis and a set of predictions on the evolution of interspecific territoriality.

2. Males of many animals guard territories that contain food needed by females of their species. The males do not consume the food, but instead use it to attract females, with whom they mate. What are the costs and benefits of this kind of mating territoriality for males *in terms of females contacted?* Use your list to predict what ecological factors should lead to this kind of mating territoriality.

3. Develop a game theoretical hypothesis to analyze the decision of some great tits to employ the nonterritorial route to breeding. Come up with a two-strategies hypothesis and a conditional strategy hypothesis. What predictions follow from your two hypotheses?

4. In Belding's ground squirrels and other mammals, young males disperse while young females remain on or near the natal territory, but this pattern is exactly reversed in most birds. Why might this be so? In producing your hypotheses, consider the fitness value of a territory to most male mammals and to most male birds [470].

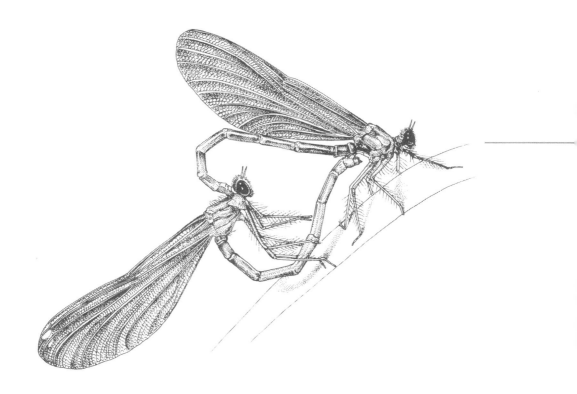

CHAPTER 12

Male and Female Reproductive Tactics

*H*OWEVER SKILLFUL AN ANIMAL IS AT REPELLING PREDATORS, foraging efficiently, or defending territories, these abilities have evolutionary consequences only if the individual succeeds in passing on more of its genes than other genetically different individuals in the same species. Reproductive behavior is therefore the central focus of natural selection—and the topic of the next three chapters. We begin by considering why the reproductive behavior of males and females often differs dramatically. Why do males so often take the initiative in courting mates, and why do females so often reject their suitors? Why do males fight with one another over females? Why do females prefer males that sport bizarre ornaments and perform strange behavioral displays? These are intriguing questions.

429

Answering them requires us to understand something about sexual selection, a category of Darwinian natural selection. Sexual selection occurs when individuals differ in reproductive success either (1) because of competition within one sex for access to mates and their gametes or (2) because one sex prefers the gametes received from certain members of the opposite sex. This chapter looks at these two elements of sexual selection, first examining how selected elements of male aggressive behavior and mating tactics appear to help individuals defeat rivals in the competition for mates. Then we shall examine the sexually selected products of past interactions between males and females. We will look at reproductive behavior as a contest in which females place demands on and create reproductive obstacles for males, whose responses create reciprocal selection on females. Thus, if we wish to understand why males do what they do, we have to understand the ultimate goals of females, and vice versa.

The Evolution of Males and Females

I was thrilled by the first active bower of a satin bowerbird (Color Plate 13) that I saw in Australia. Two beautifully constructed little fences of interwoven twigs outlined an avenue littered with blue parrot feathers and a smattering of bright yellow flowers. I waited nearby in the eucalyptus forest, hoping to see the animal that had constructed it. In due course, the magnificent satiny blue–black male, a little larger than an American robin, flew down to his bower, carrying an ordinary rubber band to add to his collection of ornaments.

I did not have the patience to wait for a female bowerbird to visit the male, but others have watched arriving females, which inspire the bower builder to bound wildly about, producing an extraordinary medley of mechanical sounds and buzzes, interspersed with imitations of the songs of other species, including the hooting of laughing kookaburras. Yet despite their remarkable vocal displays and delightful bowers, most males are rejected by visiting females, as documented by Gerald Borgia and his team of bowerbird observers [111]. The team recorded far more rejections than matings, which take place in the avenue between the bower walls [110]. After a female bowerbird copulates, she leaves to lay eggs and incubate them until they hatch, feed the hatchlings, and shepherd the fledglings around the neighborhood. Her partner stays at the bower, working slavishly on its improvement and decoration [109] while taking time out to steal decorations from or destroy the bowers of other males in his neighborhood [964]. A few males with especially well decorated bowers will copulate with more than one female (the record is 33), while many males will fail to mate even once per breeding season.

Bowerbirds are the only birds that build display structures; some species in the group construct amazingly large "playhouses" that make the satin bowerbird's twig fences look amateurish indeed (Figure 1). The exciting historical problem of how bowers originated and changed over time has been partly solved [111, 685]. However, our focus here is on the question of why satin bowerbird males and females

1 Bower building and decorating by the bowerbird *Amblyornis inornatus*. (Top) The huge and beautifully built bower of a male bowerbird. (Bottom) The male inside his bower near some of the fruit and flower ornaments he has collected. *Photographs by Will Betz, courtesy of Adrian Forsyth.*

behave differently when it comes to sex, with males doing the courting, females evaluating the courtship; males ready to mate at the drop of a hat, females much more sexually restrained and discriminating—a pattern that applies to the vast majority of animal species (Figure 2).

The Evolution of Sex Differences

Why do typical males compete for mates and typical females choose carefully among the competitors? Male competition usually arises because of the customarily strong male bias to the **operational sex ratio,** which is the ratio of sexually receptive males to receptive females in a population at any given time [354, 687]. When receptive females are scarce, they represent a limiting resource for males, which will compete to have a chance to inseminate those few females that are available.

But why is the operational sex ratio usually skewed toward males? Because the *potential* rate of reproduction generally differs between the sexes, with males capable of producing vast quantities of sperm, whereas females are constrained by

2 Male copulatory eagerness in elephant seals can be exploited by human researchers [289]. In the foreground is a dummy female elephant seal composed of urethane foam covered with fiberglass that has caught the eye of a male of her species. The male was lured to the area by a tape recording of the vocalization given by a female being mounted. When he arrived, he quickly climbed onto the scale in pursuit of the female model, which was moved by a hidden observer. Other observers could then determine the weight of the male without having to immobilize him with chemical sedatives. *Photograph by Chip Deutsch.*

the time and resources needed to make their much larger eggs, which they often care for after fertilization [211]. In all sexual species, males are (by definition) those individuals that produce sperm, the smaller gametes, usually no more than a set of genes in a package just large enough to contain the energy needed to drive the male's DNA to an egg. Even in species in which males produce relatively gigantic sperm, such as a fruit fly whose males make sperm more than ten times their body length when the sperm tails are uncoiled [943], the mass of an egg is still vastly greater than that of a sperm. Indeed, females, are (by definition) the sex that produces the larger gametes, the eggs, which in birds and most other animals are enormous relative to sperm (Figure 3). In birds, it is not uncommon for a single egg to constitute 15–20 percent of the female's body mass, and some go as high

3 A difference in parental investment. A sperm of a hamster fertilizing a hamster egg (magnified 4000 times) illustrates the difference in the amount of materials donated to a zygote by male and female. *Photograph by David M. Phillips/The Population Council.*

as 30 percent [692]. By way of contrast, a male splendid fairy-wren, although smaller than an English sparrow, may be the proud possessor of over 8 billion sperm at any one moment [1209]. Robert Montgomerie tells me that female coho salmon lay about 3500 eggs in the company of a male who releases about 400 billion sperm at the same time.

Among mammals, the same correlation between gamete size and numbers produced holds true. In a human female, only a few hundred cells develop into mature eggs [255]. In contrast, a single man could theoretically fertilize all the eggs of all the women in the world, given that each ejaculate contains several hundred million sperm. Few males of any species succeed in fertilizing more than a handful of eggs, but the huge abundance of sperm relative to eggs means that the potential rate of reproduction is greater for males than for females. Since females spend more time out of the mating loop in order to secure resources for egg production, they become a scarce commodity for males [209, 211, but see 1109, 1217].

The difference between the sexes in the size of the gamete each donates to an offspring can be expressed as a difference in **parental investment,** the time and energy and risks that a parent invests in one offspring that reduce the chance that the parent will have more offspring in the future [1200]. Robert L. Trivers invented this term to emphasize the trade-offs for parents that make contributions to offspring. On the plus side, parental investment may increase the probability that an existing offspring will survive to reproduce. But this fitness benefit may come at the cost of the parent's ability to generate additional offspring down the road.

In the animal world, the general rule is that females make larger parental investments than males. As we have seen, the differences start at the level of the gametes. When a female builds an egg, the many nutrients and cell organelles that she donates to the gamete constitute a much larger parental investment than is provided by the male to the tiny nutrient-free sperm that will fertilize it. These differences between the sexes in gametic parental investment are often amplified in other kinds of parental investment, which include the food or protection parents sometimes give to their progeny [1200] (Figure 4). In mammals, for example, females nourish their embryo(s), then care for the babies when they are born and

4 Parental investment takes many forms. (Top left) A male frog that carries his tadpole offspring on his back. (Top right) Using her mouth, a crocodile mother transports her babies to water just after they have hatched from her nest. (Bottom left) A female cicada killer wasp drags a cicada she has paralyzed with a sting to a burrow she dug, where her offspring will feed upon the unlucky cicada. (Bottom right) Male danaid butterflies feeding on a plant containing toxins, which they will store and pass on to their mates, which will use the material to protect their eggs. *Photographs by (top left) Roy McDiarmid; (top right) Jonathan Blair; (bottom left) the author; (bottom right) Michael Boppré.*

give them milk; the typical male mammal, in contrast, impregnates females and departs, never to see or interact with his offspring.

But why is it that the sexes differ in the resources they donate to a fertilized egg, with females supplying the bulk of the nutrients to the zygote and often providing additional parental care to their offspring? Geoffrey Parker and his colleagues have argued that the evolution of the two kinds of gametes, and thus the two sexes, stemmed from divergent selection that favored either (1) individuals whose gametes were good at fertilizing other gametes or (2) individuals whose gametes developed better after being fertilized [907]. Sperm, small and highly mobile, are superbly designed to move rapidly toward an egg when released by a male. Bluegill sperm, for example, dart along at up to four sperm lengths per second. The sedentary egg, large and packed with nutrients, contains materials useful for the development of the zygote following fertilization. No single kind of gamete could be equally good at both tasks, which led to the evolution of separate fertilization devices (the sperm of males) and development devices (the eggs of females)—although other factors may also be involved [221, 585].

Testing the Evolutionary Theory of Sex Differences

We have suggested that males usually compete for mates because of inequalities between the sexes in parental investment, which lead to differences in potential reproductive rates and a male-biased operational sex ratio (Figure 5). These differences favor males that compete aggressively with rivals for mates and copulate at every opportunity. In contrast, female fitness will rarely be advanced by receiving sperm from as many males as possible, favoring those females that avoid the fitness costs of superfluous matings. Thus the typical female shows great sexual restraint, closely inspecting potential partners to pick one whose genes or other contributions will maximize her fitness.

We can test this theory by finding unusual cases in which males make the larger parental investment or have the lower potential reproductive rate. In species of this sort [942], the operational sex ratio should be biased toward females, leading to female competition for mates and males that choose sexual partners carefully, producing a **sex role reversal.** In some species of fish, for example, males offer substantial parental investment, so much so that they are limited in their potential reproductive capacity compared with females [1219]. Thus, females of the pipefish *Syngnathus typhle* place their eggs in a male's brood pouch. During the male's "pregnancy," he provides nutrients and oxygen to the fertilized eggs for several weeks, during which time the average female produces enough eggs for two males. Since the sex ratio is 1:1, male pouch space is in short supply, and as expected, females compete for the opportunity to donate eggs to parental males, while males with free pouch space actively choose among mates, discriminating against small, plain females in favor of large, ornamented ones (Figure 6) [84, 1016].

Another example of a species in which maximum reproductive capacity may be greater for some females than for males is the flightless Mormon cricket, which

5   Differences between the sexes in sexual behavior arise from fundamental differences in parental investment (and nuptial gifts; see below), which affect the rate at which individuals can produce offspring. The sex that can potentially leave more descendants gains from high levels of sexual activity, whereas the other sex does not. An inequality in the number of receptive individuals of the two sexes leads to competition for mates within one sex, while the other can be choosy about its partners.

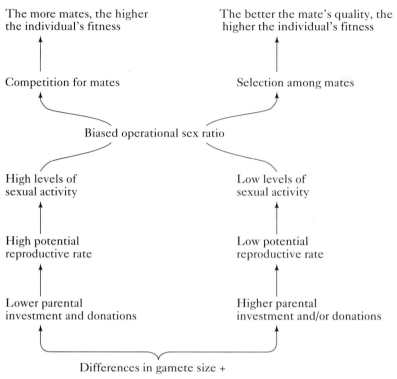

The more mates, the higher the individual's fitness

↑

Competition for mates

↑

Biased operational sex ratio

High levels of sexual activity

↑

High potential reproductive rate

↑

Lower parental investment and donations

The better the mate's quality, the higher the individual's fitness

↑

Selection among mates

↑

Low levels of sexual activity

↑

Low potential reproductive rate

↑

Higher parental investment and/or donations

Differences in gamete size +

Differences in other forms of parental investment +

Differences in resources donated directly to mates

6   Sex role reversal in a pipefish. In this species, in which males provide the greater parental investment, females compete for males and males exercise mate choice, favoring larger females and females with a larger ventral skinfold ornament. *Source: Rosenqvist [1016].*

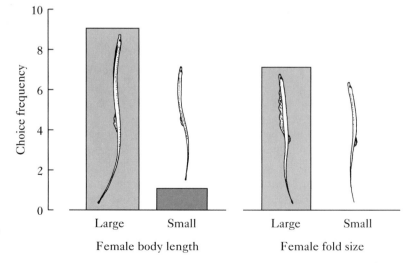

7   An edible sper-matophore. A Mormon cricket female consumes the large spermatophore that her partner donated to her during copulation. *Photograph by Darryl Gwynne.*

despite its common name has no religious affiliation and is a katydid, not a crick-et. When male Mormon crickets mate, they transfer an enormous edible sper-matophore to the female (Figure 7) [487]. Given that the spermatophore consti-tutes 25 percent of the male's body mass, most male Mormon crickets probably cannot mate more than once. In contrast, females may be able to produce sever-al clutches of eggs, provided they can induce a number of males to mate with them.

Sometimes bands of Mormon crickets march across the countryside, eating farm-ers' crops and mating as they go. In these high-density populations, males have access to many potential mates. When they begin to stridulate from a perch, announcing their readiness to mate, females come quickly to them and jostle for the opportunity to mount a male, the prelude to insertion of the male's genitalia and transfer of a spermatophore. In addition to this female sex role reversal, males often refuse to transfer a spermatophore to females that mount them. The aver-age weight of rejected females is significantly less than that of those that are per-mitted to copulate (Figure 8). By mating with heavier females, males transfer their sperm to more fecund individuals. A selective male that rejects a 3.2-gram female in favor of one weighing in at 3.5 grams fertilizes about 50 percent more eggs as a result [487]. The Mormon cricket example supports the argument that the sex with the higher potential reproductive capacity will compete for access to selec-tive members of the other sex.

The theory also predicts that if the relative reproductive capacities of the two sexes were to change within a species, the mate-acquiring roles adopted by the sexes

8 Mate choice by males. Mormon cricket males prefer to mate with heavier females. As a result of their choosiness, they gain an average of 18 extra eggs to fertilize. *Source: Gwynne [487].*

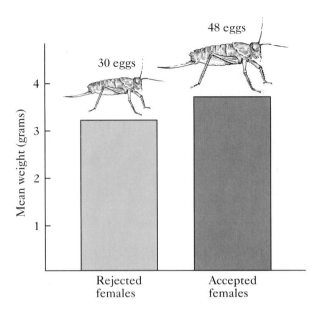

should change as well. This expectation has been examined in another katydid, a strange, thin Australian species (Color Plate 14) that in one season has only pollen-poor kangaroo paw flowers to feed upon, but at other times can forage from pollen-rich grass tree flowers. When these katydids are limited to kangaroo paw flowers, a male's large spermatophore is both difficult to produce and valuable to females as a nutrient gift. Under these conditions, males are choosy about their mates, and females fight with one another for access to receptive, spermatophore-offering males. But when grass trees come into flower and food is available in abundance, the maximum reproductive rate for males increases (they can produce spermatophores more rapid-

Table 1 A test of the evolutionary theory of sex role differences in which changes in relative reproductive rates in a katydid species lead to sex role reversals

Treatment	Mean number of		Proportion of interactions with	
	calling males	matings per female	male choice of mates	female competition for mates
Food scarce: Kangaroo paw pollen only	0.4	1.3	0.4	0.2
Food abundant: Pollen plus supplement	6.6	0.7	<0.1	0.0

Note: Data were derived from equal samples of observations of captive katydids, 24 of each sex per cage with four cages per treatment.
Source: Gwynne and Simmons [492].

ly), and the females' reproductive rate is limited by the speed with which they can turn pollen into eggs. At this time, sex roles switch to the more typical pattern, with males competing for access to females and females rejecting some males.

Darryl Gwynne and Leigh Simmons were able to generate these sex role changes experimentally by creating two groups of caged katydids, one that was fed only kangaroo paw pollen and another that received this food plus a calorie-rich supplement. Their experimental results (Table 1) show convincingly that in this case sex roles are flexible, not invariant, and that they depend on the relative reproductive rates (and operational sex ratios) of the two sexes [492].

Sexual Selection and Competition for Copulations

Whenever members of one sex compete for access to the other, while members of the sought-after sex mate in a discriminating fashion, conditions are ripe for **sexual selection.** Like almost everything else important in evolutionary theory, Charles Darwin discovered sexual selection, which he defined as "the advantage which certain individuals have over others of the same sex and species, in exclusive relation to reproduction" [264]. According to Darwin, sexual selection was an evolutionary process favoring traits such as elaborate ornaments and weaponry that enhanced an individual's ability to compete for or attract mates, even if those traits reduced survival (Figure 9).

The bizarre sexual behavior of bowerbirds, which Darwin had read about, was precisely the kind of thing that caused him to propose sexual selection as a distinct evolutionary pressure. Investments in building bowers, gathering decorations, displaying to females, and fighting with other males seem certain to reduce male survival chances to some extent. The survival costs of these traits, however, might be offset by the mating success of aggressive males with elaborate bowers and dramatic displays.

Note, however, that natural selection and sexual selection operate in fundamentally the same way. Both forms of selection require that hereditary variation among individuals in their attributes translate into differences in the number of surviving offspring produced. Although sexual selection is merely a subcategory of natural selection, the distinction Darwin made between the two processes usefully focuses attention on the selective consequences of sexual interactions *within* a species. Sexual selection helps explain traits that do not make evolutionary sense in the context of dealing with other aspects of the environment, such as predators, diseases, difficulties in finding food, and so on.

That a species' social environment can be a powerful force for evolutionary change is evident in the injurious, even life-threatening, battles that males sometimes use to determine their social status. Here we shall explain that males strive to be dominant because of the competition-for-mates component of sexual selection. But keep in mind that aggressive competition for mates is not completely restricted to males, as female pipefish and katydids illustrate. Moreover, compe-

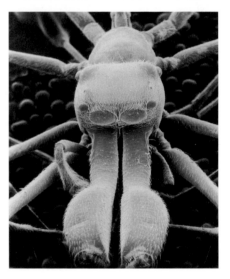

9 Sexually selected "ornaments" of males. Darwin believed that sexual selection via female choice was responsible for the evolution of elaborate plumage and remarkable displays in males in some species of birds of paradise (top left) and grouse (top right). Darwin argued that the strange horns and snouts of certain beetles (bottom left) also arose via female choice, although it is now known that males use these structures as weapons when fighting for mates. The immensely enlarged chelicerae (jaws) of this jumping spider male (bottom right) are also employed in pushing contests among males. *Photographs by (top left) Crawford Greenewalt; (top right) R. Haven Wiley; (bottom left) the author; (bottom right) Simon Pollard.*

tition for mates can be expressed more subtly than in a barroom brawl. For example, differences among males in the timing of their transition to adulthood can affect whether they get to mate with females of high reproductive value [1110]. Differences among males in their ability to keep searching for mates can also have something to say about which males meet and mate with the most females. Thus, in the European adder, the more meters wriggled by a male snake during a breeding season, the more mates he is likely to encounter (Figure 10) [752].

Still, male animals of all sorts and sizes, living and extinct [654], fight or fought with bites, kicks, antler locks, jaw grapples, head butts, even neck slams in the case of giraffes [1112], in order to establish dominance over others. For example, when two male elephant seals throw themselves at each other on a beach, 4000 or so kilograms collide in a battle that will determine a winner, the one who stays, and a loser, the one who humps away across the sand. From his records of winners and losers in a population of southern elephant seals on South Georgia Island in the Atlantic Ocean [786], T. S. McCann found that he could identify the number one male, an individual who had from 14 to 157 encounters with each of nine other males, and won them all. The number two male elicited submission from all but the top male, and so on down the line. In other words, the elephant seals fought to establish dominance over others, forming a linear dominance hierarchy, with each male in his own place.

If the ultimate significance of fighting is to acquire mates, then rank in a dominance hierarchy should correlate with male copulatory success over the breeding season. McCann's observations support this prediction (Figure 11A). Similar results have been secured for the northern elephant seal (Figure 11B) [495], and for many other animals in which the "top dogs" copulate more frequently than males of lower status [574, 576, 739].

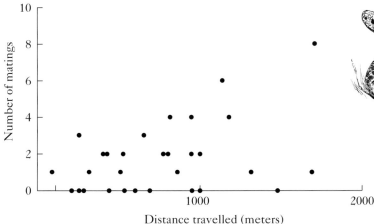

10 Sexual selection for male endurance. The more meters traveled by European adder males, the greater the number of females they mated with. *Source: Madsen et al. [752].*

11 Male dominance and mating success. (A) High-ranking males of the southern elephant seal copulate many more times than low-ranking individuals. (B) The same pattern applies to the northern elephant seal. *Sources: (A) McCann [786]; (B) Haley et al. [495].*

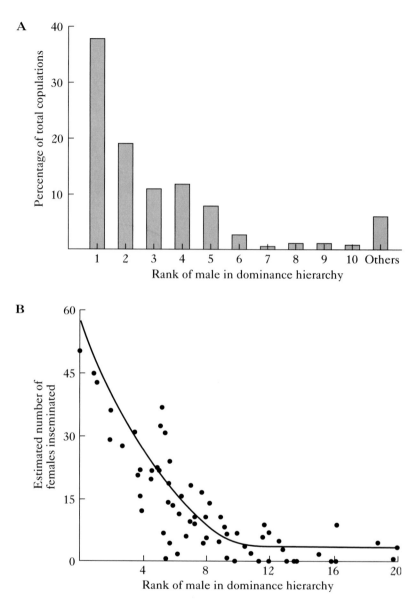

In elephant seals, larger males generally have an advantage in their beach fights [786]. The relationship between large body size and winning fights holds for a broad spectrum of animals [24], suggesting that sexual selection for fighting ability will often lead to the evolution of males with large bodies. One way of testing this possibility uses information on the extent to which males are larger than

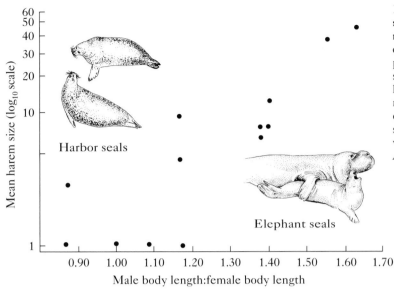

12 Opportunities for mating success and investment in male body size are positively correlated in seals and other pinnipeds. In species in which some males can acquire large harems, sexual selection has resulted in extreme sexual dimorphism. In monogamous species, males and females weigh about the same. *Source: Alexander et al. [16].*

females of their species. Richard Alexander and his co-workers realized that differences among species in the degree of **sexual dimorphism** in body size might provide a measure of the size investment males made to boost their fighting capacity [16]—although other factors besides sexual competition influence the relative sizes of the sexes [22, 523, 1028].

If a costly investment in body growth and maintenance has been sexually selected, then we can predict that there must be exceptional reproductive rewards to be gained by large males. If this is true, then males should be very large (and greatly outweigh females) in those species in which males can monopolize many females, while sexual dimorphism in body size should be least in those species whose males are limited to one or two mates per breeding season [16]. This prediction is supported by body length data from several mammalian groups, including seals and their relatives (Figure 12). Thus, in the highly sexually dimorphic elephant seal, a male may mate with as many as 100 females, whereas males and females of monogamous pinnipeds are more or less equal in length and weight.

Social Dominance and Male Fitness

Baboons are another example of a species in which males are much larger than females, and here too adult males compete violently for social dominance. Carlos Drews found that males averaged one aggressive wound—usually a bite—every 6 weeks, an injury rate about four times greater than that of females [310]. In this species, however, the value of these costly struggles for dominance is problematic, since many researchers have found that high position in a dominance hierarchy does not necessarily result in high copulatory success [e.g., 81]. For example, when Glen Hausfater counted copulations in a troop of baboons with a clear dominance hierarchy, he found that males of low and high status were equally likely to copulate with females [518]. But Hausfater had initially assumed that the more times a male mated, the greater his chances of fathering offspring. If in reality females fertilize their eggs with sperm received during only a few days in their estrous cycle, then males that copulate outside this time cannot produce offspring. It could be that high-ranking males are more likely to copulate with estrous females when it counts—that is, when their eggs can be fertilized. When Hausfater tested this possibility, he found that dominant males indeed outcopulated subordinates during the days of high female fertility [518].

Although this result suggests that dominance is correlated with the number of *effective* copulations achieved by a male baboon, the statistically significant relationship between rank and effective copulations arises because young males that have just begun to mate have very low dominance and achieve very few effective copulations [80]. Among the fully adult male members of a troop of baboons, only a weak connection exists between a male's ability to defeat rivals in direct physical confrontations and the number of offspring he produces. Among primates generally, the relationship between rank and "reproductive success" may be significant [225, 345], but it is not overwhelming, because socially subordinate males use alternative mating tactics to help them overcome physically stronger, larger rivals.

Among baboons, for example, there are several paths to sexual success in addition to securing high status. Males can and do develop "friendships" with particular females, relationships that do not depend on physical dominance, but rather on the willingness of a male to protect a given female's offspring—even if the youngster is not his own. Once a male has demonstrated that he is willing and able to provide protection for a female and her infant, that female may seek him out when she enters estrus again, even if he is not the top male in the social hierarchy of her troop [1154]. The ability of females to influence male mating success in this manner demonstrates that female baboons are not passive players when it comes to reproductive decisions.

Male baboons also form "friendships" with other males. Through these alliances, they can sometimes collectively confront a stronger rival that has acquired a partner, forcing him to give her up, even though he is socially dominant in one-on-one encounters. Thus, for example, in one troop of yellow baboons that contained eight adult males, three low-ranking males (fifth through seventh in the hierarchy) regularly formed coalitions to oppose a single higher-ranking male who was in consort with a female. In 18 of 28 cases, the single male was forced to relinquish the female to the threatening gang of subordinates [863].

Alternative Mating Tactics

Baboons are far from the only species in which subordinate males use special mating tactics in order to avoid going toe-to-toe with a stronger or more experienced opponent [31, 306, 415]. For example, small males of a Costa Rican rove beetle are at a strong disadvantage when competing with larger males for mating territories around a bit of mammal dung or on the carcass of a dead sloth. Females are drawn to these none-too-aesthetic materials in order to feed on the flies they attract. Large males chase small males away from their territories.

Sometimes, however, a small male turns around and presents the tip of his abdomen to a larger, aggressive rival. Sometimes the larger male follows the little guy, tapping the end of his abdomen in courtship, perhaps duped by a female-mimicking odor provided by the smaller male—a deceptive capacity known to occur in another rove beetle [921]. The tricky little male keeps moving slowly about the area, never permitting the larger courting male to mount and consummate their affair. Sometimes in the course of his meanderings, the courted beetle encounters a female, at which point he may court her and copulate literally under the nose of his thoroughly deceived territorial rival [396].

Another alternative mating tactic is the nonaggressive "satellite" behavior of males (Figure 13) that simply remain near others that are calling or displaying to females [168, 699, 919, 1250], or defending a resource attractive to females [658],

13 Satellite male mating tactics in the great plains toad and bighorn sheep. (Top) A non-calling male toad crouches by a caller, waiting to intercept females attracted by his signals. (Bottom) A non-territorial male bighorn sheep that has been waiting near a more dominant ram has broken through the defenses of his rival and has mounted a fleeing female. *Photographs by (top) Brian Sullivan and (bottom) Jack Hogg.*

or directly defending females (Figure 13B) [550]. In the horseshoe crab, for example, some males patrol the water off the beach, finding and grasping females heading toward the shore to lay their eggs (see Figure 19 in Chapter 13). Other males are Johnny-come-latelies that crowd around a paired couple after they reach the beach [128]. As it turns out, the attached male fertilizes at least 10 percent more eggs than a competing Johnny-come-lately [127]. If being an attached male yields more offspring than waiting for pairs to come to the beach, why do any males practice the less successful "satellite" option?

A Conditional Strategy with Alternative Mating Tactics

This question about the satellite tactic in horseshoe crabs could be asked about most examples of alternative mating tactics, since the general rule is that one tactic appears to provide greater reproductive success [479]. Thus, even though subordinate male primates can reduce their mating disadvantage through various means, dominant males nonetheless seem to have at least a slight edge in producing offspring. Likewise, a small female-mimicking rove beetle sometimes gets to mate in the presence of a larger rival, but sooner or later he is chased away, leaving the dominant male to monopolize the arriving females.

For another example of alternative tactics with unequal reproductive gains, we turn to male *Panorpa* scorpionflies, which acquire mates in three different ways: (1) some males aggressively drive other males away from dead insects, a food resource that attracts receptive females; (2) other males secrete saliva on leaves and wait for females to come and consume this nutritional gift; and (3) still others offer their mates nothing at all, but instead grab them and force them to copulate (Figure 14). To measure the reproductive success of males using these three tactics, Randy Thornhill studied caged groups of ten male and ten female *Panor-*

14 Forced copulation in *Panorpa* scorpionflies occurs when a male without a nuptial gift grabs a female and holds on until she mates with him. *Photograph by Thomas E. Moore.*

pa [1179]. Some of the male *Panorpa* were large, others medium-sized, and still others relatively small. Thornhill placed two dead crickets in each cage. The largest male scorpionflies competed for the crickets and won, and as a result, averaged nearly six copulations each. Medium-sized males generally offered mates salivary gifts, but gained only about two copulations each. Small males were unable to claim crickets, and appeared incapable of generating sufficient saliva to attract females. They employed the forced copulation route, but were relatively unsuccessful, averaging only about one copulation per male. Under the conditions of the experiment, the different tactics produced unequal fitness gains.

Let us use game theory (see Chapter 9) to analyze this situation. Since the different tactics yield different fitness returns, the separate strategies hypothesis cannot apply. Remember that an evolutionary strategy is simply a hereditarily distinct behavioral characteristic (see Chapter 11). If the differences between males using the three options were hereditary, then the one strategy that yielded the highest net benefits would eventually replace the others (see Chapter 3). Therefore, Thornhill proposed that the three tactics were options contained within a single conditional strategy, a hypothesis that requires that the differences between the phenotypes be environmentally caused (see Chapter 11). If so, males using a low-payoff option should switch to a tactic yielding higher reproductive success if the social conditions they experience make the switch possible.

To test this prediction, Thornhill removed the large males that were defending the dead crickets from the enclosure [1179]. Other males promptly abandoned their salivary mounds and moved to claim the crickets. Males that had not defended a cricket or produced a salivary gift stationed themselves by the abandoned secretions of other males. Thus, male *Panorpa* are able to adopt whichever of the three tactics returns the highest possible rate of copulations, given the current competition levels. These results clinch the case in favor of a conditional strategy as the explanation for the coexistence of three mating tactics in *Panorpa* scorpionflies.

Three Separate Strategies: Three Mating Tactics

For yet another example of a species whose males vary in the tactics they use to acquire mates, consider the marine isopod *Paracerceis sculpta*, a species that vaguely resembles the more familiar sowbugs and pillbugs—terrestrial isopods frequently found in moist spots in suburban backyards. This isopod lives in marine sponges found in the intertidal zone of the Gulf of California. If you were to open up a sufficient number of sponges, you would find females, which all look more or less alike, and males, which come in three dramatically different sizes: large (alpha), medium (beta), and small (gamma) (Figure 15). The three sizes of males also differ in their behavior. The big alpha males attempt to exclude other males from interior cavities of sponges that have one or more females living in them. If a resident alpha encounters another alpha male in a sponge, a battle ensues that may last for 24 hours before one male gives way to the other. Should an alpha

15 Three different forms of the sponge iso-pod: the large alpha male, the female-sized beta male, and the tiny gamma male. Each type not only has a different size and shape, but also uses a different tactic to acquire mates. *Source: Shuster [1097].*

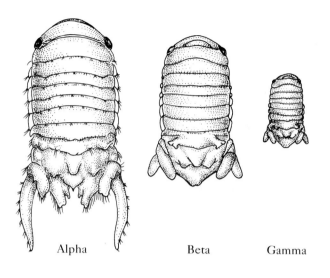

Alpha   Beta   Gamma

male find a tiny gamma male, the larger isopod simply grasps the smaller one and throws him out of the sponge. Gammas wisely avoid alphas whenever possible, lurking about and trying to sneak matings from the females living with an alpha male [1096].

When an alpha and a medium-sized beta male meet inside a sponge cavity, the beta behaves like a female, soliciting courtship, which the alpha is likely to provide—but to no good effect from the alpha's perspective. Through female mimicry, female-sized beta males coexist with much larger and stronger rivals and thereby gain access to females that the alpha male would otherwise monopolize, just as is true of the rove beetles we discussed earlier.

In this species, therefore, we have three different types of males, and one type has the potential to dominate others in male–male competition. If the three types represent three distinct strategies, then (1) the differences between them should be traceable to genetic differences and (2) the mean reproductive success of the three kinds should be equal. If, however, alpha, beta, and gamma mating behavior are three tactics controlled by a single conditional strategy, then (1) the behavioral differences should be induced by different environmental conditions, not different genes, and (2) the mean reproductive success of males using the alternative tactics need not be equal.

Stephen Shuster and his co-workers collected the information needed to test these two hypotheses [1098, 1099]. First, they showed that the size and behavioral differences between the three types appear to stem largely (with some complications) from genetic differences arising from a single gene with three alleles. Second, they asked whether the three types of males experience equal fitnesses. Shuster, like Thornhill, measured the reproductive success of the three phenotypes in a laboratory setting by placing different numbers and types of males with dif-ferent numbers of females in artificial sponges. The males he used in his experi-ment had special genetic markers, distinctive characteristics that could be passed on to their offspring, enabling Shuster to identify which male had fathered which of the baby isopods that each female eventually produced.

At the end of this experiment, Shuster found that the reproductive success of a male depended on how many females and rival males lived with him in a sponge. For example, when an alpha male and a beta male lived together with one female, the alpha isopod fathered most of the offspring. But when this male combo occu-pied a sponge with several females, the beta male outdid his rival, siring 60 per-cent of the resulting progeny, while in other combinations, gamma males outre-produced the others. When several females lived in a sponge, an alpha male could not control them all, and some mated with female-mimicking betas or secretive gammas, despite the alpha male's ability to manhandle smaller rivals.

Shuster and Michael Wade then returned to the Gulf of California to collect a large random sample of sponges, each one of which they opened to count the num-ber and type of male and female isopods present [1099]. With these data, they could estimate the reproductive success of 555 males, given the laboratory results on male reproductive success in various mixes of cohabiting individuals. When the mathematical dust had settled, alpha males had mated with an estimated 1.51 females on average, while betas checked in at 1.35 and gammas at 1.37 mates. Since these means were not significantly different, statistically speaking, Shuster and Wade concluded that the three genetically different types of males had essen-tially equal fitnesses in nature. The requirements for a three-strategies explana-tion had been met for *P. sculpta*. This species represents an exception to the gen-eral rule that conditional strategies account for behavioral variation in mating tactics within species.

Sexual Selection and Sperm Competition

In most research on male–male competition, a male's fitness is measured in terms of the number of females inseminated, since counting the number of copulations is a lot easier than identifying the number of young actually fathered by a given male. But in many animal species, fertile females mate with more than one part-ner. When this happens, whose sperm will fertilize her eggs? If some males' sperm have an advantage over those of others, just counting a male's copulations would not provide an accurate measure of his fitness.

Mechanisms of Sperm Competition

Consider what takes place in the common black-winged damselfly of eastern North America [1228]. When a female flies to a stream to lay her eggs, she may mate with a series of males over several hours as she visits two or three territories to oviposit, copulating with each owner in turn as well as mating with a nonterritorial satel-

16   Copulation in the black-winged damselfly. A territorial male (left) flies from his perch to court, grasp, and eventually copulate (right) with a female, who twists her abdomen forward to make contact with his sperm-transferring organ. *Photographs by the author.*

lite male or two. The female's behavior creates the potential for **sperm competition**—that is, competition among the sperm of different males to fertilize the same female's eggs [904]. Male damselflies have evolved an extraordinary device to give their sperm an edge in this competition.

Copulation in damselflies and dragonflies is a curious business. The male first grasps the front of the female's thorax with specialized claspers at the tip of his abdomen. A receptive female swings her abdomen under the male's body and places her genitalia over the male's penis equivalent, which occupies a place on the underside of his abdomen near the thorax (Figure 16). The male then rhythmically pumps his abdomen up and down, during which time his penis acts as a scrub brush (Figure 17), catching and drawing out any sperm already stored in the female's sperm storage organ, called a spermatheca [1229]. Jonathan Waage, the discoverer of this extraordinary sperm competition mechanism in black-winged damselflies, found that a copulating male removes between 90 and 100 percent of the competing sperm, after which he releases his own gametes. His sperm will be stored in the female's cleaned-out spermatheca for use when she fertilizes her eggs—unless she mates with another male before ovipositing, in which case his stored ejaculate will experience the same sad fate as those of his previous rivals.

The male black-winged damselfly employs an unusually efficient mechanism for eliminating the sperm of rivals stored within his partner. But this is far from the only species in which males remove or destroy sperm stored by a female

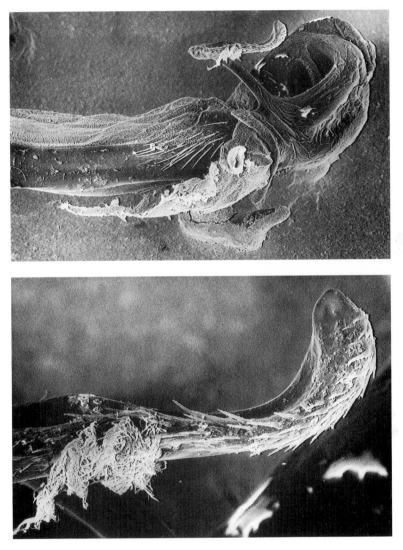

17 Sperm competition in the black-winged damselfly. The male's penis (above) has lateral horns and spines that enable him to scrub out a female's sperm storage organ before passing his own sperm to her. A close-up of a lateral horn (below) reveals rival sperm caught in its spiny hairs. *Photomicrographs by Jonathan Waage.*

[e.g., 320, 414, 630]. For example, a male dunnock—a small European songbird—may induce his partner to void sperm if she has been found near a rival male (Figure 18) [267]. Similarly, William Eberhard has proposed that some male sharks give their mates a contraceptive douche prior to ejaculating sperm into them. To this end, a shark penis has two tubes. Through one tube, a male can spray seawater at great force, perhaps washing out the female's reproductive tract before passing sperm to her via the other tube [324].

18 Sperm competition in the dunnock. A male pecks at the cloaca of his partner after finding another male near her; in response, she will eject a droplet of sperm-containing fluid. *After Davies [267].*

Mate Guarding

If some males remove or displace rival sperm from the reproductive tract of a partner, or bias egg fertilizations toward their sperm in some other way, it could well be adaptive for a male to prevent his mate from copulating with another male. In many species, males do seem to try to control their mates' sexual behavior *after* inseminating them, sometimes by sealing the female's genital opening with various secretions [62, 299]. More commonly, males remain near or in physical contact with their mates, ready to react aggressively to the arrival of other males. For example, as discussed earlier (see Chapter 8), males of some birds announce their attentiveness to their fertile mates by singing loudly, possibly to intimidate other males.

The question is, do these postcopulatory interactions actually produce fertilization benefits for the guarding male? The answer is yes, at least in some cases. For example, males of one parasitic wasp court their mates immediately *after* copulation is over, an interaction that substantially reduces the receptivity of the female to other males. Furthermore, even if the female does mate a second time, the second male's success in fertilizing eggs falls by a half when the female has received a bout of postcoital courtship from her first mate [18].

Another very different guarding tactic has the same function for males of another parasitic wasp. While copulating, pairs are often approached by another male. When the pair separates, the interloping male poses a competitive threat to the male who has just finished mating because females remain receptive for a short time after copulating. But instead of assaulting the other individual, the recently mated male adopts the receptive position of an immobile female with antennae lowered (Figure 19), which usually lures his rival into futile attempts to copulate with him. While the second male wastes his time with the female mimic, the mated female walks away, soon to become completely unreceptive to courting males [388].

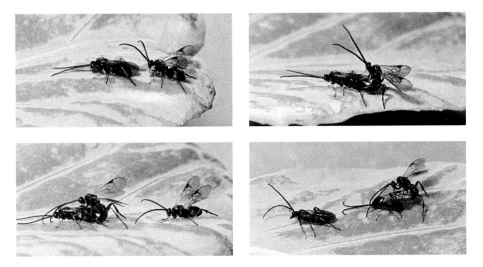

19 Female mimicry and mate guarding in a parasitic wasp. (Top left) A female crouches with antennae held low, signaling her receptivity to the male behind her. (Top right) He mounts and copulates. (Bottom left) The mating pair is approached by another male. The copulating male lowers his antennae, providing the female signal of receptivity. (Bottom right) The copulating male has released his partner. She moves off while he is mounted by the second male, who has been deceived by his male "partner's" use of female mimicry. *From Field and Keller [388].*

The fertilization benefits of mate guarding, however achieved, must be weighed against the fitness costs of the activity, especially the loss of opportunities to seek out other females as a result of staying with one already inseminated partner. Janis Dickinson measured this cost in the blue milkweed beetle by removing male beetles from the females they accompanied after copulation (Figure 20). About 25 percent of the separated males found new mates within 30 minutes. Thus, remaining on a female after inseminating her carries a considerable cost for a mate-guarding male, since he has a good chance of finding a new mate elsewhere if he would just leave his old one. On the other hand, nearly 50 percent of the females whose guarding partners were plucked from their backs also acquired new mates within 30 minutes. Since guarding males cannot easily be displaced from a female's back by rival males, they reduce the probability that an inseminated partner will mate again, giving their sperm a better chance to fertilize more of her eggs than otherwise. Dickinson calculated that if the last male to copulate with a female fertilizes even 40 percent of her eggs, he gains by giving up the search for new mates in order to guard a current one [298].

In general, the *benefits* of mate guarding *increase* as the probability rises that unguarded partners will mate again and use the sperm of later partners to fertilize their eggs. The *costs* of mate guarding *decrease* to the degree that receptive

20   Adaptive mate guarding by males of the blue milkweed beetle. The fitness of guarding males exceeds that of nonguarding individuals, provided that the last male to copulate with a female fertilizes a substantial proportion of her eggs. *Source: Dickinson [298].*

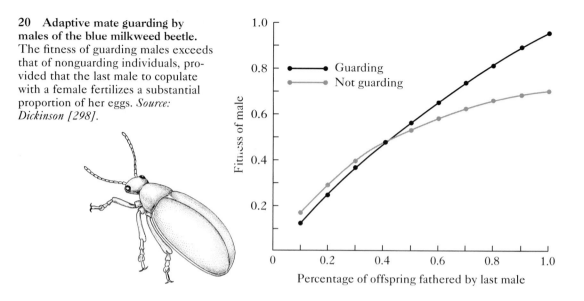

females are scarce, such that males remaining with one partner lose few chances to locate and inseminate additional mates. For example, although females of both the Idaho ground squirrel and the closely related Belding's ground squirrel often mate with more than one male, females of the Idaho ground squirrel fertilize their eggs primarily with the sperm of the *last* male to copulate with them, whereas there is a large *first*-male fertilization advantage in the other species. Furthermore, the burrows of receptive females are much farther apart in the Idaho ground squirrel than is the case for Belding's ground squirrel. You should be able to predict which of these two species exhibits mate guarding by males [1086].

In addition, males should guard their mates intensely only when their partners are receptive and have eggs that can be fertilized. White-fronted bee-eaters offer an opportunity to test this prediction. In these birds, as in many others, a male forms a cooperative relationship with a female that will last at least until the young of the season have been reared. After pairing, female bee-eaters produce a clutch of four or five eggs, laying one egg per day until the clutch is complete. Each egg is available to be fertilized for only a short period, creating a 4- to 5-day window of fertilization opportunity for males. During this time, a female's primary partner mates with her frequently. Moreover, on these days, he almost always flies along with his mate whenever she leaves the nesting area, thereby reducing by tenfold the chance that she will be harassed and forced to copulate with another male [358]. This period of close guarding of a pair-bonded partner, however, does not extend past the time when she has ova to fertilize (Figure 21). In fact, male bee-eaters switch from mate guarding to the sexual pursuit of other males' mates as soon as their primary partner finishes laying her eggs.

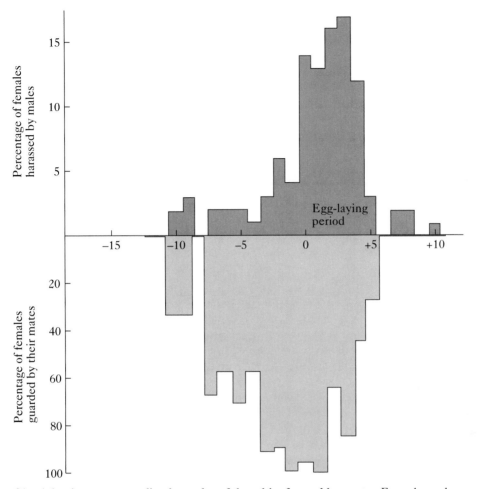

21 Adaptive mate guarding by males of the white-fronted bee-eater. Forced copulation attempts occur most often during the time when females are laying their eggs. In response to the risk of sperm competition, males remain especially close to their mates on the days just prior to and during egg-laying. *Sources: Emlen and Wrege [358], and unpublished data courtesy of Stephen Emlen and Peter Wrege.*

The same selectivity in mate guarding occurs in many other animals, with males making a costly investment in the protection of a partner *only* when she has eggs that can be fertilized [71], as demonstrated by high-ranking male baboons. Likewise, older, dominant African elephant males monopolize females only during that phase of estrus when their partners are likely to have a fertilizable egg. Early in estrus when the ovum is immature, or late in the cycle when the egg has already

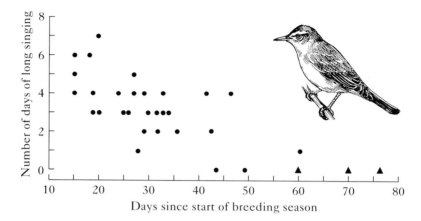

22 Seasonal change in mate-guarding by male great reed warblers. Early in the breeding season, when unmated females were still available, males sang mate-attracting "long songs" on a number of days when their current partners were still fertile, sacrificing guarding of the current mate to attract a second female. Late in the season, males sang on fewer days during their partner's fertile phase, and some males did not even attempt to attract a second mate (triangles). *Source: Hasselquist and Bensch [515].*

been fertilized, a female may copulate with younger, less socially dominant bulls, but these inseminations have little reproductive significance [952].

A cost–benefit approach to mate guarding produces the expectation that mate guarding will increase as the availability of additional mates falls. This prediction has been tested in the great reed warbler, whose males sometimes mate with more than one female [515]. In order to do so, a mated male must cease accompanying his partner and begin to sing a "long song" to attract a new partner. Attracting an unmated female is far more likely early in the breeding season, and during this period, males resume long singing several days before their current mate's fertile period has ended. As the season progresses, males guard their mates for a greater and greater proportion of the time when they have eggs to fertilize (Figure 22).

Sexual Interactions: Female Choice and Material Benefits

Our focus thus far has been on the competition among males to monopolize females and their eggs, and the often dramatic evolutionary consequences for males of this form of sexual selection. Because of the spectacle associated with outright battles between male elephant seals, or the Machiavellian nature of the stealthy sperm wars of black-winged damselflies, it is easy to forget that females have an active interest in reproduction too. In fact, females retain primary control over reproduction [1230, 1246], first, because they alone produce eggs, whose development after fertilization depends on the materials they donate to these gametes during their manufacture. What is in the eggs determines their chemical constitution and size, and may thus affect the phenotypic attributes of the young [663, 1061, 1140] (see Chapter 14). Moreover, females have a great deal to say about which sperm get to fertilize the eggs they produce [93, 328], not only by actively choosing certain males as mating partners [e.g., 159, 438], but also by sometimes permitting the sperm of certain males to fertilize their eggs while shunting to one side unwanted sperm they have received from other partners [716, 1043, 1162]. In the spotted sandpiper, for example, a species in which males care for the eggs (see Chapter 13), a female may produce a clutch of eggs for one male with whom she copulates frequently, then leave him to pair off with another male, with whom she also copulates often. But when it comes to fertilizing her second clutch of eggs, she may use sperm she received from her first partner, which she has sustained internally in tiny sperm storage tubules in her uterine wall for days or weeks [883].

The power of female control over fertilization could favor males with traits, including chemicals transferred in seminal fluids, designed to affect a female's "decisions" about which sperm to use [329]. Chemicals with this function have evolved, and some have physiologically damaging side effects on the females that receive them [630, 991], a clear case of a conflict of interest between the sexes (see below). Alternatively, selection might favor males with attributes that block or override female preferences for other males. In turn, the evolved responses of males will exert selection on female reproductive behavior and physiology, leading to reciprocal selection between the sexes, a process that should produce ever more effective operators in the arena of sexual interaction.

Female Control and Male Resources

Table 2 outlines the elements of female control over reproduction and how males may have responded evolutionarily to these features [1230]. We begin with the possibility that males sometimes transfer resources of various sorts to their mates in order to affect how females (1) produce eggs, or (2) choose copulatory partners, or (3) select sperm to fertilize their eggs. In a few cases, researchers have established that males provide their mates with nutritious **nuptial gifts** whose constituents wind up in their eggs. Thus, in a butterfly species in which males transfer an ejaculate constituting 15 percent of the male's mass, female fecundity increased 60 percent when females could mate with as many males as they wished as compared with females allowed to mate just once. The nutrients donated by males are incorporated into the extra eggs [1278], and females in nature actively search for several mates [624].

As noted earlier, male katydids often attach a large spermatophore to the genital opening of their mates, a food gift that females feed upon after copulation (Color Plate 14). Do females use these nuptial meals to make eggs? Mated female katydids that are experimentally prevented from eating the spermatophore produce fewer and smaller eggs over the next few days (Figure 23) [1108]. If this result

Table 2 Interactions between the sexes in which reproductive control is exercised primarily or exclusively by females

---

Reproductive decisions controlled primarily by females

Investment in egg: What and how much to place in egg

Mate choice: Which male or males to accept sperm from

Fertilization of egg: Which sperm to use

Investment in offspring: Which embryos to maintain; which offspring to give care to, and how much

Male traits that may influence female reproductive decisions

Transfer of resources to females: Influences investment in egg by female, or acceptance of male as mating partner, or use of male's sperm to fertilize her eggs

Elaborate courtship: Influences acceptance of male as mating partner, or use of male's sperm to fertilize her eggs

Frequent copulation with same female: Influences female to use his sperm to fertilize her eggs, increasing the benefits he gains by caring for the resulting offspring

Sexual harassment and forced copulation: Overcomes female preference for other males

Infanticide and selective paternal investment: Influences which offspring female will care for

---

Source: Modified from Waage [1230].

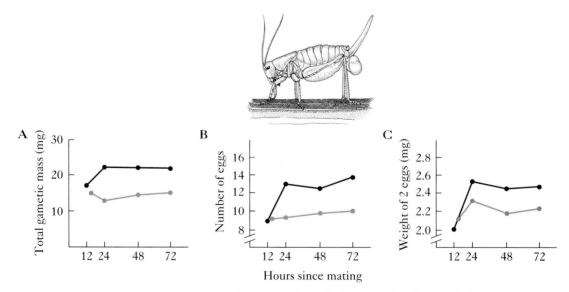

23 **The male spermatophore provides valuable materials** for female *Requena verticalis* katydids. After eating a spermatophore (solid lines), females are able to produce (A) a greater total mass of eggs, (B) more eggs, and (C) heavier eggs than females that are not allowed to do so (shaded lines). *Source: Simmons [1108].*

stems from a reduction in male-donated materials to incorporate into their eggs, then if there were some way to label the contents of a spermatophore, the label should appear in a female katydid's eggs.

To test this prediction, Darryl Gwynne and William Brown injected radioactive amino acids in the haemolymph (blood) of males of an Australian katydid. When these males mated, they transferred proteins containing these radioactive building blocks to their mates. When the mated females laid eggs, their gametes were radioactive. Thus, the proteins in spermatophores are used to make eggs in this katydid, although this result does not apply to every spermatophore-consuming insect [1286]. Interestingly, in Gwynne and Brown's study, female katydids in good condition invested relatively large amounts of male donations in their eggs. In contrast, food-stressed females diverted most of the spermatophore to their body tissues, presumably to help them stay alive long enough to reproduce again [490]. This ability of females to adjust how much of the male gift goes into their eggs is a nice demonstration of the point that females control reproduction.

Males can offer services other than calories to their mates. For example, males of the otherwise nondescript moth *Gluphisia septentrionis* spend hours drinking from mud puddles and forcefully squirting the water out their other end after having extracted sodium from the fluid. They then package the sequestered sodium within a spermatophore to be transferred to a mate; she will stock her eggs with the male's gift, supplying her larvae with a valuable ion, needed for a host of cellular biochemical processes, that is scarce in the plant matter they will consume [1123].

Or consider that male white-tailed ptarmigan devote more than a fifth of their waking hours in springtime to scanning the tundra while closely accompanying their mates. As a result of male vigilance, females can spend more time foraging, collecting the food resources that will be needed to produce a clutch of eggs [35]. Although the male may be keeping an eye open for rival males as much as for predators [132], his watchfulness contributes to his mate's ability to manufacture more eggs during her fertile period.

Nuptial gifts and other male services are valuable resources that males may offer in return for copulations, as occurs when female chimpanzees engage in sex with males that have given them a chunk of freshly killed monkey [1139]. Likewise, in the black-tipped hangingfly, females make copulation, and subsequent egg fertilizations, contingent upon receipt of a suitable nuptial gift (Figure 24). Randy Thornhill showed that females reject males that hand over unpalatable ladybird beetles. Even if the food gift is edible, and copulation begins, the duration of mating correlates with the size of the gift [1178]. If the prey is small, and the meal lasts less than 5 minutes, the female will leave without having accepted a single sperm. But if the prey cannot be polished off in less than 20 minutes, the female will depart with a full complement of the gift-giver's sperm (Figure 25).

Not only does the duration of copulation, and therefore the size of the nuptial gift, regulate how many sperm a female hangingfly will take from a male, it also determines whether she will become unreceptive and start laying her eggs. If a

24 A nuptial gift. A male hangingfly has captured a moth to offer to a female, a material benefit for his copulatory partner. *Photograph by the author.*

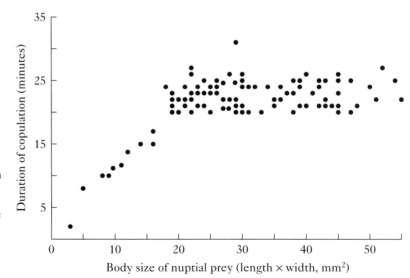

25 Sperm transfer and the size of nuptial gifts. In black-tipped hangingflies, the larger the gift, the longer the mating, and the more sperm the male is able to transfer to the female. *Source: Thornhill [1178].*

Duration of copulation (minutes)

Body size of nuptial prey (length × width, mm^2)

female has been given only a 12-minute meal, she will leave her mate prematurely, and to add injury to insult, she will seek out another male. She then feeds upon his gift and accepts his sperm, which she presumably uses to fertilize the eggs she lays before mating again. Thus, female mate choice creates sexual selection favoring males that provide large, nutritious nuptial gifts [1178].

Females of the decorated cricket use a similar mechanism to reject the sperm of males that pass a smaller-than-average spermatophore [1042]. The cricket male attaches a two-part spermatophore to his mate's genital opening during copulation. Shortly after separating from the male, the female reaches back and removes one part of the device, leaving in place the component that contains the sperm. While the female feeds on part 1, sperm migrate into her body from the other section, the sperm ampulla. The larger the nuptial gift, the longer she feeds, and the more time the sperm have to exit from the sperm ampulla. But as soon as she is done with part 1, the female turns to part 2, the ampulla, munching her way through this edible structure and demolishing any sperm that remain behind. Males that pass a skimpy spermatophore to a female fail to transfer the maximum amount of sperm, and so fertilize fewer eggs.

The kind of nuptial gift to warm the heart of an editor at the *National Enquirer* is one in which the male sacrifices himself to a cannibalistic partner as a grand finale to mating [160]. In mantids and some spiders, females do sometimes consume their mates, perhaps gaining valuable calories and nutrients in the process (see Figure 2 in Chapter 6). The $64,000 question is, can males that are eaten gain sufficient fitness benefits to favor the evolution of "voluntary" sexual suicide on their part? An alternative hypothesis is that males are eaten not because it advances their own fitness, but because females take advantage of opportunities to capture and cannibalize sexually motivated males [724, 956].

If cannibalism reduces male fitness, males should exhibit great caution when they are near females, which they do in certain mantids [687, 724]. In the case of the redback spider, however, males appear to make every effort to persuade their partners to consume them. While transferring sperm to his mate, the male redback performs a somersault that brings his body right next to the female's jaws (Figure 26). In a majority of cases, the female obliges her sexual companion and devours him. But female redbacks do not pursue or consume males during courtship, or at any other time except after the copulatory somersault [26].

Male redbacks are small, a mere 1 or 2 percent of the female's mass, barely more than a mouthful for a cannibalistic female. However, they may provide a useful snack, because hungry females are more likely to dine on a male than are well-fed females. When eaten, the deceased males derive substantial benefits from the experience, since they fertilize more of their partner's eggs posthumously than they would if they were to survive the mating. The eaten male's gains arise because cannibalistic females are less likely to mate again. Thus, the benefits to males from sexual suicide will usually outweigh the costs, since male redbacks have little chance of surviving the arduous search for a new partner [26].

26 Sexual suicide in the red-back spider. (A) The male first aligns himself facing forward on the underside of the female's abdomen while he inserts his sperm-transferring organ in her reproductive tract. (B) He then elevates his body and (C) somersaults backward into the jaws of his partner. She may oblige by consuming him while sperm transfer takes place. *Source: Forster [395].*

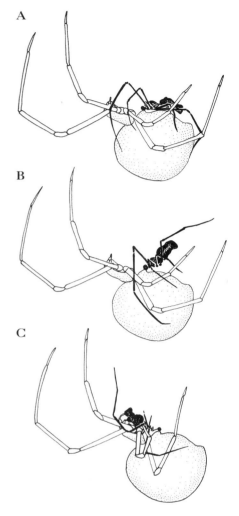

Female Choice Based on Male Appearance and Courtship

We have seen that when males have something useful to offer females (in addition to sperm), females may evaluate a male's gift, or the resources in his territory, as the basis for choosing which males to copulate with and whose sperm to use to fertilize their eggs. However, in some species, females seem as interested in the appearance or courtship behavior of males as they are in other factors (Table 3). In these cases, females typically prefer more intense and more varied courtship stimulation, favoring males with larger, more brightly colored ornaments, or more extreme rates of acoustical or tactile display, or combinations of all of the above [659]. The preference for more intense courtship stimulation may even extend to copulatory

Table 3   Examples of female mate choice based on differences among males in certain morphological and behavioral attributes

Species	Favored attribute	Reference
	Visual stimulation	
Scorpionfly	More symmetrical wings	[1180]
Barn swallow	More symmetrical and larger tail ornaments	[824]
Long-tailed widowbird	Longer tails	[23]
Wild turkey	Larger beak ornaments	[156]
Jungle fowl	Larger and redder combs and wattles	[1334]
House finch	Redder feathers	[543]
Satin bowerbird	Bowers with more ornaments	[109]
Cichlid fish	Taller display "bower"	[795]
	Acoustical stimulation	
Field cricket	Longer calling bouts	[521]
Woodhouse's toad	More frequent calls	[1158]
Ochre-bellied flycatcher	More frequent calls	[1258]
Great reed warbler	Larger repertoire	[516]
Túngara frog	Lower-pitched calls	[1031]
	Olfactory stimulation	
Mouse	Certain odors	[149, 628]
Cockroach	Certain odors	[831]
Moth	High concentrations of hydroxydanaidal in pheromone	[315, 316, 339]
	Tactile stimulation	
Sierra dome spider	Higher rates of genitalic stimulation	[1238]

courtship, with females evaluating males on the basis of the tactile stimulation they receive from the male genitalia during copulation (Figure 27) [324, 327].

For an example of mate choice based on elaborate visual courtship, consider that peahens prefer peacocks with relatively large numbers of eyespots in their immense tails, which the males spread and shake in front of potential mates [926]. The importance of these decorations was demonstrated when Marion Petrie and Tim Halliday captured some adult peacocks and removed 20 of the outermost eyespots from some males' tails. Birds so treated experienced a significant decline in mating success compared with their performance in the previous year. In contrast, controls that were captured and handled, but whose tails were left intact, were no less attractive to females after the treatment (Figure 28) [925].

If greater sensory stimulation from courting males does the trick, then experimentally augmenting a male's courtship ornaments should enhance his copulatory success. The relevant experiment was done first with the long-tailed widowbird, a species about the size of a red-winged blackbird that is endowed with a

27   Male genitalia as a test of cryptic female choice. The more elaborate penises on the left, which might provide more elaborate copulatory stimulation, are from primate species in which females regularly mate with more than one male. The simpler (less stimulating?) penises on the right are from other species, in which a single male usually copulates with a female during one estrous cycle. *After Dixson [301] and Eberhard [325].*

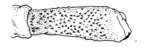

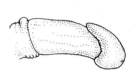

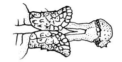

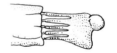

28   Removal of eyespots from a peacock's tail reduces his attractiveness to females. After 20 eyespots had been cut from their tails, males in the experimental group averaged two fewer mates in the following breeding season compared with their performance in the previous year. *Source: Petrie and Halliday [925].*

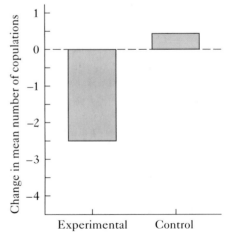

half-meter-long tail. The male flies around, displaying his magnificent tail to females passing by his grassland territory. Malte Andersson took advantage of the wonders of superglue to perform an ingenious experiment in which he captured male widowbirds, then shortened the tails of some by removing a segment of tail feather, only to glue the segment onto another bird's tail, thereby lengthening that ornament [23]. The tail-lengthened males were much more attractive to females than those that suffered the loss of a portion of their ornaments. Moreover, the tail-lengthened males also did much better than controls whose tails were cut and then put back together with dollops of superglue.

Similar experiments with barn swallows in Europe [820] and North America [1125] yielded similar results. Males with experimentally lengthened outer tail feathers attracted mates more quickly than males with experimentally shortened ornaments or control males with ornaments of intermediate length. As a result, longer-tailed males began breeding sooner than others, enabling them to rear two clutches more often than the other males [820].

The favorable response of females to heightened levels of sensory stimulation provided by a male's appearance or courtship says something about the proximate mechanisms of mate choice in these species, but leaves us wondering about the ultimate significance of female choices. What information are courting males and their elaborate ornaments providing to potential mates, and why are females sometimes persuaded by what they see, hear, smell, or feel? Several alternative theories of mate choice based on male courtship have been explored by behavioral biologists. One of these, the *material benefits theory,* applies to species in which males may provide mates with useful resources, such as nuptial gifts or food-containing territories or parental care. In species of this sort, male courtship could be an advertisement of what the male will offer a female in return for copulations and egg fertilizations.

Courtship Cues and Material Benefits

In fish species whose males provide parental care, male courtship might enable females to pick a superior caretaker for their offspring, namely, a large, aggressive, protective father [865]. To test this possibility in one species of cichlid fish, Katherine Noonan placed females in the central compartment of a three-chambered aquarium. The female cichlid could see a large male and a small male in the two end compartments on either side of her, but the two males could not see each other because of two black Plexiglas barriers in the central compartment (Figure 29). Females watched the visual courtship displays of both males, and in 16 of 20 trials, they spawned in a nest near the large male.

In the pied flycatcher, some males have jet black and white plumage, while others have a duller, browner appearance. In a three-chambered aviary (Figure 30) containing a female and two naturally dull-plumaged males, the female usually selected the male whose dull dark feathers had been artificially dyed dark black [1039]. As in the cichlid experiment, the female demonstrated her choice by begin-

29 A test of mate choice by female cichlid fish. The female in the central compartment of the aquarium can observe and approach either of the males in the end compartments, but the males cannot see each other or interact in any way, thereby eliminating male–male competition as a factor in the female's choice of a spawning nest. *Source: Noonan [865].*

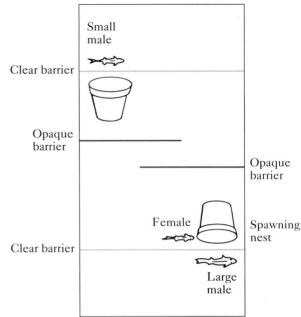

30 Mate choice by female pied flycatchers. In an experimental enclosure (shaded lines indicate netting; the solid line indicates an opaque barrier; X's indicate nest box entrances) captive females preferred to build a nest in a nest box advertised by a naturally dull-plumaged male with artificially blackened feathers rather than setting up housekeeping next to an unaltered, dull-plumaged male. *Source: Sætre et al. [1039].*

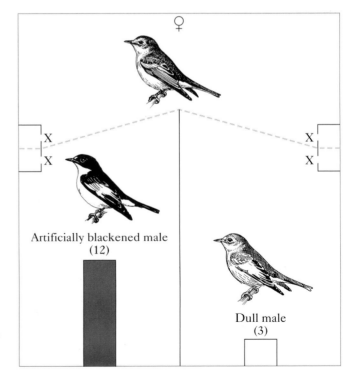

ning to build a nest in the nest box next to the compartment containing the male of her choice.

If the female flycatcher's preference for black-and-white males evolved because of direct parental care benefits offered by these males, then males with the preferred plumage should provide their offspring with more food than dull-plumaged males. In an experimental test of this prediction, the weight of nestlings fed by black-and-white males alone (after experimental removal of their mates) fell less than that of offspring cared for by dull males alone, indicating that black-and-white males were indeed better fathers [1040].

Bright plumage also carries the day with female house finches, which generally prefer males that have naturally bright red feather patches over males with paler orange or yellow patches [543, 544]. The redness of male plumage in this species may signal the nutritional state of the male [545] or his physiological condition, especially as influenced by parasite load [1176]. Individuals that have eaten large amounts of fruits and seeds and are in better condition have brighter, redder feathers laden with carotenoids, which may be derived from their diet. Here too visual assessment of plumage enables discriminating females to mate with males in better condition and with greater potential to assist them in rearing offspring than less attractive males [543].

In addition to size and color, the body symmetry of a male is sometimes important to females selecting mates. In the barn swallow, for example, females not only like males with long outer tail feathers, but they also find *symmetrical* males—those with outer tail feathers of equal length—more attractive than males with asymmetrical tails. Experiments in which the symmetry of tail ornaments was manipulated, requiring still more superglue, showed that males with symmetrical tails were chosen more quickly by females [824].

Another ingenious experimental study on the effects of symmetry employed male zebra finches that had received colored leg bands. Some males had two green bands on one leg and two orange bands on the other, an asymmetrical combination. Other males had an orange band over a green band (or green over orange) on both legs, a symmetrical arrangement. When females were given a chance to visit males in a special cage (Figure 31) where they could see the males, but not vice versa (because the two sexes were separated by one-way glass partitions), females spent significantly more time perched in compartments next to "symmetrical" males than "asymmetrical" ones (Figure 31) [1163].

Females of some species evidently can choose symmetrical males without even looking at them! When females of a Japanese scorpionfly were placed in the middle of a tubular apparatus containing pheromone-releasing males of different degrees of wing length symmetry at either end of the tube, 21 out of 25 females moved to the screen on the compartment containing the more symmetrical individual [1180]. Likewise, females of certain crickets prefer songs produced by symmetrical males [1111]. In these cases, symmetry is associated with a particular kind or quantity of pheromone, or an acoustical signal of a certain sort, so that females with certain signal preferences wind up mating with symmetrical males.

A   B

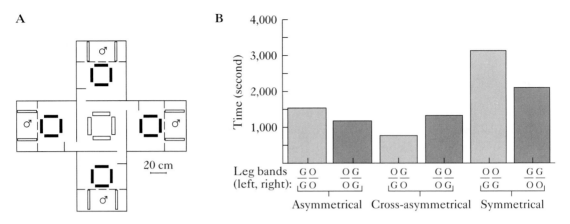

31   Female zebra finches prefer symmetrical males. (A) The cage used to test female choice. The female finch had access from a central compartment to four adjacent cages that housed one male each. The males could not see one another. The dark bars represent perches that monitored the presence of the female. (B) The mean time that females spent next to caged males with particular color band combinations (G = green color band, O = orange color band). Males with symmetrical band arrangements attracted more attention from females. *Source: Swaddle and Cuthill [1163].*

What material benefits might female scorpionflies, crickets, or swallows gain by mating with symmetrical individuals? Perhaps an asymmetrical male is a generally less effective parent or provider of nuptial gifts. Male asymmetry often arises because of environmental insults, such as severe shortages of food or parasite infestations [e.g., 947], that disrupt developmental homeostasis and so prevent equal development of both halves of the body (see Chapter 4). If body asymmetry is associated with early developmental problems [948], an adult asymmetrical male may be handicapped in helping his mates.

Sexual Interactions: Female Choice when Material Benefits Are Not Offered

In the species whose courtship we have just examined, males offer resources or parental care to their partners, so that their appearance or behavior could signal their capacity to deliver the goods. In other species, however, males do nothing to help their mates or offspring, offering no nuptial gifts, no parental care, nothing at all except their genes. If males provide only sperm to their mates, females cannot use the physical attributes of males, or their courtship behavior before, during, or after copulation, as indicators of the material benefits they will receive from their mates. So what might they gain by preferring to mate with males with elaborate ornaments or extreme courtship displays?

Three of the main theories on this point (there are others [1034]) are summarized in Table 4. The *healthy mate theory* states that male courtship and appearance inform females of a potential sexual partner's health or parasite load. If so, females could use this information to mate with males that are less likely to transmit lice, mites, fleas, or bacterial pathogens, which could harm them or their future offspring. The *good genes theory* also argues that male courtship and appearance provide information to females about the health and condition of a would-be mate, but states that the benefit from mating with a robust and healthy partner lies in the acquisition of viability-promoting genes that advance the survival chances of their offspring, usually by conferring hereditary resistance to parasitic infection or disease. In contrast, the *runaway selection theory* [34, 280], based on the work of the great evolutionist R. A. Fisher [390], proposes a different sort of genetic benefit for discriminating females. Here the benefit to the choosy female lies in the acquisition of genes that will lead her daughters to prefer attractive males and will endow her sons with attributes that will be preferred by most females—even if these traits actually reduce the survival chances of individuals that possess them.

Because the runaway selection alternative is the least intuitively obvious explanation for male courtship and female mate choice, let us briefly explain the underlying argument based on the mathematical models of Russell Lande [696] and Mark Kirkpatrick [651]. Imagine that a slight majority of the females in an ancestral population had a preference for certain male characteristics, perhaps *initially* because the preferred traits were indicative of some survival advantage enjoyed by the male. Females that mated with preferred males would produce offspring that would inherit the genes for the mate preference from their mothers *and* the genes for the attractive male character from their fathers. Sons that expressed the preferred trait would enjoy higher fitness, in part simply because they possessed the key cues that more females found attractive. In addition, daughters that found these traits attractive would gain by producing sons with the appearance or courtship traits that females liked.

Table 4 A comparison of the key features of three alternative theories for the evolution of extreme sexual ornaments and displays by courting male animals

Evolutionary mechanism	Females prefer trait that is	Primary adaptive value to choosy females
Healthy mate selection	Indicative of male health	Females (and offspring) may avoid contagious diseases and parasites
Good genes selection	Indicative of male survival chances	Sons and daughters may inherit the viability advantages of their father
Runaway sexual selection	Sexually attractive	Sons inherit the trait that makes them attractive to females; daughters inherit the majority mate preference

The idea that female mate choice genes and genes for the preferred male trait are inherited together is the basis for the concept of a runaway process in which ever more extreme female preferences and male characteristics spread together as new mutations occur. The runaway process ends only when natural selection against costly or risky displays balances sexual selection in favor of traits that are appealing to females. Thus, if peahens originally preferred peacocks with larger than average tails because such males could forage efficiently, they now might favor males with extraordinarily tails because this mating preference has taken on a life of its own, resulting in the production of sons that are exceptionally attractive to females and daughters that will choose this kind of male for their own mates. In fact, the Lande–Kirkpatrick models demonstrate that right from the start of the process, there is no need for female preferences to be directed at male traits that are utilitarian in the sense of improving survival, feeding ability, and the like. Any preexisting preference of females for certain kinds of sensory stimulation could conceivably get the process under way (see Chapter 7). As a result, traits opposed by natural selection because they reduced viability could still spread through the population due to the runaway process [651, 696]. Instead of mate choice for genes that promote the development of useful characteristics in offspring, runaway selection could yield mate choice for *arbitrary* characters that are a burden to individuals in terms of survival, a disadvantage in every sense except that females mate preferentially with males that have them!

Testing the Healthy Mate, Good Genes, and Runaway Selection Theories

Discriminating among these three alternative explanations for elaborate male courtship and female choice in species in which males do not provide material benefits has proved very difficult, in part because the three are not mutually exclusive. As just mentioned, female preferences and male traits that originated through a good genes process could then be caught up by runaway selection. Note too that males with hereditary resistance to certain ectoparasites (good gene benefits) will also be less likely to infect their partners with these ectoparasites (healthy mate benefits). Moreover, if at the end of a period of runaway selection, males had evolved extreme ornaments and elaborate displays, then only individuals in excellent physiological condition would be able to develop, maintain, and deploy their ornaments in effective displays. Males in such superb physiological condition would probably have to be highly effective foragers (good gene benefits) as well as parasite-free (healthy mate benefits), in which case females mating with such males would be unlikely to acquire a contagious disease or skin parasite, and their daughters might well get some survival-benefiting genes while their sons received the pure attractiveness genes of their fathers.

Given the difficulty in separating these three possible processes, the best we can do for the moment is to look at predictions derived from one or another theory and test them by the comparative method [e.g., 502] or by applying them to selected species. For example, good genes theory leads to several predictions about the traits of male peacocks and female choice: (1) males should differ genetically in ways related to their survival chances, (2) male behavior and ornamentation should provide accurate information on the survival value of the males' genes [660, 1323], (3) females should use this information to select partners, and (4) the off-spring of these males should benefit from their mothers' mate choice.

All four conditions apply to peacocks. Marion Petrie studied a captive but free-ranging population in a large forested English park, and found that the male peacocks killed by foxes had significantly shorter tails than their surviving companions. More-over, those taken by predators had been generally avoided by females in previous mating seasons, suggesting that females can discriminate between males with high and low survival potential, possibly on the basis of their ornamented tails [923].

Peahen preferences translate into offspring with enhanced survival chances. Young peafowl sired by highly ornamented males grow faster and survive better than the young of males that females find less attractive. Petrie established these points in a controlled breeding experiment in which she took a series of males with different degrees of ornamentation from the park and paired them in large cages with four females chosen at random from the population. The young of all the males were reared under identical conditions, weighed at intervals, and then eventually released back into the park. The sons and daughters of males with more highly ornamented tails weighed more at day 84 and were more likely to be alive after 2 years in the park than the progeny of males with fewer eyespots (Figure 32) [924].

This combination of results is consistent with the view that peahen mate pref-erences create sexual selection that is *currently* maintaining or spreading the genet-ic basis for those preferences because offspring receive good "viability-enhancing" genes from their fathers. But it is also possible that healthy male benefits are involved as well; perhaps the peahens' mate choices lead them to avoid parasitized males and so reduce their risk of acquiring damaging ectoparasites that they would

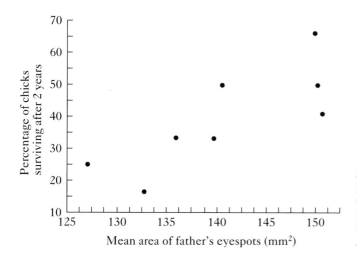

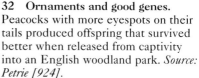

32 Ornaments and good genes. Peacocks with more eyespots on their tails produced offspring that survived better when released from captivity into an English woodland park. *Source: Petrie [924].*

pass on to their offspring, reducing their growth rate and survival chances. Moreover, a demonstration of current selective advantages associated with female preferences and male traits does not rule out the possibility that these attributes *originated* as side effects of Fisherian runaway selection in the manner described above.

The discovery that offspring growth and survival are promoted by female preferences for certain courtship characteristics in peacocks has been matched by similar findings for a cockroach species in which it is believed that only the male's genetic contribution affects female reproduction [831]. In this species, females prefer dominant males as partners, and the proximate mechanism of mate choice is based on the pheromones released by males of different social status. The offspring of preferred males develop more rapidly than those sired by males with less attractive pheromones. Rapid development enables individuals to reach maturity faster and begin producing offspring sooner, which gives them a fitness advantage over competitors slower to mature.

Improved viability of offspring may also be involved in the mating choices of female great reed warblers, whose decisions are strongly affected by the size of a male's song repertoire. The larger a male's repertoire, the higher his lifetime reproductive success as measured by the number of his young that survive to reproduce (Figure 33). In contrast, differences in territories and paternal care do not affect fitness, suggesting that by mating with males with large song repertoires, females gain genes that advance the viability of their young. Interestingly, males with large repertoires are older than average, which may explain why females that mate with them have offspring more capable of survival [516].

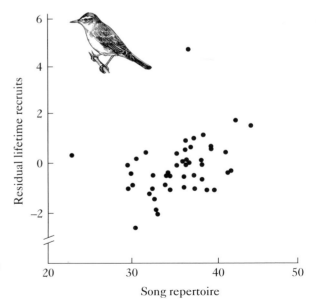

33 Song repertoire of the father and offspring survival are correlated in the great reed warbler. The larger a male's song repertoire, the greater his lifetime reproductive success, as measured by offspring that survive to the age of reproduction. *Source: Hasselquist et al. [516].*

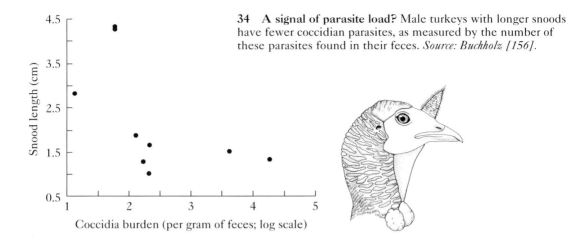

34 A signal of parasite load? Male turkeys with longer snoods have fewer coccidian parasites, as measured by the number of these parasites found in their feces. *Source: Buchholz [156].*

In these and other species, female preferences for certain male attributes may have evolved because those attributes indicate that their owners have low levels of *internal*, noncontagious parasites (as predicted from a good genes argument) or that they have low levels of *external*, contagious parasites (as predicted from the healthy mates theory). A number of interesting supportive examples exist. In male turkeys, the larger a weird beak ornament (called the snood), the lower the internal load of a certain protozoan parasite (Figure 34); females prefer males with larger snoods [156]. Likewise, the red wattles and comb of jungle fowl become paler when the birds are infected with an internal parasite, a nematode worm; females prefer males with bright red combs [715, 1334].

Anders Møller's work with barn swallows offers evidence for a possible relationship between ornament quality and heritable resistance to ectoparasites. As noted already, male swallows with long outer tail feathers enjoy a mating advantage. The length of these ornaments advertises whether males have been afflicted with mite ectoparasites in the past. Young males reared in nests that had been doused with pyrethrin, a mite killer, had much longer tail ornaments in the succeeding year than those reared in nests experimentally inoculated with mites. Blood-sucking mites seriously harm nestlings, reducing their weight and survival chances. Thus, any male that happened to possess a hereditary edge against mites would develop attractive ornaments and could pass on the genetic basis for mite resistance to his offspring. Indeed, males with attractive ornaments from unmanipulated populations do have low parasite loads, and they do produce offspring with similar advantages [822].

Sexual Interactions: Female Sperm Choice

A female's ability to control reproduction does not end with her selection of a mate on the basis of his precopulatory courtship (see Chapter 14). In many species, females mate with a number of males, which gives them the opportunity to choose one male's sperm over another's to fertilize their eggs. Thus, in flour beetles, the males whose odors females find most attractive are also the males that achieve the highest fertilization rate in competition with other males, possibly because females with multiple partners bias the fertilization of their eggs in favor of such males [712]. In the moth *Utetheisa ornatrix*, many females mate more than once, but selectively use the sperm of males that provide them with large spermatophores, which contain large quantities of the alkaloids used to protect eggs [329]. Mated females somehow sense the mass of the spermatophores they have received, and they actively move only the sperm from large spermatophores to the sperm storage organ for later use in fertilizing eggs [694, 695]. This moth, then, exhibits **cryptic female choice** *after* copulating with several males, a phenomenon that may be much more common than currently appreciated [328, 1116].

How might males respond evolutionarily to cryptic female choice? One possibility is for the male to copulate frequently with his potentially "unfaithful" partner to secure a numerical advantage in the sperm present in her reproductive tract. Timothy Birkhead and Anders Møller tested this hypothesis by predicting that pair-bonded colonial birds should be more likely to engage in frequent copulations than pairs belonging to solitary species [91]. This prediction is based on the greater likelihood that females of colonial species will acquire sperm from several males than females of solitary species, which are not surrounded by sexually active males. Birkhead and Møller tested this prediction with comparative data, using *all* the cases they could find in which two closely related species differed in the degree to which they practiced colonial breeding. For example, the red-rumped swallow (*Hirundo daurica*) nests in isolated pairs, whereas its close relative, the barn swallow (*Hirundo rustica*), nests in colonies. When many males live in close proximity to many females, the probability of extra-pair copulations should be higher. Therefore, paired males of colonial species are expected to copulate with their mates more often than paired males of solitary species do.

In fact, pairs of the colonial barn swallow copulate three times as frequently on average as pairs of the solitary red-rumped swallow [91]. The same relationship held for all other comparisons between colonial and solitary members of the same genus that Birkhead and Møller could find (an additional six pairs of comparisons). The probability that by chance alone all seven sets of comparisons would yield the expected result—namely, more frequent pair copulations in the colonial species—is less than 1 in 100. Birkhead and Møller concluded that frequent copulation between male and female partners in birds is indeed an evolved adaptation to the risk of sperm competition from rival males.

The case of repeated matings involving the same male and female offers a special illustration of the utility of considering sexual phenomena from the perspective of both sexes. Birkhead and Møller focused on what males might gain by investing time and energy in multiple matings with their long-term partners. But what about the female, whose cooperation is generally required if copulation is to succeed? In fact, female birds sometimes take the initiative, repeatedly inviting their mates to copulate [125], suggesting that females have something to gain from the practice (see Chapter 13). Perhaps a female can swamp unwanted sperm from an unfavored male by mating often with the male of her choice [583] Alternatively, frequent copulation could be a form of mate guarding by the female, who thereby keeps her partner from copulating with other females, preventing the later loss of her mate's parental care to the offspring of others. Thus, frequent copulation has potential fitness consequences for both male and female.

Sexual Interactions: Female–Male Conflict

Males often cooperate with females in an effort to influence their reproductive decisions in ways that favor the males' genes. Sometimes, however, males attempt to circumvent female choices. Conflict between the sexes may be common, since male and female interests are by no means identical. For example, males of various resource-defending species may acquire more than one mate when success in this endeavor increases the male's fitness, even though the presence of a second female depresses the reproductive chances of the first [179, 1205]. On the other side of the coin, females of a colonial seabird, the razorbill, attack their mates when they try to mate with neighboring females. By breaking up extra-pair matings, aggressive females deprive their mates of a chance to fertilize eggs outside the pair bond [1231], presumably in order to monopolize their partner's parental care.

Sexual Harassment and Forced Copulation

Conflict between the sexes is most readily apparent when males use force or harassment to overcome sexually resistant females [210]. In some water striders, for example, males often wrestle violently with females, which may somersault away (Figure 35) to avoid copulating. Persistent male water striders increase the risk that females will be captured by predatory aquatic insects that detect their prey by water surface vibrations [1020]. Likewise, males of a solitary bee pounce upon nectar-collecting females so forcefully that they often knock them to the ground. Unreceptive females can usually break free from their attackers, but frequently harassed females search for food at relatively unprofitable flowers in concealed parts of the food plant rather than foraging on the more energy-rich, but more exposed, outer flowers [1150]. Similarly, males of some birds [71, 358] and many mammals [1131], including our close relative, the chimpanzee, may literally attack certain females, either to secure an immediate copulation or to intimidate a female into mating later, but sometimes injuring or even killing the unfortunate female in the process.

Killing a potential mate obviously eliminates the possibility that a male will leave descendants with that female. However, despite this cost of the intense drive to copulate, with or without female cooperation, the trait may have overall fitness payoffs for males of some species. Female water striders that are unable to dislodge a mounted male will eventually mate with him [1021]. Persistent males also

35 Conflict between the sexes. A male water strider struggles to maintain his grasp on a female attempting to perform a backward somersault to dislodge her unwelcome partner. *Drawing by G. Marklund, from Rowe et al. [1021].*

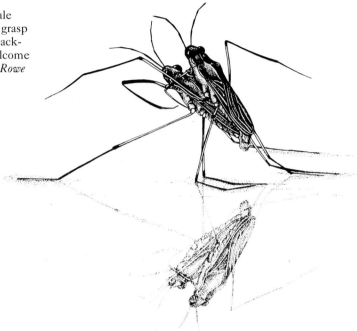

secure more matings in the Hawaiian fruit fly *Drosophila silvestris* [100]. Likewise, Fred Bercovitch found that the more times male baboons tried to mount females, the more times they eventually succeeded in inseminating them, despite a lack of female cooperation in a high proportion of cases [82].

Do females in these and other species accept persistent males because these males clearly have the energy reserves to fuel their persistence, thus demonstrating their good physiological condition and their potential to supply their mates with either good genes or material benefits, such as protection from other males? Or do females mate with persistent males simply because the costs of continued resistance reach the point at which it is less expensive to mate with the unwanted male than to spend more energy on avoiding his advances [210]?

Infanticide and Selective Male Parental Investment

The powerful ability of females to control reproduction is clearest in those species in which mothers can bias the sex ratio of the offspring they bear [559] or adjust how hard they work to bring food to their offspring [285]. In addition, females sometimes abort embryos fathered by certain males in order to produce offspring with new partners [93]. After a new stallion has taken over a band of wild mares from another male, the mares that are pregnant may abort after being forced to copulate with the new harem owner [83]. Much the same thing happens in certain mice after one male has displaced another, when exposure to the odor of the new male's urine causes pregnant females to resorb their embryonic offspring [688, 1063].

The abortion or resorption of embryos probably helps females make the best of a bad situation after they have lost or been abandoned by a previous mate and are now faced with an aggressive new male. By ending the pregnancy, a female may reduce the risk that the new male will attack her or kill her offspring. From the new male's perspective, by bringing ongoing pregnancies to an end, he speeds the onset of sexual cycling in the once pregnant females, thereby accelerating the opportunity to father offspring with them [1063].

Infanticide as a means to counteract female control of reproduction clearly occurs in those cases in which males kill young animals despite resistance from their mothers. As noted in Chapter 1, female langurs and lions try to prevent the death of their young at the jaws of the new pride masters, but they often fail. Sexual competition infanticide is widespread because it often benefits the infanticidal male, at the expense of the female who loses her infant(s). In various primate species, infanticide is believed to account for 30 to 40 percent of all infant mortality [1131]. Likewise, in an eresid spider, males cause a great deal of egg mortality by cutting egg cases out of the webs where females are guarding them. Even if the eggs hatch from a fallen egg case, the spiderlings will die without their mother's care. By destroying the egg case, an infanticidal male leaves the female no option other than to produce a replacement clutch, which the new male may fertilize and the female will care for—provided the case containing his offspring is not cut out of the web by yet another male [1058].

Evolutionary responses to the risk of infanticide include more than the adaptive acquiescence shown by female horses and mice or the physical counterattacks of female langurs and lions. In some species, pregnant females undergo a false estrus after the takeover of their group by a new male, with whom they copulate despite their inability to conceive his offspring [579, 611]. Alternatively, females may regularly mate with several males, making the paternity of their offspring difficult to determine, and thereby making attacks on their young by their various partners less likely [20, 872]. Cases of this sort illustrate powerfully that the reproductive interests of males and females do not necessarily coincide, and that members of the same species can exert intense selection on one another within the reproductive arena.

SUMMARY

1. Sexual reproduction creates a social environment of conflict and competition among individuals as each strives to maximize its genetic contribution to subsequent generations. Males usually make many small gametes and try to fertilize as many eggs as possible, while providing little or no parental care. Because females make fewer, larger gametes and often provide additional parental investment in their offspring, they usually have a lower potential rate of reproduction than males. As a result, receptive females are scarce, and males typically compete for access to them, while females can choose among many potential partners. In a few species, however, these sex roles are reversed, providing an opportunity to test the theory that the differences between the sexes stem from differences in their relative parental investments (or in their potential reproductive rates).

2. Evolution by sexual selection occurs if genetically different individuals differ in their reproductive success (a) as a result of competition within one sex for mates or (b) because of their differential attractiveness to members of the other sex. The competition-for-mates component of sexual selection is behind the evolution of many elements of male reproductive behavior, including competition for social dominance, alternative mating tactics, and mate guarding after copulation.

3. Although in a typical species, males exert selection on one another in the competition for mates, females generally have the last word on reproduction because they control the production and fertilization of eggs. Interactions between the sexes can be viewed as a mix of cooperation and conflict as males seek to win fertilizations in a game whose rules are set by the reproductive "decisions" of females.

4. In a typical species, females often choose among potential mates, creating the mate-choice component of sexual selection. Males of some species seek to win favor with females by offering them material benefits, including nuptial gifts, resources monopolized in territories, or parental care.

5. Mate choice by females occurs even in some species in which males provide no parental care or any other material benefits to females. Mate choice of this sort could arise as a result of selection by females of males whose genes will enhance the viability of their offspring (good genes theory). On the other hand, extravagant male features could spread through a population in which even arbitrary elements of male appearance or behavior became the basis for female preferences. Exaggerated variants of these elements could be selected strictly because females preferred to mate with individuals that had them (runaway selection theory). A third alternative is that female preferences for elaborate ornaments arise because males with these attributes are healthy and parasite-free (healthy mates theory). The relative importance of these various mechanisms of sexual selection remains to be determined.

6. Males in some cases may attempt to fertilize eggs by mating with females that try to reject them. The widespread occurrence of sexual harassment, forced copulation, and infanticide demonstrates the important evolutionary consequences of sexual conflict between males and females.

SUGGESTED READING

To learn more about the evolution and adaptive value of satin bowerbird sexual behavior, read two papers by Gerald Borgia [110, 112]. Tests of sex differences theory by way of sex role reversals have been reviewed by Darryl Gwynne [488] and by Amanda Vincent and her colleagues [1219]. Malte Andersson's book on sexual selection is clear and comprehensive [24], as is *The Ant and the Peacock* by Helena Cronin [246]. Anders Møller shows how many elements of the barn swallow's behavior have been shaped by sexual selection [826]. William Eberhard explores "sperm choice" by females that have mated with more than one male [328]. The consequences of sperm competition for birds are described in a book by Tim Birkhead and Anders Møller [92].

Mart Gross provides an excellent review of alternative mating tactics [479].

Jonathan Waage's articles on the black-winged damselfly provide an unusually clear description of dramatic male competition for eggs to fertilize [1228, 1229]. Randy Thornhill's studies of hangingflies show how female mate choice operates [1178, 1179].

Stevan Arnold [34] and Richard Dawkins [280] have a go at explaining runaway selection. Finally, although the question of how sexual reproduction evolved in the first place was not explored in this chapter, you can learn about it from G. C. Williams [1288] and Robert Trivers [1201], with an update from Laurence Hurst and Joel Peck [586].

DISCUSSION QUESTIONS

1. Male rats, sheep, cattle, rhesus monkeys, and humans that have copulated to satiation with one female are speedily rejuvenated if they gain access to a new female. This phenomenon is called the "Coolidge Effect," supposedly because when Mrs. Calvin Coolidge learned that roosters copulate dozens of times each day, she said, "Please tell that to the President." When the President was told, he asked, "Same hen every time?" Upon learning that roosters select a new hen each time, he said, "Please tell that to Mrs. Coolidge." Provide a sexual selectionist hypothesis for the evolution of the Coolidge Effect. Use your hypothesis to predict what kinds of male animals should *lack* the Coolidge Effect.

2. In Anders Møller's experiment on the effect of tail length on mating success in the European swallow, he made some males' tail feathers shorter by cutting them, and made some males' tail feathers longer by gluing feather sections onto their tails [820].

But he also formed a group in which he cut off parts of the males' tail feathers and then simply glued the fragments back on to produce a tail of unchanged length. What was the point of this group? And why did he randomly assign his subjects to the shortened, lengthened, and unchanged tail groups?

3. We argued that mate guarding should be common in species in which females retain their receptivity after mating and are likely to use the sperm of the last mating partner when fertilizing their eggs. But there are species, including some crab spiders, in which males remain with *immature, unreceptive* females for long periods; they will fight with other males that approach these females [304]. How can "guarding" behavior of this sort be adaptive? Produce hypotheses and allied predictions.

4. In a pipefish in which females compete for choosy males that care for the eggs they fertilize in a specialized brood pouch, males are more likely to accept eggs from courting females whose skin lacks the black spots that are caused by a parasitic trematode worm. The higher the trematode load, the lower the fecundity of the female. When Gunilla Rosenqvist and Kerstin Johansson tattooed black spots onto the skin of unparasitized females, these experimentally altered individuals experienced reduced success when courting males [1017]. What proximate and ultimate explanations did the researchers examine in this study of mate choice, and what were the predictions they tested?

5. In seahorses, males are more specialized for paternal care than any other animal. They have an elaborate pouch for brooding eggs, which they protect, aerate, and nourish for up to 6 weeks [1218]. Why, then, is it surprising that unmated males wrestle and prod other males for access to mates, whereas females lack these behavior patterns? What prediction can you make about the operational sex ratio in this species?

6. The figure below (after Gross [476]) shows three different tactics used by male bluegill sunfish for fertilizing eggs [476, 478]. (A) A territorial male guards a nest that may attract gravid females. (B) Little sneaker males wait for an opportunity to slip between a spawning pair, releasing their sperm when the territory holder does. (C) A slightly larger satellite male with the body coloration of a female hovers above a nest before slipping between the parental territorial male and his mate when the female spawns. How would you test the competing three-different-strategies versus one-conditional-strategy hypotheses to account for the existence of alternative mating tactics in this species?

A

B

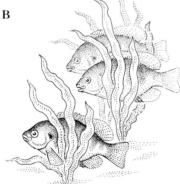

C

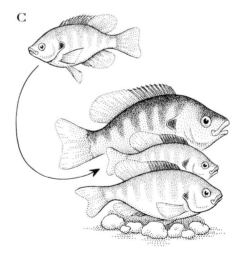

CHAPTER 13

The Evolution of Mating Systems

A
LTHOUGH THE DIFFERENCES IN THE POTENTIAL reproductive
rates of the two sexes typically lead males to compete for
females, the maximum number of mates acquired by males
and females varies considerably from species to species. In
the satin bowerbird, for example, males may copulate with
dozens of females in a breeding season, whereas in the green
catbird, a related bowerbird, males settle for a single primary
sexual partner. In a very few birds, several males share the
same female as their only mate. Female mating systems also
vary. In some birds, females have a whole retinue of partners;
in others, one female pairs with one male exclusively, while in
other species, large numbers of females pick the same male as
their sole mate in a given breeding season. Analyzing why

the number of mating partners varies among species, resulting in monogamy, polygyny, and polyandry, is the first goal of this chapter. The second goal is to examine why so many different forms of these three basic mating systems exist. For example, in some polygynous species, males secure multiple mates simply by hunting for as many receptive females as possible, while in others, territorial males cluster together at traditional arenas where they display vigorously to an audience of observant females. One theory about this diversity is that differences in key ecological factors produce different distributions of females, which in turn makes different mating tactics most productive for males. We shall use this argument to explain why males exhibit so many different kinds of mating behavior, while also considering how the active involvement of females shapes animal mating systems. This theory should be tested rigorously for entire groups of animals via the comparative method. But since this work has only just begun [1105, 1174], we will focus instead on case studies of particular species for which the link between female ecology and mating system has been investigated.

Does Monogamy Exist?

Although male mating systems vary greatly, males of most species, even species as different as black-winged damselflies and elephant seals, can be classified as **polygynous,** in that some males fertilize the eggs of many females each breeding season. Sexual selection theory (see Chapter 12) provides a possible explanation for this common pattern. Since the typical male makes no parental investment in his offspring, a male's reproductive success is usually related to how many females he inseminates—to the extent that copulations lead to egg fertilizations. As a result, males usually compete for mates, and the "winners" are polygynists. Such individuals are likely to have a relatively large number of progeny, passing on to their male offspring the genes for the tactics they used to become polygynists.

Viewed strictly in these terms, therefore, males are expected to try to acquire several mates, although many will not succeed. The real puzzles are those cases in which males, having found one partner, do not seek out others over a given breeding season, or even longer **(monogamy)**, and those even rarer instances in which the male actually shares a single long-term mate with several other males **(polyandry).**

In the discussion that follows, keep in mind that I am using the term *mating system* primarily to refer to the number of sexual partners secured by males or females during a breeding season, although other mating system classifications focus on whether males and females form long-term pair bonds. As we will see, efforts to devise a simple classification scheme of any sort for mating systems have collided with complex reality.

The Puzzle of Monogamous Males

As noted, sexual selection theory would seem to require that males attempt polygyny, and yet monogamy apparently occurs. Various explanations have been offered for why some males might not continue to hunt for more females after having found one. The *mate-guarding hypothesis* argues that monogamy may be adaptive when a female left by one male would acquire another partner, whose sperm would then fertilize her eggs [354]. Mate guarding is especially likely to be adaptive when females remain receptive after mating and when they are widely scattered and difficult to locate, conditions that apply to the redback spider, whose story we outlined in the previous chapter. A similar pattern applies to the beautiful clown shrimp, *Hymenocera picta* (Figure 1), whose males spend weeks with one female rather than attempting polygyny [1273]. As expected, males searching for mates far outnumber receptive females, which will copulate only during one short period every 3 weeks or so. Furthermore, females are sparsely distributed in their marine environment. Because of the costs and difficulty of finding one receptive female after another, males that find a potential partner wait for her to become receptive, guarding her prior to copulation in order to have access to her during the narrow window of time when her eggs can be fertilized.

Alternatives to the mate-guarding hypothesis for monogamy include the *mate-assistance hypothesis*, which states that males will remain with a single partner to help rear their offspring in environments in which male parental care can greatly promote offspring survival. The extra surviving offspring derived from paternal cooperation may be more numerous than those a male could expect to acquire through abandonment of one mate in order to search for others. Males of the monogamous seahorse *Hippocampus whitei* do indeed provide extraordinary care to

1 A monogamous shrimp. When males of the clown shrimp encounter a potential mate, they remain with her, because receptive females are scarce and widely distributed. *Photograph courtesy of Uta Seibt and Wolfgang Wickler.*

2   A pregnant male. A male seahorse (left) receives eggs from his mate, which fill his brood pouch and swell his abdomen substantially. *Drawing by S. Barker, from Vincent and Sadler [1220].*

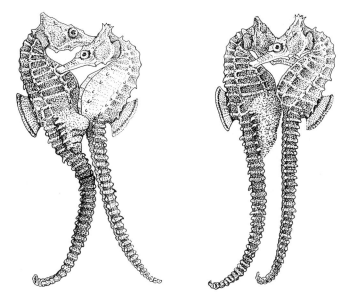

their offspring, rearing them in a sealed brood pouch during a pregnancy that lasts up to 3 weeks. Males and females form pairs that stay together over a whole series of matings, egg transfers, and pregnancies (Figure 2). Pairs even greet each other each morning before moving apart to forage separately, while ignoring others they happen to meet during the day [1220]. Since a male's brood pouch can accommodate only one clutch of eggs, males gain nothing by courting more than one female at a time. Nor is there much to gain by switching mates from one pregnancy to the next if the current mate can supply her partner with a new clutch of eggs as soon as he has reared the current brood. Many females apparently can keep their partners pregnant throughout the several-month-long breeding season. Females also have an incentive to stick with one male, since the costs of attempted polyandry would probably be high, given the very low density of seahorses, their exceedingly poor swimming power, and their vulnerability to sharp-eyed predators.

In other species, males can potentially gain by acquiring several mates, even if it harms their primary partners. In cases of this sort, we might expect females to try to block their partners' polygynous tendencies. Paired females of the burying beetle *Nicrophorus defodiens* assault their mates when they release a sex pheromone to attract another female. In this species, a male and female form a pair to bury a carcass, such as a dead mouse, which will provide food for the offspring that the female will produce after insemination by her mate. But once the carcass is buried, males may call a second female to the site. If this female were to lay her clutch of eggs on the pair's ball of dead mouse flesh, her larvae would com-

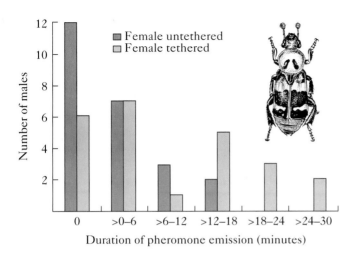

3 Female burying beetles combat male polygyny. When a paired female beetle is experimentally tethered, the amount of time that her partner spends releasing sex pheromones to attract additional mates rises dramatically. *Source: Eggert and Sakaluk [334].*

pete for food with the larvae of the first female, reducing their survival or growth rate. Thus, when the paired female smells her mate's pheromone, she hurries to him to push him from his perch or bite him firmly. These attacks reduce his ability to signal, as Anne-Katrin Eggert and Scott Sakaluk [334] showed by tethering the first female so that she could not reach a pheromone-releasing would-be polygynist. Under these circumstances, males called pheromonally for much longer periods than when they were exposed to untethered females free to enforce monogamy on their mates (Figure 3). The point here is that males may be monogamous not because it is in their best interests, but because females make it happen through *female-enforced monogamy.*

Monogamy in Mammals

Although monogamy occurs in various invertebrates, most notably the termites [854], and in some vertebrates, it is exceptionally rare in mammals, a group notable for the size of female parental investment in their offspring. Given the nature of mammalian pregnancy and milk production, sexual selection theory suggests that mammalian males, which cannot become pregnant and do not offer milk to infants, should usually try to be polygynous—and they usually do. However, as always, exceptions to this rule [e.g., 436, 1305] can be instructive in helping us to understand the evolution of the phenomenon.

Returning to the mate-assistance hypothesis, we can predict that most monogamous mammals will be ones in which males care for their offspring. Although fewer than 10 percent of all mammals feature male parental care [1305], many of these exceptions are monogamous, including some rodents and members of the dog family [354]. Male rodents can provide protection for their young against infanticidal intruders (see Chapter 6), and can also keep their young warm when they are very small [154]. These benefits may compensate monogamous males for lost mating opportunities. Male carnivores, such as wolves and foxes, can also provide helpful parental care by driving off infanticidal males, defending a feeding territory, and provisioning pups with prey.

However, the female-enforced monogamy hypothesis also predicts that monogamy will be more prevalent when males offer paternal care, because females gain by monopolizing these benefits rather than sharing them with others. For example, a wolf pack usually contains just one dominant breeding female, often a mother, sister, or aunt to the nonbreeding females [892]. Interestingly, the nonbreeders may be ovulating and courting males, but they will copulate and produce a litter only if the dominant female is removed (Figure 4). Their failure to reproduce when the dominant female is present could be to their advantage if she would be likely to kill their progeny, or if their pups would starve to death through food competition with her offspring. The fact that such reproductive suppression by subordinates in various carnivores is correlated with high energetic costs of pregnancy and postnatal care [233] is consistent with this argument. For the moment, therefore, we cannot eliminate either the mate-assistance monogamy or female-enforced monogamy hypotheses as explanations for monogamy in wolves. Nonetheless, the acceptance of nonbreeding status by subordinate females in a wolf pack keeps even the dominant male from becoming polygynous.

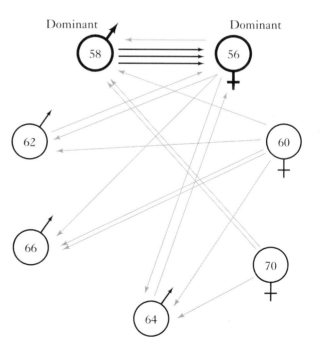

4 Monogamy in wolves. In this captive pack with three females and four males, only the dominant female (56) actually copulated with a male (58); each solid arrow represents one copulation. Although the other two females came into estrus and solicited matings (shaded arrows) from the males in the pack, they did not copulate and so did not produce litters. *Source: Packard et al. [892].*

Monogamy in Birds

Although monogamy is rare in mammals, in about 90 percent of all bird species, males and females form monogamous partnerships during a breeding season, which apparently preclude opportunities for polygyny by males (or polyandry by females) [692]. Why the difference between mammals and birds? Gordon Orians pointed out that male birds differ from most male mammals in that they can be highly effective parents. Although male birds cannot lay eggs, they can incubate them and care for the nestlings when they hatch, providing the young with food and protection against predators. In the monogamous willow ptarmigan, males do not feed the young, but they may protect the female and young from predators (Color Plate 15). Thus, a monogamous male might, through his paternal services, acquire more descendants with one assisted female than he could with several mates that he did not help [880]. The benefits of mate-assistance monogamy are especially likely to outweigh the costs in those species in which females breed synchronously, a common occurrence in birds. By the time his courtship of one female is complete, few unpaired females remain for a would-be polygynist. Under these circumstances males lose little by remaining with a current mate and her offspring.

The mate-assistance hypothesis for avian monogamy predicts that male parental care will in fact have a substantial effect on the number of surviving young reared by a female. In some birds, females lay, and pairs care for, a *single* egg, suggesting that male assistance is essential to produce *any* young. In such species, a male that divided his efforts between two or more females could wind up with no offspring at all [1303]. In some other species with larger broods, the male's parental contribution to his reproductive success has been established in various ways. In the yellow-eyed junco, observations of young fledglings reveal that these "bumblebeaks" are initially so inept at collecting and processing insect food that without 3 weeks of food supplements, they would all die of starvation [1243]. Therefore, the male's food contributions to his fledglings promote his fitness and that of his mate. Other studies have made the same point by experimentally removing males from pairs, leaving "widowed" females that generally reared fewer young that did non-widowed controls (Figure 5) [746].

Both mate assistance and mate guarding may account for the prevalence of monogamy in the tree swallow, a species with some polygynous males. Peter Dunn and Raleigh Robertson found that polygynous males actually produced fewer genetic offspring on average (0.8 fledglings) than their monogamous counterparts (3.0 fledglings), for two reasons. First, a male with two mates at two nest sites cannot guard both equally well, making it possible for his partners to be inseminated by rivals and to use their sperm to fertilize some or all of their eggs, a story that we will get to shortly. Second, a polygynous male cannot do a good job of helping both mates rear young, so that a higher proportion of his partners' offspring die before fledging [314].

Thus, monogamy could be advantageous to both males and females under some circumstances—but once again, we should consider whether females can force monogamy on their partners. Given that male birds so often provide useful ser-

5 Paternal assistance is valuable in the monogamous snow bunting. Control females often reared four or more offspring with the help of their male partners, whereas few females that had been experimentally "widowed" by removal of the male partner were able to fledge more than three chicks. *Source: Lyon et al. [746].*

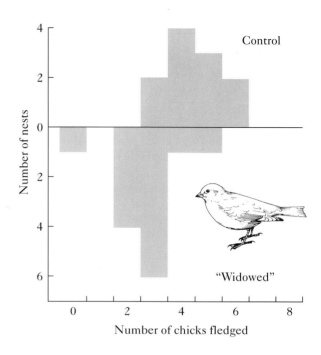

vices for their mates, females mated to monogamous males should generally enjoy much greater fitness than those mated to polygynous males, from which they would receive only a fraction of the possible paternal donation, as demonstrated by the tree swallow. If so, females should not pair with an already mated male, and they may benefit by attacking other females that approach their partner, his territory, and his resources [880].

In order to discriminate between the mate-assistance and female-enforced hypotheses for monogamy in the great tit, Mats Björklund and Björn Westman looked at the effects of male removals on female reproductive success [96]. Some females of this songbird choose to mate with males in second-rate habitats, an adaptive choice only if they gain more by mating with a monogamous male in a secondary environment than by doubling up with a male in the preferred habitat. In the preferred habitat (deciduous woods), monogamous pairs fledged an average of nearly eight young in a breeding season, with an average weight of 18.5 grams. In the secondary habitat (coniferous forest), monogamous pairs produced about seven young with an average weight of 17.5 grams. Even though pairs nesting in conifers did not do as well as those in deciduous woods, they did better than experimental pairs in the primary habitat from which the male partners were removed a few days after the hatching of the eggs. These widowed females fledged fewer than six young with a mean weight of 16.5 grams.

That widowed females of the great tit can fledge about six young even without male assistance shows that a polygynous male with two females on his territory could wind up with eleven or twelve fledglings to his credit, even if he helped neither female; thus, polygyny should be advantageous to the male. But female mate choice prevents males from becoming polygynous. Some nesting habitats are much better than others, but not so much better as to compensate a female that chose an already mated male and lost half (or all) of his assistance in feeding her young.

Extra-Pair Copulations: The Male Perspective

But what about the traditional view of birds as monogamists, with males and females alike limiting themselves to one partner per breeding season? As it turns out, many supposedly "monogamous" birds regularly mate outside the pair bond (e.g., [1264]). From the perspective of a male bird, these **extra-pair copulations,** or EPCs as they are known to acronym enthusiasts, have clear costs, including the time spent searching for matings outside the pair bond as well as the risk that his primary partner will fertilize her eggs with sperm acquired from rival males in his absence. These risks are real, judging from the tendency of male birds to guard their mates closely during the female's fertile period, while switching to pursuit of other females only after this period is over [71, 358].

Whether EPCs provide counterbalancing benefits to males depends on whether these extracurricular matings result in egg fertilizations. Evidence that they do has mounted in recent years, primarily as a result of the application of DNA fingerprinting techniques to determine paternity in wild birds. These techniques take advantage of the structure of DNA, which contains many small noncoding regions between the protein-coding genes. These noncoding regions consist of certain base sequences repeated over and over, but the number of repetitions varies greatly from individual to individual. For example, one hypervariable region in human DNA exists in at least 77 different forms (a total of only 79 persons yielded these 77 different alleles [1304]). Researchers can now break up DNA into small fragments and then locate the segments that come from particular hypervariable regions. These fragments can be labeled radioactively, separated on the basis of their molecular weight, and eventually photographed to produce a "DNA fingerprint."

The pattern of bands in a DNA fingerprint reflects the size of the fragments of DNA derived from a given hypervariable region. Thus, a comparison of the DNA fingerprints from two individuals can reveal whether they are relatives, since individuals with recent common ancestors inherit portions of the same DNA and therefore the same hypervariable regions. For example, a nestling bird's DNA fingerprint will contain bands all of which will match either bands in its mother' fingerprint or in its genetic father's fingerprint. By comparing the DNA fingerprints of a mother dunnock, her four nestlings, and the two males that copulated with her, we can tell that the "alpha" male, her primary partner, fathered nestling G because of the match between the banding patterns of their fingerprints (Figure 6). But the other nestlings (D, E, and F) have bands in their DNA fingerprints that are absent in the "alpha" male. Therefore he could not have been their father. Instead, the "beta" male sired these offspring, as shown by the match between his DNA fingerprint and those of the three nestlings [158].

6 DNA fingerprinting reveals paternity of nestlings in the dunnock by enabling researchers to compare the banding patterns of males and their putative offspring. The bands represent DNA fragments of different sizes ranging from about 2 to 20 kilobases (kb). The small black triangles point to distinctive bands present in the alpha male; the small open triangles point to bands specific to the beta male. Compare the banding patterns of the two males with those of the nestlings (D–G). Nestlings D, E, and F cannot have been fathered by the alpha male because their DNA fingerprints contain bands that are not present in his fingerprint. The banding pattern of the beta male indicates that he, not the alpha male, fathered these three nestlings. *Source: Burke et al. [158].*

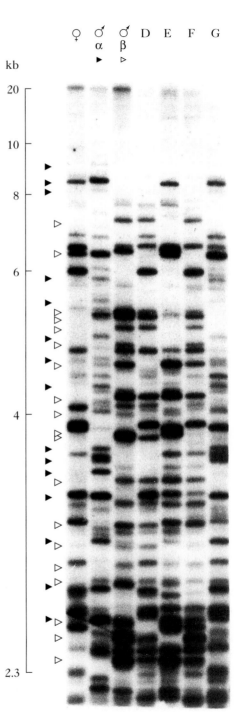

DNA fingerprinting demonstrates that EPCs often do result in increased reproductive success for male birds that copulate with females in addition to their primary partner. Extra-pair paternity often accounts for 20 percent of all nestlings in a songbird brood, and can exceed 50 percent [300, 1264].

Extra-Pair Copulations: The Female Perspective

The advantage to male birds of siring offspring by the mates of other males is clear enough, but what do females gain from EPCs? In some species, females probably do not benefit at all, judging from their vigorous resistance to "philandering" males. In other species, however, females appear to accept, or even solicit, copulations from two or more males [93, 829, 1264], evidence that they are gaining something from the activity. For example, in various populations of red-winged blackbirds, females regularly court and mate with males other than their primary partners, and they use the "extra" sperm to fertilize some of their eggs [464, 1241].

One way to classify polyandrous mating systems would be to focus on the nature of the benefits, genetic or material, that females receive from copulating with more than one male (Table 1). Among the possible genetic benefits gained by a female bird that engages in EPCs is fertility insurance. If some males have defective or

Table 1 Hypotheses on why females might benefit by mating voluntarily with more than one male

---

Genetic Benefits Polyandry

1. Egg fertilization insurance via acquisition of sufficient numbers of competent sperm [994]

2. Better sperm with genes acquired from preferred sexual partners [94, 633, 1238]

3. Genetic *variety* gained via acquisition of genetically diverse sperm [631]

   Some sperm will be genetically compatible with egg genotypes [906, 1327]

   Some sperm will come from individuals of optimal genetic distance to avoid damaging inbreeding or outbreeding [1092]

   Some sperm will be better egg fertilizers, conferring a fertilization advantage on sons [632]

   Resulting offspring will be more diverse; siblings will be less likely to engage in ecological competition [995]

   Resulting offspring will be more diverse; siblings will do better in division of labor within colony (e.g., in social insects) [984]

Material Benefits Polyandry

1. Acquisition of resources controlled by mating partners [695, 1324]

2. Acquisition of parental care from mating partners [270]

3. Improved foraging success via distraction of a male foraging competitor [1239]

4. Reduced risk of sexual harassment from nonpartners [1281]

---

infertile sperm, mating with several partners increases the chance that all of a female's eggs will be properly fertilized [92, 968]. Female red-winged blackbirds that engage in EPCs do indeed have a somewhat higher egg hatching rate, probably as a result of the more complete fertilization of their eggs [465]. Similarly, Thomas Madsen and his colleagues found that female adders that mated with several males had fewer stillborn young than females restricted to one or a very few partners. They interpreted this finding to mean that a polyandrous female improved her chances of receiving sperm from at least one high-quality male, sperm that were more likely to fertilize her eggs and whose genes were more likely to promote adequate development of her offspring [751].

The *better sperm hypothesis* also applies to females that can selectively engage in EPCs with males that offer them superior genes for their offspring. In the splendid fairy-wren, for example, a reproducing pair, often close relatives, lives on a territory with a number of subordinates, usually males. The breeding female also regularly mates with males from neighboring territories, thereby presumably increasing the genetic diversity of her young and counteracting any deleterious effects of inbreeding. Moreover, male intruders intent on EPCs are generally the dominant males in their groups. If dominant fairy-wren males do have "good genes," females could gain fitness by mating specifically with them [1025].

In a number of other birds, females that voluntarily solicit EPCs choose dominant, or older, or more impressively ornamented males as their extra partners [159, 825]. In the case of the blue tit, for example, females are far more likely to visit neighboring males whose own partner stays at home, presumably because she has such an attractive mate that she cannot acquire better sperm elsewhere (Figure 7). Attractive males whose own partners are faithful to them survive better and produce more young than less attractive males, suggesting that the males chosen for EPCs possess "good genes" [633].

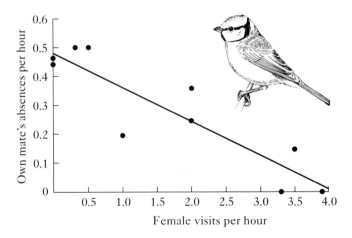

7 Selective polyandry by a songbird. In the blue tit, some females leave their home territories to visit and mate with males on other territories. Males that attract many visits have mates that stay at home, suggesting that they are high-quality partners whose mates would gain little by extra-pair copulations elsewhere. In contrast, males whose mates often leave to visit other territories are rarely visited by other females cruising for mates, suggesting that these males are considered inferior by females. *Source: Kempenaers et al. [633].*

Yet another kind of genetic benefit of polyandry falls under the *genetic compatibility hypothesis* [148, 1327, 1328]. According to this view, the genetic value of a sperm depends on the genotype of an egg, thanks to a host of factors that can affect the compatibility of the two genotypes, and thus the viability of the resulting embryo. One important factor that might contribute to the survival of an offspring is its degree of heterozygosity, especially at the MHC locus (see Chapter 4). Animals heterozygous for MHC genes may have immune responses superior to those of homozygotes, which means that females able to acquire sperm with MHC genes different from those in their eggs could produce offspring with improved disease resistance. As Jerram Brown has pointed out, female mate choice that promotes heterozygosity in offspring does not necessarily involve uniform selection for certain "good genes" that universally promote viability, but rather involves the acquisition of "different genes" that complement those possessed by the female [148].

One way for a female to acquire a variety of genes to complement the particular genotypes of her eggs is to mate with several partners. Jeanne Zeh's research on the harlequin beetle–riding pseudoscorpion, which we met in Chapter 11, shows that females that mated with several males had lower rates of embryo failure (and more surviving offspring) than females that were experimentally paired with a single male (Figure 8) [1326]. When Zeh compared the numbers of offspring sired by the same male with different females, she found no correlation. In other words, some males were not duds while others were of high quality; rather, the effect of a male's sperm on a female's reproductive success depended on the compatibility of their particular gametes, as predicted from the genetic compatibility hypothesis.

Our focus so far on genetic gains should not obscure the possibility that females can sometimes use EPCs to secure useful resources—the *more or better material ben-*

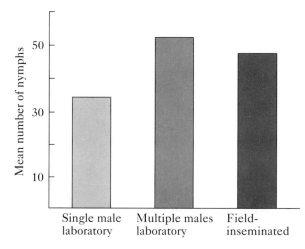

8   Female reproductive success in a polyandrous tropical pseudoscorpion. In laboratory experiments, females restricted to a single partner produced fewer nymphs than females that mated with several males. The reproductive success of females taken from the field matched that of polyandrous females in the laboratory, suggesting that females usually mate with several males under natural conditions. *Source: Zeh [1326].*

efits hypothesis. For example, female red-winged blackbirds are allowed to forage for food on the territories of the males with whom they have engaged in EPCs, whereas truly monogamous females are chased away [466]. Similarly, females of some bees must copulate with territorial males each time they enter a territory if they are to collect pollen and nectar there (Figure 9) [12]. Female hangingflies and butterflies also must mate in order to receive the nuptial gifts or valuable spermatophores provided by their partners, favoring polyandry in these insects [487, 1029]. In butterfly species whose females usually mate with a substantial number of males, the spermatophore contains much more protein than in those species whose females are monogamous or mate with few males [95]. In other words, males of polyandrous butterflies encourage females to mate with them by making it nutritionally worthwhile.

Among birds, females may solicit matings from more than one male in a breeding season to secure parental assistance from several males, as in the case of the dunnock (see below). By mating with certain of their neighbors, female red-winged blackbirds can count on these males when predators come close to their nests. Males who had had a sexual relationship with a neighboring female were much

9 Polyandry with material benefits. By mating with many males, females of this megachilid bee gain access to pollen and nectar in the males' territories. *Photograph by the author.*

more vigorous in their attacks on a stuffed magpie in the territory of this extra-pair partner than when the magpie was placed in other nearby territories [466]. On the other hand, resident males in partnerships with females that have been unfaithful are *less* likely to provide nest defense, which can reduce the average number of fledglings produced by these females compared with those who have not fertilized some or all of their eggs via extra-pair copulations [1241].

Thus, extra-pair copulations carry costs for females, including loss of the primary partner's care for their offspring, the time spent interacting with several males, increased exposure to predators, and a heightened risk of venereal disease or parasitic infection. On the other hand, as we have seen, any of several genetic or material benefits could conceivably outweigh these costs (see Table 1). The even greater puzzle is why males tolerate polyandry, which requires them to share, rather than monopolize, the reproductive output of a female [880]. However, because females control reproduction for their own benefit, their decisions may leave a male the choice of sharing or not mating at all.

Furthermore, female polyandry does not necessarily preclude male polygyny, as shown by red-winged blackbirds and others, including partridge-like tinamous, ground-nesting birds of South America whose eggs feed many predators. Male tinamous protect and incubate eggs without aid from their mates, who devote themselves to producing one clutch after another. A male sitting on a clutch, however, often attracts several more females that add their eggs to his nest, making him a polygynist while his partners practice polyandry by giving different clutches to different males [880].

Likewise, in giant water bugs, polyandrous females donate sequential clutches of eggs to several different paternal males, but some males can accommodate eggs from several females, making them polygynous [1127]. Neither polyandry nor polygyny poses a special evolutionary puzzle in these cases, given that polyandrous females gain male parental care from more than one male, while polygynous males can attract multiple mates.

Polyandry without Polygyny

A form of polyandry that does challenge sexual selection theorists is practiced by a few species in which several males mate with just one female in a breeding season. For example, up to eight males of the Galápagos hawk may pair-bond for years with one female, helping her rear a single youngster per breeding episode [374]. In this and another odd bird, the New Zealand pukeko or purple swamphen, all the males in a female's "harem" appear to have the same chance of fertilizing her eggs [600]. Thus, female mechanisms that regulate sperm competition may make it advantageous for males to cooperate by banding together to keep other males out of suitable breeding territories, which appear to be extremely scarce. Territory holders engage in an equal opportunity fertilization lottery that determines which male's sperm actually is incorporated into the resident female's egg.

In some other polyandrous species, a more or less complete sex role reversal takes place, with males assuming all or most parental responsibilities and females actually competing for access to mates. For example, female spotted sandpipers, the subject of Lewis Oring's long-term research project, behave like males in many ways [881, 884]. In addition to taking the lead in courtship, females are larger and more combative than males, and they arrive on the breeding grounds first, whereas in most birds males precede females. Once on the breeding grounds, females compete for territories, which they defend against others of their sex (Figure 10). A female's territory may attract first one and then later another male. The first male mates with the female and gets a clutch of eggs to incubate and rear on his own in the female's territory, which she continues to defend while producing a new clutch for a second mate. As a result, a few females achieve higher reproductive success than the most successful males [882], an atypical result for animals generally (see Chapter 12).

Oring believes that the understanding of *sex role reversal polyandry* in spotted sandpipers is advanced by recognizing a key historical constraint affecting this species. In all sandpiper species, females never lay more than four eggs in a clutch, and cannot adjust clutch size even if resources are abundant. Therefore they are "locked into" an inflexible four-egg clutch, a trait inherited from an ancestral sand-

10   Resource defense polyandry. Female spotted sandpipers fight for possession of territories that may attract several males to them. *Photograph by Steven Maxson, courtesy of Lewis Oring.*

piper. They can capitalize on rich food resources only by laying more than one clutch, not by increasing the number of eggs laid in any one batch [881]. To do this, however, they must acquire more than one mate to care for their sequential clutches, making this a rare case in which female fitness is limited more by access to mates than by production of gametes.

Given the fixed four-egg clutch of spotted sandpipers, polyandry may be adaptive in this species because of the confluence of several unusual ecological features [698]. First, the adult sex ratio is slightly biased toward males. Second, spotted sandpipers nest in areas with immense mayfly hatches, which provide superabundant food for females and for the young when they hatch. Third, a single parent can care for a clutch about as well as two parents, in part because the young are *precocial:* able to move about, feed themselves, and thermoregulate shortly after hatching. A broad comparative study of birds in general has shown that females are more likely to evolve short pair bonds with males and seek a series of partners when their young are precocial [1174]. This combination of excess males, abundant food, and precocial young means that female spotted sandpipers that desert their initial partners will find new ones without harming the survival chances of their first brood. Once he is deserted, a male's options are limited. Were he also to leave the eggs, they would fail to develop, and he would have to start all over again. If all females are deserters, then a male that performs the role of sole parent presumably experiences greater reproductive success than he would otherwise, even if his partner acquires another mate to assist her with a second clutch. Furthermore, the *first* partner of a female spotted sandpiper may provide her with sperm that she stores and uses much later to fertilize some or all of her second clutch of eggs (see Chapter 12). Thus, older males that arrive on the breeding grounds relatively early and pair off quickly may not lose as much from their mate's acquisition of a second copulatory partner as was once believed [883].

Female spotted sandpipers, by deserting their mates, can invest in the defense of a resource, good nesting habitat, that attracts several males to them. A similar case of polyandry is exhibited by the northern jacana, a tropical water bird. As in the spotted sandpiper, sex role reversal for this bird involves large, aggressive females that compete for possession of territories big enough to attract several males, each of which cares for one clutch of eggs by himself [609]. Females of this species may even practice ovicide, destroying the eggs of males on neighboring territories to free those males to move into their territory and accept a clutch from them [353, 1145] (see Chapter 1). Once again, given that females produce several clutches of eggs, but offer no parental care, males that exercise the sole parent option are probably doing the best they can, reproductively speaking, under these special conditions.

Thus, several different kinds of polyandry exist, with females acquiring the sperm of different males, or the nuptial presents of several males, or the parental care of several males. Even within a given category, polyandrous females can differ in the way in which they seek reproductive success. For example, females of

11 Mate defense poly-andry. A pair of red-necked phalaropes, a species in which some females guard two mates in sequence. Female phalaropes are as brightly plumaged as their mates, if not more so. *Photograph by John D. Reynolds.*

the red-necked phalarope (Figure 11) directly defend mates rather than employing the strategy of spotted sandpipers and jacanas, which is to defend a territory that attracts males [986, 987]. As individual male phalaropes arrive on the breeding grounds, females pursue and court them, jabbing rival females with their sharp beaks to keep them at bay. Once paired, a female lays her clutch in her mate's nest, and then leaves him to it. She may then find and guard another male, to whom she donates a second clutch of eggs that he alone incubates and defends.

Polygyny

Just as there is diversity in the ways in which females may achieve polyandry, males can become polygynous by various means. Consider the male mating tactics of bighorn sheep, black-winged damselflies, and satin bowerbirds (see Chapter 12). Even though males of all these species defend mating territories, what they are defending differs, providing the basis for a classification of polygyny. Bighorn rams (like female red-necked phalaropes) go where potential mates are, fighting with other males to monopolize females directly (*female defense polygyny*). Black-winged damselflies (like female spotted sandpipers) wait for mates to come to them at their resource-rich territories; female damselflies mate with them and then lay their eggs in the aquatic vegetation that the male damselfly controls (*resource defense polygyny*). Male satin bowerbirds, on the other hand, protect territories containing only a display bower, not resources that females might use (*lek polygyny*).

How can we account for the variety of mating tactics in polygynous species? Stephen Emlen and Lewis Oring have argued that the extent to which a male can monopolize mating opportunities depends on social and ecological factors that affect the distribution of females [354]. The degree to which receptive females are clumped in their environment varies as a function of such things as predation pressure and food dispersion. Females that live together or aggregate at resource patches can be monopolized economically by males, whereas scattered females cannot, because as the size of a territory grows, the costs of defending the site also increase (see Chapter 10).

Female Defense Polygyny

The Emlen–Oring theory generates a key prediction: when receptive females occur in defensible clusters, males will compete for those clusters. Qualitative support for this theory comes from the observation of many unrelated species whose females live in permanent groups defended by resident males. For example, female bighorn sheep and gorillas travel in groups, in part for protection against predators, whereas females of some lizards come together temporarily at rich foraging sites [542, 796]; males of all these vertebrates fight to live with these preexisting groups of females, becoming polygynous if they succeed in keeping other males away [506, 550].

Likewise, females of a tropical bat form groups that forage together at night (as shown by radio-tracking studies), always returning to the same spot in their home cave to roost during the day. The existence of these roosting groups is tailor-made for female defense polygyny because one male need not patrol far when guarding the cluster (Figure 12). Gary McCracken and Jack Bradbury have shown through electrophoretic analysis that successful territory holders father 60 to 90 percent of the offspring of the females in their roost, and that one male may sire as many as 50 pups during his territorial tenure [790]. Similarly, males of the Montezuma oropendola, a tropical blackbird, compete for access to clusters of nesting females, which group their long, dangling nests in certain trees; the dominant male at these

12 Female defense polygyny in a bat. The male at the left guards a roosting cluster of females. *Photograph by Gary McCracken.*

sites secures nearly 80 percent of all matings by excluding subordinates from the area [1244]. These dominant males shift from nest tree to nest tree, following females rather than defending a nesting resource, demonstrating that they deserve to be called female defense polygynists.

Equally spectacular reproductive achievements come from female defense by males in some ants of the genus *Cardiocondyla*. In these species, certain males remain inside their natal nest after becoming adults. These stay-at-home males use their saber-shaped mandibles to pierce the bodies of other males just as they metamorphose into new adults. They may also induce the workers in the nest to kill their rivals by secreting a chemical on other males that marks them for destruction by their female nestmates. The few males that survive this murderous intrigue within their colony have many freshly emerged females all to themselves, with one male capable of inseminating at least 38 queens [535].

In still other species, males may actually create and control clusters of females by herding them together. This strategy is used by certain tiny siphonoecetine amphipods that live in great numbers in the fine gravel of shallow marine bays. They typically construct elaborate cases composed of bits of pebble, fragments of mollusk shells, and the like. Males move about in their "houses" searching for females in their abodes. They "capture" single females and glue their houses to their own. A male may succeed in constructing an apartment complex containing his case cemented to the abodes of up to three females (Figure 13) [621]. He probably monitors and monopolizes the sexual activity of his apartment-mates. The point is that in all these cases, and many others, females occur in tight clusters for various reasons, enabling some males to control access to many mates.

13 **Female defense polygyny in a marine amphipod.**
(A) One individual is shown without his house. (B) A male has glued the houses of two females to his case. *Drawings courtesy of Jean Just.*

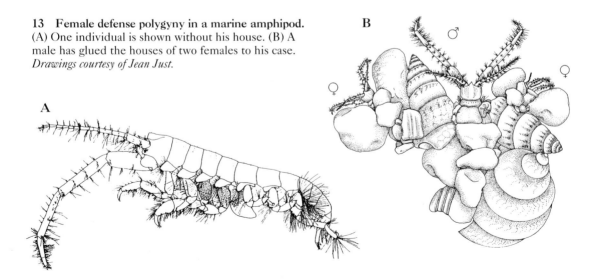

Female Defense Polygyny: The Female Perspective

Participation in polygynous mating systems has reproductive consequences for females as well as for males. In yellow-bellied marmots, females often remain near their birth site, as is typical of mammals (see Chapter 10), so that clusters of mothers, daughters, and aunts arise, presumably for improved protection against coyotes and eagles [28]. Males search for locations with one or more resident females, and those males able to find and defend a meadow that is home to two or more females practice female defense polygyny. As one might expect, the reproductive success of a male increases as the number of females living in his territory increases.

But what about female reproductive success? In one study, females living in the smallest groups had the greatest number of offspring per year (Figure 14) [309]. This result suggests that females would experience greater fitness if they were monogamous rather than being part of a female defense polygyny system.

One hypothesis to account for the failure of females to move from large to small groups is that their *lifetime* fitness is greater when they live with many others. If large groups thwart predators better, then females in large groups will live longer and reproduce during more years than monogamous females, compensating for their reduced output in any one year [344]. However, contrary to this prediction, one observer found no positive correlation between the size of a group of females and female survivorship [28].

A second hypothesis reminds us that what is adaptive for an individual depends on what its other options are [1303]. Perhaps female marmots that lived alone with a mate would indeed enjoy higher lifetime reproductive success than if they lived in a group. But in order to do this, they might have to drive off all other females, and this might be costly for a variety of reasons, particularly if arriving females were highly motivated to settle with them because of a scarcity of unoccupied habitat.

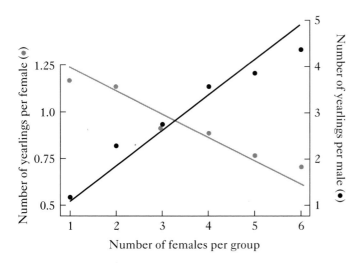

14 Polygyny versus monogamy. Annual reproductive success of male and female marmots. Polygynous males have more offspring that survive their first year than monogamous males, but females with polygynous partners have fewer surviving offspring than monogamous females. *Source: Armitage [28].*

The idea that social interactions within a species have much to say about the evolution of that species' mating system can also be illustrated with lions. You will recall that lion females form groups whose members cooperatively hunt big game, but that debate exists on whether the caloric gains from cooperative hunting are sufficient to explain the formation of prides (see Chapter 10). Female lions are long-lived animals that compete for permanent territories where they hunt and breed. Those without a territory have no chance to reproduce successfully. Therefore, lionesses may remain together in order to acquire territories and defend them against rival prides. If this hypothesis is correct, then clashes between prides over territories should be resolved in favor of the larger group. Craig Packer's team has observed 15 aggressive interactions between prides of known size, of which 13 were won by the larger pride [897].

Interactions between the sexes may also contribute to the maintenance of prides, even at the cost of short-term food shortages. By sticking together, females have some chance of protecting their cubs from infanticidal males intent on a takeover of the pride (see Chapter 1). When cub-killing males encounter a single female parent, her cubs have almost no chance of survival, whereas groups of two or more females occasionally succeed in preventing males from destroying all of their cubs. Thus, the benefits to lionesses from social living may have more to do with coping with members of their own species than with maximizing food intake [897]. The various competitive and destructive interactions among lions may produce coalitions of females, which then benefit from association with one or more polygynous males that help protect their cubs from infanticidal outsiders.

Resource Defense Polygyny

Females of many species do not live in tight clusters, but a male still may be able to defend a territory that makes him polygynous if the resources females need are clumped spatially, permitting economical defense of a resource-based territory. Male black-winged damselflies can engage in resource defense polygyny because a small patch of floating vegetation will attract many sexually receptive females and will accommodate all their eggs, which they insert one by one inside floating leaves.

A safe location for eggs constitutes a defensible resource for a host of animal species; the more of this resource a territorial male holds, the more likely he is to acquire more than one mate. For example, in an African cichlid fish, *Lamprologus callipterus*, a female deposits a clutch of eggs in an empty snail shell, then pops inside to remain with her eggs and hatchlings until they are ready to leave the nest. Territorial males of this species, which are much larger than their tiny mates, not only defend suitable nest sites, but collect shells from the lake bottom and steal them from the nests of rival males, creating middens of up to 86 shells (Figure 15). Because up to 14 females may nest simultaneously in different shells in one male's midden, the owner of a rich territory can enjoy extraordinary reproductive success [1049].

15 Resource defense polygyny in an African cichlid fish. Territorial males collect and defend small snail shells; many females are attracted to territories with many shells, where they lay their eggs and brood them until the young become independent. *Source: Sato [1049].*

In some species, the foods that females need are distributed in discrete patches. By defending valuable food centers, a male may get to mate with a number of foraging females. For example, males of the topi, an African antelope, compete in some areas for patches of grassland plain where the grass leaves are unusually green, presumably an indicator of high-quality grazing. The greener the territory, the more females are likely to visit it [52], and thus, presumably, the more copulations and egg fertilizations the territory owner can secure.

Resource Defense Polygyny: The Female Perspective

Earlier we presented the argument that monogamy in birds might result from female attempts to monopolize the parental care that males can offer their mates. In the light of this hypothesis, it is paradoxical that females of some bird species enter into polygynous relationships with males that provide parental care to their offspring. A female that pairs with a male that already has a partner is forced to share his paternal care services with his first mate. How can we account for this behavior?

A classic hypothesis, called the *polygyny threshold model*, identifies the conditions under which females can gain more by mating with an already mated male than with a current bachelor [880]. This model applies to species in which the territory of a male contains useful resources for the female and her progeny. If the quantity or quality of resources in a male's territory varies greatly from male to male, it is possible that a female that joins an already paired male on a very rich territory may reproduce better than if she were to pair with a single male on a resource-poor territory. Although a female with a polygynous partner must share his resources and paternal care with one or more other females, her share could still be greater than that of a monogamous female on a poor territory. When resources are more or less evenly distributed, however, females are expected to favor monogamous males, as we discussed earlier in the case of the great tit.

Michael Carey and Val Nolan tested the polygyny threshold model in a study of the indigo bunting, a small songbird whose parental males may attract either 0, l, or 2 mates to their territories in overgrown fields [177, 178]. Monogamous females whose relationship with a male lasted the whole breeding season had only slightly greater reproductive success (1.6 young fledged) than females that participated in a polygynous arrangement (1.3 young fledged). Because the second female in a polygynous relationship usually arrives on the breeding grounds after most males with acceptable territories have been claimed, her "expected" reproductive success should she pick an unmated male is probably very low. By joining a mated male on a "superterritory," she probably does better.

But when the same prediction was tested by Swedish observers of the pied flycatcher, they found that females in polygynous relationships had only about two-thirds the reproductive success of monogamous females on average (Figure 16)

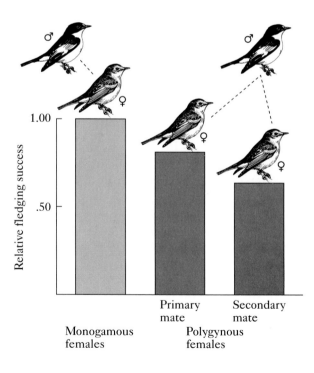

16 Does polygyny increase female fitness? Female pied flycatchers that mate with monogamous males enjoy higher reproductive success than either primary or secondary mates of polygynous males. This result suggests that females lose by mating with deceptive polygynists. *Source: Alatalo and Lundberg [7].*

[7; see also 672 for another similar case]. In this species, monogamous females receive all the parental care of the father of their offspring; in contrast, the second mate of a polygynous male gets little or no assistance from her partner. So why do females choose to mate with polygynist flycatchers?

Rauno Alatalo and his co-workers have suggested that they do not choose to do so, but are deceived by the males, who manage to maintain two separate territories, each containing a tree with a nest hole (a critical resource for female flycatchers). According to this deception hypothesis, a polygynous male can attract two mates that do not know of each other's existence because he has two territories, not one. Each female presumably lacks the information that would permit her to avoid a bad decision, and so she settles down with her mate on one of his territories, only to receive relatively little help feeding her young from her "polyterritorial" male. Even though each of his mates' average reproductive success may be lower than that of monogamous females, the *sum* of their output confers an advantage on the polygynist male at the expense of both mates [8, 5].

As noted, this deception hypothesis predicts that females should not be able to determine the pairing status of a potential mate, but Svein Dale and Tore Slagsvold have shown that mated male flycatchers spend less time singing than unmated individuals, thus providing an important potential cue for perceptive females [254]. Moreover, females looking for a mate may visit as many as seven males in a single hour [253], returning repeatedly to visit those they consider good prospects. A male that has a mate elsewhere cannot stay at his secondary territory for prolonged periods because he has to keep tabs on his primary territory as well. As a result, he will be absent during some checks by prospective mates, giving these females a chance to determine his mating status [254]. Thus, females may not be deceived when they accept an already mated male [979; 1173]. Perhaps they choose him anyway because of the high costs of finding or evaluating other potential partners [1143] or because the remaining unmated males have extremely poor territories [1172], alternative hypotheses that deserve additional testing in this and other potential cases of sexual deception.

Scramble Competition Polygyny

Although female defense and resource defense tactics by competitive males make intuitive sense when females or resources are clumped in small, defensible areas, in many other species receptive females are widely dispersed. Under these conditions, the cost–benefit ratio of mating territoriality falls, and males compete by trying to outrace rivals to receptive females; repeat winners practice *scramble competition polygyny*.

The scarce receptive females of a flightless *Photinus* firefly, for example, can appear almost anywhere over wide swaths of Florida woodland. Males of these species make no effort to be territorial. Instead they search, and search, and search some more. When James Lloyd tracked flashing males, he walked 10.9 miles in total following 199 signaling males and saw exactly two matings. Whenever he spotted a female, she signaled for only 6 minutes on average before a male found

17 Scramble competition polygyny in the nonterritorial thirteen-lined ground squirrel. Males become much more active travelers during the 2-week mating period when receptive females can be encountered here and there. *Source: Schwagmeyer [1064].*

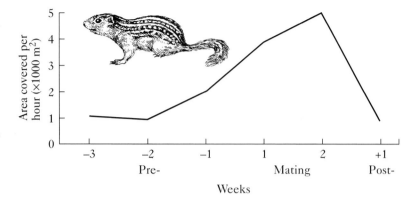

her [728]. Mating success in this species probably goes to the searchers that are the most persistent, durable, and perceptive, not the most aggressive.

The thirteen-lined ground squirrel is a vertebrate equivalent of the searching firefly; males search widely for estrous females during a 1- to 3-week mating season (Figure 17). Within this season, a female becomes receptive for a mere 4 to 5 hours. The first male to find an estrous female copulates with her. Even if she mates again, the first male will fertilize about 75 percent of her eggs, as David Foltz and Patricia Schwagmeyer [393] have shown through genetic paternity analysis of litters produced by polyandrous females. Given the first-male mating advantage and the widely scattered distribution of females, the ability to keep hunting has much to say about a male's reproductive success.

It is not just mobility that affects male fitness in this species [1064], however, but also a special kind of "intelligence." Although sexual selection rarely leads to advances in male intellect, a male ground squirrel's fitness is affected by his ability to remember where he found a female about to enter into estrus. After visiting a number of females near their scattered burrows, searching males often return to these places on the following day. When researchers experimentally removed some females from an area, the males, upon their return, spent more time searching for the missing females that were on the verge of estrus than for those that were not. Males do not simply inspect places that females use heavily, such as their burrows, but instead bias their search in favor of the spot where they actually interacted with an about-to-become-receptive female [1065, 1066]. Males with superior spatial memory can probably canvass their neighborhood more quickly and efficiently as they race to be first in line to mate with a female as soon as she can be inseminated (Figure 18).

A quite different form of scramble competition polygyny, the *explosive breeding assemblage,* occurs in species with still more highly compressed breeding seasons. For example, female horseshoe crabs come ashore to certain nesting beaches to lay their eggs on just a few nights each spring and summer when tide conditions are favorable for successful nesting. On these days, males gather off the nesting

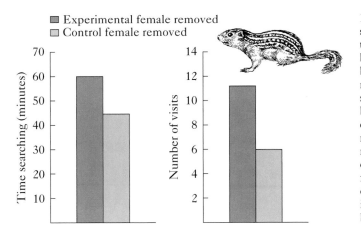

■ Experimental female removed
□ Control female removed

18 Scramble competition polygyny selects for male learning ability. Male thirteen-lined ground squirrels remember the locations of females about to become sexually receptive. When they returned to an area that had contained such a female the previous day (dark bars), but from which she had been experimentally removed, they spent more time searching for her, and returned to her home range more often, than in the case of removed females that had *not* been about to enter into estrus when the males visited them the previous day (lighter bars). *Source: Schwagmeyer [1066].*

beaches in large numbers and search for incoming females (Figure 19). When a male finds an unaccompanied female, he grasps her with special claws and goes with her onshore to fertilize her eggs as she lays them. A paired male has an advantage in the race to fertilize eggs because his grasping apparatus makes him very difficult to dislodge, allowing him to release his sperm in the optimal position for egg fertilization [126].

The mating tactics of male wood frogs likewise reflect their very brief opportunity for acquiring partners, since all the females in some populations are receptive for just one night each year. Naturally this creates intense selection on males favoring those that assemble at a pond on the mating night. Just as in horseshoe

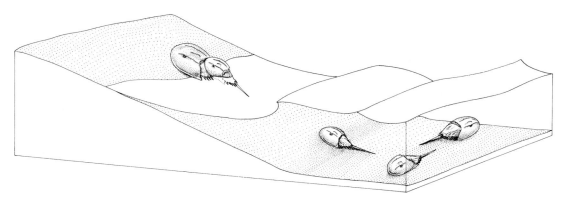

19 Scramble competition polygyny in the horseshoe crab. Males of this marine animal try to find a receptive female on her way to the beach, the better to be in position to fertilize her eggs as they are laid. *After Brockmann [126].*

crabs, the high density of rivals raises the cost of repelling opponents from a defended area. And because females are available only on this one night, a few highly aggressive, territorial males cannot monopolize a disproportionate number of females. Therefore, male wood frogs eschew territorial behavior and instead hurry about trying to encounter as many highly fecund females as possible before the one-night orgy ends [124].

Lek Polygyny

Still another mating system exhibited by polygynous males is lek polygyny, which features defense of territories too small to offer useful resources to visiting females [116]. Females come to these little territories, which are often clustered in a traditional display area, or *lek*, just to select a mate; after copulating, the female departs and may never see her partner again (Color Plate 16).

20 Lek polygyny in the white-bearded manakin. Males defend tiny display sites; females (shown here on the far left) visit the lek to select a mate from among the several males displaying there.

For example, males of the South American white-bearded manakin [717, 718], a sparrow-sized bird, each defend a single sapling or two in the forest and a bare patch of ground underneath, which the bird clears of leaves and debris to make a display court (Figure 20). The site contains nothing of value to a female, except a potential mate. As many as 70 display courts pack areas that may be only 150 meters square. The male begins his display routine with a series of rapid jumps between perches, loudly snapping his clublike wing feathers together in flight. He then pauses with body tensed before jumping to the ground with a snap and immediately back to the perch with a buzz, and then back and forth "so fast he seems to be bouncing and exploding like a firecracker" [1132]. The arrival of a female at the lek encourages many males to display simultaneously, producing an uproar. If the female is receptive and chooses a partner, she will fly to his perch for a series of mutual displays, followed by copulation. Afterward, she leaves to begin nesting, and the male remains behind to court newcomers.

Alan Lill found that in a lek with ten manakins, one male chalked up nearly 75 percent of the 438 copulations that he observed; a second male mated 56 times (13 percent), while six other males together accounted for a piddling 10 matings [717]. You will recall similar inequalities in mating success among males of the satin bowerbird [109].

Lekking is by no means restricted to birds. During the breeding season, males of the bizarre West African hammer-headed bat (Figure 21) gather in the evening along river edges at traditional display areas [115]. Each bat defends a territory 10 meters in diameter while he hangs from a perch high in a tree. The males produce loud cries that sound like "a glass being rapped hard on a porcelain sink." Receptive females fly to the lek and visit several males, each of which responds with a paroxysm of wing-flapping displays and strange vocalizations (note the convergence with manakins and satin bowerbirds). Most males are doomed to disappointment; 6 percent of the males at one lek in 1974 were responsible for nearly 80 percent of all matings.

Why Do Males Aggregate in Leks?

Why do male manakins and hammer-headed bats behave the way they do? Jack Bradbury has argued that lekking evolves only when other systems do not pay off for males, thanks to a wide and even dispersion of females. Female manakins and hammer-headed bats do not live in permanent groups, but instead travel great distances in search of widely scattered sources of food, especially figs and other tropical fruits, which are available at long and unpredictable intervals. A male that tried to defend one tree might have a long wait before it began to bear attractive fruit, and when it did, the fruit would attract hordes of consumers, which would overwhelm the territorial capacity of a single defender. Thus, the feeding ecology of females of these species makes it hard for males to monopolize them directly or indirectly. Instead, males display their merits to choosy females that come to leks to inspect them.

21   A lek-polygynous mammal: the
hammer-headed bat. *Photograph by
A. R. Devez.*

But why in most lekking species do many males congregate in small areas instead
of displaying on their own isolated territories? Of the many hypotheses proposed for
male clustering on mini-territories, we shall review three: (1) the *hotspot hypothesis*,
in which males cluster because females tend to travel along certain routes that inter-
sect at particular points, or "hotspots" [117, 962], (2) the *hotshot hypothesis*, in which
males cluster because subordinate males gather around highly attractive males in
order to have a chance to be seen by females drawn to these "hotshots," or to intercept those females [448], and (3) the *female preference hypothesis*, in which males
cluster because females prefer sites with large groups of males, where they can more
quickly, or more safely compare the quality of many potential mates [551].

One way of testing the hotspot versus hotshot hypotheses is by temporarily
removing males that have been successful in attracting females. If the hotspot
hypothesis is correct, removal of males from attractive positions will enable others to move into the favored sites. But if the hotshot hypothesis is correct, removal

of attractive males will cause the cluster of subordinates to disperse to other popular males or to leave the site altogether.

In a study of the great snipe, a European sandpiper, removal of central dominant males caused their neighboring subordinates to leave their territories. In contrast, removal of a subordinate while the central dominant snipe was in place resulted in quick replacement of the vacant territory by another subordinate. At least in this species, the position of the males most attractive to females determines where clusters of males gather, not the location per se [552, 554]. Likewise, in the unrelated black grouse, the location of the most popular territory changes from year to year within a long-lasting lek, suggesting that a popular male, rather than a particular location, most influences female behavior (Figure 22). In addition, males that occupy sites next to popular males gain slightly improved mating success, as expected from the hotshot hypothesis [998].

Comparing the hotshot and hotspot hypotheses from the female perspective, we can predict that if the hotshot hypothesis is correct, females should be attracted primarily by particular lekking males rather than the location of their display sites, as in the black grouse. In fallow deer, dominant, attractive bucks can be induced to abandon their central territories by placing sheets of black polyethylene on these sites. After moving to a new site, these hotshots still continue to attract a disproportionately large number of females, despite their shift in position [208].

Although the hotshot hypothesis seems likely in some cases, the hotspot hypothesis has received support in others. If location has much to do with lek

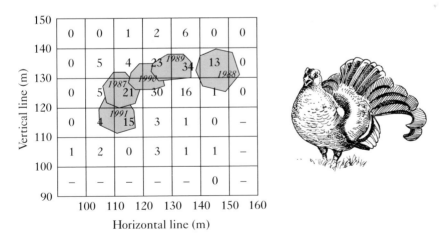

22   Hotspot or hotshot? The total number of copulations recorded in 100-square-meter sectors of a black grouse lek from 1987 to 1991. The irregular polygons show the location of the top territory for each of the five years. The shifts in preferred territory suggest that male attractiveness plays a key role in lek polygyny in this species, as required by the hotshot hypothesis. *Source: Rintamäki et al. [998].*

formation, then males of unrelated species would be expected to converge on some of the same lekking sites. In fact, lekking bees, flies, wasps, butterflies, and other insects often do occupy many of the same territories along mountain ridgelines or tropical forest streams, suggesting that these features serve as "highways" for dispersing females, thereby providing hotspots where males can wait for mates to come to them [9, 1091]. Likewise, David Westcott found leks of flycatchers, manakins, and hummingbirds in the same small areas, often at the confluences of small forest streams or where ridgelines end [1259]. This overlap in lek distributions among unrelated animals indicates that topography determines where females go, leading to aggregations of males at key intersections along travel routes.

On the other hand, in various lekking antelope and deer, males do not form leks in the places where females are most concentrated spatially, as one would predict if leks formed where females tended to travel (Figure 23) [51]. Instead, estrous females of several lekking ungulates go out of their way to visit these spots, perhaps to compare the performance of many males simultaneously (the female preference hypothesis) or to avoid interference from intrusive males that disrupt matings [855].

James Deutsch attempted a test of the female preference hypothesis in his study of the Uganda kob, an antelope whose males gather in groups to defend their display territories and court females that come to look at them. If female preferences push males into large groups, then those leks with a relatively large number

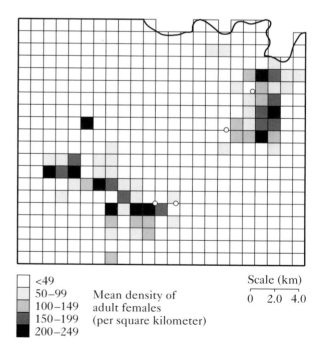

23 Female density is not correlated with lek formation in an antelope, the Kafue lechwe. The four leks in this region (open circles) are *not* located in the areas of highest female density, evidence against the hotspot hypothesis in this species. *Source: Balmford et al. [51].*

☐	<49
	50–99
	100–149
	150–199
■	200–249

Mean density of adult females (per square kilometer)

Scale (km)
0   2.0   4.0

of males should attract proportionately more females than leks with very few males. However, the operational sex ratio was the same for leks across a spectrum of sizes (Figure 24), so that males were no better off in large groups than in small ones [290]. For this species at least, the female preference hypothesis can be rejected.

For other lekking animals, such as the ruff (a sandpiper), hotshots, hotspots, and female preferences may all contribute to lek formation and male aggregation. Leks in this species are not randomly distributed, but instead are typically located near small ponds on slightly elevated ground, perhaps because females tend to visit ponds to forage and can see males better if they are on higher ground (a hotspot). Once at a suitable location, males may gain by joining others because of a female preference for groups of at least five displayers. But the benefits of joining are not equally shared by all males present, especially for groups in excess of six individuals. Larger aggregations arise because low-ranking males choose to settle near dominant resident hotshots; the mating chances of these individuals increase in groups larger than five (Figure 25) [1275, but see 553].

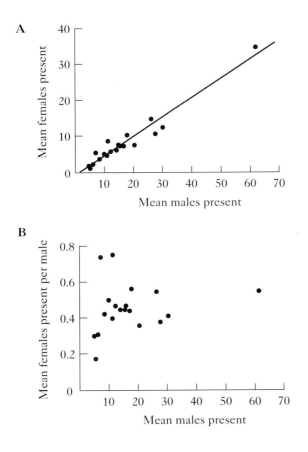

A

B

24 There is no female preference for comparing large numbers of males in the Uganda kob, an African antelope. (A) Female attendance at leks is simply proportional to the number of males displaying there. (B) Leks with many males do not attract a disproportionate number of females; therefore, the female-to-male ratio does not increase as lek size increases. Some other factor has driven lek evolution in this species. *Source: Deutsch [290].*

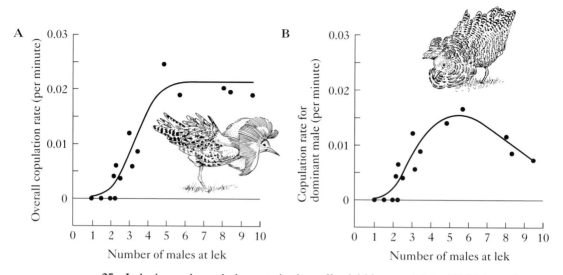

25 Lek size and copulation rate in the ruff, a lekking sandpiper. (A) Males at larger leks attract more receptive females than males at small leks. (B) At leks of six or more males, however, the number of attracted females remains the same, lowering the average reproductive success of the dominant males in attendance because some lower-ranking individuals copulate with some of the visiting females. *Source: Widemo and Owens [1275].*

Leks and Extreme Female Choice

Whatever the reason(s) for male aggregation at leks, the typical rule is that only 10 to 20 percent of the males at a given lek secure more than half the matings. After one has accounted for the effects of random processes and differences in male attendance at leks, typically 60 to 70 percent of this variance in male reproductive success remains unexplained [749]. Presumably the exceptional success of a few males arises in large measure because most females choose the same individuals. What do females gain by selecting the same males at their leks when these males appear to offer females nothing but their genes?

In the lekking black grouse, females follow the typical pattern, conferring great reproductive success on a handful of males while ignoring the rest. The most attractive males have survival rates that are twice that of the least attractive individuals, suggesting that females gain "good genes" from their choices [4], as is also possible in peacocks (see Chapter 12). However, in order to derive this genetic benefit, females must choose among males that differ in their genetic quality. Theoretical biologists have argued that any differences among males would quickly disappear in populations in which only a few males produced most of the offspring each generation, and that therefore genetic variation underlying male quality should disappear quickly in lekking species. The persistence of strong female preferences for a few males creates what is called the *lek paradox*, because it is paradoxical that females would continue to exhibit great mate selectivity in the absence of significant genetic differences among potential mates.

Ways to resolve this paradox have been suggested on several fronts. First, despite theoretical arguments to the contrary, direct measurements of genetic variation have demonstrated that a high proportion of the variation in male sexual characters in at least some lekking species does stem from genetic variation. In fact, male sexual traits in these species usually have higher genetic variation associated with them than do the corresponding traits in females. Thus, it is an empirical fact that genetic variation can indeed persist in the face of strong sexual selection by choosy females, making it possible for females to gain better genes through careful choice of their mates [951].

Moreover, John Reynolds and Mart Gross have pointed to several nongenetic gains that females might secure by choosing males carefully from a lek [989]. First, the search time costs of mate choice are reduced in systems in which groups of males conspicuously advertise their presence, facilitating speedy location of potential mates whose relative status can be quickly and accurately assessed. Furthermore, by picking dominant males as copulatory partners, females may avoid the injuries and harassment that might come from dealing with individuals that had not fully intimidated their rivals. Among fallow deer, for example, females that are being harassed move to the territories of top-ranking males, which effectively repel sexually aggressive rivals [208]. The hotshots on a lek might also be in top physical condition—a necessary requirement if a male is to dominate many other rivals—and therefore relatively free from contagious diseases.

The Mating Systems of the Dunnock

Let us now apply what we have learned about the ecology of mating systems to the dunnock, a very ordinary-looking European songbird whose mating tactics are unusually diverse—a phenomenon known for some other species as well [510, 832, 977]. Indeed, within a single dunnock population, some males may pair with one female, while others pair with two mates, or share the same female with another male (polygynandry), while neighboring territories may be occupied by several males and females in joint sexual liaison [268]. This bird illustrates the difficulty of trying to force each species into a simple mating system scheme.

In addition, the dunnock illustrates the shortcomings of treating each sex's mating system independently of the other's, since the multiplicity of arrangements in this species probably arises from interactions and conflicts of interest between males and females. First, males gain to the extent that they can control the reproductive activity of their female partner(s), improving the odds that their sperm will fertilize all of the eggs of a mate or mates. But in order to do so, a male must guard his mate(s) wherever they go. Food dispersion affects the size of the home range of female dunnocks, which eat tiny seeds and little invertebrates. When food density is low, the local females forage widely, which makes it costly for a male to defend a territory large enough to include even one female's hunting area. On the other hand, when a small area is rich in bugs and seeds, a female or two can find all they need in that small site, and a male can realistically defend the spot, monopolizing the sexual activity of one mate (monogamy) or two (polygyny) (Figure 26A).

If this scenario is correct, it should be possible to experimentally manipulate a male's potential for monogamy or polygyny by adding food to certain territories. To this end, Nicholas Davies and Arne Lundberg put out feeders with oats and mealworms in some female home ranges, and kept them stocked for months. Females with access to these supplemental foods decreased their foraging range by an average of 40 percent [275]. These females were, as predicted, more likely

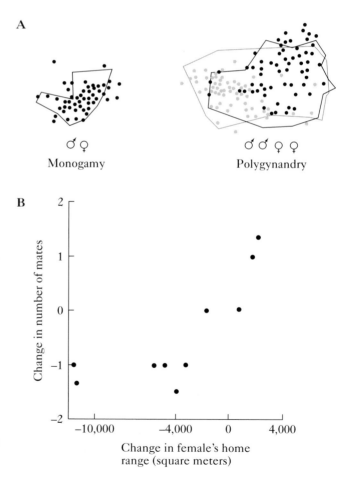

26 Multiple mating systems in the dunnock. (A) The size of unmanipulated home ranges occupied by female dunnocks varies greatly. Dots indicate sightings of banded females; lines show the areas defended by singing territorial males. On small home ranges, one male may be able to exclude others from the area, leading to monogamous or polygynous relationships. (B) Food supplements reduce the size of a female's home range greatly, in which case polyandrous females may become monogamous. However, in large home ranges where females do not receive food supplements, a single male cannot monopolize the females that live there, with the result that the females may mate with two males. *Source: Davies and Lundberg [275].*

to participate in monogamous relationships than were control females in areas without food supplements (Figure 26B).

But females are not merely passive objects of male–male competition. Their decisions are the second factor that males must contend with as they seek to maximize their egg fertilization rate. For example, females do not gain by sharing food with rival females, and they regularly attack and drive off competitors of their sex, thereby reducing the likelihood that a male will become polygynous. Females also sometimes drive off other females in order to monopolize the parental care provided by a male, who may dilute his attention to his first mate's offspring if there is a second female on his nesting territory.

Female dunnocks go further still to control reproduction on their own terms. If they live in a territory controlled by a dominant (alpha) male, they may encourage a subordinate (beta) male to remain in the area by copulating with him when the alpha male is not present to interfere with the mating. Females can recognize the songs of individual males, and they use this information to seek out multiple copulatory partners [1282]. Since female dunnocks are prepared to mate as many as 12 times per hour, and hundreds of times prior to laying a complete clutch of eggs, they make it difficult for any one male to monopolize them sexually.

The benefit to females from copulating with two males is that both will contribute parental care to their broods, provided they have copulated with the female a certain number of times. Female dunnocks act to ensure that the paternal threshold is achieved for both males by biasing their mating solicitations in favor of whichever male, alpha or beta—usually the latter—has had less time in their company (Figure 27). When both males have copulated sufficiently often, they both chip in to feed the brood that results, providing a double paternal payoff to the successfully polyandrous female [510], illustrating again that conflict between the sexes, as well as the distribution of females, affects the evolution of animal mating systems [269].

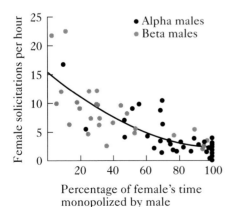

27 Female dunnocks solicit copulations more often from males that spend relatively little time monopolizing their company, whether these are the alpha or beta males in territories with two resident males. *Source: Davies et al. [273].*

SUMMARY

1. Mating systems can be defined in terms of the number of mates an individual has during a breeding season, although this simple classification scheme is at best an imperfect way to organize the complex variety of mating arrangements in nature. Sexual selection theory provides a starting point for understanding why males often attempt to be polygynous (since male fitness will usually increase with the number of mates) whereas females are more often monogamous (since female fitness is less likely to increase with multiple partners).

2. Understanding mating system diversity, however, also requires consideration of various ecological and social factors that influence male and female reproductive success in relation to monogamy, polygyny, or polyandry. For example, male monogamy may be adaptive when there are large payoffs from male parental care. Alternatively, conflict between females and males may thwart male attempts to be polygynous, even if it would be advantageous from the male perspective.

3. Likewise, there are a host of exceptions to the rule that females gain little from mating with several partners, a point made particularly clear by close examination of the sexual behavior of supposedly monogamous birds, in which both males and females engage in high levels of extra-pair copulations. Although in most birds and other animals, females can secure all the sperm they need to fertilize their eggs from a single male, female polyandry is common because of the various genetic and material benefits (including male parental care) females can gain by mating with multiple partners.

4. The role of ecological factors in the evolution of mating system diversity is highlighted by the way in which differences in the distribution of females affect the profitability of different kinds of territorial mating tactics for males. When females or the resources they need are clumped in space, female defense or resource defense polygyny evolve. If, however, females are widely dispersed or male density is high, males may engage in nonterritorial scramble competition for mates, or they may demonstrate social dominance to females through lek territoriality.

5. The dunnock illustrates the limitations of mating system classifications, while also showing that individuals can possess the behavioral plasticity to employ diverse mating tactics appropriate for different social and ecological conditions.

SUGGESTED READING

Stephen Emlen and Lewis Oring's now classic paper on the ecology of mating systems changed the analysis of mating systems [354]. Randy Thornhill and I applied their approach to insects [1181]. John Reynolds [987] argues that the Emlen–Oring approach needs major revision because females control mating and egg fertilizations to a much greater extent than Emlen and Oring assumed. Jacob Höglund and Rauno Alatalo [551] have written a useful book on leks. Nicholas Davies has summarized his studies of the dunnock in his book *Dunnock Behaviour and Social Evolution* [270].

DISCUSSION QUESTIONS

1. Construct a list of at least three hypotheses on the adaptive value to female birds of extra-pair copulations, and then make a table of predictions that follow from each hypothesis. Identify the key predictions that would enable you to discriminate among the hypotheses for a particular case. Now, what about those species in which females mate repeatedly with the *same* male, often their primary partner? Do the same hypotheses apply for this kind of multiple mating as for polyandry? Check [583] after completing your comparison.

2. Queen honeybees acquire mates by flying to certain landmarks near their hives, where they are pursued and captured by drones. A copulating honeybee drone discharges his genital apparatus (the pale structure on the right in the figure below) into his mate's abdomen, and then dies. Queens carrying these trophies are attractive to other males, and they regularly mate with several males on each of several nuptial flights. When a queen flies back to her hive, she usually returns with a male's genitalia in place. Use these observations to evaluate the following three hypotheses for

sexual suicide by monogamous drones: donation of the genitalia to a mate (1) is a mate-guarding device of males [1181], (2) helps the queen retain sperm in her reproductive tract [1309], or (3) is a form of cooperation among drones that helps them identify receptive queens in order to make sure they become fully inseminated.

3. How would mating system categories change from those presented in this chapter if mating systems were reclassified on the basis of the number of *fertilization* partners (as opposed to copulatory partners) per individual, or on the basis of the number of long-term partners per individual?

4. In the case of the burying beetle, *Nicrophorus defodiens*, discussed in this chapter, what conditions are required to make attraction of a second female costly to the first female, but beneficial to the male?

5. Apply optimal foraging theory (see Chapter 10) to the problem of female mate choice. Contrast these two options (and any others you choose to consider) for females that encounter one male after another: (1) the best-of-*n*-males tactic, in which females inspect a pool of males and then pick the top-quality individual, or (2) the threshold tactic, in which females have a minimum standard and mate whenever they encounter a male that meets that standard [601]. Under what conditions would tactic 1 be superior to tactic 2, and vice versa? How does cost–benefit analysis come into play here? How might male clustering at leks, for example, affect the payoff to females using each tactic?

CHAPTER 14

The Adaptive Tactics of Parents

ONCE AN ANIMAL HAS MATED, it faces yet another decision: namely, whether to provide parental care for any offspring that result. Most animals do not invest time and energy in efforts designed to increase the survival chances of their progeny, presumably because the costs of parenting often exceed any possible benefits. One major downside of parenting is that when an animal cares for young, it must forego some other valuable activities, such as finding food or additional mates. A male mallee fowl, for example, invests about 5 hours a day for 6 months of each year as he builds and maintains his huge "compost" nest of sand and organic material. Each time his mate lays an egg in the mound, he digs a deep pit to receive the egg, and then covers it up, shifting about 850 kilograms of compost and sand, nearly 500 times his own weight [1242]. The embryonic young developing inside the eggs benefit from their concealment from predators and the heat from the mound's decay, which the male carefully monitors and regulates. His costly parental investments presumably help the eggs he has fertilized survive to hatch, although interestingly, when the chicks appear, they are left completely to their own devices.

In this chapter, our focus will be less on establishing that parents can help their offspring than on several surprising aspects of the delivery of parental care. First, why are mothers so much more likely than fathers to take care of their offspring (Color Plate 17)? Exceptions to the rule, such as the mallee fowl, are instructive, and so we shall also examine how males sometimes gain fitness via parental care (Color Plate 18). Second, we will inspect cases in which parents care for genetic strangers, an action that seems at first glance to be thoroughly maladaptive. Third, the chapter concludes by asking why adults of some species tolerate aggression among their offspring even though some of their brood may die as a result, another intriguing Darwinian puzzle.

Why Is Parental Care More Often Provided by Females?

Anyone who has observed a nesting mallee fowl or a human family with young children knows that males have the potential to help rear the kids. However, extensive paternal care is rare indeed, and exclusive male care of the young is almost nonexistent. In most mammals, and throughout the rest of the animal kingdom, females do much more than their "fair share" of the parenting. To take just one example, female California ground squirrels with young nearby are much more likely to kick sand at rattlesnakes than are the fathers of those youngsters (see Figure 7 in Chapter 9). While male squirrels drift away, the mother squirrels continue to interact with the enemy, moderating their risk by assessing how warm and large the snake is, in part by listening to its rattles [1023], but taking risks nonetheless [1164].

Perhaps female ground squirrels have a greater interest in the welfare of their young since they have already invested so much energy in making eggs and nourishing embryos. Females of many other animals, however, abruptly terminate their parental investment after laying their eggs or giving birth to their young, indicating that past investment in an offspring does not automatically predispose females to additional parental care (but see [213, 1047]). Indeed, given the trade-offs involved in parental care, selection ought to favor added investment in an offspring *only* when its improved chances of survival (the benefit) more than compensate the parent for its loss of future reproductive opportunities (the cost).

The "Low Reliability of Paternity" Hypothesis
Given the failure of the past investment hypothesis for the prevalence of female versus male parental care, Mart Gross and Richard Shine turned their attention to three other competing hypotheses [482]. The first links the prevalence of mater-

nal care to the fact that females are more likely to be the parent of an offspring they assist than are males. If a female lays a fertilized egg or gives birth to an off-spring, this progeny will definitely have 50 percent of her genes. A male has no such assurance, especially if his partner practices internal fertilization, in which case she may have mated with another male and used his sperm to fertilize her eggs (see Chapter 13). Therefore, to the extent that a male runs the risk of caring for progeny other than his own, his potential gain from parental care falls, thus affecting the cost–benefit ratio of male parenting and making its evolution less likely than that of maternal care.

The low reliability of paternity hypothesis, although intuitively appealing, need not apply if the reliability of paternity is the same whether a male behaves pater-nally or not, as it must often be [1254]. Even if the reliability of paternity is uniformly low, a mutant paternal male can spread the hereditary basis for his behavior, pro-vided that he sufficiently improves the survival rate of those eggs he does fertilize to compensate for any resulting reduction in the number of mates he can find.

A hypothetical numerical example may help make the point. Let the proportion of eggs actually fertilized by a male when he copulates with a female be just 0.40. If the male is paternal, he will have time to find and mate with only two females (each with an average of 10 eggs), whereas if he were nonpaternal, he could acquire five mates (giving him access to 50 eggs). But if the survival of protected offspring is 50 percent versus 10 percent for eggs that lack a paternal guardian, then a paternal male will produce $0.40 \times 20 \times 0.50 = 4$ surviving offspring, whereas if he is nonpater-nal, his fitness falls to $0.40 \times 50 \times 0.10 = 2$. Multiplying both sides of the equation by the 0.40 paternity factor does not affect the relative payoffs of the two strategies.

Therefore, the evolution of paternal behavior need not be fatally inhibited by a low reliability of paternity. If, however, paternal males can improve the proba-bility of caring for their own genetic offspring by abandoning one brood in favor of trying again with another female, they can gain a great advantage over indis-criminately paternal males [1263]. In Chapter 12, we discussed one way in which some male birds may become discriminating parents, which is to make their invest-ment in brood rearing contingent upon exclusive or substantial sexual access to a mate. When a female engages in frequent extra-pair copulations in some species, she runs the risk of losing the parental assistance of her primary mate [300, 827], who may save his time and energy for another reproductive attempt with a female more likely to use his sperm to fertilize her eggs.

The "Order of Gamete Release" Hypothesis

A second hypothesis about the female bias in parental care is that females are more likely to be left holding the fertilized eggs than their male partners, enabling males not willing to invest in paternal care to "desert." By this argument, internal fertil-ization should be linked with exclusive maternal care, because after a male insem-inates a female, he can depart at once, leaving his mate to make the decision whether to provide parental care. In cases in which such care was advantageous,

Table 1   Relationship between mode of fertilization and the sex providing parental care among families of fishes and amphibians

Sex providing parental care	Families with internal fertilization	Families with external fertilization
Male	4 (14%)	75 (70%)
Female	25 (86%)	32 (30%)
Totals	29	107

Note: The data consist of the numbers of families containing one or more species in a given category; a single family may appear in more than one category.
Source: Gross and Shine [482].

the female would be "forced" to provide it, given the absence of her mate. In contrast, when fertilization is external, females often deposit their eggs before males shed their sperm, giving the females a brief moment in which to flee, leaving the males to confront the parental care dilemma.

Data on the relationship between mode of fertilization and paternal care in fishes and amphibians (Table 1) are consistent with the gamete order hypothesis. However, our confidence in this hypothesis is weakened by the results of the following test. In species in which males and females release their gametes simultaneously, the chance of the male being the parental sex should equal that of the female, because both are present when gamete release is complete. However, in a sample of 46 fish species of this sort, males were the parental sex in 36 species, significantly more than the 23 species predicted by the gamete order hypothesis. Moreover, paternal care is common in frogs, in which males release their sperm into a nest *before* the female deposits her eggs there. Finally, among some shorebirds, such as the spotted sandpiper (see Chapter 13), paternal care is also more common than maternal care, even though males donate sperm to fertilize eggs long before egg laying, giving them ample time to desert. Yet male shorebirds care for broods until they fledge in 87 percent of a sample of 96 species, whereas the corresponding figure for females is only 56 percent [1168].

Male desertion, however, may have been responsible for the evolution of exclusive maternal care in certain shorebirds in which it does occur. With respect to the sandpipers, for example, a phylogenetic analysis by Tamas Székely and John Reynolds [1168] indicates that (1) paternal care was the ancestral state, having evolved from a still more distant biparental species (Figure 1), but that (2) any number of more recent species in the genus *Calidris* have independently evolved so that females provide most or all of the brood care. Note that, for instance, the closely related semipalmated sandpiper and white-rumped sandpiper exhibit exclusive paternal care and maternal care respectively, with no current evidence for an intermediate biparental form as an ancestor. (However, the absence of such evidence

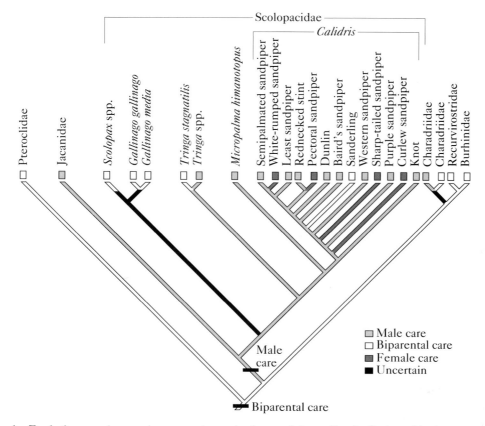

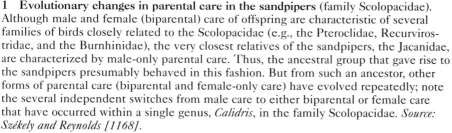

1 Evolutionary changes in parental care in the sandpipers (family Scolopacidae). Although male and female (biparental) care of offspring are characteristic of several families of birds closely related to the Scolopacidae (e.g., the Pteroclidae, Recurvirostridae, and the Burnhinidae), the very closest relatives of the sandpipers, the Jacanidae, are characterized by male-only parental care. Thus, the ancestral group that gave rise to the sandpipers presumably behaved in this fashion. But from such an ancestor, other forms of parental care (biparental and female-only care) have evolved repeatedly; note the several independent switches from male care to either biparental or female care that have occurred within a single genus, *Calidris*, in the family Scolopacidae. *Source: Székely and Reynolds [1168].*

could also stem from extinction of that biparental ancestral species or its rapid passage through the biparental phase into a modern species with one-sex care only.)

The "Association with Young" Hypothesis

Since the gamete order hypothesis cannot provide a general explanation for the observed patterns of desertion and parental care, let's consider yet another, even simpler hypothesis: because females carry the eggs, mothers are more often in a position to do something helpful for their offspring when eggs are laid or infants born [482]. In contrast, fathers usually have no direct control over the birth of their offspring, and may not even be around when their progeny appear. According to this hypothesis, even if males could provide helpful parental care, and even if it would be in their genetic interest to do so, in many cases their physical separation from their progeny makes paternal behavior impossible.

The relationship between internal fertilization and maternal care (see Table 1) is consistent with the association with young hypothesis as well as with the other two hypotheses discussed above, reinforcing the point that this result cannot be used to discriminate among the three alternatives. For a different test of the association with young hypothesis, let's consider those species of fish in which fertilization is internal, but egg laying occurs immediately after copulation. In such cases, males will still be present when their offspring are released. The association with young hypothesis predicts that males of these species should be just as likely to be paternal as males of species in which egg fertilization occurs externally. This prediction is supported.

Note, however, that the association with young hypothesis also predicts an equal number of cases of paternal and maternal care in fishes whose males and females *simultaneously* release their gametes. Since paternal care is more prevalent in this group, other forces, yet to be identified, must play a role in the evolution of parental care in these and many other species.

Exceptions to the Rule

So why do male fish provide so many exceptions to the rule that females are the sex that takes care of offspring (Figure 2)? Mart Gross and Craig Sargent have analyzed the varieties of parental care in fishes to determine what conditions beyond simple association with the young favor uniparental male care [481]. If the benefits of parental care (increased survival of the protected young) are the same for both sexes, perhaps males are paternal when the costs are lower for them than for their mates.

What are the costs of parental care for male and female fish? Male animals typically pay a mating cost for parental care: losing fitness when caring for young interferes with acquiring new mates. Indeed, as David Queller has suggested to me, this cost may be especially high for males as compared with females, since males typically have the higher potential rate of reproduction. Thus, when a male takes on the parental role, he usually gives up more than his partner does. These losses apply to male fish as well, a point nicely illustrated by male three-spined sticklebacks that are guarding their nests against foraging groups of egg-eating cannibals of their own species. When a nest guarder sees a mob coming his way, he may employ a deceptive diversionary display, moving away from his nest to dig with his snout in the lake bottom as if he were finding bottom-dwelling invertebrates [398]. The school of sticklebacks sometimes joins the "foraging" male in a

2 Paternal care in two fishes. (Left) Male sticklebacks defend the eggs laid in the nests they build. Photograph courtesy of William Rowland. (Right) Both males and females of this cichlid fish secrete a nutritious substance that their young can glean from a parent's body. *Photographs by (left) William Rowland and (right) A. van den Nieuwenhuizen.*

search for food, bypassing his nest with its vulnerable eggs or fry. The survival benefits for the deceptive male's young come at his mating expense—namely, missed opportunities to court receptive females in the foraging school.

This mating cost is real, as Susan Foster has shown in studies of male behavior in four unconnected lakes that lack cannibalistic mobs of sticklebacks. These lakes have existed only since the last glacial retreat, 22,000 years ago, and so must have been independently colonized during that time by sea-dwelling three-spined sticklebacks. Marine sticklebacks exhibit both cannibalism and diversionary displays, but in the lakes without cannibals, nest-attending males dart out to court females when feeding schools approach, rather than seeking to lead them away from their nests [399].

Most nest-defending male fish do not miss as many opportunities to mate as one might imagine, however, because their nests and the eggs within actually attract receptive females, or are located within a territory that draws mates to the paternal males [1046]. In contrast, for female fish, parenting may involve an unmit-

3   The possible relation between body size and fitness in a typical fish. If female fecundity indeed increases exponentially as body size increases, while male fitness rises only in a linear fashion, then females would pay a heavier price for lost growth than males, accounting for the frequent evolution of exclusive paternal care in fishes.

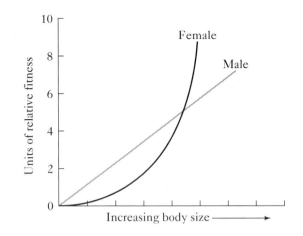

igated fertility cost in terms of lost egg production. When a female guards one brood, she cannot eat as much as she could by foraging freely, and so cannot grow as rapidly as she might otherwise. This loss in growth may be especially damaging if female fecundity increases exponentially with increasing body size, as may be typical for fishes (Figure 3). In other words, for each unit of growth lost by being parental, a female pays an especially heavy price in loss of eggs produced in the future. Males that are parental also grow more slowly than they would otherwise, but since they must be territorial to attract mates, the decrease in their growth *resulting exclusively from parental care* is slight or trivial. Thus, in fishes, paternal behavior may evolve because males pay a lower price for parenting than females do, not because they gain more from this behavior in terms of improved survival of their offspring [481].

The costs of parental behavior for males and females have been measured for a mouth-brooding cichlid, St. Peter's fish, in which either male or female may care for their young by orally incubating the fertilized eggs. In this species, both sexes lose weight when mouth brooding, since they find it difficult to eat with a mouthful of baby fish. Furthermore, the interval between spawnings increases for both parental males and females compared with individuals whose clutches are experimentally removed from their mouths (Figure 4). However, the mean nonreproductive interval is greater for brooding females (11 days) than for brooding males (7 days). Moreover, parental females produce fewer young in their next clutch than do nonparental females, whereas parental males are just as able to fertilize complete clutches of eggs in their next spawning as are nonparental males. Thus, although both sexes pay a price for parental behavior, the costs of brood care to females seem especially steep, given the immediate fall in fecundity that these individuals experience [53]—a finding that leaves unanswered the question of why some females of this species do indeed brood their young.

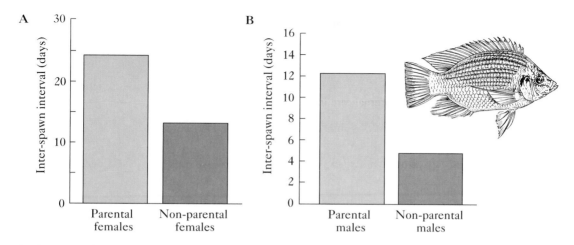

4   An actual fecundity cost of parental care in a cichlid fish. (A) The interval between spawnings for females increased greatly for those individuals that cared for a previous clutch of eggs compared with those that did not. (B) This cost applied to parental males as well, though not to the same degree (note that the scale on the vertical axis is different than that in panel A). *Source: Balshine-Earn [53].*

Why Do Male Waterbugs Do All the Parental Work?

Although some male fishes take care of their young single-handedly, exclusive male parental care is very rare among other animals, even among the insects, despite the extraordinary abundance and diversity of this group [1324]. Male giant waterbugs in the family Belostomatidae, however, do provide an example of this phenomenon. Males of some belostomatid species brood eggs that females lay out of the water on sticks and stalks of emergent aquatic vegetation (emergent brooding); males of other species take care of eggs that their mates lay on their backs (back brooding) (Figure 5). Both types of parental care involve considerable expenditures of time and energy by brooding males. In the genus *Lethocerus*, males guard their fully exposed eggs against predators and ovicidal females [588], and they regularly climb out of the water to regurgitate on the eggs, keeping them moist and alive (Color Plate 19). Back-brooding males of *Belostoma flumineum* spend much of their time keeping the eggs near the water surface, pumping their bodies to keep relatively well aerated water moving over the eggs, which they also stroke at intervals with their hind legs. In both groups, eggs that are separated from their brooding attendant fail to develop and die, demonstrating that male parental care is indispensable.

Robert Smith has explored both the history behind these unusual paternal behaviors and their adaptive significance [1129]. Since the closest relatives of the Belostomatidae, the Nepidae, are typical insects without male parental care, we can be confident that these brooding males evolved from nonbrooding ancestors

5 Parental male water-bugs brood eggs glued on their backs by their mates. Note the newly hatched nymph on the lower right. *Photograph by Robert L. Smith.*

(Figure 6). Whether emergent brooding and back brooding evolved independently from such an ancestor, or whether one preceded the other, is not known, although some evidence suggests that back brooding came later. In particular, the rare occurrence today of emergent-brooding species whose females sometimes lay their eggs on the backs of other individuals, male or female—usually when no emergent vegetation is available to them—indicates how the transition from emergent to back brooding might have occurred. Females with the tendency to lay their eggs on the backs of their mates could have reproduced in temporary ponds and pools where emergent vegetation was scarce or absent.

But why do the eggs of waterbugs require brooding? Huge numbers of aquatic insects lay eggs that do perfectly well without a parent in attendance. However, Smith notes that belostomatid eggs are much larger than the standard aquatic insect egg. A large sphere has a smaller surface area relative to its volume than a small sphere. Thus, as a large belostomatid egg develops, it has a special problem—how to move carbon dioxide out and oxygen in with sufficient rapidity to sustain the high metabolic rate required for embryonic development. Since oxygen diffuses through air much more easily than through water, laying eggs out of water—perhaps initially right at the air–water interface on stones or vegetation—can solve that problem. But this solution only creates another problem, which is the risk of desiccation that the eggs face when they are exposed to the atmosphere. The solution, brooding by males that moisten the eggs repeatedly to prevent their drying out, sets the stage for the evolutionary transition to back brooding at the air–water surface.

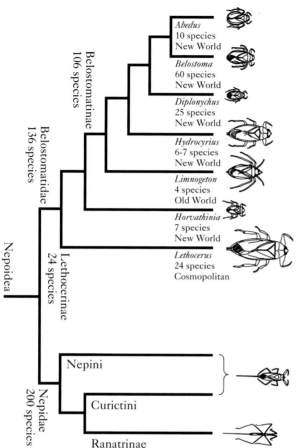

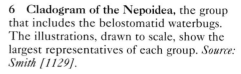

6 Cladogram of the Nepoidea, the group that includes the belostomatid waterbugs. The illustrations, drawn to scale, show the largest representatives of each group. *Source: Smith [1129].*

Wouldn't things be simpler if belostomatids were to lay small eggs, with large surface-to-volume ratios, that would not need to be brooded? To explain why waterbugs insist on producing large eggs, Smith points out that waterbugs are among the world's largest insects, almost certainly because large size is advantageous for the capture of vertebrate prey, such as fish, frogs, and tadpoles. These prey animals cannot be captured by smaller predatory insects, but large waterbugs do the job with their powerful grasping forelegs and poisonous salivary injections.

Insects do their growing during the immature stages. After the final molt to adulthood, no additional growth occurs. As an immature insect molts from one stage to the next, it acquires a new flexible cuticular skin that permits an expansion of size, up to a point. As it turns out, no immature insect can grow more than 50 or 60 percent per molt. Thus, if an insect is to grow large, one way to do so is by increasing the number of molts before making the final transition to adulthood. This option is not used by the belostomatids. No member of the belostomatid family molts more than six times, suggesting that they are locked into a five- or six-molt sequence, just as spotted sandpipers are stuck with a maximum clutch of four eggs because of the constraints of history (see Chapter 13). However, it could also be that starting small would lengthen the total development time for an immature bug, which might select against mutants that laid small eggs with the capacity for seven or eight molts.

In any event, if a belostomatid is to grow large in five or six molts, then the first instar (the nymph that hatches from the egg) must be large, because it will get to undergo only five or so 50-percent expansions. In order for the first nymph to be large, the egg must be large. In order to overcome the upper size limit for egg survival under water, large eggs must be laid with access to the air, which is where male brooding comes into play, an ancillary evolutionary development whose foundation lies in selection for large body size in conjunction with prey selection [1129].

But why the male and not the female? Here the situation parallels the fish story closely. First, emergent-brooding male waterbugs with one clutch of eggs can sometimes attract a second female, which means that remaining with one clutch is less costly for males in terms of lost mating opportunities than it appears at first glance. Second, just as is true for some fishes, the costs of parental care may be disproportionately high for females. In order to produce large clutches of large eggs, female belostomatids require far more prey than do males, which need only maintain themselves. Because brooding limits mobility and thus access to prey, parental care has greater fitness costs for females than for males, biasing selection in favor of male parental care. Admittedly, one would think that this argument would apply to many species other than belostomatids, and yet male parental care is very much the exception to the rule.

Why Are Male Burying Beetles Paternal?

Although waterbugs are about the only insects in which males alone care for their young, males of some other insects join forces with their mates to provide joint parental care for their offspring. Males of certain species of burying beetles, for example, sometimes cooperate with a female in burying a dead mouse or other small animal and constructing a "brood ball" of flesh from the corpse (see Chapter 12). The female lays a clutch of eggs near the brood ball, and when the larvae hatch out, they move onto the lump of carrion, where they feast on the food provided by their parents (see Color Plate 18).

Michelle Scott wanted to explain why males of *Nicrophorus orbicollis* sometimes exercise the parental option despite the mating cost of this decision. Do males enhance the survival of their young by feeding and grooming them? Apparently not, because in the laboratory, the mean production of young (as measured by their total weight) in broods reared by pairs was either less than or no greater than that in broods reared by single females (Table 2) [1068].

Table 2 The consequences of having a male partner on female reproductive success in the burying beetle *Nicrophorus orbicollis*

Treatment	Female only	Female with male
SMALL CARCASS (18–21 grams)		
Total weight of larvae (grams)	3.7 ± 1.0	2.4 ± 1.4
Probability of intruder killing brood	0.45	0.12
LARGE CARCASS (30–35 grams)		
Total weight of larvae (grams)	6.6 ± 0.5	6.8 ± 0.5
Probability of intruder killing brood	0.52	0.18

Source: Scott [1068, 1069, and other data].

Perhaps in nature, males stay with their mates primarily to defend the brood against predators of other species. But when Scott introduced such an enemy, a predatory beetle, into the laboratory with pairs and single-parent brooders, the pairs were no more successful in driving off the predator than were single parents. Another hypothesis bites the dust [1069].

Next, Scott considered the possibility that a major threat to broods of young larvae might come from other burying beetles rather than from predators of other species. Ovicide and infanticide do occur in *Nicrophorus orbicollis*, with intruders then using the already prepared brood ball as a food resource for their own brood [1002, 1204]. Scott showed that a single parent has some success in preventing another beetle from taking over the brood ball and destroying its offspring, but a mated pair does much better (Table 2). Pairs regularly kill and even cannibalize infanticidal intruders. Thus, the risk of infanticide may be a key factor driving the evolution of male parental care in this insect [1069] as well as in a number of monogamous mammals (see Chapter 13).

Discriminating and Nondiscriminating Parental Care

We have reviewed several hypotheses on parental care that are based on the assumption that parental animals face cost–benefit trade-offs. By considering what these fitness costs and benefits might be, and how they might vary between the sexes and in different environments, researchers have explained why females typically invest more in parental care than males do, and why there are exceptions to the rule. Let's now consider the mechanisms by which parents ensure that their own offspring, and not genetic strangers, receive the valuable parental care they have to offer.

We earlier discussed how some male birds identify broods likely to contain someone else's offspring by "remembering" how often they have copulated with the mother of those offspring (see Chapter 13). But can parental animals always identify their progeny? Consider the Mexican free-tailed bat, which migrates to

7 A crèche of Mexican free-tailed bat pups left together by their mothers while they forage outside the cave. *Photograph by Gary McCracken.*

certain caves in the American Southwest, where pregnant females form colonies in the millions. After giving birth to a single pup, a mother bat leaves her offspring clinging to the roof of the cave in a "crèche" that may contain 4000 pups per square meter (Figure 7). When the female returns to nurse her infant, she flies back to the spot where she last nursed her pup [789], but is besieged by babies in addition to her own youngster [791]. Given the shoulder-to-shoulder packing of pups, early observers believed that mothers could not possibly pick their own offspring out from the masses, and instead provided milk on a first-come, first-served basis.

But do Mexican free-tailed bat mothers really nurse indiscriminately? To find out, Gary McCracken captured and took blood samples from female bats and the pups nursing from them [788]. Using starch-gel electrophoresis, a technique somewhat similar to DNA fingerprinting that identifies variation in specific enzymes, he was able to establish whether mothers and pups had the same enzyme variants. For example, he found six forms of the enzyme superoxide dismutase, meaning that the gene coding for this enzyme was present in six different alleles in the population that he sampled.

If female bats are indiscriminate care providers, the enzyme variants of females and the pups they nurse should not be correlated. But if females tend to nurse their own offspring, then females and pups should tend to share the same alleles, and thus the same forms of superoxide dismutase and other enzymes. Despite the chaos of the bat rookery, the enzyme data indicated that females found their own pups at least 80 percent of the time [788]. More recent direct observational studies indicate that females probably do better than that, almost always recognizing their own pups through vocal and olfactory communication [47, 791]. Thus, mother free-tailed bats clearly deliver their parental care primarily to their own pups.

Offspring Recognition: Comparative Studies

The hypothesis that offspring recognition functions as a device to prevent misdirected parental care could be tested via the comparative method by looking at an appropriate sample of colonial versus solitary species. The key prediction from this hypothesis is that offspring recognition should evolve more often in colonial species, where the risk of misdirected parental care is relatively high, than in solitary species, whose parental adults are extremely unlikely to encounter babies other than their own. This prediction has not yet been tested in a statistically convincing fashion by systematically comparing closely related pairs of colonial and solitary species. However, we can illustrate the start of such a test.

Two swallows that nest in burrows in clay banks are the bank swallow, a colonial species (with a high risk of misdirected parental care), and the rough-winged swallow, a solitary species (with a low risk of misdirected care). Young bank swallows produce especially distinctive vocalizations, giving their parents a cue to use when making decisions about which individuals to feed, allowing them to distinguish between their own offspring and other fledglings that wind up in the wrong nests begging for food [69, 72, 800]. Bank swallow adults rarely make mistakes, despite the high density of nests in their colonies. The solitary rough-winged swallow, on the other hand, would never have had the chance in nature to feed another's fledglings, and should not show the same skillful recognition of offspring as its cousin, the bank swallow. To test this prediction, Michael and Inger Beecher transferred fledglings between distant roughwing burrows, and found that the parents readily accepted the transplants. Rough-winged swallows will even act as foster parents for fledgling bank swallows [69].

Two other swallows, the highly colonial cliff swallow and the less social barn swallow, also differ in the degree to which the risk of misdirected parental care could have acted in shaping their parental behavior. We would expect, therefore, that cliff swallow chicks should utter more variable and complex begging vocalizations than barn swallow nestlings. They do (Figure 8). By measuring several elements of chick calls, such as the duration and range of frequencies in these vocalizations, Mandy Medvin and her colleagues demonstrated that cliff swallow chicks produce signals containing about sixteen times as much variation as the corresponding calls of barn swallows [801].

If cliff swallow adults have experienced stronger selection to avoid feeding genetic strangers, then they should be better discriminators among chick calls than barn swallow adults. They are. In laboratory operant conditioning experiments that required adults of both species to discriminate between pairs of chick calls,

Cliff swallow   Barn swallow

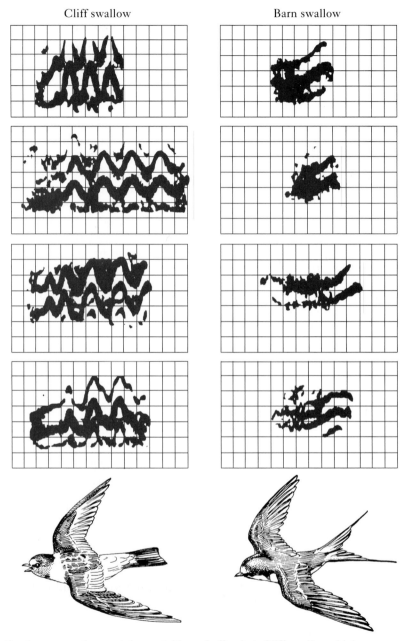

8 Call distinctiveness and parental recognition of offspring. Cliff swallow chicks produce highly structured and distinctive calls, enabling their parents to recognize them as individuals more easily than the parents of barn swallows can identify their chicks, which produce much less well-defined and more similar calls. *Source: Medvin et al. [801].*

cliff swallows reached criterion (85 percent accuracy) faster than barn swallows, suggesting that the perceptual systems of cliff swallows have evolved so as to promote accurate offspring recognition [732]. Although the proposition that colonial breeding selects for skillful recognition of offspring by parents needs broader tests, swallow behavior nevertheless supports the hypothesis.

Costly Adoption of Genetic Strangers

The cases we have reviewed support the prediction that parents should recognize their own young and discriminate against others when the risk of misdirected parental care is high. And yet the story is not always so clear-cut. For example, both herring gulls and ring-billed gulls occasionally adopt someone else's offspring, despite their colonial, ground-nesting breeding system, which exposes nesting pairs to exploitation by genetic strangers. Although early experimental studies had found that adults rejected older, mobile chicks when they were transplanted between nests [808], the aggressive responses of adults toward these transferred youngsters apparently stemmed largely from the "frightened" behavior of the displaced chicks [463]. When juveniles *voluntarily* leave their natal territory—which they sometimes do if they have been poorly fed by their parents—they do not flee from potential adopters, but instead beg for food and crouch submissively when threatened. They have some chance of being adopted, even at the advanced age of 35 days [563].

The adoption of strangers clearly qualifies as a Darwinian puzzle, for here we have parents apparently failing to act in their own best genetic interests. In order to resolve this paradox, consider the following argument. Learned recognition of one's offspring carries costs as well as benefits, notably the risk of making a mistake and not feeding, or even attacking and killing, one's own offspring. Rather than running this risk, perhaps adult herring gulls rely on the location and behavior of young gulls to make reliable, if not absolutely perfect, feeding decisions. By feeding only youngsters that they find in their nests and that behave appropriately by begging confidently, adults will usually feed only their own brood [935]. Occasionally, however, a genetic stranger is able to slip into another pair's nest and be fed, especially if it is nearly the same size and age as its hosts' biological offspring, and if it begs vigorously rather than fleeing upon the approach of the host adult [463, 563, 657]. The cost of occasional adoption is real, since parent ring-billed gulls that feed strangers rear fewer of their own chicks [153], but the probability of such an error is very low.

The hypothesis that the chance of rejecting one's own young carries higher risks than the occasional acceptance of a stranger's offspring has been applied to some more familiar cases of costly adoption, those involving specialized brood parasites such as certain cuckoos and cowbirds. These species exploit the parental devotion of other birds by depositing eggs in their nests, where the parasite hatchlings will monopolize the parental care of the host birds (see Figure 6, Chapter 5). The behavior of these parasites raises many interesting evolutionary questions, such as how the trait originated. The answer to this question may be seen in a fairly common phenomenon—namely, brood parasitism by some females within species that are not specialist parasites, which occurs in species as different as wood ducks [1078], coots [744], and swallows [927]. The occurrence of this behavior suggests that specialist parasites are derived from ancestral species in which parasitism was a sometime affair practiced against members of their own species.

What do parasitic females gain by adding eggs to another female's nest? In the coot, some of the parasites are floater females that lack nests or territories of their own. For these individuals, egg dumping appears to be a making-the-best-of-a-bad-situation tactic to salvage at least some reproductive success. But most brood parasites are territorial females with nests of their own. For these coots, brood parasitism offers a supplement to their maternal fitness if they are able to pop some surplus eggs into the nests of unwitting fellow coots. Since there are limits to how many young one female can rear with her partner, she can boost her fitness by surreptitiously enlisting the parental care of other pairs [744].

Regarding specialist brood parasites that lay eggs in other species' nests, one can ask, why do they differ in the number of species they exploit? For an answer based on historical reconstruction, we turn to Scott Lanyon's study of cowbird host preferences. Among the several parasitic cowbirds of the Americas are species that parasitize anywhere from 1 to over 200 host species. Has there been a gradual increase in the number of hosts selected by increasingly generalized cowbirds, or has there been a gradual decrease in the number of hosts parasitized by increasingly specialized cowbirds?

Lanyon used the comparative method to solve this problem [700]. He first constructed a phylogenetic diagram showing the evolutionary relationships among species or groups of species—for five species of cowbirds. These species covered the entire spectrum of host specificity (or generality, depending on your viewpoint). He developed this phylogeny (see Chapter 7) based not on behavioral traits, but rather on a totally independent character, namely, the base sequence in a gene that codes for the protein cytochrome *c*. This gene occurs in all five cowbird species, but the base sequence varies among them. The appropriate computer program can use degrees of similarity in base sequences to generate a phylogenetic tree. Starting with the sequence exhibited by an **outgroup,** the species most closely related to cowbirds, the computer creates the tree that requires the fewest steps or changes to move from the base sequence exhibited by the outgroup to that of the first species, to the next, to the next, and so on. The idea is to determine the simplest, most parsimonious way in which new species might have been derived from older ones, altering as little as possible the genetic information inherited from the ancestor of cowbirds, and retained in the outgroup.

This genetic analysis tells us that the species *Molothrus rufoaxillaris,* whose base sequence is closest to that of the outgroup, possesses the ancestral form of the cowbird cytochrome *c* gene, from which all the other forms have been derived through the accumulation of mutations. The species *Molothrus ater* and *Molothrus bonar-*

iensis have the greatest number of differences in base sequence from the outgroup, and are therefore considered to have evolved most recently (Figure 9).

When Lanyon superimposed information about host specificity on this genetically founded phylogeny, he found that the modern species judged to be closest in genetic makeup to the outgroup was an extreme specialist, whereas the intermediate forms had intermediate numbers of host species, and the most recently derived species had the greatest numbers of hosts. Lanyon interpreted this pattern to mean that the original cowbird was a specialist and that over time, species splitting off from this lineage have become more and more highly generalized, with the ability and willingness to parasitize an ever-increasing number of host species.

I believe that most students of phylogenetic analyses accept Lanyon's interpretation, but two mildly iconoclastic colleagues of mine, Walter Koenig and Robert Montgomerie, offer another view. They suggest that all brood parasites started out as generalists, but with the passage of time have become increasingly specialized. Thus, the species with the longest history of parasitism *(Molothrus rufoaxillaris)* has had time to become a specialist, whereas those that have become parasites more recently are still in the generalist phase. It ought to be possible to devise a way to rigorously discriminate between these two interpretations of history, and I look forward to the resolution of the matter.

In any event, since brood parasites are a fact of life for so many birds, the question becomes, why do host species sometimes accept eggs from genetic

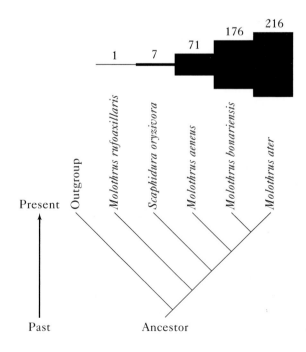

9 Evolution of brood parasitism among cowbirds. The evolutionary relationships among cowbirds, as determined by genetic analyses, are diagrammed in the form of a phylogenetic tree. Above the tree is the number of hosts parasitized by each current living species of cowbird, suggesting that the first cowbird victimized only a single host species, with increasingly generalized brood parasitism evolving subsequently. *Source: Lanyon [700].*

strangers, whether of their own or other species? At a proximate level, the accepting species probably care for the extra eggs because they are following a generally adaptive "rule of thumb" that reads: "incubate the eggs that appear in one's nest, and feed the offspring that emerge from those eggs." Such eggs will usually, though not always, be the eggs of the incubator, so that on average it pays to follow this simple rule.

Furthermore, even if it were possible for host species to recognize a parasite's eggs, the costs of doing something about them might exceed the benefits. Once again, if parents were to make incubation of eggs dependent on their learned recognition of each egg as one of their own, they might sometimes abandon or destroy their own progeny when they mistook it for a parasite's egg. Reed warblers sometimes do throw out or damage their own eggs while trying to toss a cuckoo's egg out of their nests [271].

For a small bird, whose bill cannot accommodate a parasite's egg—and there are many such host species [1012]—the only options are to abandon the clutch, either by leaving the site or by building a new nest on top of the old one, or to stay put and continue brooding the clutch along with the parasite's egg. Each option has a different mix of costs and benefits. Acceptance means rearing the cowbird, which will usually greatly reduce the number of their own offspring that the hosts can produce. But rejection requires time to find a new site and/or build a new nest, the costs of which can be high, especially for a hole-nesting species with restricted nest site availability. Lisa Petit tested the prediction that the availability of artificial nest boxes and natural nest holes would determine the likelihood that a prothonotary warbler with a parasitized nest would reject it and start over at a new site [922]. As she expected, warblers with more replacement nest sites available nearby more often abandoned their parasitized nests than warblers with few alternative sites (Figure 10).

Likewise, among yellow warblers nesting in southern Canada, individuals parasitized by cowbirds near the start of the short breeding season were likely to take action against the parasite, either by abandoning the nest or by burying the parasitized clutch under a new layer of nest material. Warblers whose nests were afflicted with cowbird eggs late in the season, however, tended to accept them, perhaps because they had too little time then to rear a new brood from scratch [157, 1070].

Even if host birds could throw out eggs laid in their nests by parasites without risk of error, the parasite might make this option most unprofitable by returning to the nest to destroy or consume the host's eggs or young if it found that its offspring had been removed. This "Mafia hypothesis" has been tested by observation and experiments with host magpies and parasitic great spotted cuckoos [1133]. As predicted by the hypothesis, magpie nests from which cuckoo eggs had been ejected suffered a significantly higher rate of predation than nests with accepted cuckoo eggs (87 percent vs. 12 percent in one sample). Furthermore, when the researchers removed the cuckoo egg from a nest that was apparently being checked by a cuckoo, and replaced the magpie's eggs as well with plasticine imitations of magpie eggs, the cuckoo approached the nest after the researchers had finished and

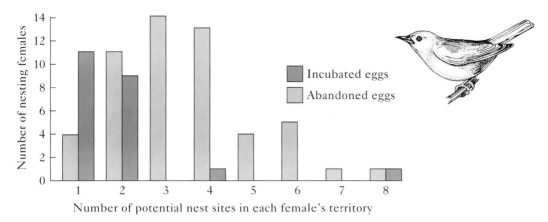

10 Nest abandonment carries variable fitness costs. Prothonotary warblers often continue incubating (darker bars) even after cowbirds have parasitized their nests *if* there are relatively few other potential nest sites available. When a warbler has access to several other nest cavities in its territory, it is more likely to abandon a parasitized nest (lighter bars). *Source: Petit [922].*

pecked the imitation eggs forcefully in an attempt to destroy them. In nature, the destruction of its clutch of eggs would force the host magpie to renest, exposing it to all the negative effects that delays in breeding have in a seasonal environment. Acceptor magpies actually have somewhat higher reproductive success, as measured in terms of fledglings produced, than do ejectors of parasite eggs [1133].

Adoption with Direct Benefits for the Adopters

Even though adoption may be a better option than abandonment of a parasitic offspring in some situations, adoption of a parasitic cowbird usually means that the host parents rear fewer offspring than they would otherwise. In some circumstances, however, adoption can actually yield absolute fitness gains for any of several reasons [319]. Consider the fathead minnow, a stream-dwelling species in which males care for their mates' eggs at nests on the undersides of rocks (Figure 11) [1046]. Appropriate nest sites are in short supply, and males compete for them, sometimes evicting an egg-tending rival in the process. If this happens, the victorious male will "adopt" the eggs already present at the nest, even though they have been fertilized by the previous owner.

The parental behavior of the adopter is costly, since he will defend the clutch against egg predators and apply antifungal secretions to the eggs. However, Craig Sargent showed that the costs of adoption were outweighed by a mating advantage for egg-guarding males, since female fathead minnows are especially attracted to parental males. Thus, a male's chance of acquiring his own mate after a takeover is enhanced if his nest has eggs in it, no matter who fertilized those eggs.

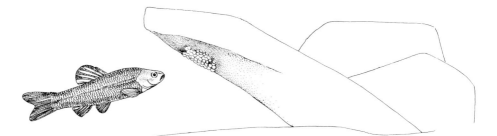

11 Adaptive adoption in the fathead minnow, whose males may care for some of the eggs at nests they acquire by territorial takeover. Having eggs at a nest attracts additional receptive females to the area.

Furthermore, Sargent discovered that males reduce the costs of adoptive egg care by reducing the size of the clutch that they inherited after the takeover. In the laboratory, when males acquired nests from other males, the number of eggs in the adopted clutch fell over 50 percent in the first 24 hours after the takeover. For a group of control males guarding eggs they had fertilized themselves, the number of eggs surviving 24 hours after the first clutch had been laid was over 85 percent. Evidently "adoptive" males prune the clutches they acquire by takeover, leaving just enough eggs to attract mates of their own while making space for new eggs and having a meal at the expense of the previous male and his partners [1046].

In reality, by adopting another's eggs, male fathead minnows get to fertilize more eggs than they would otherwise. The white-winged chough, an Australian bird that looks rather like a crow with white wing patches and a red bill, also may benefit from adopting another's offspring. The chough lives in communal groups of up to 20 individuals, all of which help to rear a single youngster (see Chapter 15 for more on cooperative breeding), a task that requires up to seven adults since it is so hard to find enough food in their extremely arid woodland environment.

Neighboring groups of choughs constantly fight with one another, destroying one another's eggs or nests if they get the chance. Occasionally, during a battle, some choughs may turn their attention to a fledgling member of the rival flock, which they attempt to separate from its group—not to harm the youngster, but rather to incorporate it into their flock. The "kidnappers" will have to feed their adopted charge for months or even years before it can feed itself on its own. In return, however, the fledgling will, if it survives long enough, become a helper at the nest of its adoptive group, providing the assistance that is essential for the parent choughs to rear young of their own. Perhaps the adopting birds kidnap the offspring of others to build up their group of helpers more quickly than they could by the slow recruitment of their own offspring into their band [533]. Or perhaps the behavior of the young birds triggers maladaptive parental behavior on the part of their "kidnappers," who are using generally adaptive traits in a fitness-reducing manner. Whatever the cause, this and other examples of "adoption" should not blind us to the fact that parents only very rarely provide care to offspring other than their own.

The Evolution of Parental Favoritism

Although parents generally reject genetic strangers in favor of their progeny, they do not always distribute their care in a completely democratic fashion. In fact, some parents may permit certain offspring to harm their siblings (Figure 12) [814]. In the great egret, for example, brothers and sisters fight for possession of the fish their parents

12 A brown booby chick dying in front of its parent, which continues to incubate its other chick. The surviving chick was responsible for the death of its sibling, which it had forced out of the nest and into the sun. *Photograph by the author.*

13 Sibling aggression in the great egret. An older, larger chick assaults a smaller nestmate. *Photograph by Douglas Mock.*

bring to the nest (Figure 13). The dominant individuals in a brood may even bludgeon their sibling rivals to death or push them out of the nest, a behavior that enables the survivors to monopolize their parents' largesse. How can **siblicide,** as Doug Mock calls it, possibly advance a parent's fitness when it leads to the loss of one or even two members from a brood of three or four?

One hypothesis is that parents do *not* gain from these actions, which have evolved because of the advantages enjoyed by winning siblings, not because of selection on parental behavior. Under special circumstances, the fitness interests of parents and offspring may clash. Consider the familial interactions of spotted hyenas, whose females deposit their newborns, usually twins, at the entrance to a small underground burrow, usually an old aardvark den. When the babies crawl in, they are safe from most predators, and free from maternal supervision as well.

Inside the burrow, the newborn hyenas often employ their fully functional canines and incisors in slashing attacks on one another (Figure 14). Some youngsters are able to kill their siblings by wounding them and by preventing them from leaving the burrow to nurse [406].

An intriguing feature of these siblicidal attacks is that they occur primarily when two sibs are of the same sex. As a result, surviving brother–brother and sister–sister pairs are very much underrepresented compared with surviving brother–sister twins [405, 406]. The ability of females to rear son–daughter pairs suggests that siblicide may reduce parental fitness. On the other hand, mother hyenas could potentially prevent their twin pups from fighting to the death by placing them in separate nursery burrows, which they sometimes do. But since they often put their pups in the same burrow, perhaps they actually use siblicide to manipulate the sex ratio of their offspring in an adaptive manner. When a high-ranking female lets her twin sons or daughters eliminate the competition for food at the outset, the survivor will be an especially vigorous, rapidly growing young animal. As an adult, the surviving son or daughter will not face potentially costly reproductive competition with its sibling. Instead, as the sole inheritor of its mother's high status, the son or daughter will have a better opportunity to secure the exceptionally great reproductive advantage enjoyed by alpha females and alpha males in this unusually socially stratified, and highly polygynous, species (see Chapter 7) [405, 407].

In other animals in which siblicide occurs, parents also could intervene if the actions of their youngsters reduced their fitness. No such intervention occurs in the case of the siblicidal juvenile egrets. Indeed, lethal sibling interactions are

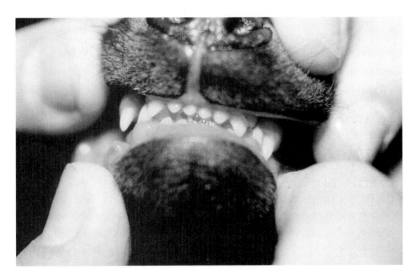

14   Siblicide in hyenas occurs when newborns use their sharp canines to slash their siblings to death. *Photograph by Laurence Frank.*

actually promoted by two earlier parental decisions about what to put in the eggs and when to begin incubating them. First, when female cattle egrets (a siblicidal species related to the great egret) produce eggs, they vary the hormones that they incorporate into the yolk. Relatively large quantities of androgen go into the first two eggs of a three-egg clutch [1062]. The more androgen contained in an egg, the more aggressive the nestling will probably be, an example of the female using her ability to influence the nature of her offspring (see Chapter 12).

Second, as soon as a female egret lays an egg, incubation begins. Because one or two days separate the laying of each egg in a three-egg clutch, the young hatch out asynchronously, with the firstborn getting a head start in growth. As a result, this chick will not only be more aggressive, but will also be much larger than the third-born chick, ensuring that the senior chick monopolizes the small fish its parents bring to the nest. Unequal feeding rates further exaggerate the size differences among siblings, creating a runt of the litter that often dies from the combined effects of starvation and assault.

Thus, parents not only tolerate siblicide, but actually seem to promote it. Why? Perhaps because parental interests are served by having the chicks eliminate those members of the brood that are unlikely to *survive to reproduce*. In most years, food will be moderately scarce, making it impossible for the adults to find enough food for all three offspring. In good years, they can, and it is then that the ability to lay three eggs sometimes pays off. Under conditions of food scarcity, a reduction in the brood accomplished by siblicide saves the parents the time and energy that would otherwise be wasted on offspring with no chance of reaching adulthood, even if their siblings left them alone.

The hypothesis that siblicide is adaptive for parent egrets has been tested experimentally by creating synchronous as well as asynchronous broods of cattle egrets [817]. Doug Mock and Bonnie Ploger created three categories of broods by shuffling newly hatched chicks among nests in a cattle egret colony: synchronous broods (all had hatched on the same day), normal asynchronous broods (chicks that had hatched at the typical 1.5-day intervals), and exaggerated asynchronous broods (chicks that had hatched 3 days apart).

If the normal hatching interval is optimal in promoting efficient brood reduction, then the number of offspring fledged per unit of parental effort should be highest for the normal asynchronous broods. This prediction was confirmed. Members of synchronous broods not only fought more and survived less well, they required more food per day than normal broods, resulting in low parental efficiency (Table 3). The same result has been recorded for the blue-footed booby, in which experimental synchronous broods of two fought more and required much more food than control broods composed of asynchronously hatched chicks [886].

Thus, cattle egret parents and others like them know (unconsciously) what they are doing when they manipulate the hormone content of their eggs and incubate them in ways that ensure differences in size and fighting ability among their chicks. Sibling rivalry and siblicide actually help parents deliver their care only to offspring

Table 3 The effect of hatching asynchrony on parental efficiency in cattle egrets

	Mean survivors per nest	Food brought to nest per day (ml)	Parental efficiency[a]
Synchronous	1.90	68.3	2.78
Normal asynchronous	2.33	53.1	4.39
Exaggerated asynchronous	2.29	65.1	3.52

[a] The number of surviving chicks produced divided by the volume of food brought to the nest per day × 100.
Source: Mock and Ploger [817].

that have a good chance of surviving to reproduce, while enabling parents to keep their food delivery costs to a minimum.

This case represents an extreme example of parental favoritism, with the adults letting their offspring sort themselves out in ways that identify those best able to provide a return on parental investment. Other animal parents may use other means to bias their parental care toward some members of a brood [1007, 1263]. For example, the begging intensity of nestling robins determines how much food their parents deliver to them [1124]. In songbirds generally, weak, ill nestlings unable to beg strongly may die if the parental "rule of thumb" reads, "stuff food into the mouth of the chick able to reach highest out of the nest and shake its body most vigorously." If weak, ill nestlings are very unlikely to become viable adults, a parental bias against them promotes efficient distribution of limited food resources, helping parents maximize the number of offspring that have a reasonable chance of surviving to adulthood.

The coot is a common black waterbird whose clownish behavior is responsible for the expression "crazy as a coot"; it is appropriate, therefore, that parent coots use a very odd system to establish favorites among their offspring. Baby coots hatch out with long, orange-tipped plumes on their backs and throats. Bruce Lyon and his colleagues suspected that these purely ornamental plumes provided signals to parents that they might use to determine which individuals to feed preferentially. To test this possibility, Lyon and company trimmed the thin orange tips from these special feathers on the backs of half of the chicks in a brood, while leaving the other members of the brood untouched. The unaltered orange-plumed chicks were fed more frequently by their parents, and they grew more rapidly as well (Figure 15).

In control broods, in which *all* of the chicks had their orange feathers trimmed, the youngsters were fed as often and survived as well as broods consisting only of untouched orange-feathered chicks. This result shows that the parents of the experimental mixed broods discriminated against the "black" chicks because they were not as strongly ornamented as their feather-intact broodmates, not because the parents failed to recognize them as their offspring. Just why coot adults use feather ornamentation to establish favorites among their offspring is unknown,

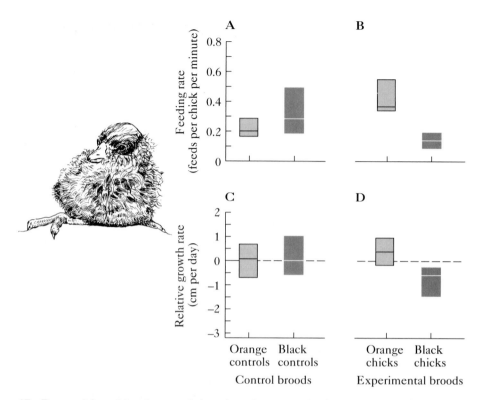

15 Parental favoritism in coots is based on the orange feather ornaments of their young. (A) Individuals in control groups composed entirely of either unaltered (orange) chicks or black chicks that had had the orange tips trimmed from their ornamental feathers were fed at the same mean rate. (B) However, in experimental broods composed of half orange and half black chicks, the ornamented individuals received more frequent feedings from parent birds. (C) The relative growth rate of chicks in both control groups was the same, but (D) ornamented orange chicks grew faster in mixed broods compared with the experimentally altered black chicks. *Source: Lyon et al. [745].*

although perhaps the degree of orange ornamentation provides a signal of the age of the chick, enabling parents to bias their feeding effort adaptively according to this criterion [745].

The general message provided by coots and cattle egrets is that parents can allocate their parental care unevenly on the basis of various features of their offspring, helping some survive, handicapping others. Cases of this sort emphasize the point that selection acts not on the number of babies produced, but on the number that survive to reproduce and pass on the hereditary traits of their parents.

SUMMARY

1. Parental care, a major form of parental investment, involves a trade-off between increasing the survival chances of one or more existing offspring and the various costs of parenting, including reduced fecundity and fewer opportunities to mate. When the circumstances are such that parental care is adaptive, females are more likely than males to provide it, perhaps because mothers are more often in a position to identify their offspring and provide useful care for them than are fathers.

2. Male fish are very unusual animals in often providing uniparental care. In many species, however, male fish may pay only a small penalty in lost reproductive opportunities for caring for eggs laid in their territories, whereas the cost of parental care to females may be much greater in immediate losses of fecundity and reductions in growth rate, which carry an additional penalty in terms of long-term reductions in fecundity.

3. An evolutionary approach to parental care yields the expectation that parents will be able to identify their offspring in order to avoid costly investments in genetic strangers. Although offspring recognition is quite widespread, particularly in colonial species, it is by no means universal, and adoption of nonoffspring occurs in many species. Multiple hypotheses exist to account for these puzzling cases, including the possibility that brood parasites select for hosts that have higher fitness on average by continuing to provide care for the parasite as well as their own offspring than they would by abandoning the parasite and their brood.

4. Another Darwinian puzzle is the indifference shown by parent egrets and some other animals to lethal aggression among their young offspring. Cases of this sort may be explained as part of a parental strategy to let offspring identify which individuals are most likely to survive, and therefore which youngsters "deserve" parental investment.

SUGGESTED READING

The Evolution of Parental Care by Timothy Clutton-Brock reviews all aspects of parental behavior from an evolutionary perspective [204]. Michael Beecher discusses why parent–offspring recognition is not universal [70]. David Westneat and Craig Sargent examine how the different interests of males and females may affect parental behavior [1262]. Marion Petrie and Anders Møller examine the costs and benefits of intraspecific brood parasitism for the parasites and their hosts [927]. Robert Montgomerie and Patrick Weatherhead employ a cost–benefit approach to the analysis of nest defense, an important element of parental care [830]. Hypotheses on siblicide are reviewed by Doug Mock and his co-workers [815].

DISCUSSION QUESTIONS

1. Gary McCracken found that although female Mexican free-tailed bats usually feed their own pups, they do make some "mistakes," which they could have avoided by leaving the pup in a spot by itself instead of in a mass of other babies [788]. Does this mean that the parental behavior of this species is less than "optimal" (see Chapter 8)? Use a cost–benefit approach to develop alternative hypotheses to account for these "mistakes."

2. When should the time, energy, and risks taken in territorial defense by a male bird be considered "parental investment" [119]?

3. You observe a bird in which females typically exhibit stronger defense of their eggs and nestlings than do their male partners. Use an optimality approach to develop testable hypotheses on why females invest more in defense than males in this species. Consider the costs and benefits of parental defense to members of the two sexes, keeping in mind that the sexes might differ in size, color, certainty that the offspring were indeed their own, and so on [830].

4. Males of the river bullhead often eat some of the eggs in the nests that they guard in Italian streams and rivers [756], as do male garibaldis, a fish that nests on reefs in the Pacific Ocean [1102]. Develop at least two different hypotheses that would explain this Darwinian puzzle. Does the finding that the eaten eggs were fertilized by their consumer in both species help you discriminate among your hypotheses?

5. Females of the royal penguin typically lay two eggs, a small first egg and a much larger second egg [1140]. In many cases, females eject the first egg shortly before or after laying the second. In what way might the evolution of ovicide by mother penguins be similar to tolerance of siblicide in cattle egrets?

CHAPTER 15

The Adaptive Value of Social Living

*T*HE INTERACTIONS BETWEEN PARENTS AND OFFSPRING
(Chapter 14) are just one of the many kinds of social behavior that can be analyzed in terms of their fitness costs and
benefits for the participants. In this chapter, we will consider
other forms of social interactions and their consequences for
individuals. Just as it is true that parental species are not
inherently superior to those whose adults do not care for their
young, species with elaborate social behavior are not somehow "better adapted" than those whose members lead largely
solitary lives. Unfortunately, people often think that the
highly complex societies of some vertebrates—most conspicuously our own—represent a crowning achievement of evolution. To counteract this mistaken impression that complex

social behavior is always superior to "primitive" sociality or solitary behavior, this chapter begins with a discussion of the *costs* of social living. The cost–benefit approach of behavioral ecology suggests that many circumstances favor the evolution of solitary living as the more adaptive mode of existence. After establishing this point, we shall examine how cooperation might evolve via several different evolutionary routes before turning to one of the most intriguing problems in social behavior, the evolution of altruism, or self-sacrificing helpful behavior. Altruism takes several different forms in the animal world, including the complete rejection of personal reproduction by sterile workers who labor for other individuals. Biologists have developed several ingenious hypotheses to account for the evolution of altruism; this chapter examines their different proposals and shows that extreme cooperation among social creatures can arise by more than one pathway.

The Costs and Benefits of Sociality

The economic analysis of animal behavior has proved remarkably useful, as we have seen with respect to communication, territoriality, mating tactics, and many other topics. We shall now apply the cost–benefit approach to living in groups (Table 1) [13]. Because humans are a social species, we like to flatter ourselves by believing that sociality must be the "most advanced" way of life. It is true that living and cooperating with others has a variety of potential benefits, some of which we have already seen. Black-headed gulls capture food more easily when they hunt in flocks than when foraging by themselves [450]. A pride of lions can better defend a hunting territory than a single lion can [897]. A cluster of male emperor penguins huddled together on the ice through the long and frigid Antarctic winter conserve their energy reserves much better than isolated penguins, which enables them to brood their eggs without eating anything for more than 100 days

Table 1 Major fitness benefits and costs of sociality

BENEFITS

Reduction in predator pressure by improved detection or repulsion of enemies or by the dilution effect (see Chapter 9)

Improved efficiency in foraging for large or evasive prey (see Chapter 10)

Improved defense of key spatial or food resources against other groups of conspecifics

Improved care of offspring through communal feeding and protection

COSTS

Increased competition within the group for food, mates, or other limited resources

Increased risk of infection by contagious diseases or parasites

Increased risk of exploitation of or interference with parental care by other group members

Source: Alexander [13].

1 Social defense against predators, as practiced by these musk oxen in their circular defensive formation. *Photograph by Ted Grant/Information Canada Phototheque.*

[19]. And social prey can often spot danger more quickly (see Chapter 9), or repel an enemy more effectively (Figure 1), or simply swamp the consumption capacity of the local predators better than solitary individuals (Figure 2) [1277].

But social species are rare, and if sociality is so advantageous, why should this be so? We can better answer this question if we first recognize that social life carries costs as well as benefits. In many groups of social foragers, for example, subordinates are exploited by dominants, as when male lions push the pride's females from a kill that the smaller females have made [1053]. In other species, subordi-

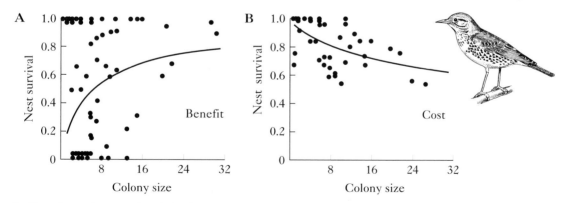

2 Benefits and costs of sociality in a songbird, the fieldfare, that nests in loose colonies in woodlands. (A) A benefit: the larger the colony, the greater the probability of nest survival over a breeding attempt. (B) A cost: the larger the colony, the lower the survival rate of nestlings, due to increased mortality caused largely by starvation. *Source: Wiklund and Andersson [1277].*

nate members of a group may pay a special price for living with others who copulate with their mates [358, 827] or practice infanticide on their offspring [578]. Or consider the ovicidal society of the acorn woodpecker [848], a bird that forms groups containing as many as three breeding females and four breeding males. The breeding females all lay their eggs in the same nest, perhaps because females that try to nest alone have all their eggs destroyed by their companions [663]. Even when two or three females settle on the same tree hole nest, the first eggs laid by one woodpecker are almost always removed and destroyed by another member of the band (Figure 3). Eventually the "cooperatively breeding" females all lay eggs on the same day, at which time they tend to stop stealing eggs and begin joint incubation of the brood. By this time, more than a third of the eggs laid by the breeders may have been destroyed.

The price of being social may fall unequally upon some members of a group, but one general and nearly inescapable disadvantage of social life is heightened

3 Reproductive interference in a social animal. One member of a breeding group of acorn woodpeckers removes an egg of a companion female from their communal nest. *Photograph courtesy of Walt Koenig.*

competition for food [5]. When many animals cluster together, the odds are that they will deplete the local supply of food, as shown by the increased rate of starvation of nestling fieldfares as the number of pairs nesting in close proximity increases in this colonial forest-breeding thrush (see Figure 2B) [1277].

Another cost of sociality that can affect dominant and subordinate alike is increased vulnerability to parasites and pathogens. Consider the colonial cliff swallow, which nests in groups ranging from a couple of breeding pairs to over 3000 individuals. These swallows build their mud nests shoulder-to-shoulder under overhanging cliffs, bridges, and culverts. Large colonies are more likely to contain at least one adult infested with a blood-sucking parasite, the swallow bug, which can make its way from that bird to others in the group. Charles and Mary Brown found that the larger the colony, the more blood-sucking swallow bugs attacked each nestling, and the less each 10-day-old nestling weighed on average (Figure 4) [141]. The Browns demonstrated that the bugs were the guilty parties by fumigating a sample of nests in an infested colony while leaving other control nests untreated with insecticide. The nestlings freed from the bugs weighed much more, and were more likely to survive, than those plagued by many parasites (Figure 5).

The deaths of heavily parasitized cliff swallow chicks graphically illustrate that social living is far from an unmitigated blessing. If sociality is to evolve, special ecological conditions are required so that the many costs of associating with others will be outweighed by certain benefits to social individuals. The primary fitness benefit of social life may be improved success in dealing with predators [13] (see Chapter 9), as shown by the social bluegill sunfish and its close relative, the solitary pumpkinseed sunfish [480]. Bluegills become social during the breeding

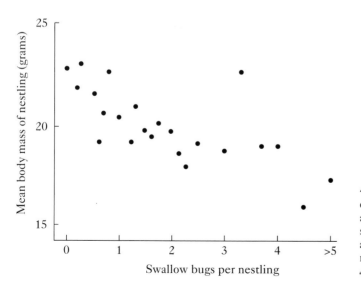

4 A cost of sociality. In large colonies of cliff swallows, nestlings are afflicted by more swallow bug parasites on average, and they weigh less as a result. Each point represents the mean mass for a sample of nestlings. *Source: Brown and Brown [141].*

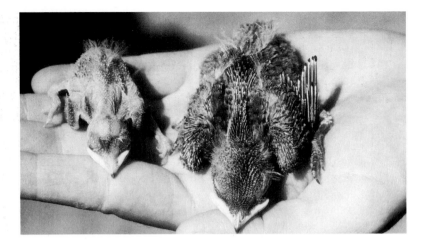

5   Effect of parasites on cliff swallow nestlings. The much larger nestling comes from an insecticide-treated nest; the stunted baby of the same age occupied a parasite-infested nest. *Source: Brown and Brown [141].*

season, when groups of 50 to 100 males cluster their pit nests on sandy lake bottoms. Each breeding male fertilizes and cares for the eggs deposited in his nest by spawning females. Although bluegills may aggregate in part because some places are better than others for nest building, colonial males almost certainly derive a benefit in terms of a reduction in predator pressure on their eggs. By defending their overlapping territories, for example, social males "cooperate" in expelling egg-eating catfish from the colony (Figure 6). Because colonial males share the expenses of driving off egg predators, they can spend more time fanning their eggs, a parental activity that reduces the chance that the eggs will become infected with a fungus. As a result, males nesting at high densities have a lower percentage of infected eggs in their nests than do solitary nesting males [223].

But social bluegills do not get their antipredator and antifungus benefits for free. A fish that nests in a group must contend with the tendency of his neighbors (and other nonnesting bluegills attracted to the group) to consume the eggs in his nest. Moreover, he faces sexual interference in courtship and spawning by "sneaker" males and female mimics (see Discussion Question 6 in Chapter 12) that gather at large colonies. These costs reduce the net benefit enjoyed by social bluegills.

In contrast to their bluegill relatives, pumpkinseed sunfish do not breed in colonies. Whereas bluegills have small, delicate mouths designed for "inhaling" small, soft-bodied insect larvae, pumpkinseeds have powerful jaws adapted for picking up, crushing, and consuming heavy-bodied molluscan prey. Thus, although a bluegill cannot pick up a snail and cart it away from the nest, pumpkinseeds are easily able to do this, and may consume their egg-loving enemy to boot. In

6 A society of bluegills. Each colonial male defends a territory bordered by the nest sites of other males, while bass (above), bullhead catfish (left), snails, and pumpkinseed sunfish (right foreground) roam the colony in search of eggs. *Drawing courtesy of Mart Gross.*

addition, a bluegill's bite does little damage to a nest-raiding bullhead catfish, but a pumpkinseed's attack packs a considerable wallop. Thus, pumpkinseeds are relatively free from nest predation, and are solitary, whereas bluegills are more vulnerable to nest predation, and are social, supporting the hypothesis that social living is adaptive *only* when special benefits counterbalance its unavoidable costs [480]. Pumpkinseed sunfish are in no way inferior to or less well adapted than bluegills because they are solitary; they simply face ecological circumstances that make solitary nesting a more adaptive response.

The Evolution of Helpful Behavior

Although mates often help each other rear their offspring (see Chapter 14), helpful behavior is not limited to parental couples. Until the mid-1960s, biologists took helpful interactions for granted because they assumed that animals should help one another for the benefit of the species as a whole. But with the recognition that group selection is less potent than individual selection in shaping behavior (see Chapter 1), helpful actions become considerably more interesting [1287]. Here, as in the parallel analysis of communication (see Chapter 8), the trick is to ascertain who benefits from a helpful action (Table 2).

Table 2 The effect on reproductive success of various social interactions.

Type of interaction	Effect on reproductive success of	
	"Donor"	Recipient
Mutualism (Cooperation)	+	+
Reciprocity	+ (delayed)	+
Altruism	−	+
Selfish behavior	+	−
Spiteful behavior	−	−

Some kinds of helpful actions generate an immediate benefit for both helper and helpee. When one lioness drives a wildebeest into a lethal ambush set by her fellow pride members [1138], the driver will also get some meat, even if she did not personally pull the antelope down and kill it. Likewise, if several male bluegills succeed in fending off a bullhead catfish that has entered the space they mutually claim and defend, the eggs guarded by all the males are safer as a result. When both helper and recipient enjoy reproductive gains from their interaction, they have engaged in **mutualism** or **cooperation,** which requires no special evolutionary explanation.

However, some kinds of helpful interactions are more puzzling. Robert Heinsohn and Craig Packer [534] found that some lionesses consistently took the lead when their group moved to challenge presumptive intruders, in this case taped roars played in their pride's domain by the researchers. Others held back, letting their pridemates take the more dangerous leading positions. Why do the leaders do "more than their share" of the risky work? Perhaps because over the very long term, they benefit from having the laggards around for other mutually beneficial activities (such as babysitting the cubs) or for emergency aid. In addition, when the pride is under direct attack, the slackers might come to its defense, since they too depend on the maintenance of a group territory for their reproductive success [704].

The lionesses that Heinsohn and Packer studied did not take turns leading their territorial patrols, but **reciprocity** or **reciprocal altruism** does occur. Robert Trivers pointed out that if an individual could help another now at relatively little cost and later receive valuable "repayment" from the helped animal, the original helper would experience a net reproductive benefit from its initial helpful action [1199].

The logic of game theory, however, suggests that a population composed of reciprocal altruists would be vulnerable to invasion by individuals that accepted help but later conveniently "forgot" to return the favor. "Cheaters" would reduce the fitness of "noncheaters" in such a system, presumably making reciprocal altru-

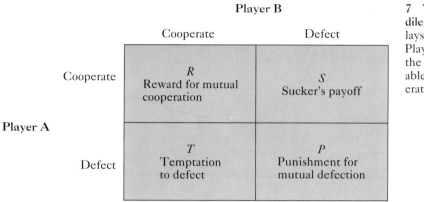

Player B

	Cooperate	Defect
Cooperate	*R* Reward for mutual cooperation	*S* Sucker's payoff
Defect	*T* Temptation to defect	*P* Punishment for mutual defection

Player A

7 The prisoner's dilemma. The diagram lays out the payoffs for Player A associated with the different options available to two potential cooperators.

ism unlikely to evolve. The problem can be illustrated with a game theoretical model called the *prisoner's dilemma* (Figure 7), based on a human situation. Imagine that a crime has been committed by two persons, who agree not to squeal on each other if caught. The police have brought them in for interrogation and have put them in separate rooms. The cops offer to let each one go free if he will implicate his pal in the crime. If suspect A accepts the tempting offer while B maintains their agreed-upon story, A gets his freedom ("*T*" in Figure 7), while B gets hit with the maximum punishment (the sucker's payoff). If together they maintain their agreement, they make it harder, though not impossible, for the police to convict either one (see "*R*" in Figure 7). And if each one fingers the other, the police will use this evidence against both and renege on their offer of freedom for the snitch, so that both A and B will be punished for their mutual "defection."

In a setting in which the payoffs for the various responses are ranked $T > R > P > S$, the optimal response for suspect A is to defect, not to cooperate, provided that $R > (S + T)/2$. Under these circumstances, if suspect B maintains their joint innocence, A gets reward *T*, which is greater than reward *R*; and if suspect B defects, defection is still the superior tactic for A, because the payoff from *P* is greater than that from *S*. By the same token, suspect B will come out ahead on average if he defects and squeals on his buddy.

This model predicts, therefore, that reciprocal cooperation should not evolve. But cases of apparent reciprocity do exist [893, 1199, 1283], perhaps because under some circumstances, two players may interact repeatedly over the long term, not just once. Robert Axelrod and W. D. Hamilton have shown that when this condition applies, individuals that use the simple behavioral rule "Do unto individual X as he did unto you the last time you met" can reap greater fitness gains than can cheaters that accept assistance but do not return the favor [37]. When multiple interactions are possible, the rewards for back-and-forth cooperation can add up, exceeding the short-term gain from a single defection. However, computer simulation contests also suggest that a "cheat if you can get away with it" approach will yield even higher cumulative returns than the "tit-for-tat" strategy just described. Among other things, such a strategy takes advantage of individuals that always cooperate, should these types happen to exist in a population [871].

Altruism and Indirect Selection

Reciprocal altruism is really a special kind of mutualism in which the helpful individual endures a short-term loss until its help is reciprocated, at which time it earns a net increase in reproductive success. In evolutionary biology, unadorned **altruism** is a term restricted to cases in which the donor really does lose reproductively over the long haul as a result of helping another produce more surviving offspring. Altruistic actions, if they exist, are an especially exciting puzzle for adaptationists because they cannot have evolved via natural selection for increased personal reproductive success.

Cases do exist in which an altruistic individual is never "repaid" by the animal that it helped, nor is the helper actually "selfishly" manipulating the recipient with its "assistance" so as to increase its own reproductive success, as sometimes happens. A brilliant explanation for "real" altruism came from W. D. Hamilton, who showed that "for-the-good-of-the-group" selection is not needed for the spread of a mutant gene that causes its bearer to behave altruistically [500]. Hamilton's explanation for altruism rests on the premise that the unconscious goal of reproducing, from an evolutionary perspective, is to propagate one's distinctive alleles. Personal reproduction achieves this ultimate goal in direct fashion. But helping genetically similar individuals—that is, one's relatives—survive to reproduce provides an indirect route to the very same end.

To understand why, we must discuss the concept of the **coefficient of relatedness** (r), the probability that two individuals both possess the same rare allele by virtue of inheriting it from a recent common ancestor. Imagine that a parent has the genotype amy^1/amy^2, and that amy^2 is a rare form of the gene. Any offspring of this individual will have a 50 percent chance of inheriting the amy^2 allele because any egg (or sperm) that the parent donates to the production of an offspring has one chance in two of containing the amy^2 allele. Thus the coefficient of relatedness in this case is 1/2, or 0.5. In contrast, r is close to 0 between this parent and nonrelatives, which almost surely lack the allele in question.

Imagine an uncle and his sister's son. The uncle possesses the amy^2 allele. The uncle and his sister have 1 chance in 2 of sharing the allele in question, and the sister has 1 chance in 2 of passing that allele on to any of her offspring. Therefore the coefficient of relatedness for an uncle and his nephew is $1/2 \times 1/2 = 1/4$, or 0.25. The coefficients of relatedness for some other relatives appear in Table 3.

With this information on r values, we can determine the fate of a rare "altruistic" allele in competition with a common "selfish" allele. The key question is whether an individual can leave more copies of its alleles by helping relatives reproduce than it could if it reproduced personally instead. Let us say that an animal

Table 3   Coefficients of relatedness (*r*) between relatives

Relationship	*r*	Relationship	*r*
Full siblings	0.5	Parent–offspring	0.5
Half-siblings	0.25	Uncle/aunt–niece/nephew	0.25
Cousin–cousin	0.125	Grandparent–grandchild	0.25

could potentially have two offspring of its own, or give up reproduction altogether and invest entirely in relatives, helping three siblings survive that would not have lived otherwise. Offspring share half their genes with a parent; siblings also share half their genes with each other. Therefore, in this example, personal reproduction yields $r \times 2 = 0.5 \times 2 = 1$ unit of genes contributed directly to the next generation, whereas altruism yields $r \times 3 = 0.5 \times 3 = 1.5$ genetic units passed on indirectly in the bodies of relatives. In this example, the altruistic route results in a greater number of shared alleles being transmitted to the next generation.

Another way of looking at this matter is to compare the selective consequences of individuals who aid others at random versus those who direct their aid to close relatives. If aid is delivered randomly, then no one form of a gene is likely to benefit more than any other. But if close relatives aid one another selectively, then any distinctive alleles they possess may survive better, helping those alleles increase in frequency compared with other forms of the gene in the population at large.

When one thinks in these terms, it becomes clear that a form of selection can occur when genetically different individuals differ in their effects on the reproductive success of close relatives. Jerram Brown calls this form of selection **indirect selection,** which he contrasts with **direct selection** for traits that raise individual reproductive success (Figure 8) [146].

A brief digression is necessary at this point to deal with yet another term, *kin selection*. Although many persons treat this term as a synonym for indirect selection, Brown points out that kin selection, as originally defined by John Maynard Smith, refers to the evolutionary effects of both parental aid given to descendant kin (offspring) and altruism directed to **nondescendant kin** (relatives other than offspring). Biologists have long recognized that parents can affect the evolutionary process by increasing the survival chances of their offspring, and that parental care will evolve when the resulting increase in the survival of the aided offspring more than compensates a parent for the loss of opportunities to produce additional offspring in the future (see Chapter 14). In genetic terms, a parent can gain via parental investment because it shares 50 percent of its genes with each offspring. By the same token, however, individuals can promote the survival of certain of their genes by helping relatives other than offspring. Altruism can be favored by the component of kin selection that Brown calls indirect selection, provided that

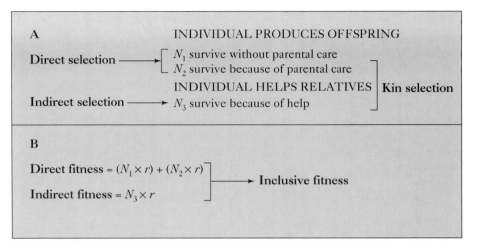

8 **The components of selection and fitness.** (A) Direct selection acts on variation in individual reproductive success. Indirect selection acts on variation in the effects individuals have on their relatives' reproductive success. (B) Direct fitness is measured in terms of personal reproductive output; indirect fitness is measured in terms of genetic gains derived by helping relatives. Inclusive fitness is the sum of the two and represents the total genetic contribution of an individual to the subsequent generation. *After Brown [146].*

the gain in genes transmitted via helped relatives more than makes up for the resulting lost opportunities for personal reproduction. The use of the term *indirect selection*, rather than *kin selection*, keeps the focus clearly on the distinction between parental effects on offspring and an aid-giver's effects on nondescendant kin, so we shall use it here [147].

Because fitness gained through personal reproduction (**direct fitness**) and through increased production of surviving nondescendant kin (**indirect fitness**) can both be expressed in identical genetic units, we can sum up an individual's total contribution of genes to the next generation, direct and indirect, creating a quantitative measure called **inclusive fitness** (see Figure 8B). We do not quantify inclusive fitness by adding up an individual's genetic representation in its offspring plus that in all its other relatives; instead we add up only its effects on gene propagation (1) directly in the bodies of surviving offspring *that owe their existence to the parent's actions, not to the efforts of others*, and (2) indirectly via nondescendant kin that owe their existence to the altruist's helpful actions. As David Queller has noted, the primary value of the concept lies in its ability to help us compare the evolutionary (genetic) consequences of competition between two alternative traits or strategies [972]. Calculating the inclusive fitness effects of strategy A versus strategy B can tell us which of the two should be adaptive under certain conditions, and which therefore can be predicted to be employed by individuals living under those conditions.

Hamilton's Rule

All of this leads us to "Hamilton's rule," which states that in order for an altruistic act to be adaptive (gene-promoting), its direct fitness cost to the altruist (the number of offspring not produced, $*c$, times the coefficient of relatedness between parent and offspring, r_c) must be less than its indirect fitness benefit (the *added* number of relatives that exist thanks to the altruist's action, $*b$, times the coefficient of relatedness between altruist and recipient(s), r_b). For example, if the genetic cost of an altruistic act were the loss of one offspring ($1 \times r = 1 \times 0.5 = 0.5$ genetic units), but the altruistic act led to the survival of three nephews that would have otherwise perished ($3 \times r = 3 \times 0.25 = 0.75$ genetic units), the altruist would receive a net genetic gain, thereby increasing the frequency of any distinctive allele associated with its altruistic behavior.

A concrete example of the utility of Hamilton's rule is offered by Heinz-Ulrich Reyer's study of the pied kingfisher [985]. These attractive African birds nest colonially in tunnels in banks by large lakes. Some males that become *primary helpers* do not breed during the first year of their adult life. Instead, they help their parents, bringing food to their mother and her nestlings while attacking predatory snakes and mongooses. Some pied kingfisher males cannot acquire a mate because males outnumber females in the population; primary helpers are drawn from this group. The key question is, are these males propagating their genes as effectively as possible by helping to raise their siblings [147]? Males that cannot find a mate do have other options: they can help unrelated nesting pairs as *secondary helpers*, or they can become *delayers*, sitting out the breeding season, waiting for the next year to try to find a mate.

To learn why primary helpers help, we need to know the costs and benefits of their actions. Primary helpers work much harder than delayers, and they also do more work than secondary helpers, which take a more relaxed approach to helping (Figure 9). The greater sacrifices of primary helpers correlate with a lower rate of return to the breeding grounds (just 54 percent) in the following year compared with secondary helpers (74 percent return) or delayers (70 percent return). Furthermore, only two in three surviving primary helpers find a mate in their second year and reproduce personally, whereas 91 percent of returning secondary helpers succeed in breeding. Many one-time secondary helpers breed with the female they helped the preceding year (10 of 27 in Reyer's sample), suggesting that improved access to a potential mate is the ultimate reason for their initial altruism.

These data enable us to calculate the cost to primary helpers in terms of reduced personal reproduction in the second year. For simplicity's sake, we shall restrict our comparison to primary helpers that help their parents rear siblings in year one with no other helpers present and then breed on their own in the second year, if they survive and find a mate, versus secondary helpers that help nonrelatives

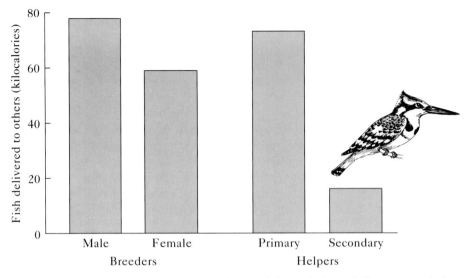

9   **Altruism and relatedness in pied kingfishers.** Primary helpers deliver more calories per day in fish to a nesting female and her offspring than do secondary helpers, which are not related to the breeders they assist. *Source: Reyer [985].*

with no other helpers present in year one and then reproduce on their own in the second year, if they survive and find a mate.

Primary helpers throw themselves into helping their parents produce offspring *at the cost of having less chance of reproducing personally in the next year.* Although primary helpers do better than delayers in the second year (0.41 versus 0.29 units of direct fitness), secondary helpers do better still (0.84 units of direct fitness) because they have a higher survival rate and greater probability of securing a partner (Table 4).

But is the cost to primary helpers of 0.43 lost units of direct fitness (0.84 – 0.41 = 0.43) in the second year offset by a gain in indirect fitness during the first year? To the extent that these males increase their parents' reproductive success, they create siblings that would not otherwise exist, indirectly propagating their genes in this fashion. In Reyer's study, the parents of a primary helper gained an *extra* 1.8 offspring on average when their son was present. Some primary helpers assisted their genetic mother and father, in which case the extra 1.8 siblings were full brothers and sisters, with a coefficient of relatedness of 0.5. But in other cases, one parent had died and the other had remated, so that the offspring produced were only half-siblings ($r = 0.25$). The average coefficient of relatedness for sons helping a breeding pair was between one-quarter and one-half ($r = 0.32$). Therefore, the average gain for helper sons was 1.8 sibs $\times$ 0.32 = 0.58 units of indirect fitness, a figure higher than the mean direct fitness loss experienced in their second year of life.

To calculate the average *inclusive* fitness for a primary helper over 2 years, we

Table 4 Calculations of inclusive fitness for male pied flycatchers exercising different behavioral options in their first year, and then trying to breed in their second year

Behavioral tactic	First year			Second year					Inclusive fitness
	y	r	f_1	o	r	s	m	f_2	$f_1 + f_2$
Primary helper	$1.8 \times 0.32 = 0.58$			$2.5 \times 0.50 \times 0.54 \times 0.60 = 0.41$					0.99
Secondary helper	$1.3 \times 0.00 = 0.00$			$2.5 \times 0.50 \times 0.74 \times 0.91 = 0.84$					0.84
Delayer	$0.0 \times 0.00 = 0.00$			$2.5 \times 0.50 \times 0.70 \times 0.33 = 0.29$					0.29

Note: Symbols: y = extra young produced by helped parents; o = offspring produced by breeding birds; r = coefficient of relatedness between the male and y or o; f_1 = fitness in year 1, f_2 = fitness in year 2; s = probability of surviving from year 1 to year 2; m = probability of finding a mate in year 2.
Source: Reyer [985].

add $0.58 + 0.41 = 0.99$ total genetic units. This figure is greater than the equivalent sum for secondary helpers ($0 + 0.84 = 0.84$). (Secondary helpers raise the reproductive output of the pairs they assist, but they are not related to the extra offspring, and so gain no *indirect* fitness benefits from their actions.)

The inclusive fitness figures calculated here do not include the units of indirect fitness gained by primary helpers as a result of improving the survival chances of their parents (see below), which would further raise their inclusive fitness [849]. Our calculations are meant to illustrate that primary helpers sacrifice future personal reproduction in exchange for *increased* numbers of nondescendant kin [985]. Because these added siblings carry some of the helpers' alleles, they provide sufficient indirect fitness gains to offset the loss in direct fitness that primary helpers experience in their next year relative to secondary helpers.

Helpers at the Nest

In effect, Reyer's study of the pied kingfisher provides support for two adaptationist hypotheses on the evolution of helping: (1) primary helpers raise their fitness indirectly through the increased production of nondescendant kin, whereas (2) secondary helpers raise their fitness directly by increasing their future chances of reproducing personally. The two kinds of male helpers behave differently, matching their brand of altruism to the kind of breeders that receive it. Such discriminating altruists in the past must have spread their distinctive mutant alleles through the population, modifying the ancestral form of helping behavior, which was probably less finely tuned at the outset. Of course, all of this assumes that the fairly small sample of data collected by Reyer over a few years accurately reflects what was going on in the species as a whole during this time; moreover, persons who accept Reyer's conclusions, as I do, assume that the fitness consequences of the various behavioral options open to male pied kingfishers have been much the same over thousands of years.

The pied kingfisher is merely one of a large number of birds and mammals now known to have helpers at the nest [147, 610, 1137]. Each case presents a separate puzzle that deserves to be analyzed in the light of a full range of hypotheses [355]. Consider Ian Jamieson's nonadaptationist argument that caring for another's offspring could occur purely as a nonadaptive side effect of other adaptive traits [598, 599]. This hypothesis proposes that helping *originated* as a result of a mutation, or changed social or ecological conditions, that caused some birds to delay their dispersal from their natal territory. The fitness consequences of delayed dispersal could have been positive for any of a number of reasons unrelated to helping relatives, favoring young birds that exercised this option, at least until changed conditions made dispersal advantageous [664]. But while remaining on their natal territory, the young birds might have been exposed to nests containing offspring of their breeding parents. At a proximate level, the visual stimulation provided by the nestlings could have triggered "helping" behavior, especially feeding of the nestlings, by the nonbreeding birds. These actions could be maintained in the population indefinitely, even if they provided no direct or indirect fitness benefits, and indeed, even if they carried a small fitness cost to the helper. All that would be required is that the helper derive reproductive benefits from delayed dispersal and its ability to be parental eventually, at which time it would feed its own offspring adaptively.

This nonadaptationist hypothesis for helping differs from various alternative adaptationist hypotheses (Table 5), especially with respect to its focus on the possible origins of "helping" and any subsequent changes that might have made it more selective, an issue related to the reconstruction of the trait's history rather than its current utility (see Chapter 7). If our goal is to discriminate among the var-

Table 5 The difference between nonadaptationist and adaptationist hypotheses on helping at the nest in birds

Historical sequence			Current function[a]
Origin of helping	Original function[a]	Subsequent modification	
NONADAPTATIONIST HYPOTHESIS			
Parental care	None	None	None
ADAPTATIONIST HYPOTHESES: DIRECT FITNESS GAINS			
Not specified	Mutualism	None	Mutualism
Not specified	Mutualism	Yes	More effective mutualism
ADAPTATIONIST HYPOTHESES: INDIRECT FITNESS GAINS			
Not specified	Altruism	None	Altruism
Not specified	Altruism	Yes	More discriminating altruism

[a] "Function" refers to the adaptive value of the trait with respect to helping individuals other than offspring.

Breeding system   Helping behavior

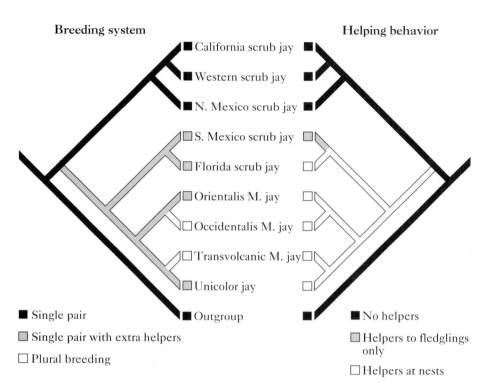

California scrub jay

Western scrub jay

N. Mexico scrub jay

S. Mexico scrub jay

Florida scrub jay

Orientalis M. jay

Occidentalis M. jay

Transvolcanic M. jay

Unicolor jay

Outgroup

■ Single pair

☐ Single pair with extra helpers

☐ Plural breeding

■ No helpers

☐ Helpers to fledglings only

☐ Helpers at nests

10   The evolution of helping at the nest in species and subspecies of jays belonging to the genus *Aphelocoma*. The Florida scrub jay is one of several members of the genus in which helping at the nest occurs, a trait that apparently originated in one ancestral jay that gave rise to all these species. *Source: Brown and Li [150].*

ious hypotheses on the possible adaptive value of helping, we need to determine whether helpers raise their *indirect* fitness or their *direct* fitness by helping at the nest. If the actions of "helpers" fail to increase the production of nondescendant kin or fail to increase their personal reproductive success, we can reject all hypotheses on the adaptive value of helping per se.

Many studies of the current utility of helping at the nest have found that helpers favor close relatives (e.g., [359, 360]) and that they reap fitness benefits as a result [147, 355]. For example, consider the Florida scrub jay, which has been studied for many years by Glen Woolfenden and John Fitzpatrick [1307]. This jay is a member of a cluster of closely related species in which nonbreeding birds help rear both nestlings and fledglings (Figure 10). These nonbreeding adults may be 2 or 3 years old, or even older; they are physiologically capable of producing offspring of their own, but instead defend the breeding pair's territory and feed the nestlings while also detecting [794] and repelling predators (Figure 11). If one

11 Cooperation among scrub jay relatives. Helpers at the nest in the Florida scrub jay provide food for the young, defense for the territory, and protection against snakes. *Source: Wilson [1294].*

measures the number of offspring fledged by pairs with and without helpers, one finds an increase, albeit a modest one (Table 6), for pairs with helpers.

But what if good territories can support helpers, while bad ones cannot? If this were the case, then the positive association between helpers and offspring fledged might not have been caused by the helpers, but rather by the better food supplies or superior nesting sites on the territories that just happened to have "helpers." To control for this possibility, Ronald Mumme experimentally captured and removed the nonbreeding helpers from some randomly selected breeding pairs, while leaving other helpers unmolested. The removal of helpers reduced the reproductive success of the experimental pairs by about 50 percent, as measured by the number of offspring known to be alive 60 days after hatching (Figure 12). Mumme concluded that helpers really do help [846].

Who are these helpers? Woolfenden and Fitzpatrick studied color-banded birds for years in order to show that helper jays are, almost without exception, the off-

Table 6 Effect of scrub jay helpers at the nest on the reproductive success of their parents and on their own inclusive fitness

	Inexperienced pairs[a]	Experienced pairs
Average number of fledglings produced with no helpers	1.03	1.62
Average number of fledglings produced with helpers	2.06	2.20
Increase due to help	1.03	0.58
Average number of helpers	1.70	1.90
Indirect contribution to fitness per helper	0.60	0.30

[a] Pairs in which one or both members were breeding for the first time.
Source: Emlen [349].

spring of the pair they help. Scrub jay groups contain no "secondary helpers" of the sort Uli Reyer found in pied kingfishers. Because helpers boost the survival chances of full and half-siblings, they immediately raise their inclusive fitness by adding 0.15–0.30 units to the indirect fitness column (see Table 6) each time they help their parents. Moreover, helper scrub jays improve the chances that their parents will live to breed again another year, as do helpers in the pied kingfisher. Improved parental survival means that helpers are responsible for still more siblings in the future; these extra siblings yield an average of about 0.30 additional indirect fitness units for helpers [849]. Thus, the total indirect fitness gains from altruistic helping can potentially exceed its costs in terms of lost direct fitness, particularly if the young birds have almost no chance of reproducing personally. Woolfenden believes that the habitat of scrub jays is saturated with territorial

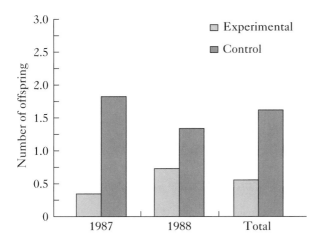

12 Helpers at the nest help parents raise more siblings in the Florida scrub jay. The graph shows numbers of offspring alive after 60 days in experimental nests that lost their helpers (lighter bars) and in unmanipulated nests (darker bars) during a 2-year experiment. *Source: Mumme [847].*

groups, which means that there are few openings for dispersing young adults. When territorial vacancies are extremely rare, helping becomes an adaptive option for subordinate young adults.

However, the extent to which available nesting habitat really is in unusually short supply for bird species with helpers at the nest is a matter of debate [714, 963]. Jan Komdeur examined the issue in an experiment with another bird species with helpers, the Seychelles warbler. He transplanted 58 birds from one island (Cousin) to two other nearby islands with suitable but unoccupied warbler habitat. In so doing, Komdeur created vacant territories on Cousin. If young Seychelles warblers remain on their natal territories because they cannot find suitable open habitat, then helpers on Cousin should have moved into the openings created by the removal of some territory holders. They did, almost immediately filling all the vacant territories.

Since the islands that received the transplants had more potential territories than warblers, Komdeur expected that the offspring of the transplanted adults would leave home promptly in order to breed in territories of their own. They did, providing further evidence that young birds help only when they have little chance of making direct fitness gains by dispersing [668].

Komdeur provided even more insight into the decision-making abilities of young birds by comparing the behavior of first-year birds on territories of *different* quality on the saturated island, Cousin. Breeding birds occupy sites that vary in size, vegetational cover, and insect supplies. By using these variables to divide warbler territories into categories of low, medium, and high value, Komdeur showed that territory quality affected the survival chances of first-year birds that remained to help their parents, as well as the indirect fitness gains they could achieve by helping their parents. Youngsters on good territories were likely to survive there as helpers while increasing the number of siblings produced by their parents, and they often stayed put. In contrast, young birds left poor natal territories, where they had little chance of making it to the next year and where they could have little positive effect on the reproductive success of their parents (Table 7).

Table 7 The effect of the quality of the natal territory on the willingness of first-year birds to delay dispersal in order to help their parents

	Territory quality		
	Low	Medium	High
Increase in fledglings produced by pairs with helpers	0.03	0.34	0.42
Proportion of helpers that survive the year	0.30	0.67	0.86
Proportion of young birds that delay dispersal	0.29	0.69	0.93

Sources: Data from Komdeur [667]; Mumme [846].

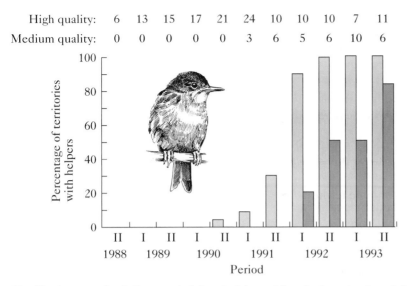

| High quality: | 6 | 13 | 15 | 17 | 21 | 24 | 10 | 10 | 10 | 7 | 11 |
| Medium quality: | 0 | 0 | 0 | 0 | 0 | 3 | 6 | 5 | 6 | 10 | 6 |

13 Territory quality influences helping decisions. After the introduction of the Seychelles warbler to Aride Island, the population grew until all the best territories were taken, after which some birds were forced to settle for medium-quality sites. Subsequently, helpers at the nest first appeared in the high-quality territories (lighter bars), and only later in some medium-quality territories (darker bars). Numbers at the top are the numbers of territories that could have had helpers. Source: *Komdeur et al. [669].*

If the birds' dispersal decisions on Cousin were influenced by the availability of open habitat and the relative quality of their natal territories, then on the once-unoccupied island of Aride, as the transplants reproduced and the growing population filled more and more of the available sites, helpers should have appeared first in the best territories, later on in medium-quality sites, and last in low-quality territories [669]. Note that this prediction is based on optimal habitat selection theory, with its assumption that individuals will make decisions that maximize their net fitness gains (see Chapter 11). Whether it is optimal for a Seychelles warbler to disperse or to stay on as a helper depends on its particular circumstances. The birds did indeed make habitat selection decisions based on the options available to them as individuals, decisions that maximized their inclusive fitness, whether achieved via direct reproduction or indirectly by helping their parents rear more siblings (Figure 13) [667, 846].

Alarm Calls and Indirect Selection

Helping at the nest is not the only type of potential altruism. In Chapter 9, we discussed the advantages to social animals of giving alarm signals upon detection of a predator. You may recall the case of the whistling Belding's ground squirrels whose alarm calls alert others to the arrival of a hunting hawk, to the mutual benefit of the signal giver and the signal receivers [1084]. Belding's ground squirrels also produce a special alarm call, a staccato whistle, when they spot a *terrestrial* predator, such as a coyote or badger (Figure 14). This alarm call also causes other squirrels to rush for safety. Is this another case of cooperation, or is it altruism? The answer has been helpfully provided by Paul Sherman, whose findings enable us to discriminate among the following alternative hypotheses [1082]:

1. *Direct selection hypotheses:* The caller enhances its personal chances for reproductive success by giving an alarm signal.

    a. *The predator confusion hypothesis:* The signal's function is to alert the caller's neighbors so that they can all run together for safety, confusing the terrestrial predator and helping everyone in the process—except the predator.

    b. *The predator deterrence hypothesis:* Although one might think that the caller risks drawing attention to itself and should slip off quietly when it spots danger, predators that know they have been detected are more likely to abandon the hunt. The caller signals to communicate with the predator to save its own skin.

    c. *The reciprocal altruism hypothesis:* A caller may give the signal at some immediate risk to itself, but later it will be more than repaid by others in the group when they return the favor.

    d. *The parental care hypothesis:* A caller gives the signal to warn its offspring, increasing their chances of survival and thus the caller's direct genetic contribution to subsequent generations.

2. *Indirect selection hypothesis:* The caller reduces its lifetime chances for reproductive success by sounding the alarm, but its altruism nevertheless raises its inclusive fitness. Parents, aunts, uncles, brothers, sisters, and cousins are alerted by the signal, generating indirect fitness benefits that outweigh the direct fitness costs of the action.

14   A Belding's ground squirrel that has spotted a terrestrial predator giving an alarm call. *Photograph by George Lepp.*

We can quickly eliminate some of the hypotheses listed above by demonstrating that alarm calling when a terrestrial predator is present puts the caller at considerable risk. Terrestrial predators are not confused (hypothesis 1*a*) or deterred from a hunt (hypothesis 1*b*) when they hear a squirrel call. Sherman and his gang of observers saw weasels, badgers, and coyotes stalk alarm callers and kill them at a rate higher than noncalling, fleeing ground squirrels. Moreover, the probability that an individual will give an alarm call is not correlated with familiarity or length of association between the caller and the animals that benefit from its signal. This result decreases the probability that alarm calling is maintained through tit-for-tat reciprocity (hypothesis 1*c*), which requires long-term associations in order to be evolutionarily stable [1083].

We are left with the parental care (1*d*) and the altruism (2) hypotheses, both of which predict that females, rather than males, will be more likely to give risky alarm calls. Female Belding's ground squirrels tend to be sedentary, and therefore a female often lives with her daughters, sisters, aunts, and nieces. Males, on the other hand, move away from the natal burrow (see Chapter 10) and do not live near offspring that they might help. According to the parental care hypothesis, females should give more alarm calls than males because only female callers are likely to be compensated for their costly help by the improved survival chances of their offspring. The altruism hypothesis suggests that females should give more calls because the loss in future reproduction that they might suffer by attracting predators to them might be more than repaid indirectly through the increased probability of survival of kin *other than offspring*.

Sherman found that females are in fact much more likely to give alarm calls when they spot a predator than are males, and that females with relatives living nearby call more frequently than females without relatives as neighbors (Table 8) [1082]. Thus, alarm calling by female Belding's ground squirrels may yield direct fitness gains, if females help their offspring escape from predators, and indirect fitness gains, to the extent that aunts, nieces, and sisters escape as well.

Table 8   Who gives alarm calls among Belding's ground squirrels?

Category of squirrel	Exposures to predator[a]	Squirrels giving alarm call (%)	Squirrels expected to call if alarms are random (%)
Males >1 year old	67	12 (18)	19 (28)
Females >1 year old with living relatives[b]	190	75 (39)	53 (28)
Females >1 year old without living relatives	168	31 (18)	46 (28)

[a]Number of times ground squirrels in each category were present when a terrestrial predator appeared, 1974–1979.

[b]Relatives consist of daughters, granddaughters, mothers, or sisters.

Source: Sherman, unpublished data.

15 Discriminating cooperation among Belding's ground squirrels. Only close relatives defend a territory together in this species, with mothers assisting daughters and sisters cooperating with sisters in the defense of space. *Source: Sherman [1083].*

	Number of pairs	Proportion of cooperative chases by pairs of this type
Mother–daughter	59	
Littermate sisters	52	
Nonlittermate sisters	36	
Grandmother–granddaughter	17	
Aunt–niece	15	
Gtgrandmother–gtgranddaughter	4	
Half-aunt–niece	9	
First cousins	10	
First cousins, once removed	3	
Nonkin	89	

The fact that females without young, but with nondescendant kin nearby, will sound the alarm suggests that indirect selection contributes to the maintenance of this behavior in the squirrel population, complementing the action of direct selection. Further evidence for the effects of indirect selection on ground squirrel behavior comes from studies by Warren Holmes and Paul Sherman, who showed that females help close female relatives as well as their offspring in territorial conflicts with intruders (Figure 15) [567]. Therefore, both direct and indirect selection probably contribute to the evolution of alarm calling in Belding's ground squirrels, whose cries can serve a parental function even while simultaneously assisting other relatives. One unit of inclusive fitness gained by keeping offspring alive is no different from one unit of inclusive fitness gained by saving the lives of assorted nondescendant kin.

Mating Cooperation among Males

Although parental care and altruism are restricted to females in Belding's ground squirrels, male animals as different as dwarf mongooses [1014], lions [88, 166], and bottle-nosed dolphins [216] specialize in a different kind of cooperation, the joint defense of groups of females (Figure 16). In lions, for example, the males in a coalition living with a pride of females hurry to confront any opponents who dare to roar in their territory. When Jon Grinnell and his co-workers played taped roars in lion country, the resident males often approached the tape recorder and attacked a stuffed dummy lion placed nearby (Figure 17). The eviction of intruder males is essential to the maintenance of the residents' control of a pride, and

16 A coalition of male lions, taking a break from defense of their mates against other coalitions that would wrest control of the pride from them if they could. *Photograph by Craig Packer and Anne Pusey.*

thus all members of a coalition benefit from a cooperative response to potential threats to this control [475].

Once some males have begun to cooperate in this manner, they select for the trait in other males, forcing individuals to sacrifice the chance to monopolize an entire pride of females in order to have any chance at all of mating. The more males in a coalition, the longer they are likely to maintain control of the females. Yet when groups of males associate with a pride, they do not share the receptive females equally. In one coalition of six lions, for example, the top male copulated 3.5 times as frequently as the low male in the hierarchy. In general, groups of four or more male lions exhibit great variation in mating success because they usually outnumber the estrous females in a pride. Less variation in reproductive success occurs in small groups, since the ratio of estrous females to males is higher. In large groups, males excluded from mating at any one time because of a shortage of females are in effect nonreproducing helpers who assist their sexually active companions in keeping rivals away from the pride. They are making the best of a bad job; unable to hold a pride by themselves, they are forced to cooperate with others to have some chance of mating at some time during the tenure of the coalition [661].

But if males in large male coalitions are forced to sacrifice direct fitness, perhaps they can gain back some compensatory indirect fitness by joining only large groups of close male relatives. To examine this possibility, Craig Packer and his associates compared the genetic relationships among males in large coalitions (four

17 Cooperation among male lions in a coalition includes the task of defending the pride against intruders. In this case, the "intruder" is a stuffed lion; the experimenters played a tape of a roaring male to attract a pair of resident males to the spot. *Source: Grinnell et al. [475].*

to six males) with those among small partnerships (two or three males) [896]. Packer's crew used DNA fingerprinting techniques (see Chapter 12) to determine the genetic relationships of pride members. As expected, they found that males in large coalitions possessed more similar DNA fingerprints on average than did males in smaller groups. Thus, duos and trios were often, though not always, unrelated individuals that formed cooperative social teams to compete with other groups, whereas large groups were clusters of brothers, half-brothers, and cousins that had stayed together after leaving their natal pride. In these large groups, a male that had little reproductive success was, to use Brian Bertram's phrase, "reproducing by proxy through his companions," because they all had a recent common ancestor and shared a relatively high proportion of the same alleles [88].

Manakin Coalitions and Cooperative Courtship

The altruists in large coalitions of lions appear to help relatives, but when genetic strangers cooperate, they usually do so for mutual reproductive advantage. Does the long-tailed manakin of Central America provide an exception to this general rule, given that males form teams of nonrelatives to court females—but only one male reaps the copulatory reward from their mutual efforts [397, 792]?

These joint efforts are extraordinary. Pairs of beautifully plumaged males, each outfitted in a stunning combination of dark blue, powder blue, and crimson, with two long, trailing tail feathers, cooperate by loudly whistling "toe-lay-doe" together as many as 300 times per hour at a display perch. What sounds like one male's "toe-lay-doe" is actually produced by the two individuals calling in almost perfect synchrony. The closer the match between the sound frequencies used in the two birds' calls, the more likely females are to visit the callers [1198]. Visiting females land on the pair's display perch, often a horizontal section of liana vine that lies a foot or so above the ground. In response, the two males dart in and land close to the prospective mate before performing an astonishing cartwheel display (Figure 18). The male closest to the female leaps into the air and then begins to come down where his partner is perching. To make room for the descending bird, the other male sidles quickly over to the jumping-off point that had been used by his companion, and then leaps up to exchange places with him. After a series of these cartwheels, one male may leap over the head of the observant female, with the other member of the pair instantly following in order to perform a new run of cartwheels.

As if this were not sufficiently dramatic, long-tailed manakins have other joint displays, including a combined slow fluttering flight that David McDonald calls the "butterfly flight." Birds that are helicoptering back and forth in this fashion may suddenly switch to the other extreme, in which they race at top speed around the perched female. Despite the length and elaborate nature of the cooperative routines of the males, female visitors only rarely signal that they are ready to mate, by beginning to jump about on the perch. At this cue, one member of the display duo discreetly leaves, while the remaining male stays to copulate. The female then

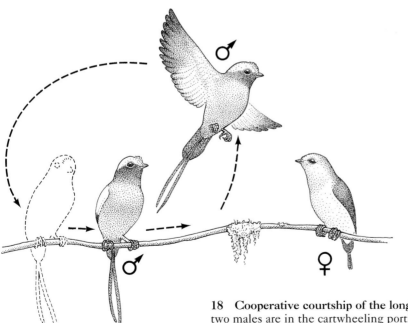

18   Cooperative courtship of the long-tailed manakin. The two males are in the cartwheeling portion of their dual display to a female, who is perched on the vine to the right.

flies off, and the mated male calls for his partner, who flies back to resume his cooperative calling and courtship duties.

By marking the attenders at display perches, McDonald and his manakin watchers found that one male is dominant to all other males that display at a given site. The top male has a favorite display companion, a beta male, which in turn is dominant to a variable number of other part-time cooperators [397, 792]. Over many years, McDonald's team recorded 263 copulations by marked males; the alpha male contributed 259 to this total, a finding that makes the reciprocal altruism hypothesis untenable.

Having eliminated the reciprocal altruism and indirect fitness hypotheses, what is left? The answer comes from McDonald's patient tracking of what happens to males over their lifetimes. He found that the dominance hierarchy at a display site changed slowly over the years, with some young lowly subordinates gradually moving up the ladder to reach the beta position. Only by becoming a beta male can a long-tailed manakin eventually inherit the alpha spot when the top male dies. Betas acquired the top position in all 11 cases in which an alpha male disappeared during McDonald's study. By displaying with the alpha male, a beta individual establishes his claim to be next, keeping other (younger?) birds at bay for years—although just why manakin hierarchies should be so stable is intriguing. When a beta male succeeds to the alpha position, his chances of copulating improve sharply, because the females that have been visiting the site generally continue to do so after his elevation. As a result, the old beta, now alpha, male's mating success correlates with that of his deceased predecessor (Figure 19) [792]. Thus, cooperation between alpha and beta males has evolved through direct selection, with an eventual reproductive payoff going to patient and long-lived subordinates.

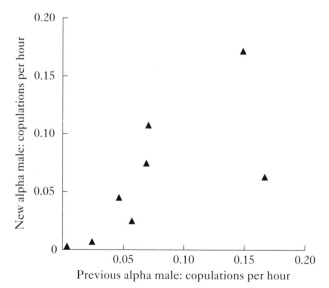

19 Cooperation with an eventual payoff in the long-tailed manakin. After the death of his alpha male partner, the beta male (now an alpha) will copulate about as frequently as his predecessor did, presumably because the females attracted to the duo in the past continue to visit the display arena when receptive. *After McDonald and Potts [792].*

The Evolution of Eusocial Behavior

Although a Martian ethologist would surely appreciate the ingenuity of earthbound researchers in devising and testing alternative hypotheses on the helpful behavior exhibited by kingfishers, lions, and long-tailed manakins, he or she would probably find being stung by a honeybee more fascinating (Figure 20). Although the pain of a bee sting is attention-getting, a Martian aware of the theory of evolution by natural selection would be even more impressed by the discovery that the bee that stung him was only one of thousands of nonreproducing individuals living together in a hive [1074]. In fact, of the twenty to forty thousand bees in a typical honeybee colony, only one is likely to reproduce. This queen bee is surrounded by a huge workforce of effectively sterile daughters that care for the brood, build new wax combs, maintain the proper temperature within the hive, collect pollen, and store honey (see Figure 4 in Chapter 4). Worker bees are much smaller than the queen of their colony, and their ovaries are correspondingly reduced—a proximate basis for their usual failure to produce eggs. Although very occasionally a worker will make and lay a few eggs, their mother, the queen bee, is so well fed by her workers, and her ovaries are so much larger, that she can manufacture hundreds of thousands of eggs in her lifetime.

20 Suicidal altruism. When a honeybee stings a vertebrate, she dies after leaving her stinger and the associated poison sac attached to the body of the victim. *Photograph by Bernd Heinrich.*

Nor are honeybees unique in forming immense colonies in which reproduction is dominated by one or a few females whose offspring are aided by sterile helpers. In various ant, bee, and wasp colonies, workers may come in several very different shapes and sizes, each type, or caste, of nonreproducing individuals specialized for a particular helpful role within the colony (Figure 21). For example, the largest and most aggressive workers in some species of ants constitute the soldier caste. Soldier ants have just one function in life: protecting the colony against dangerous invaders, such as predatory ants or raiding vertebrates, which they may sting or chop in half with shearing mandibles. Some termite soldiers also have formidable jaws, while others spray colony intruders with a sticky resinous glue (Figure 22) [1293]. Soldier insects engage in dangerous work, nowhere more obviously suicidal than in the case of the glue grenade ants that rush at enemies while constricting their abdominal muscles so violently that they burst a large abdominal gland, killing themselves, but splashing a disabling glue on the intruders [777].

How can we account for the evolution of **eusocial** (that is, caste-forming) insects? The historical sequence of events resulting in complex eusociality is made somewhat more understandable by the discovery of species intermediate in their behav-

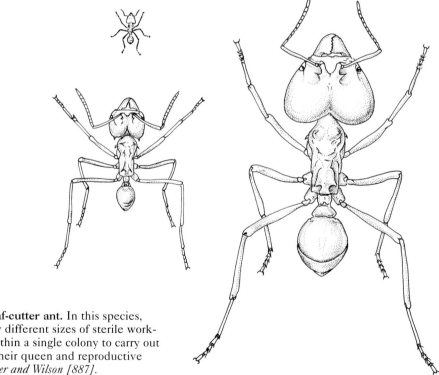

21 Castes in a leaf-cutter ant. In this species, several distinctively different sizes of sterile workers are produced within a single colony to carry out different tasks for their queen and reproductive nestmates. *After Oster and Wilson [887].*

22   Sterile worker and soldier termites. (Top) Large soldiers with cutting jaws guard a group of smaller workers. (Bottom) Soldiers of a nasute termite spray an entangling glue over a colony intruder, in this case, a fruit fly. *Photographs by (top) E. S. Ross and (bottom) E. Ernst.*

23 A solitary wasp, *Ammophila novita.* The female provisions her nests with moth larvae, like this caterpillar she is holding in her jaws, without help from any other member of her species. *Photograph by the author.*

ior between typical solitary insects, whose females provide care to their offspring all by themselves (Figure 23), and the extreme exponents of eusociality, such as honeybees, army ants, and termites. These weakly eusocial species have a colony organization that may provide clues about the transition from solitary living to complex eusociality. Among them are the social paper wasps in the genus *Polistes*.

A "typical" temperate-zone *Polistes* female emerges in the fall from a paper nest, often built under the eaves of a house (Figure 24). After mating, she will spend the winter hibernating in a crevice until spring comes, when she will leave her winter shelter and start building a papery nest of her own from chewed plant fibers. The nest contains a cluster of cells, each of which receives a single egg from the new queen. A foundress female may be joined by other overwintering females, often her sisters; this situation generates dominance contests and the formation of a hierarchy, with the alpha female reproducing more than lower-ranking individuals [981, 1152], which nevertheless help rear the larvae and protect the nest against predators, parasites [368], and nest usurpers [419].

The eggs that are laid early in the season are destined to become daughter wasps because they are fertilized by the queen wasp, using sperm stored from last fall's copulation. As these eggs hatch, the adult females feed the newborn larvae water, nectar, and fragments of insect prey. When the first female offspring emerge, they assist their mother and aunts in raising still more daughters (their sisters and cousins) rather than flying off to start new colonies in which to rear their own offspring. These workers are only a little smaller than the queen, and they can have functional ovaries, although they rarely lay many eggs.

Several broods are produced in this fashion during the colony's life, with several dozen females joining the workforce. They increase the size of the nest, add

24 Social paper wasps. Females of this *Polistes* wasp cooperate in the construction and defense of a nest and in the feeding and care of the young. One of the adult females here is the queen; the others are probably sisters of hers that do not reproduce but instead assist in the care of the eggs (visible in several upper cells) and larvae (the translucent grubs in many of the other cells). All three adult wasps are threatening the intruding photographer. *Photograph by the author.*

cells, feed the larvae, detect and drive off parasites, and sting predators and unlucky humans into retreat. Eventually, the queen produces broods of daughters, and then sons, that do not join the workers, but instead lounge around on the nest, appropriating food from their working sisters. Activity at the nest dramatically decreases as these "lazy" females and males fly off to find mates elsewhere. All the males die, but the mated females—future colony foundresses—hibernate through the winter months and resume the cycle the following spring [368].

Thus, the lives of paper wasps are dramatically different from those of solitary wasps, but the colonies they produce are not nearly as large as those of some eusocial insects, nor are the differences in size and structure among foundresses and workers nearly so dramatic. In addition, the skewing of reproduction among colony members is somewhat less pronounced, and the commitment to sterility by workers is also less firm at a paper wasp nest than in a honeybee hive. As a result, paper wasp societies suggest what an intermediate stage in the historical transition from a strictly solitary lifestyle to the exceptional eusociality exhibited by certain wasps, bees, ants, and termites might have been like.

Eusociality, Genetics, and Haplodiploidy

Even in weakly eusocial species, some females forego reproduction entirely to help others reproduce. One does not have to be a Martian to appreciate that reproductive self-sacrifice of this sort poses a fascinating evolutionary problem, one that Darwin himself struggled to solve. Darwin's solution was to point out that for the most part, each insect colony is an extended family. Therefore, by helping close relatives survive to reproduce, the sterile members of a colony in effect help maintain family traits, including the ability of reproducing individuals to generate some sterile helpers.

Darwin's theory for the evolution of sterile workers is based on what we now call indirect selection, with an indirect fitness benefit to the altruist arising from an increased production of reproducing relatives. More than 120 years after Darwin's contribution to the subject, W. D. Hamilton developed a formal cost–benefit analysis of worker altruism based firmly on the theory of indirect selection. Remember that according to Hamilton's rule, the (direct fitness) cost of altruism is calculated by counting the number of its own offspring that the altruist gave up in order to help others reproduce (*c) and multiplying this number by the degree of relatedness of parent to offspring (r_c). The (indirect fitness) benefit to the altruist is equal to the added number of reproducing relatives that the altruist helped produce (*b) multiplied by the degree of relatedness between the altruist and those it helped (r_b). When ($r_c \times$ *c) is less than ($r_b \times$ *b), altruism by the worker is adaptive because direct fitness losses are exceeded by indirect fitness gains.

Therefore, anything that increased the indirect fitness side of the equation would make the evolution of worker altruism more likely. Hamilton focused on r_b, the relatedness of the altruist to the relatives it helped. He was the first to point out that r_b could be unusually high in the Hymenoptera—the ants, bees, and wasps—because of the unusual method of sex determination in this group, which happens to be the insect order with by far the largest number of eusocial species [500]. Hymenopteran males are **haploid** (have only one set of chromosomes) because they arise from unfertilized eggs, whereas females are **diploid** (have two sets of chromosomes) because they are the product of the union of egg and sperm. Thus, if a female hymenopteran mates with just one male, all the sperm she receives will be identical, because the haploid male has only one set of chromosomes to copy when making gametes. As a result, all the daughters she produces will carry the same set of their father's genes. Any one daughter will share these genes (50 percent of her total genotype) with all her sisters. The other set of chromosomes will come from her mother. The mother's eggs are not genetically uniform because she is diploid; gamete formation in animals with two sets of chromosomes involves the production of a cell with just one set. The statistically average egg made by a female wasp, bee, or ant will have 50 percent of the same alleles carried in her other eggs. Thus, when hymenopteran eggs unite with genetically identical sperm, the resulting offspring share on average 75 percent of their alleles: 50 percent from their father and 0–50 percent from their mother (Figure 25).

A   Mother–offspring genetic relatedness

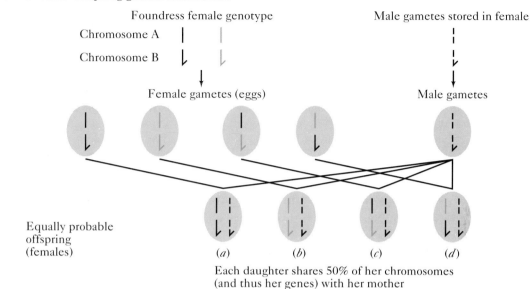

Each daughter shares 50% of her chromosomes
(and thus her genes) with her mother

B   Sister–sister genetic relatedness

Pick any daughter genotype and compare it with
the possible genotypes of her sisters

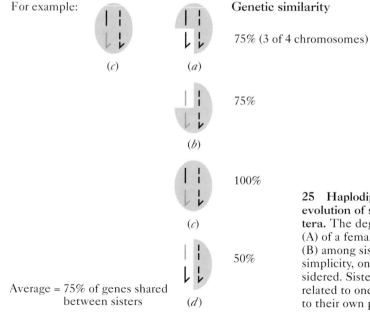

25   Haplodiploidy may play a role in the evolution of sociality in the Hymenoptera. The degree of genetic relatedness (A) of a female wasp to her offspring and (B) among sister wasps. For the sake of simplicity, only two chromosomes are considered. Sisters may be more closely related to one another than they would be to their own progeny.

Because of the haplodiploid nature of sex determination in the Hymenoptera, sisters may therefore be genetically very similar ($r = 0.75$) to one another, more so than a mother to her daughters and sons ($r = 0.50$). As a consequence of this genetic fact, ($r_c \times {}^*c$) should be less than ($r_b \times {}^*b$) more often in the Hymenoptera than in other groups, all other things being equal, which would facilitate the evolution of eusociality in the bees, ants, and wasps. The essence of Hamilton's *haplodiploid hypothesis* is that because of the genetics of sex determination in this group, worker hymenopterans can gain more inclusive fitness by helping their reproductively competent sisters (future queens) than by reproducing themselves. Provided that sisters really are closely related, indirect selection could favor females that, so to speak, put all their eggs (alleles) in a sister's basket and give up personal reproduction entirely.

The Haplodiploid Hypothesis Examined

The haplodiploid hypothesis generates some testable predictions, which we can examine. First, if a female worker in a eusocial bee, ant, or wasp colony is to cash in on her potentially high degree of relatedness with her fellow females, she should bias her help toward reproductively competent sisters rather than toward male siblings. Although sister bees, ants, and wasps share up to 75 percent of their genes in common, a sister has only 25 percent of her haploid brother's genes. Brothers do not receive any of the paternal genes that their sisters possess. The remaining half of the genome that sisters and brothers both receive from their mother ranges from 0 to 100 percent identical, averaging 50 percent; 50 percent of a half means that a sister has one-quarter of the genes of her brothers on average ($r = 0.25$).

Robert Trivers and Hope Hare recognized that these brother–sister inequalities in *r* should lead to a conflict of interest between queens and workers regarding how workers allocate their assistance [1202]. If workers share three times as many genes on average with their sisters as with their brothers, the stable investment ratio from the workers' perspective would be 3:1 in favor of sisters. In contrast, their queen mother donates 50 percent of her genes to each offspring, male or female; she presumably gains no genetic advantage by investing more resources in the production of sons ($r = 0.5$) than daughters ($r = 0.5$).

Various conflicts have indeed been observed in ant colonies, including efforts by queens to consume haploid eggs that workers have laid, eggs that would have become sons of the workers were it not for their cannibalistic grandmother [536]. In large eusocial colonies, however, if queen and workers "disagree" about the allocation of food to the queen's sons and daughters, the workers should hold the upper hand. In colonies with thousands of workers, the queen cannot possibly monitor everyone's behavior. Thus, the numerous workers could control how they care for and feed the larvae, withholding food from brothers in order to nourish sisters instead. If workers attempt to maximize their own inclusive fitness, then the combined *weight* of all the female reproductives (a measure of total resources devoted to female production) raised by the colony's workers should exceed the combined weight of the male reproductives by 3 to 1. When Trivers and Hare surveyed the literature, they found the expected skewing of investment toward females [1202].

However, workers should bias their production of reproductives toward females only if their mother mates just once. If sperm from two or more haploid males are used by a queen hymenopteran to fertilize her eggs, the resulting daughters with different fathers will not be closely related at all. Only when females have the same father will they share 75 percent of their genes in common (see Figure 25). As it turns out, some queens in a species of *Formica* ant do mate with two or even three males, while others do not. Liselotte Sundstrom realized that this species provided a wonderful opportunity to find out whether workers did indeed take their reproductive sisters' r_b into account when allocating food to future queens and kings. They did. The daughters of queens that had mated once heavily skewed their investment toward producing sister queens. But workers in colonies with multiply mated queens behaved quite differently. For them, brothers were as genetically valuable as sisters, and they did *not* bias the colony's production toward females [1160].

Ulrich Mueller has also shown that worker hymenopterans can alter their investment in colonymates according to their relatedness [845]. He experimentally manipulated colonies of a primitively eusocial bee, removing the foundress queen from some nests but leaving her alone in others. When a colony has its foundress queen, the usual asymmetry in relatedness between workers and their sisters ($r = 0.75$) and their brothers ($r = 0.25$) persists, and a bias toward female progeny is expected under the indirect selection hypothesis. But in colonies from which the foundress has been removed, a daughter assumes reproductive leadership. Under these conditions, her sister–workers are now helping her produce nieces ($r = 0.375$) and nephews ($r = 0.375$), rather than additional siblings. Thus, the relatedness asymmetry disappears, and workers ought to treat male progeny more favorably in these colonies. In fact, workers in the experimental colonies did invest more in males (the combined weight of which equaled 63 percent of the weight of all reproductives) than did workers in colonies that retained their foundress queen (in which males constituted 43 percent of the total weight).

The asymmetry in relatedness arising from haplodiploidy has also been used to explain differential treatment of siblings in certain parasitic wasps that sometimes lay two eggs on a moth caterpillar host. Each egg undergoes repeated divisions, forming a mass of about 1500 descendant eggs, which then all develop into separate offspring within the unfortunate caterpillar. When two eggs coexist on the same host, one gives rise to female larvae and the other to male larvae. However, a handful of the female eggs usually develop before all the rest, turning into precocious larvae with a selective appetite for their brothers, which they cannibalize. These precocious cannibals never metamorphose into adults, instead devoting their short lives to the removal of their brothers. Why?

26   Selective sibling rivalry in a parasitic wasp. Sterile, quick-developing (precocious) female larvae preferentially attack clusters of cells that will give rise to brothers while leaving their sisters alone. In these experiments, 33 pairs of all-male and all-female egg masses were placed in containers with a precocious larva for 30 minutes. The percentage of male and female broods attacked during this time is shown. *Source: Grbić et al. [467].*

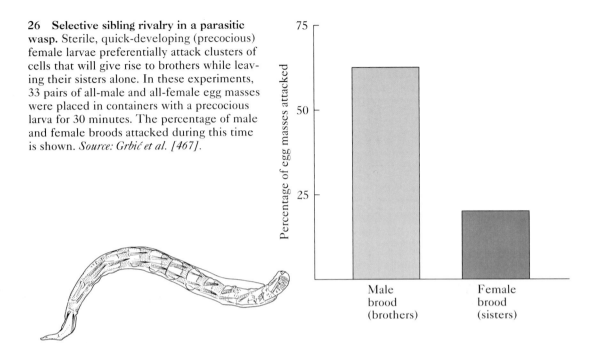

Thanks to the haplodiploid system of sex determination and the closer relatedness among sisters, female altruists have more to gain by helping their sisters survive to reproduce than by helping their brothers. They achieve this goal by eliminating some brothers and thus ensuring that more caterpillar flesh will be available for their maturing sisters, gaining indirect fitness benefits from their selective cannibalism (Figure 26) [467].

Very Close Relatedness Is Not Essential for the Evolution of Eusociality

Our attention thus far has been solely on how exceptionally close relatedness between helper and recipient could promote eusociality. However, much evidence exists to suggest that high values of r_b are *not* an absolute requirement for the evolution of eusocial systems. First, the queens of many eusocial ants, bees, and wasps mate more than once, a pattern that should tend to eliminate the close relatedness among the daughters of a queen, as we noted above. Queen honeybees, for example, often copulate with several dozen partners over a series of nuptial flights [665], a behavior that could greatly reduce the coefficient of relatedness among the workers in a colony. On the other hand, a female that mates with several males need not use the sperm from every male, and evidence exists that some females of social hymenopterans are indeed highly selective [103]. Therefore, the number of matings by a female does not permit an estimate of the degree of relatedness among her daughters. Direct measurements are required. When Kenneth Ross made these

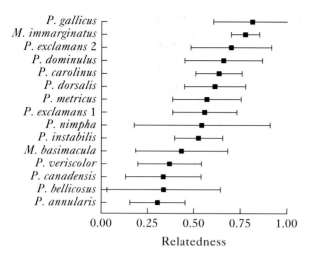

27 The mean coefficient of relatedness among female colony members in several eusocial wasp species, most of them members of the genus *Polistes*. Note that most values are well under the maximum possible (0.75). *Source: Strassmann et al. [1153].*

measurements in two species of eusocial wasps in which queens mate several times, the *maximum* average coefficient of relatedness among workers and their reproducing sisters did not exceed 0.4 [1018].

Further evidence that the workers in eusocial societies may not be especially closely related comes from the discovery that many such colonies contain more than one queen ruling conjointly, each producing eggs cared for by the worker force at large. In such a society, workers presumably help reproductive females that are not their sisters. Even in *Polistes* wasp colonies, which generally seem to be run by a single dominant female, Joan Strassmann and her colleagues have shown through genetic analyses that the actual average r of nestmates almost never reaches the 0.75 maximum value, and often is less than 0.50 (Figure 27) [1153]. These results indicate that many paper wasp queens either mate more than once or share reproductive "duties" with other females in their nests.

The point is that the haplodiploid system of sex determination does not guarantee that workers in eusocial hymenopteran societies will be very closely related, nor do fairly low levels of r prevent eusociality in these insects. In fact, other organisms that do not exhibit haplodiploid sex determination have evolved eusocial societies, showing that haplodiploidy and the potential for very high r_b values are not essential for the evolution of extreme social behavior. The termites, for example, are every bit as social as honeybees and paper wasps, despite the fact that both males and females are diploid. Termite colonies may have thousands of sterile workers (see Figure 22) that labor on behalf of a huge, bloated queen in a nest chamber set in the center of what may be an immense nest mound (see Figure 30) riddled with tunnels.

There is even one diploid vertebrate that is essentially eusocial [1088]. The bizarre-looking naked mole-rat is a little, hairless, sausage-shaped mammal (Figure 28) that lives in colonies of about 70 to 80 individuals. Each colony occupies

28 A eusocial mammal. Naked mole-rats live in colonies made up of a queen, kings, and a worker caste. *Photograph courtesy of J. U. M. Jarvis.*

its own complex maze of underground tunnels, which may total 3 kilometers in length. The immense size of their subterranean home stems from extraordinary cooperation among chain gangs of colony members, which work together to move tons of earth to the surface each year while burrowing in search of edible tubers. Yet when it comes to reproducing, breeding is restricted to a single big "queen" and several "kings" that live in a centrally located nest chamber. Females other than the queen do not even ovulate, but serve as sterile helpers at the nest, consigned to specialized support roles for the queen and kings, as are most of the males in the colony (Figure 29; Color Plate 20) [690].

But these sterile workers usually do labor on behalf of relatives. In all of the highly social animals studied to date, colonies are composed at least in part of related individuals, and as we noted above, workers may well have the ability to partition their aid in accordance with the coefficient of relatedness. Indeed, even diploid termite and naked mole-rat workers may be quite closely related to those they help because of intense inbreeding involving repeated brother–sister or son–mother matings. DNA fingerprinting has shown that the members of any given naked mole-rat colony are extremely similar genetically while differing greatly from the members of other colonies, a condition that provides high potential indirect fitness benefits to helpers within a colony [983].

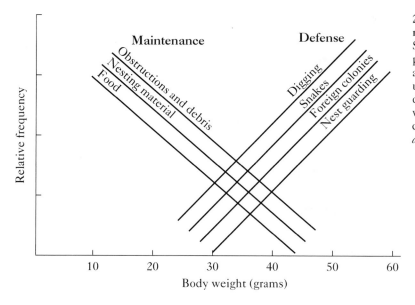

29 Division of labor in naked mole-rat colonies. Small individuals engage primarily in maintenance activities; larger individuals undertake digging and defense duties. Only the very largest members of the colony breed. *Source: Lacey and Sherman [690].*

The Ecology of Eusociality

If we take yet another look at Hamilton's rule that altruism can evolve when ($r_c \times {}^*c$) is less than ($r_b \times {}^*b$), we can see that a high r_b is not the only element in the equation. To the extent that ${}^*c$ (offspring the helper gave up to be an altruist) is low, adaptive altruism becomes more likely, provided that r_b is greater than zero. In other words, if the ecology of the species is such that young dispersing adults have very little chance of reproducing successfully, then nondispersers do not sacrifice much by foregoing personal reproduction in order to remain at their natal site to help other relatives. Under these circumstances, any indirect fitness gains need not be great in order to make altruism adaptive.

Richard Alexander has pointed out that for many social animals, especially the social insects, critical resources are highly limited outside the natal nest, which is often very difficult to establish, but once constructed, offers a safe, nearly impregnable structure (Figure 30) [13]. Under these circumstances, dispersing reproductives face the monumental and nearly hopeless task of building an entirely new nest. Consider what a foundress female of a eusocial ambrosia beetle goes through in order to establish her nest in a eucalyptus tree [641]. She will need 7 months of gnawing just to carve out the first 5 centimeters or so of what may eventually become an extensive gallery in the heartwood of the tree. Most foundresses probably are killed before they get safely deep into the tree. Once established, however, a colony can persist for decades, with helper daughters assured of a safe home in which to assist their mother in rearing reproductive males and new foundresses.

30 Australian termite colonies often live in huge, hard-packed mounds constructed by millions of workers. Costs of dispersal in eusocial insects may be high because of the low probability of finding a place as safe as the natal nest. The chance that a single termite from an established colony will successfully found a new equivalent colony is very small. *Photographs by the author.*

The average direct fitness payoff for individuals leaving an ambrosia beetle tunnel system or termite mound or naked mole-rat burrow is probably very low, but if some stay behind and help, they can do useful things for their relatives, generating relatively large numbers of future reproductives, perhaps increasing the odds that a few of these individuals will found new colonies when they eventually leave home.

One of the potentially valuable services of helpers is the defense of their mother's nest against predators and potential nest usurpers of their own species. An example of this sort of help occurs in certain gall-forming aphids. The foundress female induces a plant to form a hollow gall on a leaf petiole. Until the gall forms she is highly vulnerable, but once inside she is far safer. However, she becomes safer still after producing a first group of daughters, some of which are sterile soldiers (Figure 31). These soldiers use their specially thickened mouth parts or spiny legs to stab and pierce the larvae of ladybird beetles and other aphid consumers when they try to make their way through the narrow entrance into the gall. In experiments conducted by William Foster, an average of about 20 soldiers of *Pemphigus spirothecae* died while dispatching one predatory syrphid fly larva; in the absence of soldiers, a larva consumed all 100 nonsoldier aphids that Foster had assembled for his tests [400].

Suicidal aphid soldiers do not die in vain, because the beneficiaries of their actions are usually very close relatives, some of which have the ability to go on to reproduce. The high coefficient of relatedness between soldiers and those whose lives they save arises because they are the asexually produced offspring of their mother. A single aphid female can produce a clone by giving birth parthenogenetically to daughters, which also reproduce asexually. In colonies like this, soldiers typically defend others related to them by an *r* of 1 [593, 1146]—and you cannot get any more closely related than that. Thus the high coefficients of relatedness between social altruists and social recipients, the usefulness of the aid that altruists can supply, and the low costs of forgoing personal reproduction all work in the same direction to promote eusociality in these insects.

However, soldier aphids have been found in only about 50 of the 4500 or so described aphid species, and not even all gall-forming aphid species have these altruists. Therefore, other ecological factors must be involved in this phenomenon, perhaps those affecting the costs to a foundress female of allocating some of her daughters to the often suicidal, and necessarily nonreproductive, role of soldier [1146, 1147].

Defense of a fortress nest in wood or plant tissue on which colony members can safely feed occurs in at least one ambrosia beetle, a few aphids, and many termite species. But this is not the only possible way in which sterile workers can increase the production of their relatives, whether closely related or less so. David Queller and Joan Strassmann believe that in many of the eusocial ants, bees, and wasps, the sterile workers' most important service is gathering food for their larval relatives [973]. Whereas the fortress defenders live amid a wealth of digestible plant material, the typical ant, bee, or wasp must roam far from the nest site in search of scarce food. In so doing, it runs a gauntlet of predators. Because forager mortality rates are so high, a female that lived alone would often die before her brood achieved independence. If, however, a nesting female can enlist the aid of others, care will continue to be provided for her young even if she should die before they are reared. Under these circumstances, helpers related to the pri-

31 Altruism in aphids. Four species of aphids in which (left) obligate sterile soldiers with enlarged grasping legs and short, stabbing beaks protect their more delicate colonymates (right), which have the potential to reproduce when mature. The species are drawn at different scales. *Drawings by Christina Thalia Grant, from Stern and Foster [1146].*

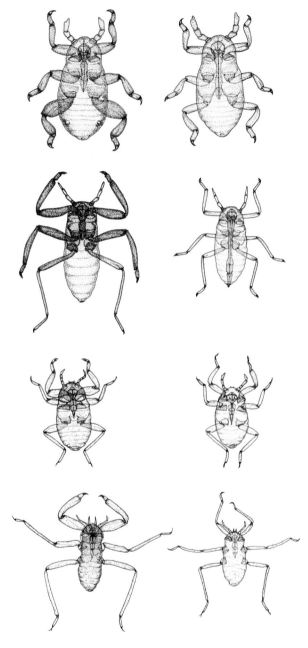

mary reproductive female gain considerable indirect fitness by bringing the brood through to adulthood [970].

Applying these categories of social altruism to vertebrates, naked mole-rats can be considered fortress defenders whose workers help prevent usurpation of the network of protective underground tunnels, while avian helpers at the nest are life insurers, providing the extended care their *altricial* (helpless newborn) siblings require even in the absence of a deceased parent.

Thus, although certain genetic factors may give the evolution of altruism a boost, ecological factors that increase the effect of helping on the survival of relatives are equally important. However, our understanding of complex sociality is still incomplete. For example, there are seven species of African mole-rats, all of which are burrowing, parental animals. The costs of leaving a safe underground burrow would seem to be high, and the benefits of helping great, for all species, given the value of the burrow, the care required by helpless youngsters, and the advantages of communal burrow construction. Therefore, we might expect to observe eusocial life in all seven species. We do not. Only one of the other species, the Damaraland mole-rat, has a social system anything like that of the naked mole-rat. In fact, these two species are the *only* burrowing mammals known to have evolved eusocial behavior [602]; even joint occupation of a burrow by several adults has been reported for only a handful of subterranean rodents [691]. These facts pose uncomfortable questions for the argument that safe fortress burrows make eusociality more likely to evolve [602]. The use of subterranean tunnels and the associated high costs of dispersal must be only part of the story behind the evolution of the eusocial lifestyle in mammals.

In general, it is easier to offer a tentative explanation for why a species *has* evolved a particular trait than for why a species has *not* evolved a particular trait. For example, researchers have accounted convincingly for the social life of the Florida scrub jay. But what about its close relative, the western scrub jay, a nonsocial species? Would members of this species also benefit from sociality? There must be years when young western scrub jays have little chance of finding a suitable vacant territory. Why haven't they evolved the ability to remain as helpers at the nest under these conditions? Much more remains to be learned about the genetic and ecological bases of altruism and social living.

SUMMARY

1. In animal societies, individuals tolerate the close presence of conspecifics despite the increased competition for limited resources and the heightened risk of disease that this entails. Under some ecological circumstances, the advantages of sociality (usually improved defense against predators) are great enough to outweigh the many and diverse costs of social living. The common view that social life is always superior to solitary life is incorrect.

2. Animals that live together may help one another in various ways. Helpful interactions have a variety of fitness consequences for the participants. Some cooperative acts (mutualisms) may immediately elevate the personal reproductive success of both cooperators. Others may be performed at some cost to one individual, but with the potential of maintaining the relationship for adaptive cooperation in the future. Still others may be performed at some cost that will be more than repaid when the recipient reciprocates in the next of a series of interactions (reciprocal altruism).

3. The kinds of interactions listed above can readily spread through a population through the action of direct (natural) selection. If, however, a helper really does permanently reduce its personal reproductive success while helping to raise the fitness of another, its altruism poses a major evolutionary puzzle. One solution to this puzzle may be that adaptive altruism raises the helper's inclusive fitness (total genetic propagation) by increasing the fitness of its nondescendant kin (kin other than offspring).

4. Altruism is far from rare in animal societies, and includes the alarm calls of some birds and mammals, help given to rear offspring that do not belong to the helper, and even the complete rejection of reproduction by sterile workers that help other individuals reproduce. In almost all cases, sterile altruists direct their aid to relatives, and they favor closer relatives over more distant ones, if they have a choice in the matter. But sterile worker castes can evolve even when the degree of relatedness is not exceptionally high, provided that the ecology of the species is such that helpers can improve the production of relatives markedly, and especially if the odds are against successful personal reproduction for individuals dispersing from a safe natal nest.

SUGGESTED READING

W. D. Hamilton's work [500] initiated a revolution in the understanding of social behavior; see also reviews by Richard Alexander [13], Mary Jane West-Eberhard [1261], and Stephen Emlen [350]. I have relied heavily on Jerram Brown's *Helping and Communal Breeding in Birds* as a guide for understanding the kinds of selection that affect the evolution of social behavior [146]. Heinz-Ulrich Reyer's study of the social pied kingfisher is superb [985], as are the papers by Jan Komdeur and colleagues on the Seychelles warbler [667, 669].

Excellent analyses of social behavior are available for ants [561] and other social insects [1077, 1293], lions [894, 1051], ground squirrels [1082], acorn woodpeckers [662], and naked mole-rats [1088].

DISCUSSION QUESTIONS

1. Some vampire bats regurgitate blood meals to other hungry individuals [1283]. What is the minimum information that you would need in order to determine whether this helpful behavior is currently maintained by indirect selection? By direct selection?

2. Let's say that in calculating the inclusive fitness of a male in a coalition of lions, you measured his direct fitness by multiplying by 0.5 the number of offspring produced by the male, and then added as his indirect fitness the total number of all offspring produced by the other members of the coalition times the mean r of those offspring to the male in question. Your calculation of his inclusive fitness would be challenged on what grounds?

3. If in a social wasp, a female could help to produce more females with an r of 0.75, why would any females reproduce personally, given that reproducers are related to their offspring by just 0.5?

4. In the ant *Veromessor pergandei*, two or three unrelated females may join forces to found a colony together [999]. They cooperate completely until the time when the first offspring begin to emerge, at which point they start fighting, with the winner ultimately killing her companion(s). Develop at least one cost–benefit hypothesis to account for the timing of the switch from cooperation to aggressive behavior.

5. This chapter examines the evolution of sterile workers strictly from the perspective of fitness gains and losses incurred by the workers themselves. But what if we view the issue from the standpoint of selection on queens? How might queens gain direct fitness benefits by "forcing" some of their offspring to be nonreproductive helpers— even if this reduced the inclusive fitness of these helpers? (See [13, 1260] after considering this question.)

CHAPTER 16

The Evolution
of Human Behavior

HUMANS BELONG TO AN ANIMAL SPECIES with an evolutionary history. It is true that we are an unusual species, but kittiwakes and hangingflies are unique and wonderful creatures too. If we can apply evolutionary thinking to kittiwakes and hangingflies, we ought to be able to do the same for humans. On the assumption that natural selection has shaped the evolution of our species, our genes should contribute to the development of adaptations. This expectation produces testable hypotheses about the possible fitness-maximizing properties of many human psychological and behavioral attributes. However, not everyone thinks we ought to employ evolutionary theory in this manner. Opponents of the proposition that our behavior has been shaped by natural selection often point to the complexity and variety of human behavior in different societies. Although cultural traditions do vary impressively, some biologists and psychologists have identified basic features of the human psyche that make sense only from an evolutionary perspective. To support this argument, this chapter presents a sample of evolutionary hypotheses about human behavior, some more speculative than others, some better tested than others. None of these ideas has been exhaustively tested against a full range of alternative explanations, but the same is true of hypotheses on the behavior of kittiwakes and hangingflies. Even so, this book has shown how evolutionary theory can identify what needs to be explained about our fellow animals while at the same time providing us with a way to do the explaining. Why should we stop there? You will not find the Final Word on human behavior here, but instead, a conceptual exercise on how to formulate and test plausible ideas about the ultimate significance of human behavior. Understanding ourselves is a worthy goal, and evolutionary biology has something interesting to say about us.

The Adaptationist Approach to Human Behavior

Human behavior is full of fascinating puzzles for evolutionary biologists. To take one small example, consider the decision to donate blood to a blood bank, a generous act performed by many undergraduates on college campuses in North America, among other places. Behavior of this sort has the appearance of altruism because the loss of blood carries a possible minor fitness cost to the donor, while providing a substantial fitness benefit to a recipient. And yet the blood recipient will almost never have a chance to say thanks to the donor.

You will recall from the preceding chapter that one ultimate hypothesis to account for helpful, self-sacrificing acts is that they generate indirect fitness gains for the helper. Blood donation, however, cannot produce such gains when the helper and recipient are not related, which is usually the case. Moreover, the action can hardly be a case of reciprocity with delayed repayment to the donor, given that someone receiving a blood transfusion rarely knows who the donor was.

A number of persons have claimed, therefore, that blood donation is immune to evolutionary analysis and is instead a kind of "pure" altruism practiced for no possible fitness advantage. According to Peter Singer, "Common sense tells us that people who give blood do so to help others, not for a disguised benefit to themselves" [1113]. Perhaps so, but because adaptationist Richard Alexander was intrigued by the challenge of explaining apparent pure altruism from an evolutionary perspective, he kept at it. His hypothesis is that the donor is "repaid," not by the recipient of his blood, but by his everyday companions, who may be impressed by someone "so altruistic that he is willing to give up a most dear possession for a perfect stranger" and therefore more inclined to cooperate with him [15].

Alexander is not arguing that the "blood donation" trait has been shaped by natural selection, but instead that our psychological mechanisms have evolved in ways that encourage us to "do good deeds" that are visible to others. Martin Daly and Margo Wilson suggest that this hypothetical proximate mechanism makes us feel better when we help others, even perfect strangers, provided that the personal cost is low. In modern Western societies, this feeling motivates us to give blood occasionally or to exhibit other inexpensive forms of culturally approved charity. In the past, when our psychological systems were evolving, our ancestors almost certainly lived in small bands of individuals whose reproductive success depended on building and maintaining cooperative relationships with one another. In such an environment, those individuals who were motivated to do small helpful acts for others may have acquired a good reputation for cooperativeness, thereby facilitating mutually beneficial alliances of the sort Alexander envisions.

Alexander's evolutionary argument yields several testable predictions, including the expectation that unpaid blood donors should almost always let some other people know that they have given blood. Bobbi Low and Joel Heinen report that students at the University of Michigan are significantly more likely to donate to fundraising drives if they receive a pin or tag that advertises their participation. Moreover, pins that announce blood donations are distributed by the American Red Cross during their blood drives [740]. A second prediction is that if people were given a choice between working with a known blood donor and someone known to refuse to give blood, most people would opt for the blood donor.

Although the evolutionary analysis of blood donation per se has just begun, other elements of human cooperation and reciprocity have been much more thoroughly explored from an evolutionary perspective [15, 222, 424]. These studies indicate that because our brains contain evolved systems that promote adaptive cooperation, humans everywhere share certain interrelated psychological attributes that promote this ultimate goal, including a desire to engage in reciprocity, a keen awareness of the social debt incurred when accepting aid from one's companions, nagging guilt about failure to pay someone back, alert detection of non-reciprocation by those we have helped, and so on. We shall not review the considerable evidence on these subjects here. Instead, the point of the blood donation example is simply that we can potentially view all of our actions, even those occurring in novel modern environments, as arising from adaptive psychological mechanisms that evolved in a particular environment in the past. If we were to listen only to what some persons consider "common sense," we would never take advantage of a theory capable of generating interesting and testable explanations for all elements of human behavior.

The Sociobiology Controversy

Alexander's examination of the blood donation problem employs the same adaptationist approach illustrated repeatedly in this book. For the purposes of building an ultimate hypothesis on blood donation, Alexander assumed that the trait was caused by a fitness-promoting proximate mechanism shaped by natural selection. This approach rarely causes intense controversy when it is applied to ground squirrels or dunnocks (but see [454, 655]). The same is not true, however, when humans are involved.

The adaptationist approach to human social behavior is often called **sociobiology,** after a book by the same name written by E. O. Wilson in 1975 [1294]. This book drew the attention of a large audience to the possibility that social behavior could be analyzed in evolutionary terms. In Wilson's audience, however, were some of his colleagues at Harvard University, including Stephen J. Gould and Richard Lewontin [17]. They and many others vehemently denounced the discussion of the evolutionary basis of human behavior in the last chapter of *Sociobiology*. Wilson and others effectively answered the critics' objections, and human sociobiology is currently flourishing [1310] under the new label **evolutionary psychology,** with several professional journals devoted exclusively to the discipline. Even so, some persons still continue to misrepresent or misunderstand the goals of evolutionary biologists. Here are five persistent misconceptions about the sociobiological approach to human behavior (see also [636]):

"Sociobiology is E. O. Wilson's own theory of human behavior."

Sociobiology is not a novel discipline, nor is its subject matter limited to human beings. Wilson simply gave an easily remembered label to what might be more ponderously called the study of social behavior from a Darwinian evolutionary perspective, or the adaptationist approach to social behavior. *Sociobiology* reviewed information gathered by many evolutionary biologists about the social behavior of everything from slime molds to human beings. In his last chapter, Wilson used an evolutionary approach to generate a number of hypotheses about the possible adaptive significance of elements of our own behavior. Many others before and after Wilson have done the same.

"No one has ever identified the genes responsible for any human behavior."

Many persons are under the mistaken impression that a sociobiologist's goal is to discover genes for various human behaviors such as altruism or aggression. It is true that an adaptationist approach to the behavior of any species rests on the assumption that the behavior of interest has evolved. In order for behavior to evolve, individuals in the past with different alleles must have exhibited different behaviors that affected their inclusive fitness. Critics of human sociobiology have claimed, therefore, that one cannot talk about the possible adaptive value of human behavior because no one has ever demonstrated precisely which genes underlie particular behavioral traits in humans.

But sociobiology is the analysis of behavior at the ultimate, not proximate, level. Proximate studies on how genes affect the development of human behavior are interesting and valuable (see Chapter 3). But one does not have to know everything there is to know about the proximate causes of a behavior in order to test hypotheses about its adaptive consequences. We know even less about the genetics of Florida scrub jay behavior and ground squirrel alarm calling than we do about the genetics of human behavior. Fortunately this lack of proximate information has not prevented evolutionary biologists from asking, and often answering, ultimate questions about the evolution of scrub jay and ground squirrel behavior. As we shall see, one can test sociobiological hypotheses about the adaptive nature of certain human characteristics without any information on the genetic or developmental basis for these characteristics.

"But humans don't do things just because they want to raise their inclusive fitness."

Some opponents of sociobiology have pointed out that humans rarely, if ever, seem to be motivated to do the things they do because of a desire for reproductive success [1041]. If you were to have asked Picasso why he wished to produce attractive paintings, or Bill why he wanted to marry Jane, Picasso and Bill would probably not incorporate the term "inclusive fitness" in their answers. But if a baby cuckoo could talk, it would not tell you that it rolled its host's eggs out of the nest "because I want to propagate as many copies of my genes as possible." No animal, cuckoo or human, needs to be aware of the ultimate reasons for its activities in order to behave adaptively. Our brain's costly decision-making mechanisms, like those of other species [513, 617], were shaped by natural selection to enhance fitness, not to provide us with the capacity to monitor the fitness consequences of each and every action. It is enough that proximate mechanisms motivate individuals to do things that raise their direct or indirect fitness. On the proximate level, we enjoy sweet foods, we fall in love, we derive satisfaction from our charitable actions, we desire approval from others, and we learn a language because we possess physiological mechanisms that make us want to do these things. Because honey tastes good, we want to eat it, and when we do, we acquire useful calories that may contribute to our survival and reproductive success without ever being aware that this is the function of our desires.

"But not all human behavior is biologically adaptive!"

Critics of sociobiology often claim that certain cultural practices, such as blood donation, circumcision, prohibitions against eating some perfectly edible foods, or a celibate priesthood, seem most unlikely to advance individual fitness. If some human practices currently reduce fitness, then sociobiology cannot be correct, according to these persons, who apparently believe that evolutionary biologists think that every aspect of every organism is currently adaptive.

But biologists use the adaptationist method simply to identify interesting puzzles and make sense of them (see Chapter 8). The method is central to the conduct of biological science because it can be used to create testable hypotheses on so many difficult-to-explain phenomena, as the example of blood donation illustrates. If a sociobiologist or adaptationist were to examine the religious practice of celibacy, he or she would be fully aware that the trait might be a maladaptive by-product of psychic mechanisms that generated adaptive behavior in our evolutionary past, but do not now in today's novel environment. Still, that person might try to produce a testable hypothesis on how the trait might paradoxically enable priests to leave more copies of their genes than if they were not celibate. Needless to say, this would be a challenge, but perhaps not an insuperable one for someone aware of indirect selection.

Even if twenty adaptationist hypotheses on celibacy were developed, there is no guarantee that any would withstand testing. This is as it should be. T. H. Huxley, the great defender of Darwinian theory, wrote, "There is a wonderful truth in [the] saying [that] next to being right in this world, the best of all things is to be clearly and definitely wrong, because you will come out somewhere" [587]. If our sociobiological hypotheses about the celibate priesthood were incorrect, effective tests would tell us so, and in the process we would learn many interesting things about celibacy and priestly behavior.

"Sociobiology is a politically reactionary doctrine that supports social injustice and inequality."

A fifth and central argument of many critics is that sociobiology provides "scientific" justification for immoral social policies [17]. These individuals claim that to say a trait is adaptive is to imply that it is both genetically determined and good, and therefore cannot and should not be changed. Noting that racist and fascist demagogues have misused biological theories in the past to promote evil political programs, they argue that human sociobiology could easily be used in this fashion. After all, if one claims that male dominance is adaptive, isn't this saying that the status quo in our society is desirable and that feminist claims fly in the face of what is genetically fixed and morally necessary?

Scientific findings can be employed in ways that often surprise, and even horrify, the investigator. Einstein's basic research on the relation between energy and matter contributed to the development of atomic weapons, much to his dismay; Darwin's theory of evolution has been misunderstood and misused by some persons to defend the principle that the rich are evolutionarily superior beings, as well as to promote unabashedly racist plans for the "improvement of the species" by selective breeding of humans.

We can hope that political perversions of evolutionary theory have been so discredited that they will not happen again. The critical point, however, is that sociobiology is a discipline that attempts to explain why social behavior exists, not to justify any given trait. This distinction is easily understood in cases involving other organisms. Biologists who study infanticide by male langurs or the lethal effects of the HIV virus are never accused of approving of infanticide or AIDS. To say that something is biologically or evolutionarily adaptive means only that it tends to elevate the inclusive fitness of individuals with the trait—and nothing more.

Moreover, a hypothesis that a behavioral ability is adaptive does not mean that the characteristic is "determined" by genes, for the very good reason that there are *no* genetically determined traits (see Chapter 3). No alleles exist for altruism or aggression or any other trait, in the sense of genes that encode these characteristics and produce them no matter what. Behavioral development depends on the interaction between the genetic recipe and the environment of the developing individual. Change the environment, and you will probably change the behavioral outcome.

A classic example in human biology involves language acquisition, a clearly adaptive ability to which there must be a genetic contribution (see Chapter 3). Among our many genes are some that code for enzymes that promote the development of certain specialized brain components that facilitate language learning. These neuronal units evolved specifically in the context of language acquisition, rather than as a side effect of some sort of generalized intelligence, as we can see from the existence of two rare human phenotypes. On the one hand, certain profoundly retarded individuals chatter away, producing completely grammatical sentences with little inherent meaning. On the other hand, some English-speaking individuals with normal to above normal intelligence have great trouble with the rules of grammar, often failing, for example, to add "-ed" to verbs when they wish to speak of past events [940] (see Chapter 2).

Of course, the genes "for" language ability interact with all sorts of environmental inputs, both in the form of the building blocks for enzyme production and neuronal development and in the form of acoustical experience. Alter the environment and the nature of the phenotype can change, as demonstrated by the diversity of human languages, which reflect the different social and acoustical environments that children experience in cultures around the world. The proposition that we possess evolved proximate mechanisms underlying a capacity for language, or aggression, or sexual jealousy, or the urge to reproduce does not condemn us to a particular expression of these characteristics for all time. For example, thanks to an environment in which effective means of birth control exist, my wife and I chose to have only two children, even though as an evolutionary biologist I believe that in some sense we exist solely to propagate the genes within us.

Explaining the Diversity of Human Cultures

Given the apparent flexibility of human behavior coupled with the extraordinary diversity of cultures around the world, how can a sociobiologist claim that human behavior has been designed by past natural selection in ways that elevate fitness? After all, there are polyandrous, polygynous, and monogamous societies, cultures in which females make important political decisions and others in which males dominate females, human groups for which warfare is a constant fact of life and other groups that never fight. Depending on where you were born, you might delight in wearing a pig's tusk through your nose, you might take pride in memorizing the Koran, you might speak Jivaro or Japanese, you might be forbidden to look at your mother-in-law, or you might be forced to have your penis cut from stem to stern at adolescence.

Although human cultures are indisputably diverse (Color Plate 21) and human behavior highly changeable, other species also exhibit considerable behavioral flexibility, thanks to the widespread evolution of conditional strategies. Recall the various reproductive, dispersal, and social options exercised by animals as different as blackbirds (Chapter 4), scorpionflies (Chapter 12), dunnocks (Chapter 13), Seychelles warblers, and lions (Chapter 15). To reinforce the point, consider the remarkably sophisticated decision-making abilities of young female white-fronted bee-eaters, which belong to an African bird species that nests in loose colonies in clay banks. Like male pied kingfishers (see Chapter 15), female white-fronted bee-eaters select one of several behavioral options at the start of their first breeding season (Figure 1). A young female bee-eater can choose to breed, or to help a breeding pair at their nest burrow, or to sit out the breeding season altogether. If an unpaired, dominant, older male courts her, the young female almost always leaves her family and natal territory to join him in nesting in a different part of the colony, particularly if he has a group of helpers to assist in feeding the offspring they will produce. This choice is likely to result in high direct fitness. But if young, subordinate males are the only potential mates available, the young female will usually refuse to set up housekeeping with one of them. Young males come with few or no helpers, and when they try to breed, they are often harassed by their fathers, who may succeed in inducing their sons to abandon their mates and return to become helpers at their fathers' nests.

A female that opts not to pair off under unfavorable conditions may choose to slip an egg in someone else's nest or become a helper at the nest in her natal territory—provided that the breeding pair there are her parents, to whom she is closely related. If one or both of her parents have died or moved, she is unlikely to provide altruistic assistance to the chicks there, which are at best distant relatives. If the indirect fitness gains from helping are likely to be small or nil, the female will simply wait, biding her time, conserving her energy for a better time in which to reproduce [360].

Thus, despite their bird-sized brains, female bee-eaters can accurately evaluate key environmental variables and make choices that usually maximize their fitness. The crucial point here is that the behavioral "flexibility" of the bee-eater is constrained in an adaptive manner. The bird secures information on the reproductive status of her relatives; she uses this information to choose among a restricted set of "options" in ways that generally promote, rather than lower, her inclusive fitness. Some evolutionary biologists have proposed that the many and varied conditional strategies that underlie our behavior do for us what the female bee-eater's conditional strategy does for her: namely, provide us with a *limited*—not open-ended—set of options and the ability to make decisions that tend to *increase*—not decrease—our inclusive fitness under the cultural conditions we happen to encounter.

Sociobiology versus Arbitrary Culture Theory on Adoption

Sociobiological theory produces hypotheses about flexible decision-making by humans that are very different from those based on the theory that our cultural activities have nothing to do with fitness maximization, a point that can be illustrated by

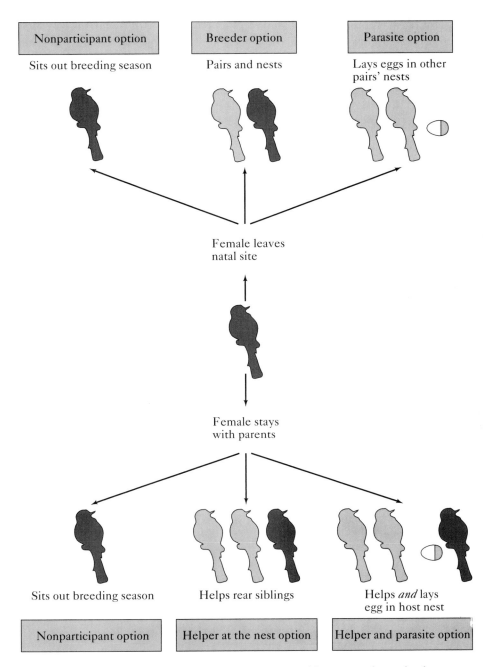

1   Complex social decisions are made by humans—and by many other animals as well. Female white-fronted bee-eaters choose adaptively among several behavioral options when they reach the age of reproduction. *Source: Emlen et al. [360].*

contrasting two approaches to adoption by humans. In 1976, the cultural anthropologist Marshall Sahlins wrote a critique of sociobiology in which he analyzed adoption in Oceania, the islands of the central Pacific Ocean, where an amazing 30 percent of the children are adoptees [1041]. Sahlins argued that the adoption-promoting practices of these island cultures had no relation whatsoever to kinship, demonstrating (he claimed) the arbitrary nature of these traditions and the irrelevance of evolutionary theory to an understanding of human behavior (**arbitrary culture theory**).

Sahlins evidently believed that only one sociobiological hypothesis existed for adoption, one based on indirect selection, which requires that children be adopted by relatives. Because some persons in Oceania adopted children who were not their kin, Sahlins concluded that evolutionary theory could not contribute to an analysis of adoption. But is adoption in Oceania really practiced in an arbitrary manner, with no positive effect on the inclusive fitness of individuals [1104]? When Joan Silk analyzed data on the relationships between moderately large samples of adopters and adoptees in 11 different cultures in Oceania [1103], she found that most adopters cared for children who were cousin equivalents or closer (minimum $r = 0.125$) (Figure 2). The highly nonrandom nature of adoption in these societies casts doubt on the arbitrary culture hypothesis while supporting an indirect fitness hypothesis for this form of human altruism.

Nonetheless, a minority of adopters in Oceania do take in the children of strangers, youngsters that they may treat with the same love and affection that parents typically supply to their own genetic offspring. These parents apparently subsidize the genes of genetic competitors. Are these exceptions to the typical adoption pattern in Oceania impossible to explain from a sociobiological perspective? No. Silk suggests that small families in some agricultural cultures might benefit from gaining adoptees, even if they were nonrelatives, if these adopted persons ultimately contributed to the family workforce, raising the economic productivi-

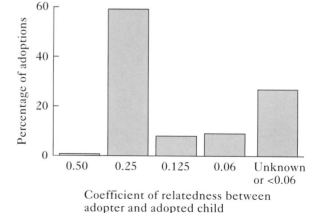

2 The indirect fitness hypothesis for adoption can be tested by examining the coefficient of relatedness between adopters and their adopted children. In 11 island societies in Oceania, adopter and adoptee were usually close relatives, yielding indirect fitness gains for the adopters. *Source: Silk [1103].*

ty of the family unit and improving the survival chances of the adopters' genetic offspring. This direct fitness hypothesis produces the prediction that small families in Oceania should have been more likely to adopt than large ones, a prediction that Silk showed was correct.

An alternative evolutionary hypothesis for adoption among nonrelatives recognizes that some decisions may be the maladaptive by-products of otherwise adaptive proximate mechanisms. For example, adoption of a nonrelative may be one consequence of a motivational system that causes adult humans to want to have children and raise a family. According to this hypothesis, although adults who adopt infant strangers may reduce their fitness, the urge to have a family and the love of children are *usually* adaptive. Because these psychological mechanisms tend to elevate fitness, they are maintained in human populations even though they *sometimes* induce people to behave maladaptively.

The maladaptive side effect hypothesis for adoption also generates testable predictions, one of which is that husbands and wives who have lost an only child or who fail to produce children themselves should be especially prone to adopt strangers. The proximate mechanism that produces the adaptive desire to be parents could cause some people to adopt substitutes for genetic offspring.

Another prediction based on this hypothesis is that adoption of nonrelatives should also sometimes occur in nonhuman animals when adults have lost their offspring and fortuitously encounter a substitute (Figure 3) [618]. Cardinals have

3 Adoption occurs in non-human animals, often when adults have just lost an offspring but encounter a substitute. Here two emperor penguins compete for "possession" of a youngster. *Photograph by Yvon Le Maho.*

been known to feed goldfish [1252], and a white whale was seen trying to lift a floating log part way out of the water, as if the log were a distressed infant that needed help to reach the surface to breathe [121]. Adopting a goldfish or a log did nothing to promote the cardinal's or the white whale's genes. But these individuals had almost certainly lost their young recently, and were employing generally adaptive parental behaviors in a nonadaptive fashion.

The risk to human beings of a maladaptive application of an adaptive psychological mechanism may be especially high in modern societies, environments that are very different from the ones in which our behavioral mechanisms evolved. In Western culture, babies are routinely made available to nonrelatives who do not know the parents of the adoptee, something that almost never occurred in the distant past. Under these novel conditions, our neuronal systems, which evolved long ago, can cause us to behave maladaptively in the present and, in so doing, reveal something about the naturally selected features of our psyches.

The abundance of testable sociobiological hypotheses on adoption (and we have not even discussed the possibility that adoption can be a form of "do-good" behavior that raises the social status of adopters) speaks to the productive nature of evolutionary theory for scientists. No single sociobiological hypothesis will explain every case of adoption, but an evolutionary approach seems far more likely to cast light on the subject that the notion that adoption by human beings is a purely nonadaptive, arbitrary manifestation of human desire and intellect.

Adaptive Decisions: Human Sexual Behavior

Having briefly illustrated how evolutionary biologists develop and test hypotheses on reciprocity and adoption, we now turn our attention to those human traits whose relationship to fitness is most obvious—namely, human sexual behavior. In analyzing the sexual tactics of men and women, we shall use the same theoretical framework that has proved so useful in understanding reproductive behavior in other animal species. As noted earlier (see Chapter 12), interactions between males and females create selection pressures on both sexes, leading to the coevolution of their respective reproductive tactics. You cannot understand what males of a species are doing without paying attention to females, and vice versa.

In the human species, as is true of most other animals, control of reproduction is primarily in the hands of females, thanks to their major physiological investments in egg production, nurturing of embryos, and feeding of infants after they are born. Although human males are also able and often willing to make large parental investments in offspring, their reproductive decisions nevertheless take place in a setting defined by female physiology and psychology. Women have eggs available for fertilization on only a few days each month, and then only if they are not pregnant or nursing dependent offspring. Prior to modern contraception, most women were pregnant or nursing for most of their reproductive lives, with the result that the ratio of females with fertilizable eggs to males with mature sperm has been heavily tilted toward males during our evolutionary history.

These basic elements of human reproductive physiology have enormous significance for our understanding of human sexual psychology and behavior, just as is true in other species with similar inequalities in potential reproductive rates. Male-biased operational sex ratios mean that from the male perspective, mates with fertilizable eggs are a highly limited resource. A scarcity of suitable females should lead to intense competition among males for mates (see [188]). In turn, the abundance of sexually active males should enable females to exercise selective mate choice, choosing sexual partners whose attributes are likely to promote their fitness. Since human males have the potential to provide material benefits to their partners and offspring, female choice can be expected to operate heavily on this criterion, with females favoring males who signal their willingness to transfer considerable resources to their mates and offspring. Since these transfers are costly to males, men are in turn expected to exhibit mate choice themselves, making investments that are likely to maximize the fitness payoff for them: namely, the prompt production of offspring that carry their genes, not those of other gentlemen. Thus, we can view human reproduction as a contest in which both cooperation and conflict come into play, a contest in which the participants exert reciprocal sexual selection on one another, with evolutionary strategies producing counter strategies that come back to shape and refine the psychological mechanisms of the interactors.

Mate Choice by Females and Its Consequences

We can test this approach to our behavior in the customary fashion by using theory to produce testable predictions. Since females are expected to set the rules for male mating competition, let's begin by analyzing female mate choice. First, do females who secure wealthy mates have higher fitness than females whose partners can provide fewer material benefits? This prediction deserves to be tested in societies like those of our ancestors, namely, cultures without modern birth control technology. In the Ache tribe of eastern Paraguay, the children of men who were good hunters were in fact more likely to survive to reproductive age than the children of less skillful hunters [626]. Likewise, several studies of traditional societies in Africa and Iran have shown a positive correlation between a woman's fitness and her husband's wealth, measured in terms of land owned or number of domestic animals in the husband's herds [106, 107, 591, 748], even though polygyny is permitted in all these groups. Women married to rich polygynists need not pay a fitness penalty, since one-half or one-third of a great deal can be more than all of a poor man's holdings—the same argument developed to explain female acceptance of polygynist mates in birds and other animals (see Chapter 13).

The correlation between wealth and reproductive success applies not just to "traditional" cultures, but also to pre–birth control societies more like our own. For example, James Boone found that during the 15th and 16th centuries, Portuguese women married to the men of highest nobility and presumed greatest wealth had more children than women whose husbands were untitled or in the military [104] (Table 1).

Given that a wealthy or resource-rich male elevates the fitness of his female partner, do females possess psychological preferences that "encourage" them to

Table 1 Reproductive performance of married men in Portuguese society in the 15th and 16th centuries

Rank	All offspring		Illegitimate offspring only	
	Number of men	Mean number of offspring	Number of men	Mean number of offspring
Royalty	96	4.75	50	0.52
Royal bureaucracy	168	4.62	43	0.36
Landed aristocrats	216	4.54	80	0.37
Untitled/military	553	2.33	127	0.23

Source: Boone [104].

acquire such mates? Among the Ache, good hunters are more likely to have extramarital affairs and produce illegitimate children than poor hunters, evidence that females in this society find a man's capacity to acquire large amounts of food to be sexually attractive. Likewise, in Renaissance Portugal, noblemen were more likely to marry more than once and more likely to produce illegitimate children than men of lower social rank, results consistent with the prediction that females will use possession of wealth as a cue in selecting a father for their children [104].

If we bear the imprint of past evolution on our psyches, then women today should also find resource control and its correlates, such as high social status, attractive features in potential mates. When women in modern Western societies are surveyed about what they are looking for in a mate, they consistently rank "good earning power" highly, whereas males place much more importance on "physical attractiveness" [161]. Why this difference between the sexes?

When we examine what men mean by "physical attractiveness," we find that preferred female traits include symmetrical facial and body features, full lips, small noses, and a waist circumference to hip circumference ratio of about 0.7 (the hourglass figure) [61, 1114, 1115]. These attributes that men find appealing are associated with developmental stability [420], a strong immune system, high estrogen levels, and youthfulness; in other words, they signal high fertility.

If males assess the physical attractiveness of potential mates [162] in order to marry or copulate with young females especially capable of producing offspring, then as males grow older, there should be an increasing disparity between their age and that of a preferred partner. To test this prediction, Douglas Kenrick and Richard Keefe turned to the *Arizona Solo* [637]. Although this newspaper is not on a par with the *New York Times*, Kenrick and Keefe found it useful because it contains numerous advertisements by men and women in search of sexual partners. Some advertisers in the *Arizona Solo* specified their own age and the acceptable minimum or maximum age that they desired in a companion. Young males in their twenties expressed an interest in partners of their own age, give or take 5 years, but males in their sixties preferred women 5 to 15 years younger than themselves. Women advertisers, on the other hand, desired partners that were 5 years younger to 10 years older, no matter whether they themselves were young or old. These patterns apply cross-culturally [162, 637] and to homosexual men as well hetero-sexual ones [638], suggesting that males possess an evolved cognitive mechanism that motivates them to seek out fertile women with relatively many reproductive years ahead of them (Figure 4). By the time men reach a marriageable age, such women will tend to be younger than they are. On average, men tend to be somewhat older than their wives [637], demonstrating the adaptive consequences of their proximate psychological mechanisms.

In contrast, the preference of women for older men indicates that resource control, not fertility, is a key ultimate factor influencing female mate preferences, since older men tend to have acquired more wealth and higher status than younger males in human societies everywhere. In keeping with this hypothesis, and contrary to arbitrary culture theory, women in cultures around the world indicate a preference

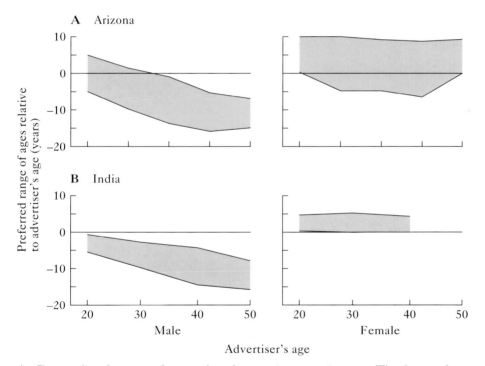

4 Cross-cultural mate preferences based on age in men and women. The data are from advertisements by men and women in (A) an Arizonan newspaper and (B) the *Times* of New Delhi, India. The advertisers indicated their own age and the maximum and minimum ages they would accept in potential partners. *Source: Kenrick and Keefe [637].*

for partners with good financial prospects [162]. The universal psychological mechanisms of human females encourage them to marry good providers.

However, it could be that women's interest in the earning power of potential mates is a purely rational response to the fact that in almost every culture males control their society's economy, making it difficult for a female to achieve economic security on her own. If this hypothesis explains why females favor wealthy males, then women who are themselves well off and not dependent on a partner's resources should place much less importance on male earning power. Contrary to this prediction, several surveys have shown that women with relatively high expected incomes actually put *more* emphasis, not less, on the financial status of prospective mates [1197, 1276]. Moreover, when David Buss surveyed female mate preferences across cultures with great differences in the economic equality of men and women, he found that the degree of inequality did not correlate with the strength of the preference for wealthy husbands [162].

The consequence of an evolved female preference for relatively wealthy partners ought to be higher fitness for males who possess greater income, a prediction that, as we have already noted, has received support from studies of traditional cultures. But what about modern societies in which poor couples have as many or more surviving children than rich ones (Figure 5) [1222]? If equality of fitness in Western cultures is the result of modern birth control technology, an evolutionary novelty, then there might well be no correlation between a man's wealth and the number of his offspring. When Daniel Perusse examined the relationship between male income and reproductive success in modern Quebec, he confirmed that

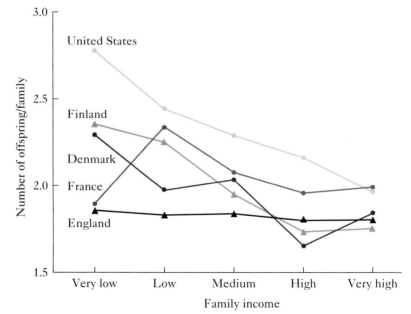

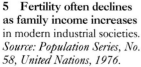

5 Fertility often declines as family income increases in modern industrial societies. *Source: Population Series, No. 58, United Nations, 1976.*

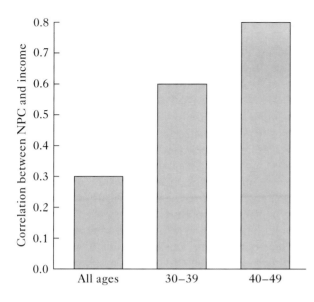

6   **Correlation between income and number of potential conceptions (NPC) in the preceding year for unmarried Canadian men of different ages.** Income and NPC are positively correlated, especially for older men. *Source: Perusse [920].*

wealthier men did not have more children. But would there have been a positive correlation had these men been living in a pre-contraceptive society?

From the data Perusse's respondents supplied to him on the number and frequency of their copulations in the preceding year, Perusse was able to estimate the *number of potential conceptions* (NPC) a male would have been responsible for, had he and his partner(s) abstained from birth control. NPC increases with the number of partners and the frequency with which a male copulates with each partner because conception is a product of the frequency of copulation. Male mating success, as measured by NPC, was highly correlated with male income, especially for unmarried males (Figure 6). Perusse concluded that unmarried Canadian men often attempt to mate with many women, but their ability to do so is heavily influenced by their wealth and status, thanks to the female preference for rich partners. Thus, the striving for high income and status exhibited by males in modern societies can be considered the adaptive product of past selection by choosy females, which occurred in environments in which potential conceptions had an excellent chance of being actual ones [920].

Conflict between the Sexes

Females that copulate with wealthy males will not maximize their fitness gains unless they receive substantial resources from their mates over the long haul. The duration of the male's commitment and the amount of resources he is willing to transfer can be a matter of contention between would-be partners. Although males may gain fitness by giving prolonged assistance to one mate, they can also potentially produce a greater total number of offspring if they have several wives or extramarital partners. Throughout human evolution, polygyny has almost certainly been practiced by men able to afford multiple partners, judging from the widespread occurrence of culturally sanctioned polygyny in historical times [852]. Extramarital affairs could also be part of the ancestral reproductive pattern; they have the potential to increase the fitness of an adulterous male substantially at little cost to him, provided he transfers few resources to those women who bear his illegitimate children. Women, in contrast, can never escape the physiological and time costs of nurturing embryos and newborns, which should tend to reduce the potential fitness to be gained by females from copulating with several males (but see below). Married women in modern societies are indeed more likely to be faithful than their husbands [1167].

The point is that women and men are not expected to have the same interest in acquiring many copulatory partners because of these basic differences in the minimum parental investment in an offspring and thus in the payoffs of polygamy for the two sexes. Donald Symons has suggested, therefore, that males have evolved a distinctive sexual psychology that motivates them to seek out multiple copulatory partners under the appropriate cost–benefit conditions. The value that males place on sexual variety for its own sake should, he predicts, be absent or much reduced in women because of evolved differences in the operation of male and female brains [1167].

Much evidence supports this prediction. Men commonly patronize female prostitutes, whereas women almost never pay men to copulate with them. Males, not females, support a huge pornography industry in Western societies because men, not women, are willing to pay even to look at nude women. Note that in modern societies these aspects of male behavior are surely *maladaptive;* prostitutes almost universally employ effective birth control or undergo abortions when pregnant, and payment for pornographic materials is unlikely to do much for a man's reproductive success. The prostitution and pornography industries take advantage of the male psyche. But widespread maladaptive responses can tell us what people really care about, giving us insight into the nature of the human psyche, which evolved prior to modern birth control technologies and the publication of *Playboy* magazine [258].

These proposed differences in psychic functioning can be tested in yet another way. Symons notes that in Western societies, the behavior of male homosexuals differs markedly from that of lesbians. Not only is male homosexuality much more common than the female variety, but males until recently were likely to have a progression of partners, whereas their female counterparts far more often had long-lasting, stable relationships. Symons's argument is that male homosexuals were free to express the proximate mechanisms that cause all males to have an interest in sexual variety [1167].

David Buss and David Schmitt illuminated these differences between the sexes simply by asking a sample of college undergraduates to state the ideal number of sexual partners they would like to have. The men in his study wanted many more mates than the women did (Figure 7A). Moreover, when Buss and Schmitt asked their subjects to evaluate the likelihood that they would be willing to have sexu-

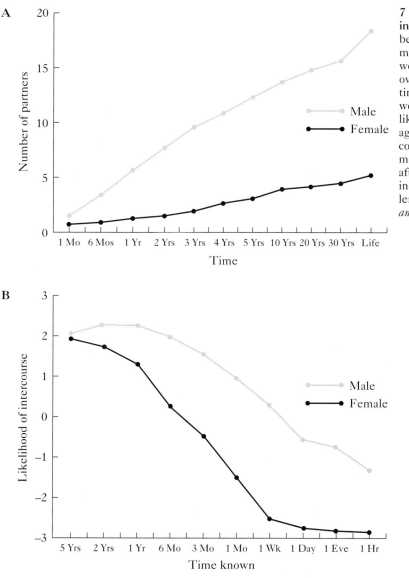

7 Sex differences in mating tactics. (A) The number of sexual partners that men and women claim they would ideally like to have over different periods of time. (B) Men's and women's estimates of the likelihood that they would agree to have sexual intercourse with an attractive member of the opposite sex after having known that individual for varying lengths of time. *Source: Buss and Schmidt [164].*

al intercourse with a desirable potential mate after having known this person for periods ranging from 1 hour to 5 years, the differences between men and women were equally dramatic (Figure 7B): "After knowing a potential mate for just 1 hour, men are slightly disinclined to consider having sex, but the disinclination is not strong. For most women, sex after just 1 hour is a virtual impossibility"[164].

8 Sex differences in mate selectivity. College men differ from college women in the minimum intelligence that they say they would require in a casual sexual partner. However, men and women have similar standards with respect to the minimum intelligence they would require in a marriage partner. *Source: Kenrick et al. [640].*

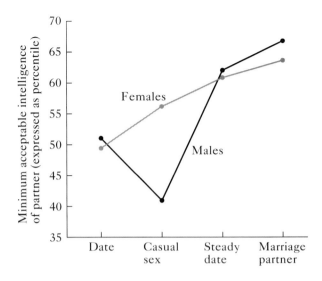

The lower standards of males, at least when it comes to casual sex, are evident in the responses of college students when asked about their absolute *minimum* requirements vis-a-vis the intelligence of persons with whom they would have different kinds of sexual relationships. Men, but not women, were prepared to have sex with partners of far below average intelligence, provided that the encounter carried no commitment (Figure 8) [640].

The enthusiasm of males for low-cost sexual opportunities has also been documented by two social psychologists who sent confederates, an attractive young man and an attractive young woman, on the following mission. They were to approach strangers of the opposite sex on a college campus, asking some of them "Would you go to bed with me tonight?" Not one woman agreed to the proposition, but 75 percent of the men said, "Yes." Remember that these gentlemen had met the woman in question all of a minute before [199].

Now it is possible that all the women in this study who said "No" did so because they sensibly feared becoming pregnant or did not wish to risk injury in a sexual encounter with a male stranger. If so, then homosexual women should have no such inhibitions about casual sex, since sexual interactions between two women cannot result in pregnancy and are very unlikely to lead to physical assault. But homosexual women are no more interested in having multiple partners than are heterosexual females [40].

Coercive Sex

The greater eagerness of males for sex in general is also apparent in the readiness of some to engage in coercive sex, including criminal rape. Although rapists are often severely punished, rape occurs in every culture studied to date [899].

Many persons have tried to explain the phenomenon, including Susan Brownmiller, in her favorably reviewed and still influential book *Against Our Will* [155]. In her view, rapists act on behalf of all men to instill fear into all women, the better to intimidate and control them, thus keeping them "in their place."

This intimidation hypothesis implies that some males are willing to take the risks associated with rape, which is a capital crime in many cultures, in order to provide a benefit for the rest of male society. This argument suffers from all the logical problems inherent in "for-the-good-of-the-group" hypotheses (with the added difficulty that groups composed of only one sex cannot be the focus of any realistic sort of group selection), but we can test it anyway. If the evolved function of the trait is to subjugate all women, then the rapist element in male society can be predicted to target older, dominant women (or young women who aspire to positions of power) to demonstrate the penalty that comes from stepping outside the traditional subordinate role. This prediction is not met. Most rape victims are young, *poor* women [1184]. The intimidation hypothesis, therefore, cannot be correct.

An alternative evolutionary hypothesis proposed by Randy and Nancy Thornhill is that rape is an optional tactic in a conditional sexual strategy [1184]. According to the Thornhills, sexual selection has favored males with the capacity to commit rape *under some conditions* as a means of fertilizing eggs and leaving descendants. In this view, rape in humans is analogous to forced copulation in *Panorpa* scorpionflies (see Chapter 12), in which males unable to offer nuptial gifts in return for matings use the low-gain, last-chance tactic of trying to force females to copulate with them. According to the rape as adaptive tactic hypothesis, human males unable to attract willing sexual partners might also rape as a reproductive option of last resort.

The idea that rape might serve a sexual function has angered persons who feel that such a claim excuses rapists [e.g., 381], but the hypothesis is an attempt at explanation, not a justification, for the behavior. Others have dismissed such hypotheses on the grounds that rapists are motivated by a desire to attack, injure, or humiliate women. But this is a proximate hypothesis, not a true alternative to the ultimate hypothesis that rape has evolutionary consequences. At the proximate level, some rapists might well be driven by a desire to attack women violently, but if this behavior sometimes resulted in the fertilization of their victims, one could legitimately discuss its ultimate reproductive function—again, without approving of the rapist's actions. And raped women do sometimes become pregnant [900, 1184], even in modern societies in which many women employ chemical birth control technology. In the past, when reliable birth control pills and abortion procedures were not available, rapists would have had a higher probability of fathering children through forced sex.

Another ultimate hypothesis to account for rape proposes that the behavior is a *maladaptive* by-product of those well-documented elements of the male psyche that lead to quick sexual arousal, a desire for variety in sexual partners, and an interest in impersonal sex. The plausibility of this hypothesis is enhanced by the observation that males of many nonhuman animals engage in sexual activity that

Table 2 Alternative ultimate hypotheses on why some human males commit rape and some allied predictions

Hypothesis	Predictions
Rape is a nonsexual, purely violent act (proximate hypothesis) designed to subjugate all women for the benefit of all men (ultimate hypothesis)	The populations of rapists and their victims will be similar to populations of criminals and victims in other violent crimes
	Rapists will target women in positions of power or those who aspire to these positions
Rape is an adaptive component of a conditional male sexual strategy	Victims of rape will tend to be young and fertile, and will sometimes become pregnant as a result of the rape
	Rapists are likely to be unmarried, poor men
Rape is an incidental by-product of an otherwise adaptive male sexual strategy	Victims of rape will tend to be young and fertile
	Rapists will have unusually high sex drive

cannot possibly result in offspring, such as the copulatory mounting of weaned pups by male elephant seals [1015]. As applied to humans, the rape as by-product hypothesis requires that the motivating systems regulating male sexuality have a net positive effect on fitness, even though coercive copulation usually reduces the fitness of its practitioners [900].

Discriminating between the rape as by-product and rape as adaptive tactic hypotheses is difficult, given that they generate many of the same predictions (Table 2) [900, 1185]. For example, both hypotheses produce the prediction that rape victims will be young, relatively fertile women with a relatively high probability of becoming pregnant. This prediction is apparently correct (Figure 9), but does not enable us to discriminate between the two hypotheses.

The rape as adaptive tactic hypothesis predicts that young, poor men will commit rape at a disproportionately high frequency. Although many apprehended rapists are indeed young, poor, unmarried men [1184], men of this sort are more likely to be arrested and prosecuted for any criminal act than are men of high social rank. If rape by married men of power and prestige is underreported, then we cannot rely entirely on official crime statistics to test the adaptive tactic hypothesis.

The rape as by-product hypothesis produces the prediction that rapists will have unusually high levels of sexual activity with consenting as well as nonconsenting partners, contrary to the adaptive tactic hypothesis. Some evidence supporting this prediction exists [693, 900], but as is true of many issues in human sociobiology, the results of testing competing hypotheses on coercive sex are not conclusive.

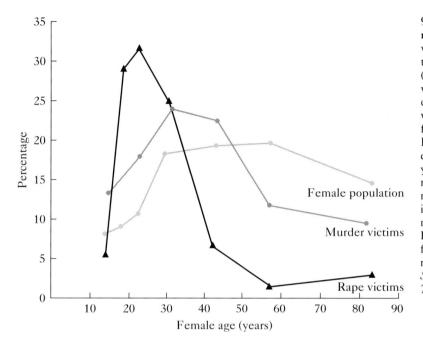

9 **Testing hypotheses on rape.** If rape were motivated purely by the intent to attack women violently (a proximate hypothesis), we would expect that the distribution of rape victims would match that of female murder victims. Instead, rape victims are especially likely to be young (fertile) women, a result consistent with ultimate hypotheses proposing that rape is linked to male reproductive tactics. Data on rape victims come from 1974–1975 police reports for 26 U.S. cities. *Source: Thornhill and Thornhill [1184].*

Female Control of Paternity

Males do not always behave in ways that advance the fitness of their sexual partners, rapists being an extreme example. However, since fertilization is internal, females retain considerable control over the paternity of their offspring, which they can influence in a variety of ways. In particular, women can actively solicit sperm from more than one male, engaging in the human equivalent of the extra-pair copulations for which female birds have become famous (see Chapter 13). If their husbands or primary partners then care for offspring fathered by other males, these women impose a heavy fitness penalty on the men they have cuckolded, given the costly nature of parental care.

Sociobiological hypotheses abound on why women might sometimes cheat on their primary partners. The fitness gains could include the receipt of extra resources or superior protection for their offspring from the other man; alternatively, mating with more than one male may eventually enable the woman to exchange an inferior primary partner for a superior one (from her perspective). Thus, we would expect women involved in extramarital affairs to choose wealthier or more emotionally committed partners than their current spouses, whom they may subsequently divorce [1167].

Given that women sometimes are sexually involved with more than one male, can they bias fertilization of their eggs in favor of the individual who would be the superior father? One possibility is that females use copulatory orgasms to control the paternity of their offspring, a hypothesis that has been disputed by those who note that female orgasm is not necessary for conception. These persons argue that orgasms caused by stimulation of the clitoris are a nonadaptive, incidental side effect of the homology of the clitoris with the penis. Both organs have their developmental origins in the same embryonic tissues; the penis arises as a result of exposure to testosterone, a hormone absent in the female embryo (see Chapter 4). Thus, the clitoris can be viewed as a rudimentary structure that is maintained merely because of the utility of the penis, a structure of great adaptive value for males, just as selection for adaptive female nipples may maintain the genes that also express themselves in the development of useless male nipples [457, 1167]. The fact that females do not experience orgasm on every copulation is consistent with this hypothesis, which is founded on the premise that the clitoris is an incomplete structure without a direct reproductive function.

However, others have proposed and tested adaptationist hypotheses on the clitoris and the orgasms it can, but does not always, produce. If orgasm is a means by which females can selectively affect the paternity of their offspring, then we can predict that orgasm should not occur on every copulation, which is true, as just mentioned. More importantly, we can predict that orgasm should increase the likelihood of conception. Some researchers have claimed that orgasms can increase sperm retention in the vagina, drawing the ejaculate toward the cervix, in effect helping move sperm closer to their ultimate target, an unfertilized egg higher in the reproductive tract [45]. If true, this result raises the possibility that females might unconsciously adjust the timing or occurrence of their orgasms to favor preferred partners [46].

More evidence in favor of female paternity choice via orgasm has been supplied by Randy Thornhill and his co-workers, who found that the single best predictor of female orgasm is the body symmetry of the female's sexual partner. Body symmetry in this study was determined by measuring the dimensions of right and left ears, feet, ankles, hands, wrists, and elbows and calculating the sum of the *differences* between the two sides of the body. Men vary considerably in their symmetry scores, which are correlated with the frequency of copulatory orgasm by their partners [1183].

Needless to say, it would be helpful to have these findings replicated by other researchers. If human females, like female barn swallows, really do give symmetrical partners a fertilization advantage, what do they gain from their choices? Thornhill and his colleagues suggest that this mechanism favors males with a developmental history free from resource shortages, diseases, or parasites. Male children afflicted by any of these problems might have difficulty maintaining developmental homeostasis (see Chapter 4), and if so, they would be marked by various bilateral asymmetries in the body and face [1183]. If this is true, then males blessed with symmetrical bodies and faces are those likely to possess good immune systems with effective means of combating the developmentally disruptive effects of parasites and diseases. Therefore, symmetrical males may not only be less likely to infect a partner with parasites or contagious illnesses, they may also confer the genetic basis of their ability to combat environmental stressors on their offspring

[1240]. Alternatively, if body symmetry is linked to a high degree of heterozygosity, then by mating with a symmetrical male, a female promotes heterozygosity in her offspring [148], especially at the MHC loci, with the resulting immune system benefits discussed earlier [1245] (see Chapter 4). Moreover, symmetrical males of high heterozygosity and allied high immunocompetence would survive better (especially in the challenging environments of our ancient ancestors), and so would potentially provide their mates and offspring with longer periods of assistance and parental care.

Sperm Competition and Mate Guarding

Women's ability to influence the paternity of their offspring through extra-pair copulations and selective orgasms creates selection pressure on men to counter the possibility that their partners will fertilize their eggs with sperm from other men. Robert Smith has proposed that certain aspects of male anatomy and physiology help men to engage in sperm competition [1128]. Human testes, for example, weigh more than those of gorillas [506], and human ejaculates contain many more sperm as well. Gorillas are much larger animals than humans, but they have a mating system in which one male completely monopolizes access to several females. In such a system the risk that a fertilizable female will receive ejaculates from more than one male is almost nil. Therefore, delivery of large quantities of sperm to a mate carries no special advantage for male gorillas, whereas the quantity of sperm transferred to a human female may affect a male's chances of fertilizing an egg, given the possibility that his partner will receive sperm from another male within a few days.

In some other animals in which sperm competition is intense, males adjust the number of sperm donated to a partner in relation to the risk that she has received sperm from a rival male [416]. Do human males do the same (Figure 10)? Robin Baker and Mark Bellis reported that the more time a couple spent together between copulations, the lower the number of sperm in the man's ejaculate (which was cap-

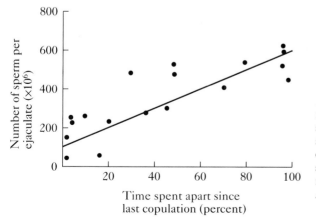

10 Sperm competition in humans. The number of sperm in an ejaculate increases as a function of the time a man and woman have spent away from each other since their last copulation, perhaps to help the male swamp any rival ejaculates his partner may have received in his absence. *Source: Baker and Bellis [44].*

tured in a condom for later sperm counts by dedicated sex researchers). The rationale here is that when a male is constantly in the company of his mate, the chance that his partner has had sex with a rival is low. Under these circumstances, little can be gained by providing extra sperm to swamp any gametes that she may have received from another male. However, when a male is away from his partner for much of the time, the probability of sperm competition increases (Figure 11), favoring an added investment in sperm transferred during copulation [44].

Martin Daly suggests an interesting, and as yet untested, alternative hypothesis for the increased sperm deliveries of men who have been out of sight of their partners for some time. He wonders whether such males might have had more social interactions with female nonpartners, which the male psyche might be designed to interpret as potential opportunities for extra-pair copulations. Even looking at large numbers of women might do the trick, causing the brain to order the testes to produce extra quantities of sperm in case a potential EPC should become reality. If so, males stimulated by nonpartners might eventually transfer the extra sperm to their regular partner when they got back together. (Perhaps you may wish to design a study that would discriminate between these two alternative interpretations of the data presented in Figure 10.)

Another response to the risk of sperm competition, however, is to prevent the introduction of rival ejaculates into one's mate in the first place. As we have just noted, men that remain close to a partner reduce the probability that she will mate

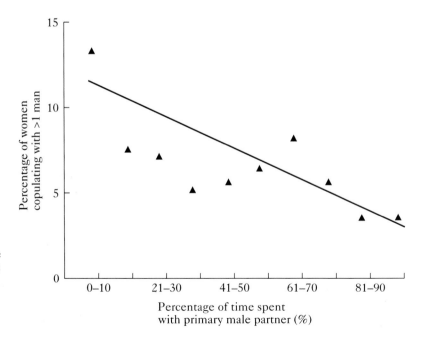

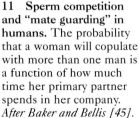

11 Sperm competition and "mate guarding" in humans. The probability that a woman will copulate with more than one man is a function of how much time her primary partner spends in her company. *After Baker and Bellis [45].*

with another man (see Figure 11) [45]. In this light, marriage itself may be a cultural institution that serves the function of mate guarding from the male perspective.

Everywhere men aspire to monopolize or restrict access to the sexual favors of their mates, although as we have seen, they do not necessarily succeed. Marriage rules institutionalize these ambitions. Although one sometimes hears about "sexually permissive" societies in which complete sexual freedom is the norm, the notion that such cultures actually exist appears to have been a (wistful?) misinterpretation on the part of outside observers. In *all* cultures studied to date, adultery committed by a woman, or even suspicion of it, is considered an offense against her husband and often precipitates violence [260]. A cuckolded husband may legally murder his wife or her lover in many societies [262], revealing another dark side of the coevolution of male and female reproductive tactics in the human species.

Adaptive Decisions: The Human Family

Ultimately the sexual preferences and practices of men and women translate into the production of children. As noted, both men and women can provide parental care for their offspring, but there is great variation among individuals with respect to how much care they actually offer a given offspring. Evolutionary theory has been used to develop a broad range of testable hypotheses about the behavior of individuals living in family groups, whether the families are composed of Seychelles warblers, white-fronted bee-eaters, or human beings [351, 352].

Partitioning Parental Care

Sociobiologists have predicted that married men should be especially careful about the allocation of their paternal investment, providing it freely only to progeny likely to be their genetic offspring. As we have just noted, males that learn of a wife's adultery are likely to have a strong emotional reaction, sometimes leading to violence. They are also likely to seek a divorce, with its consequent reduction or elimination of paternal care. In contrast, adultery by a husband is much less likely to be cited by women in Western societies as the reason they desire a divorce [1167]. Male concern about paternity is so obsessive in many cultures that husbands of rape victims may divorce their unfortunate wives, an action accepted or even encouraged by a diversity of religious groups and legal codes [155].

Some societies, such as the cultures of Oceania, are more sexually permissive than our own. Even though the extent of sexual freedom there may have been greatly overestimated by some Western observers, the average certainty of paternity in these cultures is probably relatively low. The standard practice in cultures of this sort is for a man to withhold some or all paternal care from the children of his wife and instead help the children of his sister (Figure 12). Richard Alexander has shown that if a male can expect to father only one of four children produced by his wife, he would share 5/32 of his genes with his nephews and nieces on his sister's side, but only 4/32 of his genes with his wife's children (Figure 13).

12 Paternal care by uncles. Uncles invest heavily in their sisters' children in many societies, including the Ifaluk in the western Caroline Islands. Here several Ifaluk children, who live with their unmarried uncles, butcher a pig under the supervision of their guardians. *Photograph by Laura Betzig.*

Given the material benefits that sisters can derive from brothers under such a system, they and their parents may gain by "encouraging" males to adopt the uncle–father role [260, 512]. Although the uncle–father role is adaptive for males only when certainty of paternity is very low, the correlation between sexual permissiveness and male investment in sisters' children nevertheless supports the hypothesis that paternity issues affect the evolution of child care by men [14].

Stepfathers are another category of men who are placed in the position of caring for children who are not their own. To test the prediction that stepfathers will favor their own genetic offspring over stepchildren, Mark Flinn examined relationships within some Trinidadian families in which a stepfather lived with children of his own as well as those his wife had by another man. In these families, significantly more conflict occurred between stepfathers and their stepchildren than between these men and the children acknowledged to be their genetic offspring (Figure 14) [391]. Furthermore, children were far more likely to leave a household with a stepfather to move in with relatives (their grandparents, for example) than were children growing up in a household without a resident stepfather.

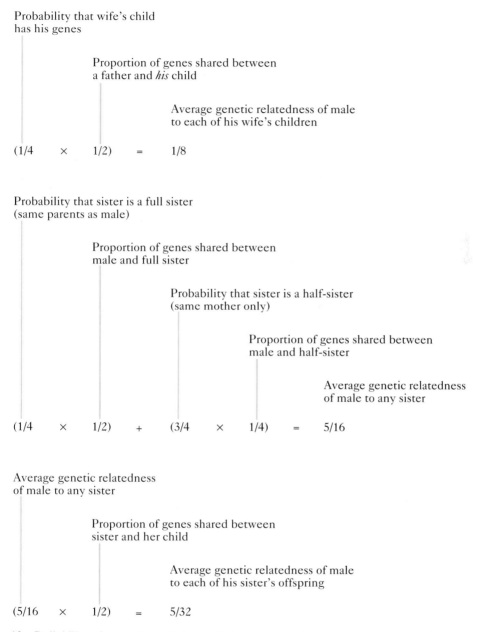

Probability that wife's child
has his genes

Proportion of genes shared between
a father and *his* child

Average genetic relatedness of male
to each of his wife's children

(1/4   ×   1/2)   =   1/8

Probability that sister is a full sister
(same parents as male)

Proportion of genes shared between
male and full sister

Probability that sister is a half-sister
(same mother only)

Proportion of genes shared between
male and half-sister

Average genetic relatedness
of male to any sister

(1/4   ×   1/2)   +   (3/4   ×   1/4)   =   5/16

Average genetic relatedness
of male to any sister

Proportion of genes shared between
sister and her child

Average genetic relatedness of male
to each of his sister's offspring

(5/16   ×   1/2)   =   5/32

13   Reliability of paternity and the coefficient of relatedness between a male and the offspring of his wife and of his sister. These calculations assume that his wife's children have a 1 in 4 chance of being his genetic offspring. Under these circumstances, a male's average coefficient of relatedness to his wife's offspring can be lower than his relatedness to his sister's offspring.

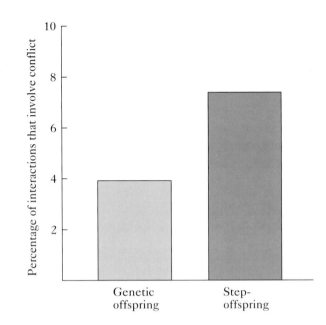

14 Stepparenting and conflict in families. Interactions involving conflict between men and their stepchildren occurred significantly more often in one Trinidadian village than did conflicts between these same men and their genetic offspring. *Source: Flinn [391].*

The converse of parental solicitude is child abuse, which was not observed by Flinn in his Trinidadian study, but occurs with discouraging regularity in our society. Martin Daly and Margo Wilson have hypothesized that child abuse is a maladaptive side effect of the psychological mechanisms that encourage humans to bias their parental care toward their own genetic offspring [256, 258, 261]. They argue that our evolved brain promotes certain kinds of behavior that *on average* lead to increased fitness, but may occasionally lead some individuals to behave in ways that reduce their fitness.

Daly and Wilson tested their parental preference hypothesis by predicting that stepparents should contribute a disproportionately large fraction of the cases of criminal child abuse in Western societies. They found that in Hamilton, Ontario, children 4 years old or younger were *40 times more likely* to suffer abuse in families with a stepparent than in families with both genetic parents present! Note that for both categories of parents in this study, the absolute likelihood that a child would be abused was small (Figure 15), but the *relative* risk was far greater for children in households with a stepparent [256, 258]. Evidently the psychological mechanisms that promote selective parental care can cause emotional difficulties for adults that embark on the stepparent role. The vast majority cope, but a few do not.

Helping Children Marry

We have seen that adults care for children in a selective fashion, usually favoring genetic offspring over nonrelatives. But parents do not always treat their sons

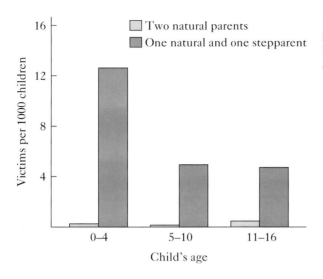

15 Child abuse is far more likely to occur in households with a stepparent than in households with two genetic parents. *Source: Daly and Wilson [256].*

and their daughters identically when it comes to dishing out parental benefits. An example involves the material sacrifices parents make to help their offspring acquire spouses. In some societies, the groom and his family are called upon to donate various resources such as cattle or money or labor—the bridewealth—to the bride's family (Figure 16); in other groups, the family of the bride sends their daughter off to marry with a special donation—the dowry—to her new husband or his family.

If bridewealth or dowry payments were purely arbitrary traditions, the accidental products of cultural evolution, then we would predict that the two forms of payments should be equally represented among cultures worldwide. They are not [425]. (Before reading further, you may wish to use sexual selection theory to predict whether bridewealth or dowry should be more commonly practiced among the world's cultures.)

Because men typically compete for access to females, sexual selection theory yields the prediction that bridewealth payments should be far more common than dowries. This is indeed the case: bridewealth payments have been recorded in 66 percent of the 1267 societies described in the *Ethnographic Atlas*, a comprehensive compendium of cultural traditions worldwide [852], whereas dowry is standard practice in just 3 percent of these societies. Bridewealth payment occurs particularly frequently in polygynous cultures, occurring in more than 90 percent of those societies classified under the heading "general polygyny," in which more than 20 percent of married men have more than one wife (Table 3). Note that most societies fall under this heading. When some males monopolize several females, marriageable females become an especially scarce and valuable commodity. One way to secure multiple wives, and so achieve exceptional reproductive success, is

16 Bridewealth payments in many traditional African cultures are made in cattle. (Top) A bridewealth herd of cattle being driven from a waterhole. (Bottom) Four Kipsigis young women in traditional leather clothes shortly after their initiation into adult status. The marriage of each woman, which occurred within a few months after the photograph was taken, required a bridewealth payment. *Photographs by (top) Monique Borgerhoff-Mulder and (bottom) Philip arap Bii.*

Table 3 The relationship between the mating systems of human cultures, the occurrence of bridewealth payments, and the occurrence of an inheritance system that favors sons

	Bridewealth payment		Inheritance system	
Mating system	No	Yes	Even	Sons favored
Monogamy	62%	38%	42%	58%
Limited polygyny	46%	53%	20%	80%
General polygyny	9%	91%	3%	97%

Note: Data come from Murdock's *Ethnographic Atlas* [852], which codes for the cultural components appearing here. The percentages are based on 112 monogamous cultures, 290 that practice limited polygyny (less than 20 percent of males are polygynous), and 448 that practice general polygyny (more than 20 percent of married men have more than one wife). When sons are favored in an inheritance system, they receive all or almost all of the parents' property.
Source: Hartung [511].

to "purchase" them by providing a payment to their parents or other relatives. To do so often requires a substantial outlay.

Wealthy men in polygynous societies can produce a great many children. Under these conditions, parents can potentially secure more descendants if they direct their wealth into the hands of a son, who can then use it to acquire a number of wives. Biased parental investment may even occur posthumously if wealthy parents distribute an inheritance to a son or sons, enabling their male offspring to become successful polygynists [108]. These sons can produce many more grandchildren for the deceased parent than a daughter whose reproductive success can never exceed the number of embryos she can personally produce and nurture. John Hartung found the predicted correlation between the practice of polygyny in a society and the occurrence of inheritance rules that favor sons, concentrating wealth in the hands of offspring who may then have the economic capacity to become polygynous (see Table 3) [511].

Guy Cowlishaw and Ruth Mace confirmed that this pattern was real after controlling for the nonindependence of cultures. They dealt with the statistical problem of how to treat information from several cultures that may have inherited similar practices as a result of sharing a recent common cultural ancestor. First, they constructed cladograms of cultures based on linguistic information; these diagrams identified which cultures are closely related and which are not. They then superimposed the mating system of each culture on the cladogram to identify those whose mating systems had changed to polygyny from a monogamous predecessor, and vice versa. Independent cultural changes to polygyny were much more likely to be associated with inheritance favoritism toward sons than were changes to monogamy [226].

Even in supposedly monogamous Western societies, rich men may have opportunities for unusual copulatory success (as demonstrated by Perusse's study of Canadian men) because their wealth makes them attractive to women. If parents in modern societies retain an ancestrally selected bias that causes them to favor offspring with high reproductive potential, we can predict that even today very wealthy parents will be inclined to bias their legacies in favor of sons rather than daughters, a prediction that has received support [1126] (Figure 17).

In contrast to the prevalence of biased parental investment in males, societies in which parents give significantly more to their daughters are rare, as one might expect given that females typically are in demand as marriage partners. However, Lee Cronk has found one tribal society, the Mukogodo of Kenya, in which parents often provide more food and more medical care (from a local Catholic mission clinic) to their daughters than to their sons [247]. The Mukogodo have only recently abandoned their traditional hunting–gathering lifestyle in favor of an economy built on sheep and goat herding. However, their herds are small, and their standing with other pastoral tribes in the area is very low. A son of an impoverished Mukogodo family is unlikely to acquire a large enough flock to pay the bride price for a wife from any of the surrounding tribes. In contrast, a Mukogodo daughter

17 Inheritance decisions. A test of the expectation that wealthy Canadian parents will bias their inheritance distribution toward their sons, who are more likely than daughters to convert exceptional wealth into exceptional reproductive success. Parents with the largest wills indeed favored sons in this study. *Source: Smith et al. [1126].*

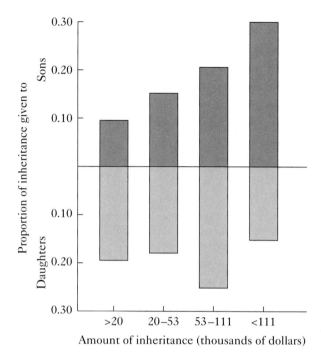

has a good chance of marrying a member of a higher-status tribe, since polygyny is standard among these groups and women are in short supply. As a result of the greater ease of marriage for daughters than sons, the average number of offspring of a Mukogodo daughter is nearly 4, whereas a son's direct fitness is only about 3. This inequality favors families that have more daughters than sons, an outcome that is achieved through neglect of young boys relative to young girls.

This case nicely illustrates the point that human behavior is adaptively flexible, not arbitrarily or capriciously or infinitely variable. Whatever psychological mechanisms control parental solicitude toward offspring, these evolved systems permit parents to favor sons under some circumstances, daughters under others, while encouraging equal treatment under still others. The option chosen tends to enhance parental fitness under local conditions. The *differences* between the daughter-favoring Mukogodo and those nearby herding cultures in which sons receive preferential treatment are surely not genetic. Instead, these parental differences reflect our ability to use evolved conditional strategies to select among a limited set of options, choosing the one with the highest fitness payoff in a given environment, just as female white-fronted bee-eaters unconsciously select the best fitness-raising option in their variable social environment.

Table 4 The relationship between dowry payments and the degree of social stratification in a society

Society	Nonstratified		Stratified	
	Polygynous	Monogamous	Polygynous	Monogamous
Dowry absent	624	99	263	45
Dowry present	1	2	5	27

Note: Stratified societies are those coded by Murdock [852] as containing an elite element or having complex social stratification. Monogamous societies include the few polyandrous ones as well.
Source: Gaulin and Boster [425].

In the Mukogodo and most tribal groups, men pay a price to acquire a bride. Why do a few contrary cultures sanction payments that help a woman secure a husband? One answer is that in monogamous societies in which males typically invest materially in their children, parents that help their daughter "buy" the right kind of man gain fitness as a result. Males in socially stratified monogamous cultures vary greatly in their status and access to wealth. In such cultures, women married to "elite" males should generally enjoy a reproductive advantage because a monogamous husband's wealth will not be divided among a bevy of wives, but instead will go to his only wife and her children. To the extent that wealth and high status translate into reproductive success, a woman's parents may gain by competing with the parents of other families for an "alpha" husband, even if this requires that they themselves offer a material inducement to the right male or to his family. Steven Gaulin and James Boster found that the tradition of dowry payments is about 50 times more likely to occur in a socially stratified monogamous culture than in a polygynous or nonstratified society [425] (Table 4). This finding supports the hypothesis that dowry practices are not a random cultural artifact, but part of an adaptive parental strategy that favors selective assistance of daughters.

Thus, we have yet another example of how sociobiologists, far from being flummoxed by cultural diversity, can make use of it to test evolutionary hypotheses about human behavior. The successes of sociobiology demonstrate that if our goal is the satisfaction of curiosity about ourselves, the Darwinian approach has a lot to tell us. There is no guarantee, however, that an evolutionary understanding of our behavior will be put to uses that most people would consider good. The brain mechanisms that push us to solve mysteries about ourselves and our world may lead to discoveries that will in fact be used to worsen, rather than alleviate, the social and military crises that characterize modern life. But whatever our wishes, the fact that we are an evolved animal species is not going to change, and so we might as well understand the significance of this fact, if only to give ourselves insight into what Pogo meant when he declared, "We have met the enemy and it is us."

SUMMARY

1. Human beings are an evolved animal species. Human sociobiology, also known as evolutionary psychology, employs an evolutionary approach to generate testable hypotheses based on the assumption that natural selection has shaped our species' behavior.

2. The study of human sociobiology has been marked by intense controversy, in part because some persons have misunderstood the goals and foundations of the discipline. Contrary to the claims of some critics, sociobiologists are not motivated by any particular political agenda, nor are sociobiological hypotheses based on the premise that elements of human behavior are both genetically determined and morally desirable.

3. Sociobiological hypotheses are designed to explain, not to justify, our behavior; testing these hypotheses can help us understand why we have psychological mechanisms that motivate us to behave in certain ways. Unlike arbitrary culture theory, which proposes that human behavior is the arbitrary product of cultural traditions constrained only by the limits of our imagination, sociobiological theory views human behavior as the product of conditional strategies involving limited options that tend to increase, rather than decrease, the fitness of individuals.

4. Many sociobiological hypotheses have been advanced for elements of human reproductive behavior, which has been shaped by cooperation and conflict between the sexes. On the one hand, both males and females can potentially gain fitness through parental care of their offspring. On the other hand, opportunities for conflict of interest also exist, since male reproductive potential is higher than that of females and since certainty of maternity is higher than certainty of paternity. Females run the risk of losing support for their offspring; males run the risk of caring for offspring fathered by men other than themselves. These possibilities appear to have heavily influenced the evolution of the proximate psychological mechanisms that underlie much of human sexual and parental behavior.

5. To the extent that we are curious about the evolution of human behavior, a sociobiological approach offers avenues of exploration that other disciplines do not. Although there are problems in employing this approach, the same can be said about any scientific endeavor that touches on human concerns.

SUGGESTED READING

The debate on sociobiology began with E. O. Wilson; see the final chapter of *Sociobiology* [1294] and a commentary on sociobiology and politics[1295], as well as an article by his opponents [17]. Subsequently, a large number of other people have chimed in, with Stephen J. Gould [e.g., 456, 458] and Philip Kitcher [655] in the forefront. Commentaries by various academics on a paper by Randy and Nancy Thornhill on rape reveal the continuing misunderstandings about human sociobiology and the intensity of feeling that the debate still generates [1185]. Doug Kenrick has written an amusing and accurate portrayal of these misunderstandings and their often ill-informed advocates [636]. Comments by anthropologists about the use of arbitrary culture theory appear in two books [143, 409]. Proponents of an evolutionary approach to human behavior include Richard Alexander [14, 15], Helena Cronin [246], and Martin Daly and Margo Wilson [259, 260]. Human sexual psychology is the subject of two entertaining books, one by Donald Symons [1167], and another by David Buss [163]. This topic has also been reviewed by Buss and David Schmitt [164], and by Randy Thornhill and Steven Gangestad [1182]. The theory that our psychological mechanisms have been designed by natural selection has been brilliantly examined by a nonscientist, Robert Wright, in *The Moral Animal* [1310]. Most of the classic papers on the evolutionary analysis of human behavior can be found, along with updates and critiques, in *Human Nature*, edited by Laura Betzig [89]. Readers can find many interesting articles on human behavior in the journal *Ethology and Sociobiology*, now called *Human Behavior and Evolution*.

DISCUSSION QUESTIONS

1. Marshall Sahlins has argued that sociobiology is contradicted because people in most cultures do not even have words to express fractions. Without fractions, a person cannot possibly calculate coefficients of relatedness, and without this information (Sahlins claims) people cannot determine how to behave in order to maximize their indirect fitness [1041]. Is Sahlins right? Has he delivered a serious blow to sociobiological theory?

2. Philip Kitcher states that "socially relevant science," such as sociobiology, demands "higher standards of evidence" because if a mistake is made (a hypothesis presented as confirmed when it is false), the societal consequences may be especially severe. For example, a hypothesis that men are more disposed to seek political power and high status in business and science than women is dangerous because it "threaten[s] to stifle the aspirations of millions" [655]. How do you think a sociobiologist could effectively answer Kitcher's claim?

3. Using the same logic that Richard Alexander employed in analyzing the evolutionary basis of the uncle–father role, make predictions about the relative willingness of aunts and uncles to invest in the nephews and nieces produced by their female siblings versus their male siblings. How would you test these predictions [428]?

4. In writing about the widespread occurrence of genocide, Stephen J. Gould reviews the adaptationist hypothesis that the capacity for large-scale murder evolved as a result of intense competition for resources or mates between small bands during our evolutionary history. Gould dismisses this hypothesis on the grounds that evolutionary theory has nothing novel to say about this aspect of human behavior. "An evolutionary speculation can only help if it teaches us something we don't know already—if, for example, we learned that genocide was biologically enjoined by certain genes, or even that a positive propensity, rather than a mere capacity, regulated our murderous potentiality. But the observational facts of human history speak against determination and only for potentiality. Each case of genocide can be matched with numerous incidents of social benevolence; each murderous clan can be paired with a pacific clan" [458]. Evaluate Gould's argument critically in the light of what you know about (1) the proximate–ultimate distinction, (2) conditional strategies, and (3) the distinction between adaptationist and arbitrary culture hypotheses.

5. Some persons believe that men in Western societies prefer sexual partners who are younger than they are because they have been taught this arbitrary cultural convention from an early age. What pre- diction follows from this hypothesis with respect to the dating preferences of teen- age males for females of different ages? What conclusion follows from the data in the figure below [639]?

- • Most attractive (minus own age)
- • Maximum difference preferred
- ▲ Minimum difference preferred

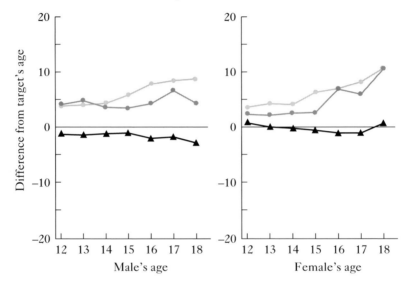

Glossary

Action potential   The neural signal; a self-regenerating change in membrane electrical charge that travels the length of a nerve cell.

Adaptation   A characteristic that confers higher inclusive fitness on individuals than any other existing alternative exhibited by other individuals within the population; a trait that has spread, or will spread, or is being maintained in a population as a result of natural selection.

Adaptationist   A behavioral biologist who develops and tests hypotheses on the possible adaptive value of particular traits.

Adaptive value   The contribution that a trait or gene makes to inclusive fitness.

Allele   A form of a gene; different alleles typically code for distinctive variants of the same enzyme.

Altruism   Helpful behavior that raises the recipient's direct fitness while lowering the donor's direct fitness.

Arbitrary culture theory   The view that human behavior is the arbitrary product of whatever cultural traditions we happen to be exposed to in our societies; thus, our actions are not expected to be explicable in evolutionary terms.

Artificial selection   *See* Selection.

Brood parasite   An animal that exploits the parental care of individuals other than its offspring's parents.

Causal question   In the scientific method, a question about the cause of a natural phenomenon.

Central pattern generator   A group of cells in the central nervous system that can produce a particular pattern of signals that triggers a functional behavioral response.

Circadian rhythm   A roughly 24-hour cycle of behavior that expresses itself independent of environmental changes.

Circannual rhythm   A annual cycle of behavior that expresses itself independent of environmental changes.

Cladogram   A diagram of the evolutionary relationships among species or groups of species.

Coefficient of relatedness   The probability that an allele present in one individual will be present in a close relative; the proportion of the total genotype of one individual present in another as a result of shared ancestry.

Communication   The cooperative transfer of information from a signaler to a receiver.

Comparative method   A procedure for testing evolutionary hypotheses based on disciplined comparisons among species with known evolutionary relationships.

Conditional strategy   *See* Strategy.

Convergent evolution   The independent acquisition over time of similar characteristics in two or more unrelated species through similar selection pressures.

Cooperation   A mutually helpful interaction.

Critical period   A phase in an animal's life when certain experiences are particularly likely to have a potent developmental effect.

Cryptic female choice   The ability of a female in receipt of sperm from more than one male to choose which sperm to use to fertilize her eggs.

Cumulative selection *See* Selection.

Darwinian puzzle A trait that appears to reduce the fitness of individuals that possess it.

Developmental homeostasis The capacity of developmental mechanisms within individuals to produce adaptive traits despite the potentially disruptive effects of mutant genes and suboptimal environmental conditions.

Dilution effect Safety in numbers that comes from swamping the ability of local predators to consume prey.

Diploid Having two copies of each gene in one's genotype.

Direct fitness *See* Fitness.

Direct selection *See* Selection.

Display A stereotyped action used as a communication signal by individuals.

Divergent evolution The evolution of differences among closely related species due to differing selection pressures in their different environments.

Dominance hierarchy A social ranking within a group in which some individuals give way to others, often conceding useful resources to others without a fight.

Entrainment The effect of changing environmental cues on the mechanisms controlling circadian (or circannual) rhythms, leading to a synchronization of an individual's activity patterns with local conditions.

Ethology The study of the proximate mechanisms and adaptive value of animal behavior.

Eusocial Of or relating to societies that contain specialized nonreproducing castes that assist the reproductive members of the society.

Evolutionarily stable strategy *See* Strategy.

Evolutionary psychology The study of the adaptive value of psychological mechanisms, especially of human beings; a key component of sociobiology.

Extra-pair copulation (EPC) A mating by a male or female with an individual other than their primary partner in a seemingly monogamous species.

Female defense polygyny *See* Polygyny.

Fitness A measure of the genes contributed to the next generation by an individual, often stated in terms of the number of offspring produced by that individual that survive to reproduce.

Direct fitness The genes contributed to the next generation by an individual via personal reproduction in the bodies of its own offspring.

Indirect fitness The genes contributed to the next generation by an individual indirectly by helping nondescendant kin, in effect creating relatives that would not have existed without the help of the individual.

Inclusive fitness The sum of an individual's direct and indirect fitness.

Fitness benefit That aspect of a trait that tends to raise the inclusive fitness of individuals.

Fitness cost That aspect of a trait that tends to reduce the inclusive fitness of individuals.

Fixed action pattern An innate, highly stereotyped response that is triggered by a well-defined, simple stimulus; once the pattern is activated, the response is performed in its entirety.

Free-running cycle The cycle of activity of an individual that is expressed in a constant environment.

Frequency-dependent selection *See* Selection.

Game theory An optimality approach to the study of adaptive value in which the payoffs to an individual of a particular behavioral tactic are dependent on what the other members of the group are doing.

Genetic mosaic An individual whose tissues are a mix of different genotypes.

Genotype The genetic constitution of an individual; may refer to the alleles of one gene possessed by the individual or to its complete set of genes.

"Good genes" theory The argument that mate choice advances individual fitness because it provides the offspring of choosy individuals with genes that advance their chances of survival.

Group selection *See* Selection.

Haploid Having only one copy of each gene in one's genotype, as, for example, the sperm and eggs of diploid organisms.

Heritability The proportion of the total variance in the phenotypes in a population that arises because of genetic variance among individuals, or V_g divided by $V_g + V_e$, where V_g = phenotypic variance caused by genetic differences among individuals and V_e = phenotypic variance among individuals that is environmentally induced.

Home range An area that an animal occupies but does not defend, in contrast to a territory, which is defended.

Honest signal A signal that accurately conveys information to receivers about the signaler's fighting ability or quality as a potential mate.

Hypothesis A possible explanation for what causes something to occur.

Ideal free distribution The distribution of nonterritorial individuals in space when they are free to make decisions that maximize their individual fitness.

Illegitimate receiver An individual that uses information gained from the signals of another to advance its own fitness at a cost to the signaler.

Illegitimate signaler An individual that produces a signal that may deceive another into responding in ways that advance the signaler's fitness at a cost to the receiver.

Imprinting A form of learning in which individuals exposed to certain key stimuli, usually early in life, form an association with an object and may later show sexual behavior toward similar objects.

Inbreeding depression The tendency of the descendants of individuals that mate with close relatives to have lower fitness than non-inbred members of their species.

Inclusive fitness *See* Fitness.

Indirect fitness *See* Fitness.

Indirect selection *See* Selection.

Innate releasing mechanism A hypothetical, neural mechanism responsible for controlling an innate response to a sign stimulus.

Instinct A behavior pattern that reliably develops in most individuals, promoting a functional response to a releaser the first time the action is performed.

Interneuron A neuron that relays messages either from receptor neurons to the central nervous system (a sensory interneuron) or from the central nervous system to neurons commanding muscle cells (a motor interneuron).

Kin discrimination The capacity of an individual to react differently to others based on their degree of genetic relatedness to that individual.

Kin selection *See* Selection.

Learning A durable and usually adaptive change in an individual's behavior traceable to a specific experience in that individual's life.

Lek A traditional display site that females visit to select a mate from among males displaying on small territories lacking resources useful to the females.

Mobbing behavior A behavior in which prey closely approach and attempt to harass a predator.

Monogamy A mating system in which one male mates with one female in a breeding season.

Mutualism A mutually beneficial relationship or cooperative interaction.

Natural selection *See* Selection.

Neuron A nerve cell.

Nondescendant kin Relatives other than offspring.

Nuptial gift A food item transferred by a male to a female just prior to or during copulation.

Operant conditioning A kind of learning based on trial and error, in which an action, or operant, becomes more frequently performed if it is rewarded.

Operational sex ratio The ratio of receptive males to receptive females over a given period.

Optimality theory A theory based on the assumption that the attributes of organisms are optimal, that is, better than others in terms of cost–benefit ratio; the theory is used to generate hypotheses about the possible adaptive value of traits in terms of their fitness payoffs to individuals.

Outgroup The species or group of species most closely related to a cluster of related species or groups of species of interest.

Parental investment Costly parental activities that increase the likelihood of survival for some existing offspring, but that reduce the parent's chances of producing offspring in the future.

Phenotype Any measurable trait of an individual that arises from the interaction of the individual's genes and its environment.

Phenotype matching A proximate mechanism of kin discrimination in which an individual's behavior toward another is based on how similar they are in some way, such as odor or appearance.

Pheromone A volatile chemical released by an individual as a scent signal for another.

Photoperiod The number of hours of light in a 24-hour period.

Phylogeny An evolutionary genealogy of the relationships among a number of species or clusters of species.

Phylogenetic tree A diagram of the evolutionary relationships among species.

Polyandry A mating system in which a female has several partners in a breeding season.

Polygyny A mating system in which a male fertilizes the eggs of several females in a breeding season.

Polymorphism The coexistence of two or more distinctive forms or traits within a species.

Prediction In the scientific method, an expected result that should be observed of a particular hypothesis is true.

Proximate cause An immediate underlying cause based on the operation of internal mechanisms possessed by an individual.

Receptive field That portion of a receptor surface that is monitored by a higher-order neuron.

Reciprocity Reciprocal altruism in which a helpful action is repaid at a later date by the recipient of assistance.

Releaser A sign stimulus given by individual as a social signal to another.

Reproductive success The number of surviving offspring produced by an individual; direct fitness.

Resource defense polygyny *See* Polygyny.

Runaway selection *See* Selection.

Satellite male A male that waits near another male to intercept females drawn to the signals produced by that male or attracted by the resources defended by that male.

Scientific conclusion In the scientific method, a hypothesis that has been tested and rejected or accepted.

Scramble competition polygyny *See* Polygyny.

Search image A perceptual screening mechanism used by predators to search visually for cryptic, edible prey.

Selection The effect of differences among individuals in their ability to transmit copies of their genes to the next generation.

    Artificial selection A process that is identical to natural selection, except that humans control the reproductive success of alternative types within the selected population.

    Cumulative selection The effect of repeated bouts of natural selection, resulting in the accumulation of many small adaptive changes in an evolving population, which can add up to a large evolutionary change over time.

Direct selection *See* Natural selection.

Frequency-dependent selection A form of natural selection in which the fitness of a type depends upon its relative frequency in the population.

Group selection Selection that occurs when groups differ in their collective attributes and those differences affect the survival chances of those groups.

Indirect selection A form of natural selection that occurs when individuals differ in their effects on the survival of nondescendant kin, creating differences in the indirect fitness of the individuals interacting with this category of kin.

Kin selection A form of natural selection that occurs when individuals differ in ways that affect their parental care or helping behavior, and thus the survival of their own offspring or of nondescendant kin.

Natural selection ("direct selection") The process that produces evolutionary change when individuals differ in heritable traits that are correlated with differences in their individual reproductive success

Runaway selection A form of sexual selection that occurs when female mating preferences for certain male attributes create a positive feedback loop favoring both males with these attributes *and* females that prefer them.

Sexual selection A form of natural selection that occurs when individuals differ in their ability to compete with others for mates or to attract members of the opposite sex.

Selfish herd A group of individuals whose members use others as living shields against predators.

Sensory exploitation A situation in which a signaler is able to tap into a preexisting sensitivity or bias in the perceptual system of a receiver, thereby gaining an advantage in transmitting a message to that receiver.

Sex role reversal   A situation in which females compete for access to males, which may choose selectively among potential mates.

Sexual dimorphism   A difference between males and females of the same species.

Sexual selection   *See* Selection.

Siblicide   The killing of a sibling by a brother or sister.

Sign stimulus   The effective component of an action or object that triggers a fixed action pattern.

Sociobiology   A discipline that uses evolutionary theory as a foundation for the study of social behavior; often used to refer to studies of this sort involving human beings.

Sperm competition   The competition between males that determines whose sperm will fertilize a female's eggs when both males' sperm have been accepted by that female.

Stimulus filtering   The capacity of neurons and neural networks to ignore stimuli that could potentially elicit a response from them.

Strategy   A genetically distinct set of rules for behavior exhibited by individuals.

Conditional strategy   A set of rules that provides for different tactics under different environmental conditions; the inherited behavioral capacity to be flexible in response to certain cues or situations.

Evolutionarily stable strategy   That set of rules of behavior that when adopted by a certain proportion of the population cannot be replaced by any alternative strategy.

Supernormal stimulus   A sign stimulus that is more effective in eliciting a response than naturally occurring actions or objects.

Synapse   The point of near contact between one nerve cell and another.

Tactic   A behavior pattern that is specified by an evolved strategy.

Territory   An area that an animal defends against intruders.

Test   In the scientific method, actual results that permit one to evaluate a hypothesis by comparing the evidence against the predicted result.

Ultimate cause   The evolutionary, historical reason why something is the way it is.

Bibliography

1 Able, K. P. 1993. Orientation cues used by migratory birds: A review of cue-conflict experiments. *Trends in Ecology and Evolution* 10:367–371.

2 Able, K. P. 1996. The debate over olfactory navigation by homing pigeons. *Journal of Experimental Biology* 199:121–124.

3 Adkins-Regan, E. 1981. Hormone specificity, androgen metabolism, and social behavior. *American Zoologist* 21:257–271.

4 Alatalo, R. V., A. Carlson, A. Lundberg, and S. Ulfstrand. 1981. The conflict between male polygamy and female monogamy: The case of the pied flycatcher *Ficedula hypoleuca. American Naturalist* 117:738–753.

5 Alatalo, R. V., D. Eriksson, L. Gustafsson, and K. Larsson. 1987. Exploitation competition influences the use of foraging sites by tits: Experimental evidence. *Ecology* 68:284–290.

6 Alatalo, R. V., J. Höglund, and A. Lundberg. 1991. Lekking in the black grouse—a test of male viability. *Nature* 352:155–156.

7 Alatalo, R. V., and A. Lundberg. 1984. Polyterritorial polygyny in the pied flycatcher *Ficedula hypoleuca*—evidence for the deception hypothesis. *Annales Zoologici Fennici* 21:217–228.

8 Alatalo, R. V., A. Lundberg, and K. Ståhlbrandt. 1984. Female mate choice in the pied flycatcher *Ficedula hypoleuca. Behavioral Ecology and Sociobiology* 14:253–262.

9 Alcock, J. 1987. Leks and hilltopping in insects. *Journal of Natural History* 21:319–328.

10 Alcock, J. 1996. Provisional rejection of three alternative hypotheses on the maintenance of a size dichotomy in males of Dawson's burrowing bee, *Amegilla dawsoni* (Apidae, Apinae, Anthophorini). *Behavioral Ecology and Sociobiology* 39:181–188.

11 Alcock, J., and W. J. Bailey. 1995. Acoustical communication and the mating system of the Australian whistling moth *Hecatesia exultans* (Noctuidae: Agaristidae). *Journal of Zoology* 237:337–352.

12 Alcock, J., G. C. Eickwort, and K. R. Eickwort. 1977. The reproductive behavior of *Anthidium maculosum* and the evolutionary significance of multiple copulations by females. *Behavioral Ecology and Sociobiology* 2:385–396.

13 Alexander, R. D. 1974. The evolution of social behavior. *Annual Review of Ecology and Systematics* 5:325–383.

14 Alexander, R. D. 1979. *Darwinism and Human Affairs.* University of Washington Press, Seattle, WA.

15 Alexander, R. D. 1987. *The Biology of Moral Systems.* Aldine de Gruyter, New York.

16 Alexander, R. D., J. L. Hoogland, R. D. Howard, K. M. Noonan, and P. W. Sherman. 1979. Sexual dimorphism and breeding systems in pinnipeds, ungulates, primates and humans. In *Evolutionary Biology and Human Social Behavior: An Anthropological Perspective*, N. A. Chagnon and W. Irons (eds.). Duxbury Press, North Scituate, MA.

17 Allen, G. E. et al. 1976. Sociobiology—Another biological determinism. *BioScience* 26:183–186.

18 Allen, G. R., D. J. Kazmer, and R. F. Luck. 1994. Post-copulatory male behaviour, sperm precedence and multiple mating in a solitary parasitoid wasp. *Animal Behaviour* 48:635–644.

19 Ancel, A., H. Visser, Y. Handrich, D. Masman, and Y. Le Maho. 1997. Energy saving in huddling penguins. *Nature* 385:304–305.

20 Andelman, S. J. 1987. Evolution of concealed ovulation in vervet monkeys (*Cercopithecus aethiops*). *American Naturalist* 129:785–799.

21 Anderson, J. B., and L. P. Brower. 1996. Freeze-protection of overwintering monarch butterflies in Mexico: Critical role of the forest as a blanket and an umbrella. *Ecological Entomology* 21:107–116.

22 Anderson, R. A., and L. J. Vitt. 1990. Sexual selection versus alternative causes of sexual dimorphism in teiid lizards. *Oecologia* 84:145–157.

23 Andersson, M. 1982. Female choice selects for extreme tail length in a widowbird. *Nature* 299:818–820.

24 Andersson, M. 1994. *Sexual Selection.* Princeton University Press, Princeton, NJ.

25 Andersson, M., F. Gotmark, and C. G. Wiklund. 1981. Food information in the black-headed gull, *Larus ridibundus. Behavioral Ecology and Sociobiology* 9:199–202.

26 Andrade, M. C. B. 1996. Sexual selection for male sacrifice in the Australian redback spider. *Science* 271:70–72.

27 Andrew, R. J. 1962. Evolution of intelligence and vocal mimicking. *Science* 137:585–589.

28 Armitage, K. B. 1986. Marmot polygyny revisited: The determinants of male and female reproductive strategies. In *Ecological Aspects of Social Evolution: Birds and Mammals,* D. I. Rubenstein and R. W. Wrangham (eds.). Princeton University Press, Princeton, NJ.

29 Armstrong, D. P. 1992. Correlation between nectar supply and aggression in territorial honeyeaters: Causation or coincidence? *Behavioral Ecology and Sociobiology* 30:95–102.

30 Armstrong, E. 1983. Relative brain size and metabolism in mammals. *Science* 220:1302–1304.

31 Arnold, S. J. 1976. Sexual behavior, sexual interference and sexual defense in the salamanders *Ambystoma maculatum, Ambystoma tigrinum* and *Plethodon jordani. Zeitschrift für Tierpsychologie* 42:247–300.

32 Arnold, S. J. 1980. The microevolution of feeding behavior. In *Foraging Behavior: Ecology, Ethological and Psychological Approaches,* A. Kamil and T. Sargent (eds.). Garland STPM Press, New York.

33 Arnold, S. J. 1982. A quantitative approach to antipredator performance: Salamander defense against snake attack. *Copeia* 1982:247–253.

34 Arnold, S. J. 1983. Sexual selection: The interface of theory and empiricism. In *Mate Choice,* P. P. G. Bateson (ed.). Cambridge University Press, Cambridge.

35 Artiss, T., and K. Martin. 1995. Male vigilance in white-tailed ptarmigan, *Lagopus leucurus:* Mate guarding or predator detection? *Animal Behaviour* 49:1249–1258.

36 Aviles, L. 1993. Interdemic selection and the sex ratio—a social spider perspective. *American Naturalist* 142:320–345.

37 Axelrod, R., and W. D. Hamilton. 1981. The evolution of cooperation. *Science* 211:1390–1396.

38 Ayala, F. J., and C. A. Campbell. 1974. Frequency-dependent selection. *Annual Review of Ecology and Systematics* 5:115–138.

39 Bachmann, G. C. 1993. The effect of body condition on the trade-off between vigilance and foraging in Belding's ground squirrels. *Animal Behaviour* 46:233–244.

40 Bailey, J. M., S. Gaulin, Y. Agyei, and B. A. Gladue. 1994. Effects of gender and sexual orientation on evolutionarily relevant aspects of human mating psychology. *Journal of Personality and Social Psychology* 66:1081–1093.

41 Bailey, J. M., and R. C. Pillard. 1991. A genetic study of male sexual orientation. *Archives of General Psychiatry* 48:1089–1096.

42 Baker, M. C., and M. A. Cunningham. 1985. The biology of bird-song dialects. *Behavioral and Brain Sciences* 8:85–133.

43 Baker, M. C., D. B. Thompson, and G. L. Sherman. 1981. Neighbour/stranger song discrimination in white-crowned sparrows. *Condor* 83:265–267.

44 Baker, R. R., and M. A. Bellis. 1989. Number of sperm in human ejaculates varies in accordance with sperm competition theory. *Animal Behaviour* 37:867–869.

45 Baker, R. R., and M. A. Bellis. 1993. Human sperm competition: Ejaculate adjustment by males and the function of masturbation. *Animal Behaviour* 46:861–885.

46 Baker, R. R., and M. A. Bellis. 1993. Human sperm competition: Ejaculation manipulation by females and a function for the female orgasm. *Animal Behaviour* 46:887–909.

47 Balcombe, J. P. 1990. Vocal recognition of pups by mother Mexican free-tailed bats, *Tadarida brasiliensis mexicana. Animal Behaviour* 39:960–966.

48 Balda, R. P. 1980. Recovery of cached seeds by a captive *Nucifraga caryocatactes*. *Zeitschrift für Tierpsychologie* 52:331–346.

49 Balda, R. P., and A. C. Kamil. 1992. Long-term spatial memory in Clark's nutcracker, *Nucifraga columbiana*. *Animal Behaviour* 44:761–769.

50 Baldaccini, N. E., S. Benvenuti, V. Fiaschi, and F. Papi. 1975. Pigeon navigation: Effects of wind deflection at home cage and homing behavior. *Journal of Comparative Physiology* 99:177–186.

51 Balmford, A., J. C. Deutsch, R. J. C. Nefdt, and T. Clutton-Brock. 1993. Testing hotspot models of lek evolution: Data from three species of ungulates. *Behavioral Ecology and Sociobiology* 33:57–65.

52 Balmford, A., A. M. Rosser, and S. D. Albon. 1992. Correlates of female choice in resource-defending antelope. *Behavioral Ecology and Sociobiology* 31:107–114.

53 Balshine-Earn, S. 1995. The costs of parental care in Galilee St Peter's fish, *Sarotherodon galilaeus*. *Animal Behaviour* 50:1–7.

54 Baptista, L. F. 1996. Nature and its nurturing in avian vocal development. In *Ecology and Evolution of Acoustic Information in Birds*, D. E. Kroodsma and E. H. Miller (eds.). Cornell University Press, Ithaca, NY.

55 Baptista, L. F., and C. K. Catchpool. 1989. Vocal mimicry and interspecific aggression in songbirds: Experiments using white-crowned sparrow imitation of song sparrow song. *Behaviour* 109:247–257.

56 Baptista, L. F., and S. L. L. Gaunt. 1994. Historical perspectives: Advances in studies of avian sound communication. *Condor* 96:817–830.

57 Baptista, L. F., and M. L. Morton. 1988. Song learning in montane white-crowned sparrows: From whom and when. *Animal Behaviour* 36:1753–1764.

58 Baptista, L. F., and L. Petrinovich. 1984. Social interaction, sensitive phases and the song template hypothesis in the white-crowned sparrow. *Animal Behaviour* 32:172–181.

59 Baptista, L. F., and L. Petrinovich. 1986. Song development in the white-crowned sparrow: Social factors and sex differences. *Animal Behaviour* 34:1359–1371.

60 Baptista, L., and P. W. Trail. 1992. The role of song in the evolution of passerine diversity. *Systematic Biology* 41:242–247.

61 Barber, N. 1995. The evolutionary psychology of physical attractiveness: Sexual selection and human morphology. *Ethology and Sociobiology* 16:395–424.

62 Barker, D. M. 1994. Copulatory plugs and paternity assurance in the nematode *Caenorhabditis elegans*. *Animal Behaviour* 48:147–156.

63 Basolo, A. L. 1990. Female preference predates the evolution of the sword in swordtail fish. *Science* 250:808–810.

64 Basolo, A. L. 1991. Male swords and female preferences. *Science* 253:1426–1427.

65 Basolo, A. L. 1995. Phylogenetic evidence for the role of a pre-existing bias in sexual selection. *Proceedings of the Royal Society of London B* 259:307–311.

66 Bass, A. H. 1996. Shaping brain sexuality. *American Scientist* 84:352–363.

67 Baum, D. A., and A. Larson. 1991. Adaptation reviewed: A phylogenetic methodology for studying character macroevolution. *Systematic Zoology* 40:1–18.

68 Baylies, M. K., T. A. Bargiello, F. R. Jackson, and M. W. Young. 1987. Changes in abundance or structure of the *per* gene product can alter periodicity of the *Drosophila* clock. *Nature* 326:390–392.

69 Beecher, M. D. 1982. Signature systems and kin recognition. *American Zoologist* 22:477–490.

70 Beecher, M. D. 1992. Successes and failures of parent-offspring recognition in animals. In *Kin Recognition*, P. G. Hepper (ed.). Cambridge University Press, Cambridge.

71 Beecher, M. D., and I. M. Beecher. 1979. Sociobiology of bank swallows: Reproductive strategy of the male. *Science* 205:1282–1285.

72 Beecher, M. D., M. B. Medvin, P. K. Stoddard, and P. Loesche. 1986. Acoustic adaptations for parent-offspring recognition in swallows. *Experimental Biology* 45:179–183.

73 Beecher, M. D., P. K. Stoddard, S. E. Campbell, and C. L. Horning. 1996. Repertoire matching between neighboring song sparrows. *Animal Behaviour* 51:917–923.

74   Behrmann, M., G. Winocur, and M. Moscovitch. 1992. Dissociation between mental imagery and object recognition in a brain-damaged patient. *Nature* 359:636–637.

75   Beletsky, L. D., and G. H. Orians. 1987. Territoriality among male red-winged blackbirds. I. Site fidelity and movement patterns. *Behavioral Ecology and Sociobiology* 20:21–34.

76   Beletsky, L. D., and G. H. Orians. 1989. Territoriality among male red-winged blackbirds. III. Testing hypotheses of territorial dominance. *Behavioral Ecology and Sociobiology* 24:333–339.

77   Benkman, C. W. 1990. Foraging rates and the timing of crossbill reproduction. *Auk* 107:376–386.

78   Bentley, D., and R. R. Hoy. 1974. The neurobiology of cricket song. *Scientific American* 231 (Aug):34–44.

79   Benzer, S. 1973. Genetic dissection of behavior. *Scientific American* 229 (Dec):24–37.

80   Bercovitch, F. B. 1986. Male rank and reproductive activity in savanna baboons. *International Journal of Primatology* 7:533–550.

81   Bercovitch, F. B. 1991. Social stratification, social strategies, and reproductive success in primates. *Ethology and Sociobiology* 12:315–333.

82   Bercovitch, F. B. 1995. Female cooperation, consortship maintenance, and male mating success in savanna baboons. *Animal Behaviour* 50:137–149.

83   Berger, J. 1983. Induced abortion and social factors in wild horses. *Nature* 303:59–61.

84   Berglund, A., G. Rosenqvist, and I. Svensson. 1986. Mate choice, fecundity and sexual dimorphism in two pipefish species (Syngnathidae). *Behavioral Ecology and Sociobiology* 19:301–307.

85   Berthold, P. 1991. Genetic control of migratory behaviour in birds. *Trends in Ecology and Evolution* 6:254–257.

86   Berthold, P., A. J. Helbig, G. Mohr, and U. Querner. 1992. Rapid microevolution of migratory behaviour in a wild bird species. *Nature* 360:668–670.

87   Berthold, P., and F. Pulido. 1994. Heritability of migratory activity in a natural bird population. *Proceedings of the Royal Society of London B* 257:311–315.

88   Bertram, B. C. R. 1978. Kin selection in lions and evolution. In *Growing Points in Ethology*, P. P. G. Bateson and R. A. Hinde (eds.). Cambridge University Press, New York.

89   Betzig, L. (ed.). 1997. *Human Nature, A Critical Reader.* Oxford University Press, New York.

90   Bildstein, K. L. 1983. Why white-tailed deer flag their tails. *American Naturalist* 121:709–715.

91   Birkhead, T. R., and A. P. Møller. 1992. A pairwise comparative method as illustrated by copulation frequency in birds. *American Naturalist* 139:644–656.

92   Birkhead, T. R., and A. P. Møller. 1992. *Sperm Competition in Birds: Evolutionary Causes and Consequences.* Academic Press, London.

93   Birkhead, T. R., and A. P. Møller. 1993. Female control of paternity. *Trends in Ecology and Evolution* 8:100–103.

94   Birkhead, T. R., A. P. Møller, and W. J. Sutherland. 1993. Why do females make it so difficult for males to fertilize their eggs? *Journal of Theoretical Biology* 161:51–60.

95   Bissoondath, C. J., and C. Wiklund. 1995. Protein content of spermatophores in relation to monandry/polyandry in butterflies. *Behavioral Ecology and Sociobiology* 37:365–372.

96   Bjorklund, M., and B. Westman. 1986. Adaptive advantages of monogamy in the great tit (*Parus major*): An experimental test of the polygyny threshold model. *Animal Behaviour* 34:1436–1440.

97   Black, A. H. 1971. The direct control of neural processes by reward and punishment. *American Scientist* 59:236–245.

98   Blest, A. D. 1957. The evolution of protective displays in the Saturnoidea and Sphingidae (Lepidoptera). *Behaviour* 11:257–309.

99   Blest, A. D. 1957. The function of eye-spot patterns in the Lepidoptera. *Behaviour* 11:209–256.

100   Boake, C. R. B., and A. Hoikkala. 1995. Courtship behaviour and mating success of wild-caught *Drosophila silvestris* males. *Animal Behaviour* 49:1303–1313.

101 Boggess, J. 1984. Infant killing and male reproductive strategies in langurs (*Presbytis entellus*). In *Infanticide: Comparative and Evolutionary Perspectives*, G. Hausfater and S. B. Hrdy (eds.). Aldine, Chicago.

102 Bolles, R. C. 1973. The comparative psychology of learning: The selection association principle and some problems with "general" laws of learning. In *Perspectives in Animal Behavior*, G. Bermant (ed.). Scott, Foresman & Company, Glenview, IL.

103 Boomsma, J. J., and F. L. W. Ratnieks. 1996. Paternity in eusocial Hymenoptera. *Philosophical Transactions of the Royal Society of London B* 351:947–975.

104 Boone, J. L. III. 1986. Parental investment and elite family structure in preindustrial states: A case study of late medieval-early modern Portuguese genealogies. *American Anthropologist* 88:859–878.

105 Borg-Karlson, A.-K. 1990. Chemical and ethological studies of pollination in the genus *Ophrys* (Orchidaceae). *Phytochemistry* 29:1359–1387.

106 Borgerhoff Mulder, M. 1987. Kipsigis bridewealth payments. In *Human Reproductive Behavior: A Darwinian Perspective*, L. L. Betzig, M. Borgerhoff Mulder, and P. Turke (eds.). Cambridge University Press, Cambridge.

107 Borgerhoff Mulder, M. 1987. Resources and reproductive success in women with an example from the Kipsigis of Kenya. *Journal of Zoology* 213:489–505.

108 Borgerhoff Mulder, M. 1988. Reproductive consequences of sex-biased inheritance. In *Comparative Socioecology of Mammals and Man*, V. Standen and R. Foley (eds.). Blackwell, London.

109 Borgia, G. 1985. Bower quality, number of decorations and mating success of male satin bowerbirds (*Ptilonorhynchus violaceus*). *Animal Behaviour* 33:266–271.

110 Borgia, G. 1986. Sexual selection in bowerbirds. *Scientific American* 254 (June):92–100.

111 Borgia, G. 1995. Comparative behavioral and biochemical studies of bowerbirds and the evolution of bower building. In *Biodiversity II*, M. Reaka, D. Wilson, and E. O. Wilson (eds.). Smithsonian Institution, Washington, D.C.

112 Borgia, G. 1995. Why do bowerbirds build bowers? *American Scientist* 83:542–547.

113 Bouchard, T. J., Jr., D. T. Lykken, M. McGue, N. L. Segal, and A. Tellegen. 1990. Sources of human psychological differences: The Minnesota study of twins reared apart. *Science* 250:223–228.

114 Bouchard, T. J., Jr., and M. McGue. 1981. Familial studies of intelligence: A review. *Science* 212:1055–1059.

115 Bradbury, J. W. 1977. Lek mating behavior in the hammer-headed bat. *Zeitschrift für Tierpsychologie* 45:225–255.

116 Bradbury, J. W. 1982. The evolution of leks. In *Natural Selection and Social Behavior*, R. D. Alexander and D. W. Tinkle (eds.). Chiron Press, New York.

117 Bradbury, J. W., S. L. Vehrencamp, and R. M. Gibson. 1989. Dispersion of displaying male sage grouse. I. Patterns of temporal variation. *Behavioral Ecology and Sociobiology* 24:1–14.

118 Breedlove, S. M. 1992. Sexual dimorphism in the nervous system. *Journal of Neuroscience* 12:4133–4142.

119 Breitwisch, R. 1989. Mortality patterns, sex ratios, and parental investment in monogamous birds. *Current Ornithology* 6:1–50.

120 Breitwisch, R., and G. H. Whitesides. 1987. Directionality of singing and non-singing behaviour of mated and unmated northern mockingbirds, *Mimus polyglottos*. *Animal Behaviour* 35:331–339.

121 Bremmer, F. 1986. White whales on holiday. *Natural History* 95 (Jan):40–49.

122 Brenowitz, E. A. 1991. Altered perception of species-specific song by female birds after lesions of a forebrain nucleus. *Science* 251:303–305.

123 Brenowitz, E. A. 1994. Flexibility and constraint in the evolution of animal communication. In *Flexibility and Constraint in Behavioral Systems*, R. J. Greenspan and C. P. Kyriacou (eds.). John Wiley & Sons, New York.

124 Breven, K. A. 1981. Mate choice in the wood frog, *Rana sylvatica*. *Evolution* 35:707–722.

125 Briskie, J. V. 1992. Copulation patterns and sperm competition in the polyandrous Smith's longspur. *Auk* 109:563–575.

126 Brockmann, H. J. 1990. Mating behavior of horseshoe crabs, *Limulus polyphemus. Behaviour* 114:206–220.

127 Brockmann, H. J., T. Colson, and W. Potts. 1994. Sperm competition in horseshoes crabs (*Limulus polyphemus*). *Behavioral Ecology and Sociobiology* 35:153–160.

128 Brockmann, H. J., and D. Penn. 1992. Male mating tactics in the horseshoe crab, *Limulus polyphemus. Animal Behaviour* 44:653–665.

129 Brockway, B. F. 1964. Social influences on reproductive physiology and ethology of budgerigars (*Melopsittacus undulatus*). *Animal Behaviour* 12:493–501.

130 Brockway, B. F. 1965. Stimulation of ovarian development and egglaying by male courtship vocalizations in budgerigars (*Melopsittacus undulatus*). *Animal Behaviour* 13:575–578.

131 Brodie, E. D., Jr. 1977. Hedgehogs use toad venom in their own defense. *Nature* 268:627–628.

132 Brodsky, L. M. 1988. Mating tactics of male rock ptarmigan, *Lagopus mutus:* A conditional mating strategy. *Animal Behaviour* 36:335–342.

133 Brooks, D. R., and D. A. McLennan. 1991. *Phylogeny, Ecology and Behavior: A Research Program in Comparative Biology.* University of Chicago Press, Chicago.

134 Brower, J. V. 1958. Experimental studies of mimicry in some North American butterflies. 1. The monarch, *Danaus plexippus*, and viceroy, *Limenitis archippus. Evolution* 12:3–47.

135 Brower, J. V., and L. P. Brower. 1962. Experimental studies of mimicry. 6. The reaction of toads (*Bufo terrestris*) to honeybees (*Apis mellifera*) and their dronefly mimics (*Eristalis vinetorum*). *American Naturalist* 96:297–307.

136 Brower, L. P. 1985. Foraging dynamics of bird predators on overwintering monarch butterflies in Mexico. *Evolution* 39:852–868.

137 Brower, L. P. 1996. Monarch butterfly orientation: Missing pieces of a magnificent puzzle. *Journal of Experimental Biology* 199:93–103.

138 Brower, L. P., and W. H. Calvert. 1984. Chemical defence in butterflies. In *The Biology of Butterflies*, R. I. Vane-Wright and P. R. Ackery (eds.). Academic Press, London.

139 Brower, L. P., W. N. Ryerson, J. L. Coppinger, and S. C. Glazier. 1968. Ecological chemistry and the palatability spectrum. *Science* 161:1349–1351.

140 Brown, C. H. 1982. Ventriloquial and locatable vocalizations in birds. *Zeitschrift für Tierpsychologie* 59:338–350.

141 Brown, C. R., and M. B. Brown. 1986. Ectoparasitism as a cost of coloniality in cliff swallows (*Hirundo pyrrhonota*). *Ecology* 67:1206–1218.

142 Brown, C. R., and M. B. Brown. 1992. Ectoparasitism as a cause of natal dispersal in cliff swallows. *Ecology* 73:718–723.

143 Brown, D. E. 1991. *Human Universals.* Temple University Press, Philadelphia, PA.

144 Brown, J. L. 1975. *The Evolution of Behavior.* W. W. Norton, New York.

145 Brown, J. L. 1982. The adaptationist program. *Science* 217:884–886.

146 Brown, J. L. 1987. *Helping and Communal Breeding in Birds: Ecology and Evolution.* Princeton University Press, Princeton, NJ.

147 Brown, J. L. 1987. Testing inclusive fitness theory with social birds. In *Animal Societies: Theories and Facts*, Y. Ito, J. L. Brown, and J. Kikkawa (eds.). Japan Scientific Societies Press, Tokyo.

148 Brown, J. L. 1997. A theory of mate choice based on heterozygosity. *Behavioral Ecology* 8:60–65.

149 Brown, J. L., and A. Eklund. 1994. Kin recognition and the major histocompatibility complex: An integrative review. *American Naturalist* 143:435–461.

150 Brown, J. L., and S.-H. Li. 1995. Phylogeny of social behavior in *Aphelocoma* jays: A role for hybridization? *Auk* 112:464–472.

151 Brown, J. L., and G. H. Orians. 1970. Spacing patterns in mobile animals. *Annual Review of Ecology and Systematics* 1:239–262.

152 Brown, J. R., H. Ye, R. T. Bronson, P. Dikkes, and M. E. Greenberg. 1996. A defect in nurturing in mice lacking the immediate early gene *fosB. Cell* 86:297–309.

153 Brown, K. M., M. Woulfe, and R. D. Morris. 1995. Pattern of adoption in ring-billed gulls: Who is really winning the inter-generational conflict? *Animal Behaviour* 49:321–331.

154 Brown, R. E. 1993. Hormonal and experiential factors influencing parental behaviour in male rodents: An integrative approach. *Behavioural Processes* 30:1–28.

155 Brownmiller, S. 1975. *Against Our Will: Men, Women and Rape.* Simon & Schuster, New York.

156 Buchholz, R. 1995. Female choice, parasite load and male ornamentation in wild turkeys. *Animal Behaviour* 50:929–943.

157 Burgham, M. C. J., and J. Picman. 1989. Effect of brown-headed cowbirds on the evolution of yellow warbler anti-parasite strategies. *Animal Behaviour* 38:298–308.

158 Burke, T., N. B. Davies, M. W. Bruford, and B. J. Hatchwell. 1989. Parental care and mating behaviour of polyandrous dunnocks *Prunella modularis* related to paternity by DNA fingerprinting. *Nature* 338:249–251.

159 Burley, N. T., D. A. Enstrom, and L. Chitwood. 1994. Extra-pair relations in zebra finches: Differential male success results from female tactics. *Animal Behaviour* 48:1031–1041.

160 Buskirk, R. E., C. Frolich, and K. G. Ross. 1984. The natural selection of sexual cannibalism. *American Naturalist* 123:612–625.

161 Buss, D. M. 1987. Sex differences in human mate selection criteria: An evolutionary perspective. In *Sociobiology and Psychology: Ideas, Issues, and Applications,* C. Crawford, M. Smith, and D. Krebs (eds.). Lawrence Erlbaum Associates, Hillsdale, NJ.

162 Buss, D. M. 1989. Sex differences in human mate preferences: Evolutionary hypotheses tested in 37 cultures. *Behavioral and Brain Sciences* 12:1–14.

163 Buss, D. M. 1994. *The Evolution of Desire.* Basic Books, New York.

164 Buss, D. M., and D. P. Schmitt. 1993. Sexual strategies theory: An evolutionary perspective on human mating. *Psychological Review* 100:204–232.

165 Byers, B. E., and D. E. Kroodsma. 1992. Development of two song categories by chestnut-sided warblers. *Animal Behaviour* 44:799–810.

166 Bygott, J. D., B. C. R. Bertram, and J. P. Hanby. 1979. Male lions in large coalitions gain reproductive advantage. *Nature* 282:839–841.

167 Byne, W. 1994. Evidence against the biological evidence. *Scientific American* 270 (5):50–55.

168 Cade, W. 1980. Alternative male reproductive strategies. *Florida Entomologist* 63:30–45.

169 Cade, W. 1981. Alternative male strategies: Genetic differences in crickets. *Science* 212:563–564.

170 Cain, A. J. 1989. The perfection of animals. *Biological Journal of the Linnean Society* 36:3–29.

171 Calvert, W. H., and L. P. Brower. 1986. The location of monarch butterfly (*Danaus plexippus* L.) overwintering colonies in Mexico in relation to topography and climate. *Journal of the Lepidopterists' Society* 40:164–187.

172 Calvert, W. H., L. E. Hedrick, and L. P. Brower. 1979. Mortality of the monarch butterfly (*Danaus plexippus* L.): Avian predation at five overwintering sites in Mexico. *Science* 204:847–851.

173 Camhi, J. M. 1984. *Neuroethology.* Sinauer Associates, Sunderland, MA.

174 Caple, G., R. P. Balda, and W. R. Willis. 1983. The physics of leaping animals and the evolution of preflight. *American Naturalist* 121:455–476.

175 Caraco, T. 1979. Time budgeting and group size: A test of theory. *Ecology* 60:618–627.

176 Caraco, T., and L. L. Wolf. 1975. Ecological determinants of group sizes of foraging lions. *American Naturalist* 109:343–352.

177 Carey, M., and V. Nolan, Jr. 1975. Polygyny in indigo buntings: A hypothesis tested. *Science* 190:1296–1297.

178 Carey, M., and V. Nolan, Jr. 1979. Population dynamics of indigo buntings and the evolution of avian polygyny. *Evolution* 33:1180–1192.

179 Carlsson, B.-G. 1991. Recruitment of mates and deceptive behavior by male Tengmalm's owls. *Behavioral Ecology and Sociobiology* 28:321–328.

180 Caro, T. M. 1986. The functions of stotting: A review of the hypotheses. *Animal Behaviour* 34:649–662.

181 Caro, T. M. 1986. The functions of stotting in Thomson's gazelles: Some tests of the predictions. *Animal Behaviour* 34:663–684.

182 Caro, T. M. 1995. Pursuit-deterrence revisited. *Trends in Ecology and Evolution* 10:500–503.

183 Caro, T. M., L. Lombardo, A. W. Goldizen, and M. Kelly. 1995. Tail-flagging and other antipredator signals in white-tailed deer: New data and synthesis. *Behavioral Ecology* 6:442–450.

184 Carpenter, F. L., D. C. Paton, and M. H. Hixon. 1983. Weight gain and adjustment of feeding territory size in migrant hummingbirds. *Proceedings of the National Academy of Sciences* 80:7259–7263.

185 Carr, A. 1967. Adaptive aspects of the scheduled travel of *Chelonia*. In *Animal Orientation and Navigation*, R. M. Storm (ed.). Oregon State University Press, Corvallis.

186 Catania, K. C., and J. H. Kaas. 1996. The unusual nose and brain of the star-nosed mole. *BioScience* 46:578–586.

187 Catchpole, C. K., and P. J. B. Slater. 1995. *Bird Song, Biological Themes and Variations.* Cambridge University Press, Cambridge.

188 Chagnon, N. A. 1988. Life histories, blood revenge, and warfare in a tribal population. *Science* 239:985–991.

189 Chai, P., and R. B. Srygley. 1990. Predation and the flight, morphology, and temperature of Neotropical rain-forest butterflies. *American Naturalist* 135:748-765.

190 Chen, J.-S., and A. Amsel. 1980. Recall (versus recognition) of taste and immunization against aversive taste anticipations based on illness. *Science* 209:831–833.

191 Cheney, D. L., and R. M. Seyfarth. 1985. Vervet monkey alarm calls: Manipulation through shared information? *Behaviour* 93:150–166.

192 Cheney, D. L., and R. M. Seyfarth. 1990. *How Monkeys See the World.* University of Chicago Press, Chicago.

193 Chilton, G., M. R. Lein, and L. F. Baptista. 1990. Mate choice by female white-crowned sparrows in a mixed-dialect population. *Behavioral Ecology and Sociobiology* 27:223–227.

194 Chivers, D. P., G. E. Brown, and R. J. F. Smith. 1996. The evolution of chemical alarm signals: Attracting predators benefits alarm signal givers. *American Naturalist* 148:649–659.

195 Chivers, D. P., and R. J. F. Smith. 1995. Fathead minnows (*Pimephales promelas*) learn to recognize chemical stimuli from high-risk habitats by the presence of alarm substance. *Behavioral Ecology* 6:155–158.

196 Christy, J. H. 1995. Mimicry, mate choice, and the sensory trap hypothesis. *American Naturalist* 146:171–181.

197 Clark, D. L., and B. G. Galef. 1995. Prenatal influences on reproductive life history strategies. *Trends in Ecology and Evolution* 10:151–153.

198 Clark, L. 1990. Starlings as herbalists: Countering parasites and pathogens. *Parasitology Today* 6:358–360.

199 Clark, R. D., and E. Hatfield. 1989. Gender differences in receptivity to sexual offers. *Journal of Psychology and Human Sexuality* 2:39–55.

200 Clayton, D. H., and N. D. Wolfe. 1993. The adaptive significance of self-medication. *Trends in Ecology and Evolution* 8:60–63.

201 Clayton, N. C., and J. R. Krebs. 1994. Hippocampal growth and attrition in birds affected by experience. *Proceedings of the National Academy of Science* 91:7410–7414.

202 Clayton, N. C., J. C. Reboreda, and A. Kacelnik. 1998. Seasonal changes of hippocampus volume in parasitic cowbirds. *Behavioral Processes* 41:237–243.

203 Clode, D. 1993. Colonially breeding seabirds: Predators or prey? *Trends in Ecology and Evolution* 8:336–338.

204 Clutton-Brock, T. H. 1991. *The Evolution of Parental Care.* Princeton University Press, Princeton, NJ.

205 Clutton-Brock, T. H., and S. D. Albon. 1979. The roaring of red deer and the evolution of honest advertisement. *Behaviour* 69:145–170.

206 Clutton-Brock, T. H., S. D. Albon, R. M. Gibson, and F. E. Guinness. 1979. The logical stag: Adaptive aspects of fighting in red deer. *Animal Behaviour* 27:211–225.

207 Clutton-Brock, T. H., and P. H. Harvey. 1984. Comparative approaches to investigating adaptation. In *Behavioural Ecology: An Evolutionary Approach* (2nd edition), J. R. Krebs and N. B. Davies (eds.). Blackwell, Oxford.

208 Clutton-Brock, T. H., M. Hiraiwa-Hasegawa, and A. Robertson. 1989. Mate choice on fallow deer leks. *Nature* 340:463–465.

209 Clutton-Brock, T. H., and G. A. Parker. 1992. Potential reproductive rates and the operation of sexual selection. *Quarterly Review of Biology* 67:437–456.

210 Clutton-Brock, T. H., and G. A. Parker. 1995. Sexual coercion in animal societies. *Animal Behaviour* 49:1345–1365.

211 Clutton-Brock, T. H., and A. C. J. Vincent. 1991. Sexual selection and the potential reproductive rates of males and females. *Nature* 351:58–60.

212 Colbert, E. H. 1955. *Evolution of the Vertebrates*. John Wiley & Sons, New York.

213 Coleman, R. M., and M. R. Gross. 1991. Parental investment theory: The role of past investment. *Trends in Ecology and Evolution* 6:404–406.

214 Collins, J. P., and J. E. Cheek. 1983. Effect of food and density on development of typical and cannibalistic salamander larvae in *Ambystoma tigrinum nebulosum*. *American Zoologist* 23:77–84.

215 Collins, S. A., C. Hubbard, and A. M. Houtman. 1994. Female mate choice in the zebra finch—the effect of male beak color and male song. *Behavioral Ecology and Sociobiology* 35:21–26.

216 Connor, R. C., R. A. Smolker, and A. F. Richards. 1992. Two levels of alliance formation among male bottlenose dolphins (*Tursiops* sp.). *Proceedings of the National Academy of Sciences* 89:987–990.

217 Conover, M. R. 1994. Stimuli eliciting distress calls in adult passerines and response of predators and birds to their broadcast. *Behaviour* 131:19–37.

218 Cooney, R., and A. Cockburn. 1995. Territorial defence is the major function of female song in the superb fairy-wren, *Malurus cyaneus*. *Animal Behaviour* 49:1635–1647.

219 Cooper, W. E. 1995. Foraging mode, prey chemical discrimination, and phylogeny in lizards. *Animal Behaviour* 50:973–985.

220 Cooper, W. E., Jr., and L. J. Vitt. 1985. Bluetails and autotomy: Enhancement of predation avoidance in juvenile skinks. *Zeitschrift für Tierpsychologie* 70:265–276.

221 Cosmides, L. M., and J. Tooby. 1981. Cytoplasmic inheritance and intragenomic conflict. *Journal of Theoretical Biology* 89:83–129.

222 Cosmides, L. M., and J. Tooby. 1992. Cognitive adaptations for social exchange. In *The Adapted Mind*, J. Barlow, L. Cosmides, and J. Tooby (eds.). Oxford University Press, New York.

223 Côté, I. M., and M. R. Gross. 1993. Reduced disease in offspring: A benefit of coloniality in sunfish. *Behavioral Ecology and Sociobiology* 33:269–274.

224 Court, G. S. 1996. The seal's own skin game. *Natural History* 105(8):36–41.

225 Cowlishaw, G., and R. I. M. Dunbar. 1991. Dominance rank and mating success in male primates. *Animal Behaviour* 41:1045–1056.

226 Cowlishaw, G., and R. Mace. 1996. Cross-cultural patterns of marriage and inheritance: A phylogenetic approach. *Ethology and Sociobiology* 17:87–98.

227 Cox, G. W. 1985. The evolution of avian migration systems between temperate and tropical regions of the New World. *American Naturalist* 126:451–474.

228 Craig, C. L. 1992. Aerial web-weaving spiders: Linking molecular and organismal processes in evolution. *Trends in Ecology and Evolution* 7:270–273.

229 Craig, C. L., and G. D. Bernard. 1990. Insect attraction to ultraviolet-reflecting spider webs and web decorations. *Ecology* 71:616–623.

230 Craig, C. L., G. D. Bernard, and J. A. Coddington. 1994. Evolutionary shifts in the spectral properties of spider silks. *Evolution* 48:287–296.

231 Craig, C. L., R. S. Weber, and G. D. Bernard. 1996. Evolution of predator-prey systems: Spider foraging plasticity in response to the visual ecology of prey. *American Naturalist* 147:205–229.

232 Craig, J. L., and M. E. Douglas. 1986. Resource distribution, aggressive asymmetries and variable access to resources in the nectar feeding bellbird. *Behavioral Ecology and Sociobiology* 18:231–240.

233 Creel, S. R., and N. M. Creel. 1991. Energetics, reproductive suppression and obligate communal breeding in carnivores. *Behavioral Ecology and Sociobiology* 28:263–270.

234 Creel, S. R., and N. M. Creel. 1995. Communal hunting and pack size in African wild dogs, *Lycaon pictus*. *Animal Behaviour* 50:1325–1339.

235 Cresswell, W. 1994. Flocking is an effective anti-predation strategy in redshanks, *Tringa totanus*. *Animal Behaviour* 47:433–442.

236 Crews, D. 1975. Psychobiology of reptilian reproduction. *Science* 189:1059–1065.

237 Crews, D. 1984. Gamete production, sex hormone secretion, and mating behavior uncoupled. *Hormones and Behavior* 18:22–28.

238 Crews, D. (ed.). 1987. *Psychobiology of Reproductive Behavior: An Evolutionary Perspective*. Prentice-Hall, Englewood Cliffs, NJ.

239 Crews, D. 1991. Trans-seasonal action of androgen in the control of spring courtship behavior in male red-sided garter snakes. *Proceedings of the National Academy of Sciences* 88:3545–3548.

240 Crews, D. 1992. Behavioral endocrinology and reproduction: An evolutionary perspective. *Oxford Reviews of Reproductive Biology* 14:303–370.

241 Crews, D. 1992. Diversity of hormone-behavior relations. In *Introduction to Behavioral Endocrinology*, J. Becker, M. Breedlove, and D. Crews (eds.). MIT Press/Bradford Books, Cambridge, MA.

242 Crews, D., and N. Greenberg. 1981. Function and causation of social signals in lizards. *American Zoologist* 21:273–294.

243 Crews, D., V. Hingorani, and R. J. Nelson. 1988. Role of the pineal gland in the control of annual reproductive behavioral and physiological cycles in the red-sided garter snake (*Thamnophis sirtalis parietalis*). *Journal of Biological Rhythms* 3:293–302.

244 Crews, D., and M. C. Moore. 1986. Evolution of mechanisms controlling mating behavior. *Science* 231:121–125.

245 Crnokrak, P., and D. A. Roff. 1995. Fitness differences associated with calling behaviour in the two wing morphs of male sand crickets, *Gryllus firmus*. *Animal Behaviour* 50:1475–1481.

246 Cronin, H. 1991. *The Ant and the Peacock*. Cambridge University Press, Cambridge.

247 Cronk, L. 1993. Parental favoritism toward daughters. *American Scientist* 81:272–279.

248 Cullen, E. 1957. Adaptations in the kittiwake to cliff nesting. *Ibis* 99:275–302.

249 Cumming, J. M. 1994. Sexual selection and the evolution of dance fly mating systems (Diptera: Empididae; Empidinae). *Canadian Entomologist* 126:907–920.

250 Curio, E. 1978. The adaptive significance of avian mobbing. I. Teleonomic hypotheses and predictions. *Zeitschrift für Tierpsychologie* 48:175–183.

251 Cuthill, I. C., and A. Kacelnik. 1990. Central place foraging: A reappraisal of the "loading effect." *Animal Behaviour* 40:1087–1101.

252 Cuthill, I. C., and W. A. Macdonald. 1990. Experimental manipulation of the dawn and dusk chorus in the blackbird *Turdus merula*. *Behavioral Ecology and Sociobiology* 26:209–216.

253 Dale, S., H. Rinden, and T. Slagsvold. 1992. Competition for a mate restricts mate search of female pied flycatchers. *Behavioral Ecology and Sociobiology* 30:165–176.

254 Dale, S., and T. Slagsvold. 1994. Polygyny and deception in the pied flycatcher: Can females determine male mating status? *Animal Behaviour* 48:1207–1217.

255 Daly, M., and M. Wilson. 1983. *Sex, Evolution and Behavior* (2nd edition). Willard Grant Press, Boston.

256 Daly, M., and M. Wilson. 1985. Child abuse and other risks of not living with both parents. *Ethology and Sociobiology* 6:197–210.

257 Daly, M., and M. Wilson. 1987. Children as homicide victims. In *Child Abuse and Neglect: Biosocial Dimensions*, R. Gelles and J. Lancaster (eds.). Aldine de Gruyter, New York.

258 Daly, M., and M. Wilson. 1987. Evolutionary psychology and family violence. In *Sociobiology and Psychology*, C. Crawford, M. Smith, and D. Krebs (eds.). Lawrence Erlbaum Associates, Hillsdale, NJ.

259 Daly, M., and M. Wilson. 1988. *Homicide*. Aldine de Gruyter, Chicago.

260 Daly, M., and M. Wilson. 1992. The man who mistook his wife for a chattel. In *The Adapted Mind*, J. Barkow and L. Cosmides (eds.). Oxford University Press, New York.

261 Daly, M., and M. Wilson. 1994. Some differential attributes of lethal assaults on small children by stepfathers versus genetic fathers. *Ethology and Sociobiology* 15:207–217.

262 Daly, M., M. Wilson, and S. J. Weghorst. 1982. Male sexual jealousy. *Ethology and Sociobiology* 3:11–27.

263 Darwin, C. 1859. *On the Origin of Species*. Murray, London

264 Darwin, C. 1871. *The Descent of Man and Selection in Relation to Sex*. Murray, London.

265 Davies, N. B. 1977. Prey selection and social behaviour in wagtails (Aves: Motacillidae). *Journal of Animal Ecology* 46:37–57.

266 Davies, N. B. 1978. Territorial defence in the speckled wood butterfly (*Pararge aegeria*): The resident always wins. *Animal Behaviour* 26:138–147.

267 Davies, N. B. 1983. Polyandry, cloaca-pecking and sperm competition in dunnocks. *Nature* 302:334–336.

268 Davies, N. B. 1985. Cooperation and conflict among dunnocks, *Prunella modularis*, in a variable mating system. *Animal Behaviour* 33:628–648.

269 Davies, N. B. 1989. Sexual conflict and the polygamy threshold. *Animal Behaviour* 38:226–234.

270 Davies, N. B. 1992. *Dunnock Behaviour and Social Evolution*. Oxford University Press, Oxford.

271 Davies, N. B., and M. de L. Brooke. 1988. Cuckoos versus reed warblers: Adaptations and counteradaptations. *Animal Behaviour* 36:262–284.

272 Davies, N. B., and T. R. Halliday. 1978. Deep croaks and fighting assessment in toads *Bufo bufo*. *Nature* 275:683–685.

273 Davies, N. B., I. R. Hartley, B. J. Hatchwell, and N. E. Langmore. 1996. Female control of copulations to maximize male help: A comparison of polygynandrous alpine accentors, *Prunella collaris*, and dunnocks, *P. modularis*. *Animal Behaviour* 51:27–47.

274 Davies, N. B., and A. I. Houston. 1981. Owners and satellites: The economics of territory defense in the pied wagtail, *Motacilla alba*. *Journal of Animal Ecology* 50:157–180.

275 Davies, N. B., and A. Lundberg. 1984. Food distribution and a variable mating system in the dunnock, *Prunella modularis*. *Journal of Animal Ecology* 53:895–912.

276 Davis-Walton, J., and P. W. Sherman. 1994. Sleep arrhythmia in the eusocial naked mole-rat. *Naturwissenschaften* 81:272–275.

277 Dawkins, R. 1977. *The Selfish Gene*. Oxford University Press, New York.

278 Dawkins, R. 1980. Good strategy or evolutionarily stable strategy? In *Sociobiology: Beyond Nature/Nurture?* G. W. Barlow and J. Silverberg (eds.). Westview Press, Boulder, CO.

279 Dawkins, R. 1982. *The Extended Phenotype*. W. H. Freeman, San Francisco.

280 Dawkins, R. 1986. *The Blind Watchmaker*. W. W. Norton, New York.

281 Dawkins, R., and J. Krebs. 1978. Animal signals: Information or manipulation? In *Behavioural Ecology: An Evolutionary Approach* (1st edition), J. R. Krebs and N. B. Davies (eds.). Blackwell, Oxford.

282 de Belle, J. S., A. J. Hilliker, and M. B. Sokolowski. 1989. Genetic localization of *foraging* (*for*): A major gene for larval behavior in *Drosophila melanogaster*. *Genetics* 123:157–163.

283 de Belle, J. S., and M. B. Sokolowski. 1987. Heredity of *rover/sitter*: Alternative foraging strategies of *Drosophila melanogaster* larvae. *Heredity* 59:73–83.

284 DeCoursey, P. J., and J. Buggy. 1989. Circadian rhythmicity after neural transplant to hamster third ventricle: Specificity of suprachiasmatic nuclei. *Brain Research* 500:263–275.

285 de Lope, F., and A. P. Møller. 1993. Female reproductive effort depends on the degree of ornamentation of their mates. *Evolution* 47:1152–1160.

286 Desmond, A., and J. Moore. 1991. *Darwin, The Life of a Tormented Evolutionist.* W. W. Norton, New York.

287 Dethier, V. G. 1962. *To Know A Fly.* Holden-Day, San Francisco.

288 Dethier, V. G. 1976. *The Hungry Fly: A Physiological Study of the Behavior Associated with Feeding.* Harvard University Press, Cambridge, MA.

289 Deutsch, C. J., M. P. Haley, and B. J. Le Boeuf. 1990. Reproductive effort of male northern elephant seals: Estimates from mass loss. *Canadian Journal of Zoology* 68:2580–2593.

290 Deutsch, J. C. 1994. Uganda kob mating success does not increase on larger leks. *Behavioral Ecology and Sociobiology* 34:451–459.

291 DeVoogd, T. J. 1991. Endocrine modulation of the development and adult function of the avian song system. *Psychoneuroendocrinology* 16:41–66.

292 DeWolfe, B. B., L. F. Baptista, and L. Petrinovich. 1989. Song development and territory establishment in Nuttall's white-crowned sparrow. *Condor* 91:397–407.

293 Dewsbury, D. A. 1992. On the problems studied in ethology, comparative psychology, and animal behavior. *Ethology* 92:89–107.

294 Dhondt, A. A., and J. Schillemans. 1983. Reproductive success of the great tit in relation to its territorial status. *Animal Behaviour* 31:902–912.

295 Dial, B. E., and L. C. Fitzpatrick. 1983. Lizard tail autotomy: Function and energetics of postautotomy tail movement in *Scincella lateralis. Science* 219:391–393.

296 Diamond, J. 1992. *The Third Chimpanzee.* HarperCollins Publishers, New York

297 Diamond, J. M., E. Cooper, C. Turner, and L. Macintyre. 1976. Trophic regulation of nerve sprouting. *Science* 193:371–377.

298 Dickinson, J. L. 1995. Trade-offs between postcopulatory riding and mate location in the blue milkweed beetle. *Behavioral Ecology* 6:280–286.

299 Dickinson, J. L., and R. L. Rutowski. 1989. The function of the mating plug in the chalcedon checkerspot butterfly. *Animal Behaviour* 38:154–162.

300 Dixon, A., D. Ross, S. L. C. O'Malley, and T. Burke. 1994. Paternal investment inversely related to degree of extra-pair paternity in the reed bunting. *Nature* 371:698–700.

301 Dixson, A. F. 1987. Observations on the evolution of genitalia and copulatory behavior in primates. *Journal of Zoology* 213:423–443.

302 Dodd, J., and J. M. Jessell. 1988. Axon guidance and the patterning of neuronal projections in vertebrates. *Science* 242:692–699.

303 Dodson, G. N. 1997. Resource defense mating system in antlered flies, *Phytalmia* spp. (Diptera: Tephritidae). *Annals of the Entomological Society of America* 90:80–88.

304 Dodson, G. N., and M. W. Beck. 1993. Precopulatory guarding of penultimate females by male crab spiders, *Misumenoides formosipes. Animal Behaviour* 46:951–959.

305 Doherty, J. A., and H. C. Gerhardt. 1983. Hybrid tree frogs: Vocalizations of males and selective phonotaxis of females. *Science* 220:1078–1080.

306 Dominey, W. J. 1980. Female mimicry in male bluegill sunfish—A genetic polymorphism? *Nature* 284:546–548.

307 Domjan, M., and B. Burkhard. 1986. *The Principles of Learning and Behavior* (2nd edition). Brooks/Cole Publishing, Monterey, CA.

308 Dow, H., and S. Fredga. 1983. Breeding and natal dispersal of the goldeneye, *Bucephala clangula. Journal of Animal Ecology* 52:681–695.

309 Downhower, J. F., and K. B. Armitage. 1971. The yellow-bellied marmot and the evolution of polygamy. *American Naturalist* 105:355–370.

310 Drews, C. 1996. Contests and patterns of injuries in free-ranging male baboons (*Papio cynocephalus*). *Behaviour* 133:443–474.

311 Duffey, S. S. 1970. Cardiac glycosides and distastefulness: Some observations on the palatability spectrum of butterflies. *Science* 169:78–79.

312 Dumont, J. P. C., and R. M. Robertson. 1986. Neuronal circuits: An evolutionary perspective. *Science* 233:849–853.

313 Dunn, J. 1976. How far do early differences in mother-child relations affect later development? In *Growing Points in Ethology*, P. P. G. Bateson and R. A. Hinde (eds.). Cambridge University Press, Cambridge.

314 Dunn, P. O., and R. J. Robertson. 1993. Extra-pair paternity in polygynous tree swallows. *Animal Behaviour* 45:231–239.

315 Dussourd, D. E., C. A. Harvis, J. Meinwald, and T. Eisner. 1991. Pheromonal advertisement of a nuptial gift by a male moth (*Utetheisa ornatrix*). *Proceedings of the National Academy of Sciences* 88:9224–9227.

316 Dussourd, D. E., K. Ubik, C. Harvis, J. Resch, J. Meinwald, and T. Eisner. 1988. Biparental defensive endowment of eggs with acquired plant alkaloid in the moth *Utetheisa ornatrix*. *Proceedings of the National Academy of Sciences* 85:5992–5996.

317 Duvall, D., M. B. King, and K. J. Gutzwiller. 1985. Behavioral ecology and ethology of the prairie rattlesnake. *National Geographic Research* 1:80–111.

318 Dyer, F. C., and J. L. Gould. 1983. Honey bee navigation. *American Scientist* 71:587–597.

319 Eadie, J. M., F. P. Kehoe, and T. D. Nudds. 1988. Pre-hatch and post-hatch brood amalgamation in North American Anatidae: A review of hypotheses. *Canadian Journal of Zoology* 66:1709–1721.

320 Eady, P. E. 1995. Why do male *Callosobruchus maculatus* beetles inseminate so many sperm? *Behavioral Ecology and Sociobiology* 36:25–32.

321 Eason, P., and S. J. Hannon. 1994. New birds on the block: New neighbors increase defensive costs for territorial male willow ptarmigan. *Behavioral Ecology and Sociobiology* 34:419–426.

322 East, M. L., H. Hofer, and W. Wickler. 1993. The erect "penis" is a flag of submission in a female-dominated society: Greetings in Serengeti spotted hyenas. *Behavioral Ecology and Sociobiology* 33:355–370.

323 Eberhard, W. G. 1980. The natural history and behavior of the bolas spider *Mastophora dizzydeani* sp. n. (Araneidae). *Psyche* 87:143–169.

324 Eberhard, W. G. 1985. *Sexual Selection and Animal Genitalia*. Harvard University Press, Cambridge, MA.

325 Eberhard, W. G. 1990. Animal genitalia and female choice. *American Scientist* 78:134–141.

326 Eberhard, W. G. 1990. Function and phylogeny of spider webs. *Annual Review of Ecology and Systematics* 21:341–372.

327 Eberhard, W. G. 1993. Evaluating models of sexual selection: Genitalia as a test case. *American Naturalist* 142:564–571.

328 Eberhard, W. G. 1996. *Female Control: Sexual Selection by Cryptic Female Choice*. Princeton University Press, Princeton, NJ.

329 Eberhard, W. G., and C. Cordero. 1995. Sexual selection by cryptic female choice on male seminal products—a new bridge between sexual selection and reproductive physiology. *Trends in Ecology and Evolution* 10:493–496.

330 Eckert, E. D., T. J. Bouchard, Jr., J. Bohlen, and L. L. Heston. 1986. Homosexuality in monozygotic twins reared apart. *British Journal of Psychiatry* 148:421–425.

331 Edmunds, M. 1974. *Defence in Animals*. Longman Group, Harlow, Essex.

332 Eens, M., R. Pinxten, and R. F. Verhegen. 1991. Male song as a cue for mate choice in the European starling. *Behaviour* 116:210–238.

333 Eens, M., R. Pinxten, and R. F. Verheyen. 1992. Song learning in captive European starlings, *Sturnus vulgaris*. *Animal Behaviour* 44:1131–1141.

334 Eggert, A.-K., and S. K. Sakaluk. 1995. Femalecoerced monogamy in burying beetles. *Behavioral Ecology and Sociobiology* 37:147–154.

335 Ehrman, L., and P. A. Parsons. 1976. *The Genetics of Behavior*. Sinauer Associates, Sunderland, MA.

336 Eimas, P. D. 1975. Speech perception in early infancy. In *Infant Perceptions from Sensation to Cognition*, L. B. Cohen and P. Salapatek (eds.). Academic Press, New York.

337 Eisner, T. E. 1966. Beetle spray discourages predators. *Natural History* 75 (Feb):42–47.

338 Eisner, T. E. 1970. Chemical defense against predation in arthropods. In *Chemical Ecology*, E. Sondheimer and J. B. Simeone (eds.). Academic Press, New York.

339 Eisner, T., and J. Meinwald. 1995. The chemistry of sexual selection. *Proceedings of the National Academy of Sciences* 92:50–55.

340 Eisner, T., and S. Nowicki. 1983. Spider web protection through visual advertisement: Role of the stabilimentum. *Science* 219:185–186.

341 Eisner, T., R. Ziegler, J. L. McCormick, M. Eisner, E. R. Hoebeke, and J. Meinwald. 1994. Defensive use of an acquired substance (carminic acid) by predaceous insect larvae. *Experientia* 50:610–615.

342 Elgar, M. A. 1986. The establishment of foraging flocks in house sparrows: Risk of predation and daily temperature. *Behavioral Ecology and Sociobiology* 19:433–438.

343 Elgar, M. A., and N. Wedell. 1996. Role-reversed risky copulation. *Trends in Ecology and Evolution* 11:189–190.

344 Elliott, P. F. 1975. Longevity and the evolution of polygamy. *American Naturalist* 109:281–287.

345 Ellis, L. 1995. Dominance and reproductive success among nonhuman animals: A cross-species comparison. *Ethology and Sociobiology* 16:257–333.

346 Elzanowski, A., and P. Wellnhofer. 1992. A new link between theropods and birds from the Cretaceous of Mongolia. *Nature* 359:821–823.

347 Emlen, J. T., and R. L. Penney. 1966. The navigation of penguins. *Scientific American* 218 (Oct):104–113.

348 Emlen, S. T. 1975. Migration: Orientation and navigation. In *Avian Biology*, D. S. Farner and J. R. King (eds.). Academic Press, New York.

349 Emlen, S. T. 1978. Cooperative breeding. In *Behavioural Ecology: An Evolutionary Approach* (1st edition), J. R. Krebs and N. B. Davies (eds.). Blackwell, Oxford.

350 Emlen, S. T. 1991. Evolution of cooperative breeding in birds and mammals. In *Behavioural Ecology: An Evolutionary Approach* (3rd edition), J. R. Krebs and N. B. Davies (eds.). Blackwell Scientific, Oxford.

351 Emlen, S. T. 1994. Benefits, constraints and the evolution of the family. *Trends in Ecology and Evolution* 9:282–285.

352 Emlen, S. T. 1995. An evolutionary theory of the family. *Proceedings of the National Academy of Sciences* 92:8092–8099.

353 Emlen, S. T., N. J. Demong, and D. J. Emlen. 1989. Experimental induction of infanticide in female wattled jacanas. *Auk* 106:1–7.

354 Emlen, S. T., and L. W. Oring. 1977. Ecology, sexual selection and the evolution of mating systems. *Science* 197:215–223.

355 Emlen, S. T., H. K. Reeve, P. W. Sherman, P. H. Wrege, F. L. W. Ratnieks, and J. Shellman- Reeve. 1991. Adaptive versus nonadaptive explanations of behavior: The case of alloparental helping. *American Naturalist* 138:259–270.

356 Emlen, S. T., J. D. Rising, and W. L. Thompson. 1975. A behavioral and morphological study of sympatry in the indigo and lazuli buntings of the great plains. *Wilson Bulletin* 87:145–179.

357 Emlen, S. T., W. Wiltschko, N. J. Demong, R. Wiltschko, and S. Berian. 1976. Magnetic direction finding: Evidence for its use in migratory indigo buntings. *Science* 193:505–508.

358 Emlen, S. T., and P. H. Wrege. 1986. Forced copulations and intra-specific parasitism: Two costs of social living in the white-fronted bee-eater. *Ethology* 71:2–29.

359 Emlen, S. T., and P. H. Wrege. 1988. The role of kinship in helping decisions among white-fronted bee-eaters. *Behavioral Ecology and Sociobiology* 23:305–315.

360 Emlen, S. T., P. H. Wrege, and N. J. Demong. 1995. Making decisions in the family: An evolutionary perspective. *American Scientist* 83:148–157.

361 Endler, J. A. 1986. *Natural Selection in the Wild*. Princeton University Press, Princeton, NJ.

362 Endler, J. A. 1987. Predation, light intensity and courtship behaviour in *Poeocilia reticulata* (Pisces: Poeciliidae). *Animal Behaviour* 35:1376–1385.

363 Endler, J. A. 1991. Interactions between predators and prey. In *Behavioural Ecology: An Evolutionary Approach* (3rd edition), J. R. Krebs and N. B. Davies (eds.). Blackwell Scientific, Oxford.

364 Epstein, R., R. P. Lanza, and B. F. Skinner. 1980. Symbolic communication between two pigeons (*Columba livia domestica*). *Science* 207:543–545.

365 Erikkson, D., and L. Wallin. 1986. Male bird song attracts females—a field experiment. *Behavioral Ecology and Sociobiology* 19:297–299.

366 Evans, H. E. 1966. *Life on a Little Known Planet*. Dell, New York.

367 Evans, H. E. 1973. *Wasp Farm*. Anchor Press, Garden City, NY.

368 Evans, H. E., and M. J. W. Eberhard. 1970. *The Wasps*. University of Michigan Press, Ann Arbor.

369 Evans, M. R., and B. J. Hatchwell. 1992. An experimental study of male adornment in the scarlet-tufted malachite sunbird: I. The role of pectoral tufts in territorial defence. *Behavioral Ecology and Sociobiology* 29:413–420.

370 Ewald, P. W., and S. A. Rohwer. 1982. Effects of supplemental feeding on timing of breeding, clutch-size, and polygyny in red-winged blackbirds. *Science* 250:1394–1397.

371 Ewer, R. F. 1973. *The Carnivores*. Cornell University Press, Ithaca, NY.

372 Ewert, J.-P. 1974. The neural basis of visually guided behavior. *Scientific American* 230 (Mar):34–42.

373 Ewert, J.-P. 1980. *Neuro-Ethology*. Springer-Verlag, New York.

---

374 Faaborg, J., P. G. Parker, L. DeLay, T. J. de Vries, J. C. Bednarz, S. M. Paz, J. Naranjo, and T. A. Waite. 1995. Confirmation of cooperative polyandry in the Galapagos hawk (*Buteo galapagoensis*). *Behavioral Ecology and Sociobiology* 36:83–90.

375 Fahrbach, S. E., and G. E. Robinson. 1995. Behavioral development in the honey bee: Toward the study of learning under natural conditions. *Learning & Memory* 2:199–224.

376 Falls, J. B. 1988. Does song deter territorial intrusion in white-throated sparrows (*Zonotrichia albicollis*)? *Canadian Journal of Zoology* 66:206–211.

377 Farentinos, R. C., P. J. Capretta, R. E. Kepner, and V. M. Littlefield. 1981. Selective herbivory in tassel-eared squirrels: Role of monoterpenes in ponderosa pines chosen as feeding trees. *Science* 213:1273–1275.

378 Farley, C. T., and C. R. Taylor. 1991. A mechanical trigger for the trot-gallop transition in horses. *Science* 253:306–308.

379 Farner, D. S. 1964. Time measurement in vertebrate photoperiodism. *American Naturalist* 95:375–386.

380 Farner, D. S., and R. A. Lewis. 1971. Photoperiodism and reproductive cycles in birds. *Photophysiology* 6:325–370.

381 Fausto-Sterling, A. 1985. *Myths of Gender*. Basic Books, New York.

382 Feduccia, A. 1993. Evidence from claw geometry indicating arboreal habits of *Archaeopteryx*. *Science* 259:790–793.

383 Feduccia, A., and H. B. Tordoff. 1970. Feathers of *Archaeopteryx*: Asymmetric vanes indicate aerodynamic function. *Science* 203:1021–1022.

384 Feener, D. H., L. F. Jacobs, and J. O. Schmidt. 1996. Specialized parasitoid attracted to a pheromone of ants. *Animal Behaviour* 51:61–66.

385 Fernald, R. D. 1993. Cichlids in love. *The Sciences* 33(4):27–31.

386 Ferveur, J.-F., K. F. Störtkuhl, R. F. Stocker, and R. J. Greenspan. 1995. Genetic feminization of brain structures and changed sexual orientation in male *Drosophila*. *Science* 267:902–904.

387 Fewell, J. H., and M. L. Winston. 1992. Colony state and regulation of pollen foraging in the honey bee, *Apis mellifera*. *Behavioral Ecology and Sociobiology* 30:387–393.

388 Field, S. A., and M. A. Keller. 1993. Alternative mating tactics and female mimicry as post-copulatory mate-guarding behaviour in the parasitic wasp *Cotesia rubecula*. *Animal Behaviour* 46:1183–1189.

389 Fink, L. S., and L. P. Brower. 1981. Birds can overcome the cardenolide defence of monarch butterflies in Mexico. *Nature* 291:67–70.

390 Fisher, R. A. 1930. *The Genetical Theory of Natural Selection*. Clarendon Press, Oxford.

391 Flinn, M. V. 1988. Step-parent/step-offspring interactions in a Caribbean village. *Ethology and Sociobiology* 9:335–369.

392 Folstad, I., and A. J. Kartar. 1992. Parasites, bright males, and the immunocompetence handicap. *American Naturalist* 139:603–622.

393 Foltz, D. W., and P. L. Schwagmeyer. 1989. Sperm competition in the thirteen-lined ground squirrel: Differential fertilization success under field conditions. *American Naturalist* 133:257–265.

394 Ford, E. B. 1955. *Moths*. Collins, London.

395 Forster, L. M. 1992. The stereotyped behaviour of sexual cannibalism in *Latrodectus hasselti* Thorell (Araneae: Theridiidae), the Australian redback spider. *Australian Journal of Zoology* 40:1–11.

396 Forsyth, A., and J. Alcock. 1990. Female mimicry and resource defense polygyny by males of a tropical rove beetle, *Leistotrophus versicolor* (Coleoptera: Staphylinidae). *Behavioral Ecology and Sociobiology* 26:325–330.

397 Foster, M. S. 1977. Odd couples in manakins: A study of social organization and cooperative breeding in *Chiroxiphia linearis*. *American Naturalist* 111:845–853.

398 Foster, S. A. 1994. Inference of evolutionary pattern: Diversionary displays of three-spined sticklebacks. *Behavioral Ecology* 5:114–121.

399 Foster, S. A., and S. A. Cameron. 1996. Geographic variation in behavior: A phylogenetic framework for comparative studies. In *Phylogenies and the Comparative Method*, E. Martins (ed.). Oxford University Press, New York.

400 Foster, W. A. 1990. Experimental evidence for effective and altruistic colony defence against natural predators by soldiers of the gall-forming aphid *Pemphigus spyrothecae* (Hemiptera: Pemphigidae). *Behavioral Ecology and Sociobiology* 27:421–430.

401 Foulkes, N. S., G. Duval, and P. Sassone-Corsi. 1996. Adaptive inducibility of CREM as transcriptional memory of circadian rhythms. *Nature* 381:83–85.

402 Francis, R. C. 1995. Evolutionary neurobiology. *Trends in Ecology and Evolution* 10:276–281.

403 Francis, R. C., K. Soma, and R. D. Fernald. 1993. Social regulation of the brain-pituitary-gonadal axis. *Proceedings of the National Academy of Sciences* 90:7794–7798.

404 Frank, L. G. 1986. Social organization of the spotted hyena *Crocuta crocuta*. II. Dominance and reproduction. *Animal Behaviour* 34:1510–1527.

405 Frank, L. G. 1997. Evolution of genital masculinization: Why do female hyaenas have such a large "penis?" *Trends in Ecology and Evolution* 12:58–62.

406 Frank, L. G., S. E. Glickman, and P. Licht. 1991. Fatal sibling aggression, precocial development, and androgens in neonatal spotted hyenas. *Science* 252:702–704.

407 Frank, L. G., H. E. Holekamp, and L. Smale. 1995. Dominance, demographics and reproductive success in female spotted hyenas: A long-term study. In *Serengeti II: Research, Management, and Conservation of an Ecosystem*, A. R. E. Sinclair and P. Arcese (eds). University of Chicago Press, Chicago.

408 Frank, L. G., M. L. Weldele, and S. E. Glickman. 1995. Masculinization costs in hyaenas. *Nature* 377:584–585.

409 Freedman, D. 1983. *Margaret Mead and Samoa*. Harvard University Press, Cambridge.

410 Fretwell, S. D., and H. K. Lucas, Jr. 1969. On territorial behavior and other factors influencing habitat distribution in birds. I. Theoretical development. *Acta Biotheoretica* 19:16–36.

411 Frey-Roos, F., P. A. Brodmann, and H.-U. Reyer. 1995. Relationships between food resources, foraging patterns, and reproductive success in the water pipit, *Anthus sp. spinoletta*. *Behavioral Ecology* 6:287–295.

412 Fullard, J. H. 1979. Behavioral analyses of auditory sensitivity in *Cycnia tenera* Hübner (Lepidoptera: Arctiidae). *Journal of Comparative Physiology* 129:79–83.

413 Fullard, J. H., and J. E. Yack. 1993. The evolutionary biology of insect hearing. *Trends in Ecology and Evolution* 8:248–252.

414 Gack, C., and K. Peschke. 1994. Spermathecal morphology, sperm transfer and a novel mechanism of sperm displacement in the rove beetle, *Aleochara curtula* (Coleoptera, Staphylinidae). *Zoomorphology* 114:227–237.

415 Gadgil, M. 1972. Male dimorphism as a consequence of sexual selection. *American Naturalist* 106:574–580.

416 Gage, M. J. G., and R. R. Baker 1991. Ejaculate size varies with socio-sexual situation in an insect. *Ecological Entomology* 16:331–337.

417 Galea, L. A. M., M. Kavaliers, and K.-P. Ossenkopp. 1996. Sexually dimorphic spatial learning in meadow voles *Microtus pennsylvanicus* and deer mice *Peromyscus maniculatus*. *Journal of Experimental Biology* 199:195–200.

418 Galef, B. G., Jr., and H. C. Kaner. 1980. Establishment and maintenance of preferences for maternal and artificial olfactory stimuli in juvenile rats. *Journal of Comparative and Physiological Psychology* 94:588–595.

419 Gamboa, G. J. 1978. Intraspecific defense: Advantage of social cooperation among paper wasp foundresses. *Science* 199:1463–1465.

420 Gangestad, S. W., R. Thornhill, and R. A. Yeo. 1994. Facial attractiveness, developmental stability, and fluctuating asymmetry. *Ethology and Sociobiology* 15:73–86.

421 Ganzhorn, J. U. 1990. Towards the map of the homing pigeon? *Animal Behaviour* 40:65–78.

422 Garcia, J., and F. R. Ervin. 1968. Gustatory-visceral and telereceptor-cutaneous conditioning: Adaptation in internal and external milieus. *Communications in Behavioral Biology (A)* 1:389–415.

423 Garcia, J., W. G. Hankins, and K. W. Rusiniak. 1974. Behavioral regulation of the milieu interne in man and rat. *Science* 185:824–831.

424 Gaulin, S. J. C. 1998. Cross-cultural patterns and the search for evolved psychological mechanisms. In *Characterizing Human Psychological Adaptation*, M. Daly (ed.). John Wiley, Chichester.

425 Gaulin, S. J. C., and J. S. Boster. 1990. Dowry as female competition. *American Anthropologist* 92:994–1005.

426 Gaulin, S. J. C., and R. W. FitzGerald. 1986. Sex differences in spatial ability: An evolutionary hypothesis and test. *American Naturalist* 127:74–88.

427 Gaulin, S. J. C., and R. W. FitzGerald 1989. Sexual selection for spatial-learning ability. *Animal Behaviour* 37:322–331.

428 Gaulin, S. J. C., D. H. McBurney, and S. L. Brakeman-Wartell. 1997. Matrilateral biases in the investment of aunts and uncles: A consequence and measure of paternity uncertainty. *Human Nature* 8:139–151.

429 Gaunt, S., and L. F. Baptista. 1997. Social interaction and vocal development in birds. In *Social Influences on Vocal Development*, C. Snowdon and M. Hausberger (eds.). Cambridge University Press, Cambridge.

430 Gavin, T. A., and E. K. Bollinger. 1988. Reproductive correlates of breeding-site fidelity in bobolinks (*Dolichonyx oryzivorous*). *Ecology* 69:96–103.

431 Gerhardt, H. C. 1983. Communication and the environment. In *Animal Behaviour. 2. Communication*, T. R. Halliday and P. J. B. Slater (eds.). W. H. Freeman, New York.

432 Geschwind, N. 1970. The organization of language and the brain. *Science* 170:940–944.

433 Getting, P. A. 1983. Mechanisms of pattern generation underlying swimming in *Tritonia*. II. Network reconstruction. *Journal of Neurophysiology* 49:1017–1035.

434 Getting, P. A. 1989. A network oscillator underlying swimming in *Tritonia*. In *Neuronal and Cellular Oscillators*, J. W. Jacklet (ed.). Dekker, New York.

435 Getty, T. 1996. Mate selection by repeated inspection: More on pied flycatchers. *Animal Behaviour* 51:739–745.

436 Getz, L. L., and S. C. Carter. 1996. Prairie-vole partnerships. *American Scientist* 84:56–62.

437 Ghysen, A. 1978. Sensory neurones recognize defined pathways in *Drosophila* central nervous system. *Nature* 274:869–872.

438 Gibson, R. M., and T. A. Langen. 1996. How do animals choose their mates? *Trends in Ecology and Evolution* 11:468–470.

439 Gill, F. B., and B. G. Murray. 1972. Song variation in sympatric blue-winged and golden-winged warblers. *Auk* 89:625–643.

440 Gill, F. B., and L. L. Wolf. 1975. Economics of feeding territoriality in the golden-winged sunbird. *Ecology* 56:333–345.

441 Gill, F. B., and L. L. Wolf. 1975. Foraging strategies and energetics of East African sunbirds at mistletoe flowers. *American Naturalist* 109:491–510.

442 Gill, F. B., and L. L. Wolf. 1978. Comparative foraging efficiencies of some montane sunbirds in Kenya. *Condor* 80:391–400.

443 Gittleman, J. L., and P. H. Harvey. 1980. Why are distasteful prey not cryptic? *Nature* 286:149–150.

444 Gjershaug, J. O., T. Järvi, and E. Røskaft. 1989. Marriage entrapment by "solitary" mothers: A study of male deception by female pied flycatchers. *American Naturalist* 133:273–276.

445 Glander, K. E. 1981. Feeding patterns in mantled howling monkeys. In *Foraging Behavior: Ecological, Ethological and Psychological Approaches*, A. C. Kamil and T. D. Sargent (eds.). Garland Press, New York.

446 Glickman, S. E., L. G. Frank, P. Licht, T. Yalckinkaya, P. K. Siiteri, and J. Davidson. 1993. Sexual differentiation of the female spotted hyena: One of nature's experiments. *Annals of the New York Academy of Sciences* 662:135–159.

447 Goodman, C. S., and N. C. Spitzer. 1979. Embryonic development of identified neurones: Differentiation from neuroblast to neurons. *Nature* 280:208–214.

448 Gosling, L. M., and M. Petrie. 1990. Lekking in topi: A consequence of satellite behaviour by small males at hotspots. *Animal Behaviour* 40:272–287.

449 Götmark, F. 1990. A test of the information-centre hypothesis in a colony of sandwich terns *Sterna sandvicensis*. *Animal Behaviour* 39:487–495.

450 Götmark, F., D. W. Winkler, and M. Andersson. 1986. Flock-feeding on fish schools increases individual success in gulls. *Nature* 319:589–591.

451 Gould, J. L. 1975. Honey bee recruitment. *Science* 189:685–693.

452 Gould, J. L. 1982. Why do honey bees have dialects? *Behavioral Ecology and Sociobiology* 10:53–56.

453 Gould, S. J. 1981. Hyena myths and realities. *Natural History* 90 (Feb):16–24.

454 Gould, S. J. 1984. Only his wings remained. *Natural History* 93 (Sept):10–18.

455 Gould, S. J. 1986. Evolution and the triumph of homology, or why history matters. *American Scientist* 74:60–69.

456 Gould, S. J. 1986. Review of *Vaulting Ambition*. *New York Review of Books* 33 (Sept 25):47–ff.

457 Gould, S. J. 1987. Freudian slip. *Natural History* 96(2):14–21.

458 Gould, S. J. 1996. The diet of worms and the defenestration of Prague. *Natural History* 105(9):18–ff.

459 Gould, S. J., and R. C. Lewontin. 1981. The spandrels of San Marco and the Panglossian paradigm: A critique of the adaptationist programme. *Proceedings of the Royal Society of London B* 205:581–598.

460 Grafen, A. 1990. Biological signals as handicaps. *Journal of Theoretical Biology* 144:517–546.

461 Grant, B. S., D. F. Owen, and C. A. Clarke. 1996. Parallel rise and fall of melanic peppered moths in America and Britain. *Journal of Heredity* 87:351–357.

462 Grant, J. W. A., C. A. Chapman, and K. S. Richardson. 1992. Defended versus undefended home range size of carnivores, ungulates and primates. *Behavioral Ecology and Sociobiology* 31:149–162.

463 Graves, J. A., and A. Whiten. 1980. Adoption of strange chicks by herring gulls, *Larus argentatus*. *Zeitschrift für Tierpsychologie* 54:267–278.

464 Gray, E. M. 1996. Female control of offspring paternity in a western population of red-winged blackbirds (*Agelaius phoeniceus*). *Behavioral Ecology and Sociobiology* 38:267–278.

465 Gray, E. M. 1997. Do red-winged blackbirds benefit genetically from seeking copulations with extra-pair males? *Animal Behaviour* 53:605–623.

466 Gray, E. M. 1997. Female red-winged blackbirds accrue material benefits from copulating with extra-pair males. *Animal Behaviour* 53:625–639.

467 Grbić, M., P. J. Ode, and M. R. Strand. 1992. Sibling rivalry and brood sex ratios in polyembryonic wasps. *Nature* 360:254–256.

468 Greene, E. 1987. Individuals in an osprey colony discriminate between high and low quality information. *Nature* 329:239–241.

469 Greene, E., L. T. Orsak, and D. W. Whitman. 1987. A tephritid fly mimics the territorial displays of its jumping spider predators. *Science* 236:310–312.

470 Greenwood, P. J. 1980. Mating systems, philopatry, and dispersal in birds and mammals. *Animal Behaviour* 28:1140–1162.

471 Griffin, D. R. 1958. *Listening in the Dark*. Yale University Press, New Haven.

472 Griffin, D. R., F. A. Webster, and C. R. Michael. 1960. The echolocation of flying insects by bats. *Animal Behaviour* 8:141–154.

473 Grillner, S., and P. Wallen. 1985. Central pattern generators for locomotion, with special

reference to vertebrates. *Annual Review of Neuroscience* 8:233–261.

474 Grillner, S., P. Wallen, and L. Brodin. 1991. Neuronal networks generating locomotor behavior in lamprey: Circuitry, transmitters, membrane properties, and simulation. *Annual Review of Neuroscience* 14:169–199.

475 Grinnell, J., C. Packer, and A. E. Pusey. 1995. Cooperation in male lions: Kinship, reciprocity or mutualism? *Animal Behaviour* 49:95–105.

476 Gross, M. R. 1982. Sneakers, satellites, and parentals: Polymorphic mating strategies in North American sunfishes. *Zeitschrift für Tierpsychologie* 60:1–26.

477 Gross, M. R. 1987. Evolution of diadromy in fishes. *American Fisheries Society Symposium* 1:14–25.

478 Gross, M. R. 1991. Evolution of alternative reproductive strategies: Frequency-dependent sexual selection in male bluegill sunfish. *Philosophical Transactions of the Royal Society of London B* 332:59–66.

479 Gross, M. R. 1996. Alternative reproductive strategies and tactics: Diversity within sexes. *Trends in Ecology and Evolution* 11:92–98.

480 Gross, M. R., and A. M. MacMillan. 1981. Predation and the evolution of colonial nesting in bluegill sunfish (*Lepomis macrochirus*). *Behavioral Ecology and Sociobiology* 8:163–174.

481 Gross, M. R., and R. C. Sargent. 1985. The evolution of male and female parental care in fishes. *American Zoologist* 25:807–822.

482 Gross, M. R., and R. Shine. 1981. Parental care and mode of fertilization in ectothermic vertebrates. *Evolution* 35:775–793.

483 Gurney, M. E., and M. Konishi. 1980. Hormone-induced sexual differentiation of brain and behavior in zebra finches. *Science* 208:1380–1383.

484 Gwinner, E., and J. Dittami. 1990. Endogenous reproductive rhythms in a tropical bird. *Science* 249:906–908.

485 Gwinner, E., T. Rödl, and H. Schwabl. 1994. Pair territoriality of wintering stonechats: Behavior, function and hormones. *Behavioral Ecology and Sociobiology* 34:321–327.

486 Gwinner, E., and W. Wiltschko. 1980. Circannual changes in migratory orientation of the garden warbler, *Sylvia borin*. *Behavioral Ecology and Sociobiology* 7:73–78.

487 Gwynne, D. T. 1981. Sexual difference theory: Mormon crickets show role reversal in mate choice. *Science* 213:779–780.

488 Gwynne, D. T. 1991. Sexual competition among females: What causes courtship-role reversal? *Trends in Ecology and Evolution* 6:118–121.

489 Gwynne, D. T. 1995. Phylogeny of the Ensifera (Orthoptera): A hypothesis supporting multiple origins of acoustical signalling, complex spermatophores and maternal care in crickets, katydids, and weta. *Journal of Orthoptera Research* 4:203–218.

490 Gwynne, D. T., and W. D. Brown. 1994. Mate feeding, offspring investment, and sexual differences in katydids (Orthoptera: Tettigoniidae). *Behavioral Ecology* 5:267–272.

491 Gwynne, D. T., and D. C. F. Rentz. 1983. Beetles on the bottle: Male buprestids mistake stubbies for females (Coleoptera). *Journal of the Australian Entomological Society* 22:79–80.

492 Gwynne, D. T., and L. W. Simmons. 1990. Experimental reversal of courtship roles in an insect. *Nature* 346:172–174.

493 Hahn, T. P. 1995. Integration of photoperiodic and food cues to time changes in reproductive physiology by an opportunistic breeder, the red crossbill, *Loxia curvirostra* (Aves: Carduelinae). *Journal of Experimental Zoology* 272:213–226.

494 Hahn, T. P., J. C. Wingfield, R. Mullen, and P. J. Deviche. 1995. Endocrine bases of spatial and temporal opportunism in arctic-breeding birds. *American Zoologist* 35:259–273.

495 Haley, M. P. 1994. Resource-holding power asymmetries, the prior residence effect, and reproductive payoffs in male northern elephant seal fights. *Behavioral Ecology and Sociobiology* 34:427–434.

496 Haley, M. P., C. J. Deutsch, and B. J. Le Boeuf. 1994. Size, dominance and copulatory success in male northern elephant seals, *Mirounga angustirostris*. *Animal Behaviour* 48:1249–1260.

497 Hall, J. C. 1977. Portions of the central nervous system controlling reproductive behavior in *Drosophila melanogaster. Behavior Genetics* 7:291–312.

498 Hall, J. C. 1994. The mating of the fly. *Science* 264:1702–1714.

499 Hall, J. C., R. J. Greenspan, and W. A. Harris. 1982. *Genetic Neurobiology.* MIT Press, Cambridge, MA.

500 Hamilton, W. D. 1964. The evolution of social behavior. *Journal of Theoretical Biology* 7:1–52.

501 Hamilton, W. D. 1971. Geometry for the selfish herd. *Journal of Theoretical Biology* 31:295–311.

502 Hamilton, W. D., and M. Zuk. 1982. Heritable true fitness and bright birds: A role for parasites. *Science* 218:384–387.

503 Hamilton, W. J., and G. H. Orians. 1965. Evolution of brood parasitism in altricial birds. *Condor* 67:361–382.

504 Hamner, W. M. 1964. Circadian control of photoperiodism in the house finch demonstrated by interrupted-night experiments. *Nature* 203:1400–1401.

505 Hanby, J. P., and J. D. Bygott. 1987. Emigration of subadult lions. *Animal Behaviour* 35:161–169.

506 Harcourt, A. H., P. H. Harvey, S. G. Larson, and R. V. Short. 1981. Testis weight, body weight and breeding system in primates. *Nature* 293:55–57.

507 Harlow, H. F., and M. K. Harlow. 1962. Social deprivation in monkeys. *Scientific American* 207 (Nov):136–146.

508 Harlow, H. F., M. K. Harlow, and S. J. Suomi. 1971. From thought to therapy: Lessons from a primate laboratory. *American Scientist* 59:538–549.

509 Harper, D. G. C. 1991. Communication. In *Behavioural Ecology: An Evolutionary Approach* (3rd edition), J. R. Krebs and N. B. Davies (eds.). Blackwell Scientific, Oxford.

510 Hartley, I. R., N. B. Davies, B. J. Hatchwell, A. Desrochers, D. Nebel, and T. Burke. 1995. The polygynandrous mating system of the alpine accentor, *Prunella collaris.* II. Multiple paternity and parental effort. *Animal Behaviour* 49:789–803.

511 Hartung, J. 1982. Polygyny and inheritance of wealth. *Current Anthropology* 23:1–12.

512 Hartung, J. 1985. Matrilineal inheritance: New theory and analysis. *Behavioral and Brain Sciences* 8:661–685.

513 Harvey, P. H., and J. R. Krebs. 1990. Comparing brains. *Science* 249:140–146.

514 Harvey, P. H., and M. D. Pagel. 1991. *The Comparative Method in Evolutionary Biology.* Oxford University Press, London.

515 Hasselquist, D., and S. Bensch. 1991. Trade-off between mate guarding and mate attraction in the polygynous great reed warbler. *Behavioral Ecology and Sociobiology* 28:187–193.

516 Hasselquist, D., S. Bensch, and T. von Schantz. 1996. Correlation between male song repertoire, extra-pair paternity and offspring survival in the great reed warbler. *Nature* 381:229–232.

517 Hasson, O., R. Hibbard, and G. Ceballos. 1989. The pursuit deterrent function of tail-wagging in the zebra-tailed lizard (*Callisaurus draconoides*). *Canadian Journal of Zoology* 67:1203–1209.

518 Hausfater, G. 1975. Dominance and reproduction in baboons (*Papio cynocephalus*): A quantitative analysis. *Contributions in Primatology* 7:1–150.

519 Hausfater, G., and S. B. Hrdy (eds.). 1984. *Infanticide: Comparative and Evolutionary Perspectives.* Aldine, Chicago.

520 Hebblethwaite, M. L., and W. M. Shields. 1990. Social influences on barn swallow foraging in the Adirondacks: A test of competing hypotheses. *Animal Behaviour* 39:97–104.

521 Hedrick, A. V. 1986. Female preferences based on male calling in a field cricket. *Behavioral Ecology and Sociobiology* 19:73–77.

522 Hedrick, A. V., and S. E. Riechert. 1989. Genetically based variation between two spider populations in foraging behavior. *Oecologia* 80:533–539.

523 Hedrick, A. V., and E. J. Temeles. 1989. The evolution of sexual dimorphism in animals: Hypotheses and tests. *Trends in Ecology and Evolution* 4:131–135.

524 Hegner, R. E., and S. T. Emlen. 1987. Territorial organization of the white fronted bee-eater in Kenya. *Ethology* 76:189–222.

525 Heiligenberg, W. 1991. The neural basis of behavior: A neuroethological view. *Annual Review of Neuroscience* 14:247–267.

526 Heinrich, B. 1979. *Bumblebee Economics.* Harvard University Press, Cambridge, MA.

527 Heinrich, B. 1984. *In a Patch of Fireweed.* Harvard University Press, Cambridge, MA.

528 Heinrich, B. 1988. Winter foraging at carcasses by three sympatric corvids, with emphasis on recruitment by the raven, *Corvus corax. Behavioral Ecology and Sociobiology* 23:141–156.

529 Heinrich, B. 1989. *Ravens in Winter.* Summit Books, New York.

530 Heinrich, B., and G. A. Bartholomew. 1979. The ecology of the African dung beetle. *Scientific American* 241 (Nov):146–156.

531 Heinrich, B., and S. L. Collins. 1983. Caterpillar leaf damage, and the game of hide-and-seek with birds. *Ecology* 64:592–602.

532 Heinrich, B., and J. Marzluff. 1995. Why ravens share. *American Scientist* 83:342–349.

533 Heinsohn, R. 1995. Raid of the red-eyed chicknappers. *Natural History* 104(2):44–51.

534 Heinsohn, R. G., and C. Packer. 1995. Complex cooperative strategies in group-territorial African lions. *Science* 269:1260–1262.

535 Heinze, J., and B. Hölldobler. 1993. Fighting for a harem of queens: Physiology of reproduction in *Cardiocondyla* male ants. *Proceedings of the National Academy of Sciences* 90:8412–8414.

536 Heinze, J., B. Hölldobler, and C. Peeters. 1994. Conflict and cooperation in ant societies. *Naturwissenschaften* 81:489–497.

537 Helbig, A. J. 1991. Inheritance of migratory direction in a bird species: A cross-breeding experiment with SE- and SW-migrating blackcaps (*Sylvia atricapilla*). *Behavioral Ecology and Sociobiology* 28:9–12.

538 Hennessy, D. F., and D. H. Owings. 1978. Snake species discrimination and the role of olfactory cues in the snake-directed behavior of the California ground squirrel. *Behaviour* 65:115–124.

539 Henry, C. S. 1972. Eggs and rapagula of *Ululodea* and *Ascaloptynx* (Neuroptera: Ascalaphidae): A comparative study. *Psyche* 79:1–22.

540 Heske, J. 1990. Why do horse lubbers roost in the bushes? *Southwestern Naturalist* 35:455–458.

541 Heston, L. L., and J. Shields. 1968. Homosexuality in twins: A family study and a registry study. *Archives of General Psychiatry* 18:149–160.

542 Hews, D. K. 1993. Food resources affect female distribution and male mating opportunities in the iguanian lizard *Uta palmeri. Animal Behaviour* 46:279–291.

543 Hill, G. E. 1990. Female house finches prefer colourful males: Sexual selection for a condition-dependent trait. *Animal Behaviour* 40:563–572.

544 Hill, G. E. 1991. Plumage coloration is a sexually selected indicator of male quality. *Nature* 350:337–339.

545 Hill, G. E. 1995. Ornamental traits as indicators of environmental health. *BioScience* 45:25–31.

546 Hinde, R. A., and N. Tinbergen. 1958. The comparative study of species-specific behavior. In *Behavior and Evolution*, A. Roe and G. G. Simpson (eds.). Yale University Press, New Haven, CT.

547 Hirth, D. H., and D. R. McCullough. 1977. Evolution of alarm signals in ungulates with special reference to white-tailed deer. *American Naturalist* 111:31–42.

548 Hitchcock, C. L., and D. F. Sherry. 1990. Long-term memory for cache sites in the black-capped chickadee. *Animal Behaviour* 40:701–712.

549 Hoffmann, A. A. 1988. Heritable variation for territorial success in two *Drosophila melanogaster* populations. *Animal Behaviour* 36:1180–1189.

550 Hogg, J. T. 1984. Mating in bighorn sheep: Multiple creative male strategies. *Science* 225:526–529.

551 Höglund, J., and R. V. Alatalo. 1995. *Leks.* Princeton University Press, Princeton, NJ.

552 Höglund, J., and A. Lundberg. 1987. Sexual selection in a monomorphic lek-breeding bird: Correlates of male mating success in the great snipe *Gallinago media. Behavioral Ecology and Sociobiology* 21:211–216.

553 Höglund, J., R. Montgomerie, and F. Widemo. 1993. Costs and consequences of variation in the size of ruff leks. *Behavioral Ecology and Sociobiology* 32:31–40.

554 Höglund, J., and J. G. M. Robertson. 1990. Spacing of leks in relation to female home ranges, habitat requirements and male attractiveness in the great snipe (*Gallinago media*). *Behavioral Ecology and Sociobiology* 26:173–180.

555 Högstedt, G. 1983. Adaptation unto death: Function of fear screams. *American Naturalist* 121:562–570.

556 Holden, C. 1980. Identical twins reared apart. *Science* 207:1323–1328.

557 Holekamp, K. E. 1984. Natal dispersal in Belding's ground squirrels (*Spermophilus beldingi*). *Behavioral Ecology and Sociobiology* 16:21–30.

558 Holekamp, K. E., and P. W. Sherman. 1989. Why male ground squirrels disperse. *American Scientist* 77:232–239.

559 Holekamp, K. E., and L. Smale. 1995. Rapid change in offspring sex ratios after clan fission in the spotted hyena. *American Naturalist* 145:261–278.

560 Hölldobler, B. 1971. Communication between ants and their guests. *Scientific American* 224 (Mar):86–95.

561 Hölldobler, B., and E. O. Wilson. 1990. *The Ants*. Harvard University Press, Cambridge, MA.

562 Hölldobler, B., and E. O. Wilson. 1994. *Journey to the Ants*. Harvard University Press, Cambridge, MA.

563 Holley, A. J. F. 1984. Adoption, parent-chick recognition, and maladaptation in the herring gull *Larus argentatus*. *Zeitschrift für Tierpsychologie* 64:9–14.

564 Holmes, W. G. 1984. Predation risk and foraging behavior of the hoary marmot in Alaska. *Behavioral Ecology and Sociobiology* 15:293–302.

565 Holmes, W. G. 1986. Identification of paternal half-siblings by captive Belding's ground squirrels. *Animal Behaviour* 34:321–327.

566 Holmes, W. G. 1995. The ontogeny of littermate preferences in juvenile golden-mantled ground squirrels: Effects of rearing and relatedness. *Animal Behaviour* 50:309–322.

567 Holmes, W. G., and P. W. Sherman. 1982. The ontogeny of kin recognition in two species of ground squirrels. *American Zoologist* 22:491–517.

568 Holmes, W. G., and P. W. Sherman. 1983. Kin recognition in animals. *American Scientist* 71:46–55.

569 Hoogland, J. L., and P. W. Sherman. 1976. Advantages and disadvantages of bank swallow (*Riparia riparia*) coloniality. *Ecological Monographs* 46:33–58.

570 Hori, M. 1993. Frequency-dependent natural selection in the handedness of scale-eating cichlid fish. *Science* 260:216–219.

571 Horn, A. G., M. L. Leonard, and D. M. Weary. 1995. Oxygen consumption during crowing by roosters: Talk is cheap. *Animal Behaviour* 50:1171–1175.

572 Hotta, Y., and S. Benzer. 1979. Courtship in *Drosophila* mosaics: Sex-specific foci for sequential action patterns. *Proceedings of the National Academy of Sciences* 73:4154–4158.

573 Hou, L., L. D. Martin, Z. Zhou, and A. Feduccia. 1996. Early adaptive radiation of birds: Evidence from fossils from northeastern China. *Science* 274:1164–1166.

574 Hovi, M., R. V. Alatalo, and P. Siikamäki. 1995. Black grouse leks on ice: Female mate sampling by incitation or male competition? *Behavioral Ecology and Sociobiology* 37:283–288.

575 Howard, J. J. 1987. Diet selection by the leafcutting ant *Atta cephalotes*—The role of nutrients, water, and secondary chemistry. *Ecology* 68:503–515.

576 Howard, R. D. 1978. The evolution of mating strategies in bullfrogs, *Rana catesbiana*. *Evolution* 32:850–871.

577 Howlett, R. J., and M. E. N. Majerus. 1987. The understanding of industrial melanism in the peppered moth (*Biston betularia*) (Lepidoptera: Geometridae). *Biological Journal of the Linnean Society* 30:31–44.

578 Hrdy, S. B. 1977. Infanticide as a primate reproductive strategy. *American Scientist* 65:40–49.

579 Hrdy, S. B. 1977. *The Langurs of Abu*. Harvard University Press, Cambridge, MA.

580 Huang, Z.-Y., and G. E. Robinson. 1992. Honeybee colony integration: Worker-worker interactions mediate hormonally regulated plasticity in division of labor. *Proceedings of the National Academy of Sciences* 89:11726–11729.

581 Hubel, D. H., and T. N. Wiesel. 1970. The period of susceptibility to the physiological effects of unilateral eye closure in kittens. *Journal of Physiology* 206:419–436.

582 Hughes, J. J., and D. Ward. 1993. Predation risk and distance to cover affect foraging behaviour in Namib Desert gerbils. *Animal Behaviour* 46:1243–1245.

583 Hunter, F. M., M. Petrie, M. Otronen, T. Birkhead, and A. P. Møller. 1993. Why do females copulate repeatedly with one male? *Trends in Ecology and Evolution* 8:21–26.

584 Hunter, M. L., and J. R. Krebs. 1979. Geographical variation in the song of the great tit (*Parus major*) in relation to ecological factors. *Journal of Animal Ecology* 48:759–785.

585 Hurst, L. D., and W. D. Hamilton. 1992. Cytoplasmic fusion and the nature of sexes. *Proceedings of the Royal Society of London B* 247:189–194.

586 Hurst, L. D., and J. R. Peck. 1996. Recent advances in understanding of the evolution and maintenance of sex. *Trends in Ecology and Evolution* 11:46–52.

587 Huxley, T. H. 1910. *Lectures and Lay Sermons.* E. P. Dutton, New York.

588 Ichikawa, N. 1995. Male counterstrategy against infanticide of the female giant water bug *Lethocerus deyrollei* (Hemiptera: Belostomatidae). *Journal Of Insect Behavior* 8:181–188.

589 Immelmann, K. 1969. Song development in the zebra finch and other estrildid finches. In *Bird Vocalizations*, R. A. Hinde (ed.). Cambridge University Press, Cambridge.

590 Ioalé, P., M. Nozzolini, and F. Papi. 1990. Homing pigeons do extract directional information from olfactory stimuli. *Behavioral Ecology and Sociobiology* 26:301–306.

591 Irons, W. 1979. Cultural and biological success. In *Evolutionary Biology and Human Social Behavior,* N. Chagnon and W. Irons (eds.). Duxbury Press, North Scituate, MA.

592 Isbell, L. A., D. L. Cheney, and R. M. Seyfarth. 1993. Are immigrant vervet monkeys, *Cercopithecus aethiops,* at greater risk of mortality than residents? *Animal Behaviour* 45:729–734.

593 Ito, Y. 1989. The evolutionary biology of sterile soldiers in aphids. *Trends in Ecology and Evolution* 4:69–73.

594 Jackson, R. R. 1992. Eight-legged tricksters. *BioScience* 42:590–598.

595 Jackson, R. R., and R. S. Wilcox. 1990. Aggressive mimicry, prey-specific predatory behaviour and predator-recognition in the predator-prey interactions of *Portia fimbriata* and *Euryattus* sp., jumping spiders from Queensland. *Behavioral Ecology and Sociobiology* 26:111–119.

596 Jacobs, L. F. 1996. Sexual selection and the brain. *Trends in Ecology and Evolution* 11:82–86.

597 Jacobs, L. F., S. J. C. Gaulin, D. F. Sherry, and G. E. Hoffman. 1990. Evolution of spatial cognition: Sex-specific patterns of spatial behavior predict hippocampal size. *Proceedings of the National Academy of Sciences* 87:6349–6352.

598 Jamieson, I. G. 1989. Behavioral heterochrony and the evolution of birds' helping at the nest: An unselected consequence of communal breeding? *American Naturalist* 133:394–406.

599 Jamieson, I. G. 1991. The unselected hypothesis for the evolution of helping behavior: Too much or too little emphasis on natural selection? *American Naturalist* 138:271–282.

600 Jamieson, I. G., J. S. Quinn, P. A. Rose, and B. N. White. 1994. Shared paternity among non-relatives is a result of an egalitarian mating system in a communally breeding bird, the pukeko. *Proceedings of the Royal Society of London B* 257:271–277.

601 Janetos, A. C. 1980. Strategies of female mate choice: A theoretical analysis. *Behavioral Ecology and Sociobiology* 7:107–112.

602 Jarvis, J. U. M., and N. C. Bennett. 1993. Eusociality has evolved independently in two genera of bathyergid mole-rats—but occurs in no other subterranean mammal. *Behavioral Ecology and Sociobiology* 33:253–260.

603 Jaycox, E. R. 1970. Honey bee foraging behavior: Responses to queens, larvae, and extracts of larvae. *Annals of the Entomological Society of America* 63:1689–1694.

604 Jaycox, E. R., and S. G. Parise. 1980. Homesite selection by Italian honey bee swarms, *Apis mellifera ligustica* (Hymenoptera: Apidae). *Journal of the Kansas Entomological Society* 53:171–178.

605 Jaycox, E. R., and S. G. Parise. 1981. Homesite selection by swarms of black-bodied honey bees, *Apis mellifera caucasia* and *A. m. carnica*. *Journal of the Kansas Entomological Society* 54:697–703.

606 Jeanne, R. L. 1970. Chemical defense of brood by a social wasp. *Science* 168:1465–1466.

607 Jeanne, R. L., H. A. Downing, and D. C. Post. 1983. Morphology and function of sternal glands in polistine wasps (Hymenoptera: Vespidae). *Zoomorphology* 103:149–184.

608 Jenkins, P. F. 1978. Cultural transmission of song patterns and dialect development in a free-living bird population. *Animal Behaviour* 26:50–78.

609 Jenni, D. A., and G. Collier. 1972. Polyandry in the American jacana. *Auk* 89:743–765.

610 Jennions, M. D., and D. W. Macdonald. 1994. Cooperative breeding in mammals. *Trends in Ecology and Evolution* 9:89–93.

611 Jeppsson, B. 1986. Mating by pregnant water voles (*Arvicola terrestris*): A strategy to counter infanticide by males? *Behavioral Ecology and Sociobiology* 19:293–296.

612 Jiménez, J. A., K. A. Hughes, G. Alaks, L. Graham, and R. C. Lacy. 1994. An experimental study of inbreeding depression in a natural habitat. *Science* 266:271–273.

613 Johns, T. 1990. *With Bitter Herbs They Shall Eat It: Chemical Ecology and the Origins of Human Diet and Medicine.* University of Arizona Press.

614 Johnson, C. H., and J. W. Hasting. 1986. The elusive mechanisms of the circadian clock. *American Scientist* 74:29–37.

615 Johnson, L. S., and L. H. Kermott. 1991. The functions of song in male house wrens (*Troglodytes aedon*). *Behaviour* 116:190–209.

616 Johnson, L. S., and W. A. Searcy. 1996. Female attraction to male song in house wrens (*Troglodytes aedon*). *Behaviour* 133:357–366.

617 Johnston, T. D. 1982. Selective costs and benefits in the evolution of learning. *Advances in the Study of Behavior* 12:65–106.

618 Jouventin, P., C. Barbraud, and M. Rubin. 1995. Adoption in the emperor penguin, *Aptenodytes forsteri. Animal Behaviour* 50:1023–1029.

619 Joyce, F. J. 1993. Nesting success of rufous-naped wrens (*Campylorhynchus ruifnucha*) is greater near wasp nests. *Behavioral Ecology and Sociobiology* 32:71–78.

620 Judd, T. M., and P. W. Sherman. 1996. Naked mole-rats recruit colony mates to food sources. *Animal Behaviour* 52:957–969.

621 Just, J. 1988. Siphonoecetinae (Corophiidae) 6: A survey of phylogeny, distribution, and biology. *Crustaceana*, Supplement 13:193–208.

622 Kacelnik, A. 1984. Central place foraging in starlings (*Sturnus vulgaris*) I. Patch residence time. *Journal of Animal Ecology* 53:283–299.

623 Kagan, J., and R. E. Klein. 1973. Cross-cultural perspectives on early development. *American Psychologist* 28:947–961.

624 Kaitala, A., and C. Wiklund. 1994. Polyandrous female butterflies forage for matings. *Behavioral Ecology and Sociobiology* 35:385–388.

625 Kalko, E. K. V. 1995. Insect pursuit, prey capture and echolocation in pipistrelle bats (Microchiroptera). *Animal Behaviour* 50:861–880.

626 Kaplan, H., and K. Hill. 1985. Hunting ability and reproductive success among male Ache foragers: Preliminary results. *Current Anthropology* 26:131–133.

627 Katz, P. S., and W. N. Frost. 1995. Intrinsic neuromodulation in the *Tritonia* swim CPG: The serotonergic dorsal swim interneurons act presynaptically to enhance transmitter release from interneuron C2. *Journal of Neuroscience* 15:6035–6045.

628 Kavaliers, M., and D. D. Colwell. 1995. Odours of parasitized males induce aversive responses in female mice. *Animal Behaviour* 50:1161–1169.

629 Keeton, W. T. 1974. The orientational and navigational basis of homing in birds. *Advances in the Study of Behavior* 5:47–132.

630 Keller, L. 1995. All's fair when love is war. *Nature* 373:190–191.

631 Keller, L., and H. K. Reeve. 1994. Genetic variability, queen number, and polyandry in social Hymenoptera. *Evolution* 48:694–794.

632 Keller, L., and H. K. Reeve. 1995. Why do females mate with multiple males? The sexually selected sperm hypothesis. *Advances in the Study of Behavior* 24:291–315.

633 Kempenaers, B., G. R. Verheyen, M. Van der Broeck, T. Burke, C. Van Broeckhoven, and A. A. Dhondt. 1992. Extra-pair paternity results from female preference for high quality males in the blue tit. *Nature* 357:494–496.

634 Kendrick, K. 1990. Through a sheep's eye. *New Scientist* 125 (12 May):62–65.

635 Kennedy, M., H. G. Spencer, and R. D. Gray. 1996. Hop, step and gape: Do social displays of the Pelecaniformes reflect phylogeny? *Animal Behaviour* 51:273–291.

636 Kenrick, D. T. 1995. Evolutionary theory versus the confederacy of dunces. *Psychological Inquiry* 6:56–61.

637 Kenrick, D. T., and R. C. Keefe. 1992. Age preferences in mates reflect sex differences in reproductive strategies. *Behavioral and Brain Sciences* 15:75–133.

638 Kenrick, D. T., R. C. Keefe, A. Bryan, A. Barr, and S. Brown. 1995. Age preferences and mate choice among homosexuals and heterosexuals: A case for modular psychological mechanisms. *Journal of Personality and Social Psychology* 69:1166–1172.

639 Kenrick, D. T., R. C. Keefe, C. Gabrielidis, and J. S. Cornelius. 1996. Adolescents' age preferences for dating partners: Support for an evolutionary model of life-history strategies. *Child Development* 67:1499–1511.

640 Kenrick, D. T., E. K. Sadalla, G. Groth, and M. R. Trost. 1990. Evolution, traits, and the stages of human courtship: Qualifying the parental investment model. *Journal of Personality* 58:97–116.

641 Kent, D. S., and J. A. Simpson. 1992. Eusociality in the beetle *Australoplatypus incompertus* (Coleoptera: Curculionidae). *Naturwissenschaften* 79:86–87.

642 Kenward, R. E. 1978. Hawks and doves: Factors affecting success and selection in goshawk attacks on wild pigeons. *Journal of Animal Ecology* 47:449–460.

643 Kermott, L. H., L. S. Johnson, and M. S. Merkle. 1991. Experimental evidence for the function of mate replacement and infanticide by males in a north-temperate population of house wrens. *Condor* 93:630–636.

644 Kessel, E. L. 1955. Mating activities of balloon flies. *Systematic Zoology* 4:97–104.

645 Ketterson, E. D., and V. Nolan. 1992. Hormones and life histories: An integrative approach. *American Naturalist* 140, Supplement:s33–S62.

646 Kettlewell, H. B. D. 1955. Selection experiments on industrial melanism in the Lepidoptera. *Heredity* 9:323–343.

647 Kiepenheuer, J. 1985. Can pigeons be fooled about the actual release site position by presenting them information from another site? *Behavioral Ecology and Sociobiology* 18:75–82.

648 Kiepenheuer, J., M. F. Neumann, and H. G. Wallraff. 1993. Home-related and home-independent orientation of displaced pigeons with and without access to environmental air. *Animal Behaviour* 45:169–182.

649 King, A. P., and M. J. West. 1983. Epigenesis of cowbird song—a joint endeavour of males and females. *Nature* 305:704–706.

650 King, A. P., and M. J. West. 1983. Female perception of cowbird song: A closed developmental program. *Developmental Psychobiology* 16:335–342.

651 Kirkpatrick, M. 1982. Sexual selection and the evolution of female choice. *Evolution* 36:1–12.

652 Kirkpatrick, M., and M. J. Ryan. 1991. The evolution of mating preferences and the paradox of the lek. *Nature* 350:33–38.

653 Kirn, J. R., and T. J. DeVoogd. 1989. The genesis and death of vocal control neurons during sexual differentiation in the zebra finch. *Journal of Neuroscience* 9:3176–3187.

654 Kitchener, A. 1987. Fighting behaviour of the extinct Irish elk. *Modern Geology* 11:1–28.

655 Kitcher, P. 1985. *Vaulting Ambition*. MIT Press, Cambridge, MA.

656 Klump, G. M., E. Kretzschmar, and E. Curio. 1986. The hearing of an avian predator and its avian prey. *Behavioral Ecology and Sociobiology* 18:317–324.

657 Knudsen, B., and R. M. Evans. 1986. Parent-young recognition in herring gulls (*Larus argentatus*). *Animal Behaviour* 34:77–80.

658 Kodric-Brown, A. 1986. Satellites and sneakers: Opportunistic male breeding tactics in pupfish (*Cyprinodon pecoensis*). *Behavioral Ecology and Sociobiology* 19:425–432.

659 Kodric-Brown, A. 1993. Female choice of multiple male criteria in guppies: Interacting effects of dominance, coloration and courtship. *Behavioral Ecology and Sociobiology* 32:415–420.

660 Kodric-Brown, A., and J. H. Brown. 1984. Truth in advertising: The kinds of traits favored by sexual selection. *American Naturalist* 124:309–323.

661 Koenig, W. D. 1981. Coalitions of male lions: Making the best of a bad job? *Nature* 293:413–414.

662 Koenig, W. D., and R. L. Mumme. 1987. *Population Ecology of the Cooperatively Breeding Acorn Woodpecker*. Princeton University Press, Princeton, NJ.

663 Koenig, W. D., R. L. Mumme, M. T. Stanback, and F. A. Pitelka. 1995. Patterns and consequences of egg destruction among joint-nesting acorn woodpeckers. *Animal Behaviour* 50:607–621.

664 Koenig, W. D., F. A. Pitelka, W. J. Carmen, R. L. Mumme, and M. T. Stanback. 1992. The evolution of delayed dispersal in cooperative breeders. *Quarterly Review of Biology* 67:111–150.

665 Koeniger, G. 1986. Mating sign and multiple mating in the honeybee. *Bee World* 67:141–150.

666 Kojima, J. 1983. Defense of the pre-emergence colony against ants by means of a chemical barrier in *Ropalidia fasciata* (Hymenoptera, Vespidae). *Japanese Journal of Ecology* 33:213–223.

667 Komdeur, J. 1992. Importance of habitat saturation and territory quality for evolution of cooperative breeding in the Seychelles warbler. *Nature* 358:493–495.

668 Komdeur, J. 1992. Influence of territory quality and habitat saturation on dispersal options in the Seychelles warbler: An experimental test of the habitat saturation hypothesis for cooperative breeding. *Acta XX Congressus Internationalis Ornithologici*, 1325–1332.

669 Komdeur, J., A. Huffstadt, W. Prast, G. Castle, R. Mileto, and J. Wattle. 1995. Transfer experiments of Seychelles warblers to new islands: Changes in dispersal and helping behaviour. *Animal Behaviour* 49:695–708.

670 Konishi, M. 1965. The role of auditory feedback in the control of vocalization in the white-crowned sparrow. *Zeitschrift für Tierpsychologie* 22:770–783.

671 Konishi, M. 1985. Birdsong: From behavior to neurons. *Annual Review of Neuroscience* 8:125–170.

672 Korpimaki, E. 1991. Poor reproductive success of polygynously mated female Tengmalm's owls: Are better options available? *Animal Behaviour* 41:49–60.

673 Kramer, M. G., and J. H. Marden. 1997. Almost airborne. *Nature* 385:403–404.

674 Krebs, J. R. 1971. Territory and breeding density in the great tit, *Parus major* L. *Ecology* 52:2–22.

675 Krebs, J. R. 1982. Territorial defence in the great tit (*Parus major*): Do residents always win? *Behavioral Ecology and Sociobiology* 11:185–194.

676 Krebs, J. R., and A. Kacelnik. 1991. Decision-making. In *Behavioural Ecology: An Evolutionary Approach* (3rd edition), J. R. Krebs and N. B. Davies (eds.). Blackwell Scientific Publications, Oxford.

677 Kroodsma, D. E. 1996. Ecology of passerine song development. In *Ecology and Evolution of Acoustic Communication in Birds*, D. E. Kroodsma and E. H. Miller (eds.). Cornell University Press, Ithaca, NY.

678 Kroodsma, D. E., and B. E. Byers. 1991. The function(s) of bird song. *American Zoologist* 31:109–114.

679 Kroodsma, D. E., and R. A. Canady. 1985. Differences in repertoire size, singing behavior, and associated neuroanatomy among marsh wren populations have a genetic basis. *Auk* 102:439–446.

680 Kroodsma, D. E., and E. H. Miller (eds.). 1996. *Ecology and Evolution of Acoustic Communication in Birds*. Cornell University Press, Ithaca, NY.

681 Kruuk, H. 1964. Predators and anti-predator behaviour of the black-headed gull *Larus ridibundus*. *Behaviour Supplements* 11:1–129.

682 Kruuk, H. 1972. *The Spotted Hyena.* University of Chicago Press, Chicago.

683 Kruuk, H. 1976. Feeding and social behavior of the striped hyaena (*Hyaena vulgaris* Desmaret). *East African Wildlife Journal* 14:91–111.

684 Kukalová-Peck, J. 1978. Origin and evolution of insect wings and their relation to metamorphosis, as documented by the fossil record. *Journal of Morphology* 156:53–126.

685 Kusmierski, R., G. Borgia, R. H. Crozier, and B. H. Y. Chan. 1993. Molecular information on bowerbird phylogeny and the evolution of exaggerated male characters. *Journal of Evolutionary Biology* 6:737–752.

686 Kvarnemo, C., and I. Ahnesjö. 1996. The dynamics of operational sex ratios and competition for mates. *Trends in Ecology and Evolution* 11:404–408.

687 Kynaston, S. E., P. McErlain-Ward, and P. J. Mill. 1994. Courtship, mating behaviour and sexual cannibalism in the praying mantis, *Sphodromantis lineola. Animal Behaviour* 47:739–741.

688 Labov, J. B. 1981. Pregnancy blocking in rodents: Adaptive advantages for females. *American Naturalist* 118:361–371.

689 Labov, J. B., V. W. Huck, R. W. Elwood, and R. J. Brooks. 1985. Current problems in the study of infanticidal behavior of rodents. *Quarterly Review of Biology* 60:1–20.

690 Lacey, E. A., and P. W. Sherman. 1991. Social organization of naked mole-rat colonies: Evidence for divisions of labor. In *The Biology of the Naked Mole-Rat*, P. W. Sherman, J. U. M. Jarvis, and R. D. Alexander (eds.). Princeton University Press, Princeton, NJ.

691 Lacey, E. A., and P. W. Sherman. 1996. Cooperative breeding in naked mole-rats: Implications for vertebrate and invertebrate sociality. In *Cooperative Breeding in Mammals,* N. G. Solomon and J. A. French (eds.). Cambridge University Press, Cambridge.

692 Lack, D. 1968. *Ecological Adaptations for Breeding in Birds.* Methuen, London.

693 Lalumière, M., L. Chalmers, V. L. Quinsey, and M. C. Seto. 1996. A test of the mate deprivation hypothesis of sexual coercion. *Ethology and Sociobiology* 17:299–318.

694 LaMunyon, C. W., and T. Eisner. 1993. Postcopulatory sexual selection in an arctiid moth (*Utetheisa ornatrix*). *Proceedings of the National Academy of Sciences* 90:4689–4692.

695 LaMunyon, C. W., and T. Eisner. 1994. Spermatophore size as determinant of paternity in an arctiid moth (*Utetheisa ornatrix*). *Proceedings of the National Academy of Sciences* 91:7081–7084.

696 Lande, R. 1981. Models of speciation by sexual selection of polygenic traits. *Proceedings of the National Academy of Sciences* 78:3721–3725.

697 Landolt, P. J., O. H. Molina, R. R. Heath, K. Ward, B. D. Dueben, and J. G. Millar. 1996. Starvation of cabbage looper moths (Lepidoptera: Noctuidae) increases attraction to male pheromone. *Annals of the Entomological Society of America* 89:459–465.

698 Lank, D. B., L. W. Oring, and S. J. Maxson. 1985. Mate and nutrient limitation of egg-laying in a polyandrous shorebird. *Ecology* 66:1513–1524.

699 Lank, D. B., C. M. Smith, O. Hanotte, T. Burke, and F. Cooke. 1995. Genetic polymorphism for alternative mating behaviour in lekking male ruff *Philomachus pugnax. Nature* 378:59–62.

700 Lanyon, S. M. 1992. Interspecific brood parasitism in blackbirds (Icterinae): A phylogenetic perspective. *Science* 255:77–79.

701 Lauder, G. V., A. M. Leroi, and M. R. Rose. 1993. Adaptations and history. *Trends in Ecology and Evolution* 8:294–297.

702 Lawrence, E. S. 1985. Evidence for search image in blackbirds *Turdus merula* L.: Long-term learning. *Animal Behaviour* 33:1301–1309.

703 Le Boeuf, B. J. 1974. Male-male competition and reproductive success in elephant seals. *American Zoologist* 14:163–176.

704 Legge, S. 1996. Cooperative lions escape the Prisoner's Dilemma. *Trends in Ecology and Evolution* 11:2–3.

705 Lemon, W. C. 1991. Fitness consequences of foraging behaviour in the zebra finch. *Nature* 352:153–155.

706 Lemon, W. C., and R. H. Barth. 1992. The effects of feeding rate on reproductive success in the zebra finch, *Taeniopyga guttata. Animal Behaviour* 44:851–857.

707 Lendrem, D. W. 1986. *Modelling in Behavioural Ecology: An Introductory Text.* Croom Helm, London.

708 Leopold, A. S. 1977. *The California Quail.* University of California Press, Berkeley.

709 LeVay, S. 1991. A difference in hypothalamic structure between heterosexual and homosexual men. *Science* 253:1034–1037.

710 LeVay, S. 1996. *Queer Science: The Use and Abuse of Research into Homosexuality.* MIT Press, Cambridge, MA.

711 Levey, D. J., and F. G. Stiles. 1992. Evolutionary precursors of long-distance migration: Resource availability and movement patterns in Neotropical landbirds. *American Naturalist* 140:447–476.

712 Lewis, S. M., and S. N. Austad. 1994. Sexual selection in flour beetles: The relationship between sperm precedence and male olfactory attractiveness. *Behavioral Ecology* 5:219–224.

713 Licht, P. 1973. Influence of temperature and photoperiod on the annual ovarian cycle of the lizard *Anolis carolinensis. Copeia* 1973:465–472.

714 Ligon, J. D., S. H. Ligon, and H. A. Ford. 1991. An experimental study of the bases of male philopatry in the cooperative breeding superb fairy-wren *Malurus cyaneus. Ethology* 87:134–148.

715 Ligon, J., R. Thornhill, M. Zuk, and K. Johnson. 1990. Male-male competition, ornamentation and the role of testosterone in sexual selection in red jungle fowl. *Animal Behaviour* 40:367–373.

716 Lijfeld, J. T., P. O. Dunn, R. J. Robertson, and P. T. Boag. 1993. Extra-pair paternity in monogamous tree swallows. *Animal Behaviour* 45:213–229.

717 Lill, A. 1974. Sexual behavior of the lek-forming white-bearded manakin (*Manacus manacus trinitatis* Hartert). *Zeitschrift für Tierpsychologie* 36:1–36.

718 Lill, A. 1974. Social organization and space utilization in the lek-forming white-bearded manakin, *M. manacus trinitatis* Hartert. *Zeitschrift für Tierpsychologie* 36:513–530.

719 Lima, S. L. 1985. Maximizing efficiency and minimizing time exposed to predators: A trade-off in the black-capped chickadee. *Oecologia* 66:60–67.

720 Lima, S. L. 1995. Collective detection of predatory attack by social foragers: Fraught with ambiguity? *Animal Behaviour* 50:1097–1108.

721 Lima, S. L., and L. M. Dill. 1990. Behavioral decisions made under the risk of predation: A review and prospectus. *Canadian Journal of Zoology* 68:619–640.

722 Lincoln, G. A., F. Guinness, and R. V. Short. 1972. The way in which testosterone controls the social and sexual behavior of the red deer stag (*Cervus elaphus*). *Hormones and Behavior* 3:375–396.

723 Lindauer, M. 1961. *Communication among Social Bees.* Harvard University Press, Cambridge, MA.

724 Liske, E., and W. J. Davis. 1987. Courtship and mating behaviour of the Chinese praying mantid, *Tenodera aridifolia sinensis. Animal Behaviour* 35:1524–1537.

725 Littlejohn, M. J. 1965. Premating isolation in the *Hyla ewingi* complex (Anura: Hylidae). *Evolution* 19:234–243.

726 Lloyd, J. E. 1965. Aggressive mimicry in *Photuris:* Firefly *femmes fatales. Science* 149:653–654.

727 Lloyd, J. E. 1975. Aggressive mimicry in *Photuris* fireflies: Signal repertoires by *femmes fatales. Science* 197:452–453.

728 Lloyd, J. E. 1986. Firefly communication and deception: "Oh, what a tangled web." In *Deception: Perspectives on Human and Nonhuman Deceit*, R. W. Mitchell and N. S. Thompson (eds.). SUNY Press, Albany, NY.

729 Lockard, R. B. 1978. Seasonal change in the activity pattern of *Dipodomys spectabilis. Journal of Mammalogy* 59:563–568.

730 Lockard, R. B., and D. H. Owings. 1974. Seasonal variation in moonlight avoidance by bannertail kangaroo rats. *Journal of Mammalogy* 55:189–193.

731 Locke, J. L. 1994. Phases in the child's development of language. *American Scientist* 82:436–445.

732 Loesche, P., P. K. Stoddard, B. J. Higgins, and M. D. Beecher. 1991. Signature versus perceptual adaptations for individual vocal recognition in swallows. *Behaviour* 118:15–25.

733 Loher, W. 1972. Circadian control of stridulation in the cricket *Teleogryllus commodus*

Walker. *Journal of Comparative Physiology* 79:173–190.

734 Loher, W. 1979. Circadian rhythmicity of locomotor behavior and oviposition in female *Teleogryllus commodus*. *Behavioral Ecology and Sociobiology* 5:383–390.

735 Loher, W., and L. J. Orsak. 1985. Circadian patterns of premating behavior in *Teleogryllus oceanicus* Le Guillou under laboratory and field conditions. *Behavioral Ecology and Sociobiology* 16:223–231.

736 Lore, R., and K. Flannelly. 1977. Rat societies. *Scientific American* 236 (May):106–116.

737 Lorenz, K. Z. 1952. *King Solomon's Ring*. Crowell, New York.

738 Lorenz, K. Z. 1970. Companions as factors in the bird's environment. In *Studies on Animal and Human Behavior*, K. Z. Lorenz (ed.). Harvard University Press, Cambridge, MA.

739 Lott, D. F. 1979. Dominance relations and breeding rate in mature male American bison. *Zeitschrift für Tierpsychologie* 49:418–432.

740 Low, B. S., and J. T. Heinen. 1993. Population, resources, and environment: Implications of human behavioral ecology for conservation. *Population and Environment* 15:7–41.

741 Lunau, K. 1992. Evolutionary aspects of perfume collection in male euglossine bees (Hymenoptera) and of nest deception in bee-pollinated flowers. *Chemoecology* 3:65–73.

742 Lundberg, P. 1988. The evolution of partial migration in birds. *Trends in Ecology and Evolution* 3:172–176.

743 Lynch, C. B. 1980. Response to divergent selection for nesting behavior in *Mus musculus*. *Genetics* 96:757–765.

744 Lyon, B. E. 1993. Conspecific brood parasitism as a flexible female reproductive tactic in American coots. *Animal Behaviour* 46:911–928.

745 Lyon, B. E., J. M. Eadie, and L. D. Hamilton. 1994. Parental choice selects for ornamental plumage in American coot chicks. *Nature* 371:240–243.

746 Lyon, B. E., R. D. Montgomerie, and L. D. Hamilton. 1987. Male parental care and monogamy in snow buntings. *Behavioral Ecology and Sociobiology* 20:377–382.

747 Mace, R. 1987. The dawn chorus in the great tit *Parus major* is directly related to female fertility. *Nature* 330:745–746.

748 Mace, R. 1996. Biased parental investment and reproductive success in Gabbra pastoralists. *Behavioral Ecology and Sociobiology* 38:75–82.

749 Mackenzie, A., J. D. Reynolds, V. J. Brown, and W. J. Sutherland. 1995. Variation in male mating success on leks. *American Naturalist* 145:633–652.

750 MacLusky, N. J., and F. Naftolin. 1981. Sexual differentiation of the central nervous system. *Science* 211:1294–1302.

751 Madsen, T., R. Shine, J. Loman, and T. Håkansson. 1992. Why do female adders copulate so frequently? *Nature* 355:440–442.

752 Madsen, T., R. Shine, J. Loman, and T. Håkansson. 1993. Determinants of mating success in male adders, *Vipera berus*. *Animal Behaviour* 45:491–499.

753 Magurran, A. E., D. W. Irving, and P. A. Henderson. 1996. Is there a fish alarm pheromone? A wild study and critique. *Proceedings of the Royal Society of London B* 263:1551–1556.

754 Malcolm, S. B. 1987. Monarch butterfly migration in North America: Controversy and conservation. *Trends in Ecology and Evolution* 2:135–139.

755 Mangel, M., and C. Clark. 1988. *Dynamic Modelling in Behavioral Ecology*. Princeton University Press, Princeton, NJ.

756 Marconato, A., A. Bisazza, and M. Fabris. 1993. The cost of parental care and egg cannibalism in the river bullhead, *Cottus gobio* L. (Pisces, Cottidae). *Behavioral Ecology and Sociobiology* 32:229–238.

757 Marden, J. H. 1995. How insects learned to fly. *The Sciences* 35(6):26–30.

758 Marden, J. H., and P. Chai. 1991. Aerial predation and butterfly design: How palatability, mimicry, and the need for evasive flight constrain mass allocation. *American Naturalist* 138:15–36.

759 Marden, J. H., and M. G. Kramer. 1994. Surface-skimming stoneflies: A possible intermediate stage in insect flight evolution. *Science* 266:427–430.

760 Marden, J. H., and M. G. Kramer. 1995. Locomotor performance of insects with rudimentary wings. *Nature* 377:332–334.

761 Marden, J. H., and R. A. Rollins. 1994. Assessment of energy reserves by damselflies engaged in aerial contests for mating territories. *Animal Behaviour* 48:1023–1030.

762 Marden, J. H., and J. K. Waage. 1990. Escalated damselfly territorial contests and energetic wars of attrition. *Animal Behaviour* 39:954–959.

763 Maret, T. J., and J. P. Collins. 1994. Individual responses to population size structure: The role of size variation in controlling expression of a trophic polyphenism. *Oecologia* 100:279–285.

764 Margulis, S. W., and J. Altmann. 1997. Behavioural risk factors in the reproduction of inbred and outbred oldfield mice. *Animal Behaviour* 54:397–408.

765 Margulis, S. W., W. Saltzman, and D. H. Abbott. 1995. Behavioural and hormonal changes in female naked mole-rats (*Heterocephalus glaber*) following removal of the breeding female from a colony. *Hormones and Behavior* 29:227–247.

766 Marler, C. A., and M. C. Moore. 1989. Time and energy costs of aggression in testosterone-implanted free-living male mountain spiny lizards (*Sceloporus jarrovi*). *Physiological Zoology* 62:1334–1350.

767 Marler, C. A., and M. C. Moore. 1991. Supplementary feeding compensates for testosterone-induced costs of aggression in male mountain spiny lizards, *Sceloporus jarrovi*. *Animal Behaviour* 42:209–219.

768 Marler, C. A., G. Walsberg, M. L. White, and M. Moore. 1995. Increased energy expenditure due to increased territorial defense in male lizards after phenotypic manipulation. *Behavioral Ecology and Sociobiology* 37:225–232.

769 Marler, P. 1955. Characteristics of some animal calls. *Nature* 176:6–8.

770 Marler, P. 1970. Birdsong and speech development: Could there be parallels? *American Scientist* 58:669–673.

771 Marler, P., and W. J. Hamilton. 1966. *Mechanisms of Animal Behavior.* Wiley, New York.

772 Marler, P., and M. Tamura. 1964. Culturally transmitted patterns of vocal behavior in sparrows. *Science* 146:1483–1486.

773 Marr, D. 1982. *Vision.* W. H. Freeman, San Francisco.

774 Marshall, J. T., Jr. 1964. Voice communication and relationships among brown towhees. *Condor* 66:345–356.

775 Martins, E. (ed.). 1996. *Phylogenies and the Comparative Method in Animal Behavior.* Oxford University Press, New York.

776 Marzluff, J. M., B. Heinrich, and C. S. Marzluff. 1996. Raven roosts are mobile information centres. *Animal Behaviour* 51:89–103.

777 Maschwitz, U., and E. Maschwitz. 1974. Platzende Arbeiterinnen: Eine neue Art der Feindabwehr bei sozialen Hautflüglern. *Oecologia* 14:289–294.

778 Mason, R. T. 1993. Chemical ecology of the red-sided garter snake, *Thamnophis sirtalis parietalis. Brain, Behavior and Evolution* 41:261–268.

779 Mason, W. A. 1978. Social experience and primate cognitive development. In *The Development of Behavior: Comparative and Evolutionary Aspects*, G. M Burghardt and M. Bekoff (eds.). Garland STPM Press, New York.

780 Mather, M. H., and B. D. Roitberg. 1987. A sheep in wolf's clothing: Tephritid flies mimic spider predators. *Science* 236:308–310.

781 May, M. 1991. Aerial defense tactics of flying insects. *American Scientist* 79:316–329.

782 Mayr, E. 1961. Cause and effect in biology. *Science* 134:1501–1506.

783 Mayr, E. 1963. *Animal Species and Evolution.* Harvard University Press, Cambridge, MA.

784 Mayr, E. 1983. How to carry out the adaptationist program? *American Naturalist* 121:324–334.

785 McAllister, L. B., R. H. Scheller, E. R. Kandel, and R. Axel. 1983. *In situ* hybridization to study the origin and fate of identified neurons. *Science* 222:800–808.

786 McCann, T. S. 1981. Aggression and sexual activity of male southern elephant seals, *Mirounga leonina. Journal of Zoology* 195:295–310.

787 McComb, K. E. 1987. Roaring by red deer stags advances date of oestrous in hinds. *Nature* 330:648–649.

788 McCracken, G. F. 1984. Communal nursing in Mexican free-tailed bat maternity colonies. *Science* 223:1090–1091.

789 McCracken, G. F. 1993. Locational memory and female-pup reunions in Mexican free-tailed bat maternity colonies. *Animal Behaviour* 45:811–813.

790 McCracken, G. F., and J. W. Bradbury. 1981. Social organization and kinship in the polygynous bat *Phyllostomus hastatus*. *Behavioral Ecology and Sociobiology* 8:11–34.

791 McCracken, G. F., and M. K. Gustin. 1991. Nursing behavior in Mexican free-tailed bat maternity colonies. *Ethology* 89:305–321.

792 McDonald, D. B., and W. K. Potts. 1994. Cooperative display and relatedness among males in a lek-mating bird. *Science* 266:1030–1032.

793 McEwen, B. S. 1981. Neural gonadal steroid actions. *Science* 211:1303–1311.

794 McGowan, K. J., and G. E. Woolfenden. 1989. A sentinel system in the Florida scrub jay. *Animal Behaviour* 37:1000–1006.

795 McKaye, R., S. M. Louda, and J. R. Stauffer, Jr. 1990. Bower size and male reproductive success in a cichlid fish lek. *American Naturalist* 135:597–613.

796 M'Closkey, R. T., K. A. Baia, and R. W. Russell. 1987. Defense of mates: A territory departure rule for male tree lizards following sex-ratio manipulation. *Oecologia* 73:28–31.

797 McNab, B. K., and J. F. Eisenberg. 1989. Brain size and its relation to the rate of metabolism in mammals. *American Naturalist* 133:157–167.

798 McNicol, D., Jr., and D. Crews. 1979. Estrogen/progesterone synergy in the control of female sexual receptivity in the lizard *Anolis carolinensis*. *General and Comparative Endocrinology* 38:68–74.

799 Mech, L. D. 1970. *The Wolf: The Ecology and Behavior of an Endangered Species*. Doubleday (Natural History Press), Garden City, New York.

800 Medvin, M. B., and M. D. Beecher. 1986. Parent-offspring recognition in the barn swallow (*Hirundo rustica*). *Animal Behaviour* 34:1627–1639.

801 Medvin, M. B., P. K. Stoddard, and M. D. Beecher. 1993. Signals for parent-offspring recognition: A comparative analysis of the begging calls of cliff swallows and barn swallows. *Animal Behaviour* 45:841–850.

802 Meier, R. P. 1991. Language acquisition by deaf children. *American Scientist* 79:60–70.

803 Meire, P. M., and A. Ervynck. 1986. Are oystercatchers (*Haemoptopus ostralegus*) selecting the most profitable mussels (*Mytilus edulis*)? *Animal Behaviour* 34:1427–1435.

804 Mello, C. V., D. S. Vicario, and D. F. Clayton. 1992. Song presentation induces gene expression in the songbird's forebrain. *Proceedings of the National Academy of Sciences* 89:6818–6822.

805 Meylan, A. B., B. W. Bowen, and J. C. Avise. 1990. A genetic test of the natal homing versus social facilitation models for green turtle migration. *Science* 248:724–727.

806 Michelsen, A., B. B. Andersen, W. H. Kirchner, and M. Lindauer. 1989. Honeybees can be recruited by a mechanical model of a dancing bee. *Naturwissenschaften* 76:277–280.

807 Michelsen, A., B. B. Andersen, J. Storm, W. H. Kirchner, and M. Lindauer. 1992. How honeybees perceive communication dances, studied by means of a mechanical model. *Behavioral Ecology and Sociobiology* 30:143–150.

808 Miller, D. E., and J. T. Emlen, Jr. 1975. Individual chick recognition and family integrity in the ring-billed gull. *Behaviour* 52:124–144.

809 Miller, L. A. 1971. Physiological responses of green lacewings (*Chrysopa*, Neuroptera) to ultrasound. *Journal of Insect Physiology* 17:491–506.

810 Miller, L. A. 1983. How insects detect and avoid bats. In *Neuroethology and Behavioral Physiology*, F. Huber and H. Markl (eds.). Springer-Verlag, Berlin.

811 Miller, L. A., and J. Olesen. 1979. Avoidance behavior in green lacewings. I. Behavior of free flying green lacewings to hunting bats and ultrasound. *Journal of Comparative Physiology* 131:113–120.

812 Millington, S. J., and T. D. Price. 1985. Song inheritance and mating patterns in Darwin's finches. *Auk* 102:342–346.

813 Mittwoch, U. 1996. Sex-determining mechanisms in animals. *Trends in Ecology and Evolution* 11:63–67.

814 Mock, D. W. 1984. Siblicidal aggression and resource monopolization in birds. *Science* 225:731–733.

815 Mock, D. W., H. Drummond, and C. H. Stinson. 1990. Avian siblicide. *American Scientist* 78:438–449.

816 Mock, D. W., T. C. Lamey, and D. B. A. Thompson. 1988. Falsifiability and the information centre hypothesis. *Ornis Scandinavica* 19:231–248.

817 Mock, D. W., and B. J. Ploger. 1987. Parental manipulation of optimal hatch asynchrony in cattle egrets: An experimental study. *Animal Behaviour* 35:150–160.

818 Moiseff, A., G. S. Pollack, and R. R. Hoy. 1978. Steering responses of flying crickets to sound and ultrasound: Mate attraction and predator avoidance. *Proceedings of the National Academy of Sciences* 75:4052–4056.

819 Møller, A. P. 1988. False alarm calls as a means of resource usurpation in the great tit *Parus major. Ethology* 79:25–30.

820 Møller, A. P. 1988. Female choice selects for male sexual tail ornaments in the monogamous swallow. *Nature* 332:640–642.

821 Møller, A. P. 1990. Changes in the size of avian breeding territories in relation to the nesting cycle. *Animal Behaviour* 40:1070–1079.

822 Møller, A. P. 1990. Effects of a haemotophagous mite on the barn swallow (*Hirundo rustica*): A test of the Hamilton and Zuk hypothesis. *Evolution* 44:771–784.

823 Møller, A. P. 1991. Why mated songbirds sing so much: Mate guarding and male announcement of mate fertility status. *American Naturalist* 138:994–1014.

824 Møller, A. P. 1992. Female swallow preference for symmetrical male sexual ornaments. *Nature* 357:238–240.

825 Møller, A. P. 1992. Frequency of female copulations with multiple males and sexual selection. *American Naturalist* 139:1089–1101.

826 Møller, A. P. 1994. *Sexual Selection and the Barn Swallow.* Oxford University Press, Oxford.

827 Møller, A. P., and T. R. Birkhead. 1993. Cuckoldry and sociality: A comparative study of birds. *American Naturalist* 142:118–140.

828 Money, J., and A. A. Ehrhardt. 1972. *Man and Woman, Boy and Girl.* Johns Hopkins University Press, Baltimore, MD.

829 Montgomerie, R., and R. Thornhill. 1989. Fertility advertisement in birds: A means of inciting male-male competition? *Ethology* 81:209–220.

830 Montgomerie, R. D., and P. J. Weatherhead. 1988. Risks and rewards of nest defence by parent birds. *Quarterly Review of Biology* 63:167–187.

831 Moore, A. J. 1994. Genetic evidence for the "good genes" process of sexual selection. *Behavioral Ecology and Sociobiology* 35:235–242.

832 Moore, A. J., N. L. Reagan, and K. F. Haynes. 1995. Conditional signalling strategies: Effects of ontogeny, social experience and social status on the pheromonal signal of male cockroaches. *Animal Behaviour* 50:191–202.

833 Moore, J., and R. Ali. 1984. Are dispersal and inbreeding avoidance related? *Animal Behaviour* 32:94–112.

834 Moore, M. C. 1986. Elevated testosterone levels during nonbreeding-season territoriality in fall-breeding lizard, *Sceloporus jarrovi. Journal of Comparative Physiology A* 158:159–163.

835 Moore, M. C., and B. Kranz. 1983. Evidence for androgen independence of male mounting behavior in white-crowned sparrows (*Zonotrichia leucophrys gambelii*). *Hormones and Behavior* 17:414–423.

836 Mooring, M. S., and B. L. Hart. 1995. Differential grooming rate and tick load of territorial male and female impala, *Aepyceros melampus. Behavioral Ecology* 6:94–101.

837 Morton, E. S. 1975. Ecological sources of selection on avian sounds. *American Naturalist* 109:17–34.

838 Morton, E. S. 1977. On the occurrence and significance of motivation-structural rules in some bird and mammal sounds. *American Naturalist* 111:855–869.

839 Moses, R. A., and J. S. Millar. 1994. Philopatry and mother-daughter associations in bushy-tailed woodrats: Space use and reproductive success. *Behavioral Ecology and Sociobiology* 35:131–140.

840 Moskowitz, B. A. 1978. The acquisition of language. *Scientific American* 239 (Nov):91–108.

841 Moss, C. 1988. *Elephant Memories*. William Morrow, New York.

842 Moss, R., R. Parr, and X. Lambin. 1994. Effects of testosterone on breeding density, breeding success and survival of red grouse. *Proceedings of the Royal Society of London B* 258:175–180.

843 Mountjoy, D. J., and R. E. Lemon. 1991. Song as an attractant for male and female European starlings, and the influence of song complexity on their response. *Behavioral Ecology and Sociobiology* 28:97–100.

844 Mountjoy, D. J., and R. E. Lemon. 1996. Female choice for complex song in the European starling: A field experiment. *Behavioral Ecology and Sociobiology* 38:65–72.

845 Mueller, U. G. 1991. Haplodiploidy and the evolution of facultative sex ratios in a primitively eusocial bee. *Science* 254:442–444.

846 Mumme, R. L. 1992. Delayed dispersal and cooperative breeding in the Seychelles warbler. *Trends in Ecology and Evolution* 7:330–331.

847 Mumme, R. L. 1992. Do helpers increase reproductive success? An experimental analysis in the Florida scrub jay. *Behavioral Ecology and Sociobiology* 31:319–328.

848 Mumme, R. L., W. D. Koenig, and F. A. Pitelka. 1983. Reproductive competition in the communal acorn woodpecker: Sisters destroy each other's eggs. *Nature* 306:583–584.

849 Mumme, R. L., W. D. Koenig, and F. L. W. Ratnieks. 1989. Helping behaviour, reproductive value, and the future component of indirect fitness. *Animal Behaviour* 38:331–343.

850 Munn, C. A. 1986. Birds that "cry wolf". *Nature* 319:143–145.

851 Munn, C. A. 1991. Macaw biology and ecotourism, or "when a bird in the bush is worth two in the hand." In *New World Parrots in Crisis*, S. R. Beisinger and N. F. R. Snyder (eds.). Smithsonian Institution Press, Washington, D.C.

852 Murdock, G. P. 1967. *Ethnographic Atlas*. Pittsburgh University Press, Pittsburgh, PA.

853 Murray, B. G., Jr. 1989. A critical review of the transoceanic migration of the blackpoll warbler. *Auk* 106:8–17.

854 Nalepa, C. A., and S. C. Jones. 1991. Evolution of monogamy in termites. *Biological Reviews* 66:83–87.

855 Nefdt, R. J. C. 1995. Disruptions of matings, harassment and lek-breeding in Kafue lechwe antelope. *Animal Behaviour* 49:411–418.

856 Nelson, D. A., P. Marler, and A. Palleroni. 1995. A comparative approach to vocal learning: Intraspecific variation in the learning process. *Animal Behaviour* 50:83–97.

857 Nelson, R. J. 1995. *An Introduction to Behavioral Endocrinology*. Sinauer Associates, Sunderland, MA.

858 Newsome, W. T., K. H. Britten, C. D. Salzman, and J. A. Movshon. 1990. Neuronal mechanisms of motion perception. *Cold Spring Harbor Symposia on Quantitative Biology* 55:697–705.

859 Newsome, W. T., and E. B. Paré. 1988. Selective impairment of motion perception following lesions of the middle temporal visual area (MT). *Journal of Neuroscience* 8:2201–2211.

860 Newton, I. 1993. Age and site fidelity in female sparrowhawks, *Accipter nisus*. *Animal Behaviour* 46:161–168.

861 Newton, P. N. 1986. Infanticide in an undisturbed forest population of hanuman langurs, *Presbytis entellus*. *Animal Behaviour* 34:785–789.

862 Nilsson, L. A. 1992. Orchid pollination biology. *Trends in Ecology and Evolution* 7:255–259.

863 Noë, R., and A. A. Sluijter. 1990. Reproductive tactics of male savanna baboons. *Behaviour* 113:117–170.

864 Nolen, T. G., and R. R. Hoy. 1984. Phonotaxis in flying crickets: Neural correlates. *Science* 226:992–994.

865 Noonan, K. C. 1983. Female choice in the cichlid fish *Cichlasoma nigrofasciatum*. *Animal Behaviour* 31:1005–1010.

866 Norberg, U. M. 1985. Evolution of vertebrate flight: An aerodynamic model for the transition from gliding to active flight. *American Naturalist* 126:303–327.

867 Nordeen, K. W., and E. J. Nordeen. 1988. Projection neurons within a vocal motor pathway are born during song learning in zebra finches. *Nature* 334:149–151.

868 Norris, K. S. 1967. Some observations on the migration and orientation of marine mammals. In *Animal Orientation and Navigation*, R. M. Storm (ed.). Oregon State University Press, Corvallis.

869 Nottebohm, F. 1969. The song of the chingolo (*Zonotrichia capensis*) in Argentina: Description and evaluation of a system of dialects. *Condor* 71:299–315.

870 Nottebohm, F., and M. E. Nottebohm. 1971. Vocalization and breeding behaviour of surgically deafened ring doves (*Streptopelia risoria*). *Animal Behaviour* 19:313–327.

871 Nowak, M., and K. Sigmund. 1993. A strategy of win-stay, lose-shift that outperforms tit-for-tat in the Prisoner's Dilemma game. *Nature* 364:56–58.

872 O'Connell, S. M., and G. Cowlishaw. 1994. Infanticide avoidance, sperm competition and mate choice: The function of copulation calls in female baboons. *Animal Behaviour* 48:687–694.

873 O'Loghlen, A. L., and S. I. Rothstein. 1993. An extreme example of delayed vocal development: Song learning in a population of wild brown-headed cowbirds. *Animal Behaviour* 46:293–304.

874 O'Loghlen, A. L., and S. I. Rothstein. 1995. Culturally correct song dialects are correlated with male age and female song preferences in wild populations of brown-headed cowbirds. *Behavioral Ecology and Sociobiology* 36:251–260.

875 Olson, D. J., A. C. Kamil, R. P. Balda, and P. J. Nims. 1995. Performance of four seed-caching corvid species in operant tests of nonspatial and spatial memory. *Journal of Comparative Psychology* 109:173–181.

876 O'Neill, K. M. 1983. Territoriality, body size, and spacing in males of the bee wolf *Philanthus basilaris* (Hymenoptera; Sphecidae). *Behaviour* 86:295–321.

877 O'Neill, W. E., and N. Suga. 1979. Target range-sensitive neurons in the auditory cortex of the mustache bat. *Science* 203:69–72.

878 O'Riain, M. J., J. U. M. Jarvis, and C. G. Faulkes. 1996. A dispersive morph in the naked mole-rat. *Nature* 380:619–621.

879 Orians, G. H. 1962. Natural selection and ecological theory. *American Naturalist* 96:257–264.

880 Orians, G. H. 1969. On the evolution of mating systems in birds and mammal. *American Naturalist* 103:589–603.

881 Oring, L. W. 1985. Avian polyandry. *Current Ornithology* 3:309–351.

882 Oring, L. W., M. A. Colwell, and J. M. Reed. 1991. Lifetime reproductive success in the spotted sandpiper (*Actitis macularia*): Sex differences and variance components. *Behavioral Ecology and Sociobiology* 28:425–432.

883 Oring, L. W., R. C. Fleischer, J. M. Reed, and K. E. Marsden. 1992. Cuckoldry through stored sperm in the sequentially polyandrous spotted sandpiper. *Nature* 359:631–633.

884 Oring, L. W., and M. L. Knudson. 1973. Monogamy and polyandry in the spotted sandpiper. *The Living Bird* 11:59–73.

885 Orr, M. R. 1992. Parasitic flies (Diptera: Phoridae) influence foraging rhythms and caste division of labor in the leaf-cutter ant, *Atta cephalotes* (Hymenoptera: Formicidae). *Behavioral Ecology and Sociobiology* 30:395–402.

886 Osorno, J. L., and H. Drummond. 1995. The function of hatching asynchrony in the blue-footed booby. *Behavioral Ecology and Sociobiology* 37:265–274.

887 Oster, G. F., and E. O. Wilson. 1978. *Caste and Ecology in the Social Insects.* Princeton University Press, Princeton, NJ.

888 Ostrom, J. H. 1974. *Archaeopteryx* and the origin of flight. *Quarterly Review of Biology* 49:27–47.

889 Owens, D. D., and M. J. Owens. 1979. Communal denning and clan associations in brown hyenas (*Hyaena brunnea*, Thunberg) in the central Kalahari Desert. *African Journal of Ecology* 17:35–44.

890 Owens, I. P. F., and R. V. Short. 1995. Hormonal basis of sexual dimorphism in birds: Implications for sexual selection theory. *Trends in Ecology and Evolution* 10:44–47.

891 Owings, D. H., and R. G. Coss. 1977. Snake mobbing by California ground squirrels: Adaptive variation and ontogeny. *Behaviour* 62:50–69.

892 Packard, J. M., U. S. Seal, L. D. Mech, and E. D. Plotka. 1985. Causes of reproductive failure in two family groups of wolves (*Canis lupus*). *Zeitschrift für Tierpsychologie* 69:24–40.

893 Packer, C. 1977. Reciprocal altruism in *Papio anubis. Nature* 265:441–443.

894 Packer, C. 1986. The ecology of sociality in felids. In *Ecological Aspects of Social Evolution*, D. I. Rubenstein and R. W. Wrangham (eds.). Princeton University Press, Princeton, NJ.

895 Packer, C. 1994. *Into Africa*. University of Chicago Press, Chicago.

896 Packer, C., D. A. Gilbert, A. E. Pusey, and S. J. O'Brien. 1991. A molecular genetic analysis of kinship and cooperation in African lions. *Nature* 351:562–565.

897 Packer, C., D. Scheel, and A. E. Pusey. 1990. Why lions form groups: Food is not enough. *American Naturalist* 136:1–19.

898 Page, T. L. 1985. Clocks and circadian rhythms. In *Comprehensive Insect Physiology, Biochemistry, and Pharmacology*, G. A. Kerkut and L. I. Gilbert (eds.). Pergamon Press, New York.

899 Palmer, C. T. 1989. Is rape a cultural universal? A re-examination of the ethnographic data. *Ethnology* 28:1–16.

900 Palmer, C. T. 1991. Human rape: Adaptation or by-product? *Journal of Sex Research* 28:365–386.

901 Palmer, J. D. 1976. *An Introduction to Biological Rhythms*. Academic Press, New York.

902 Papi, F. 1975. La navigazione dei colombi viaggiatori. *Le Scienze* 78:66–75.

903 Papi, F. 1986. Pigeon navigation: Solved problems and open questions. *Monitore Zoologici Italiana* 20:471–517.

904 Parker, G. A. 1970. Sperm competition and its evolutionary consequences in the insects. *Biological Reviews* 45:525–567.

905 Parker, G. A. 1974. Assessment strategy and the evolution of fighting behaviour. *Journal of Theoretical Biology* 47:223–243.

906 Parker, G. A. 1992. Snakes and female sexuality. *Nature* 355:395–396.

907 Parker, G. A., R. R. Baker, and V. G. F. Smith. 1972. The origin and evolution of gamete dimorphism and the male-female phenomenon. *Journal of Theoretical Biology* 36:529–553.

908 Parker, P. G., T. A. Waite, B. Heinrich, and J. M. Marzluff. 1994. Do common ravens share food bonanzas with kin? DNA fingerprinting evidence. *Animal Behaviour* 48:1085–1093.

909 Parmigiani, S., and F. S. vom Saal (eds.). 1994. *Infanticide and Parental Care*. Harwood AcademicPress, Chur, Switzerland.

910 Pärt, T. 1991. Is dawn singing related to paternity insurance? The case of the collared flycatcher. *Animal Behaviour* 41:451–456.

911 Paton, D. C., and H. A. Ford. 1983. The influence of plant characteristics and honeyeater size on levels of pollination in Australian plants. In *Handbook of Experimental Pollination Biology*, C. E. Jones and R. J. Little (eds.). Van Nostrand Reinhold, New York.

912 Payne, R. B. 1982. Ecological consequences of song matching: Breeding success and intraspecific song mimicry in indigo buntings. *Ecology* 63:401–411.

913 Payne, R. B., and L. L. Payne. 1993. Song copying and cultural transmission in indigo buntings. *Animal Behaviour* 46:1045–1065.

914 Peakall, R. 1990. Responses of male *Zaspilothynnus trilobatus* Turner wasps to females and the sexually deceptive orchid it pollinates. *Functional Ecology* 4:159–167.

915 Penfield, W. 1970. Consciousness, memory, and man's conditioned reflexes. In *On the Biology of Learning*, K. H. Pribram (ed.). Harcourt Brace Jovanovich, New York.

916 Pengelly, E. T., and S. J. Asmundson. 1974. Circannual rhythmicity in hibernating animals. In *Circannual Clocks*, E. T. Pengelley (ed.). Academic Press, New York.

917 Perrett, D. I., and E. T. Rolls. 1983. Neural mechanisms underlying the visual analysis of faces. In *Advances in Vertebrate Neuroethology*, J.-P. Ewert, R. R. Capranica, and D. J. Ingle (eds.). Plenum Press, New York.

918 Perrigo, G., W. C. Bryant, and F. S. vom Saal. 1990. A unique neural timing system prevents male mice from harming their own offspring. *Animal Behaviour* 39:535–539.

919 Perrill, S. A., H. C. Gerhardt, and R. Daniel. 1978. Sexual parasitism in the green tree frog (*Hyla cinerea*). *Science* 200:1179–1180.

920 Perusse, D. 1993. Cultural and reproductive success in industrial societies: Testing the relationship at the proximate and ultimate levels. *Behavioral and Brain Sciences* 16:267–283.

921 Peschke, K. 1987. Male aggression, female mimicry and female choice in the rove beetle, *Aleochara curtula*. *Ethology* 75:265–284.

922 Petit, L. J. 1991. Adaptive tolerance of cow-bird parasitism by prothonotary warblers: A consequence of site limitation? *Animal Behaviour* 41:425–432.

923 Petrie, M. 1992. Peacocks with low mating success are more likely to suffer predation. *Animal Behaviour* 44:585–586.

924 Petrie, M. 1994. Improved growth and survival of offspring of peacocks with more elaborate trains. *Nature* 371:598–599.

925 Petrie, M., and T. Halliday. 1994. Experimental and natural changes in the peacock's (*Pavo cristatus*) train can affect mating success. *Behavioral Ecology and Sociobiology* 35:213–217.

926 Petrie, M., T. Halliday, and C. Sanders. 1991. Peahens prefer peacocks with elaborate trains. *Animal Behaviour* 41:323–332.

927 Petrie, M., and A. P. Møller. 1991. Laying eggs in others' nests: Intraspecific brood parasitism in birds. *Trends in Ecology and Evolution* 6:315–320.

928 Pfennig, D. W. 1992. Polyphenism in spadefoot toad tadpoles as a locally adjusted evolutionarily stable strategy. *Evolution* 46:1408–1420.

929 Pfennig, D. W., and J. P. Collins. 1993. Kinship affects morphogenesis in cannibalistic salamanders. *Nature* 362:836–838.

930 Pfennig, D. W., G. J. Gamboa, H. K. Reeve, J. S. Reeve, and I. D. Ferguson. 1983. The mechanism of nestmate discrimination in social wasps (*Polistes*, Hymenoptera: Vespidae). *Behavioral Ecology and Sociobiology* 13:299–305.

931 Pfennig, D. W., and P. W. Sherman. 1994. Kin recognition. *Scientific American* 272 (6):68–73.

932 Pfennig, D. W., P. W. Sherman, and J. P. Collins. 1994. Kin recognition and cannibalism in polyphenic salamanders. *Behavioral Ecology* 5:225–232.

933 Pierce, G. J., and J. G. Ollason. 1987. Eight reasons why optimal foraging theory is a complete waste of time. *Oikos* 49:111–118.

934 Pierce, N. E., and P. S. Mead. 1981. Parasitoids as selective agents in the symbiosis between lycaenid butterfly larvae and ants. *Science* 211:1185–1187.

935 Pierotti, R., and E. C. Murphy. 1987. Intergenerational conflicts in gulls. *Animal Behaviour* 35:435–444.

936 Pietrewicz, A. T., and A. C. Kamil. 1977. Visual detection of cryptic prey by blue jays (*Cyanocitta cristata*). *Science* 195:580–582.

937 Pietrewicz, A. T., and A. C. Kamil. 1979. Search image formation in the blue jay *Cyanocitta cristata*. *Science* 204:1332–1333.

938 Pietsch, T. W., and D. B. Grobecker. 1978. The compleat angler: Aggressive mimicry in the antennariid anglerfish. *Science* 201:369–370.

939 Pinker, S. 1991. Rules of language. *Science* 253:530–535.

940 Pinker, S. 1994. *The Language Instinct.* W. Morrow and Co., New York.

941 Pitcher, T. 1979. He who hesitates lives: Is stotting antiambush behavior? *American Naturalist* 113:453–456.

942 Pitnick, S. 1993. Operational sex ratios and sperm limitation in populations of *Drosophila pachea*. *Behavioral Ecology and Sociobiology* 33:383–391.

943 Pitnick, S., T. A. Markow, and G. S. Spicer. 1995. Delayed male maturity is a cost of producing large sperm in *Drosophila*. *Proceedings of the National Academy of Sciences* 92:10614–10618.

944 Plaisted, K. C., and N. J. Mackintosh. 1995. Visual search for cryptic stimuli in pigeons: Implications for the search image and search rate hypotheses. *Animal Behaviour* 50:1219–1232.

945 Plomin, R., and D. C. Rowe. 1978. Genes, environment and development of temperament in young human twins. In *The Development of Behavior,* G. M. Burghardt and M. Bekoff (eds.). Garland STPM Press, New York.

946 Pohl-Apel, G., and R. Sossinka. 1984. Hormonal determination of song capacity in females of the zebra finch: Critical phase of treatment. *Zeitschrift für Tierpsychologie* 64:330–336.

947 Polak, M. 1993. Parasites increase fluctuating asymmetry of male *Drosophila nigrospiracula*: Implications for sexual selection. *Genetica* 89:255–265.

948 Polak, M., and R. Trivers. 1994. The science of symmetry in biology. *Trends in Ecology and Evolution* 9:122–124.

949 Pollak, G., D. Marsh, R. Bodenhamer, and A. Souther. 1977. Echo-detecting characteristics of neurons in inferior colliculus of unanesthetized bats. *Science* 196:675–677.

950 Pollard, S. D. 1994. Consequences of sexual selection on feeding in male jumping spiders (Araneae: Salticidae). *Journal of Zoology* 234:203–208.

951 Pomiankowski, A., and A. P. Møller. 1995. A resolution of the lek paradox. *Proceedings of the Royal Society of London B* 260:21–29.

952 Poole, J. H. 1989. Mate guarding, reproductive success and female choice in African elephants. *Animal Behaviour* 37:842–849.

953 Porter, R. H., V. J. Tepper, and D. M. White. 1981. Experiential influences on the development of huddling preferences and "sibling" recognition in spiny mice. *Developmental Psychobiology* 14:375–382.

954 Potts, W. K., C. J. Manning, and E. K. Wakeland. 1991. Mating patterns in seminatural populations of mice: Influence by MHC genotype. *Nature* 352:619–621.

955 Powell, G. V. N. 1974. Experimental analysis of the social value of flocking by starlings (*Sturnus vulgaris*) in relation to predation and foraging. *Animal Behaviour* 22:501–505.

956 Prete, F. R. 1995. Designing behavior: A case study. *Perspectives in Ecology* 11:255–277.

957 Prior, K. A., and P. J. Weatherhead. 1991. Turkey vultures foraging at experimental food patches: A test of information transfer at communal roosts. *Behavioral Ecology and Sociobiology* 28:385–390.

958 Pritchard, P. C. H. 1976. Post-nesting movements of marine turtle (Cheloniidae and Dermochelyidae) tagged in the Guianas. *Copeia* 1976:749–754.

959 Proctor, H. C. 1991. Courtship in the water mite *Neumania papillator:* Males capitalize on female adaptations for predation. *Animal Behaviour* 42:589–598.

960 Proctor, H. C. 1992. Sensory exploitation and the evolution of male mating behaviour: A cladistic test. *Animal Behaviour* 44:745–752.

961 Provine, R. R. 1986. Yawning as a stereotyped action pattern and releasing stimulus. *Ethology* 72:109–122.

962 Pruett-Jones, S. G. 1988. Lekking versus solitary display: Temporal variations in dispersion in the buff-breasted sandpiper. *Animal Behaviour* 36:1740–1752.

963 Pruett-Jones, S. G., and M. J. Lewis. 1990. Sex ratio and habitat limitation promote delayed dispersal in superb fairy-wrens. *Nature* 348:541–542.

964 Pruett-Jones, S., and M. Pruett-Jones. 1994. Sexual competition and courtship disruptions: Why do male bowerbirds destroy each other's bowers? *Animal Behaviour* 47:607–620.

965 Pusey, A. E., and C. Packer. 1987. The evolution of sex-biased dispersal in lions. *Behaviour* 101:275–310.

966 Pusey, A. E., and C. Packer. 1994. Infanticide in lions. In *Infanticide and Parental Care*, S. Parmigiani and F. S. vom Saal (eds.). Harwood Academic Press, Chur, Switzerland.

967 Pusey, A. E., and M. Wolf. 1996. Inbreeding avoidance in animals. *Trends in Ecology and Evolution* 11:201–206.

968 Pyle, D. W., and M. H. Gromko. 1978. Repeated mating by female *Drosophila melanogaster:* The adaptive importance. *Experientia* 34:449–450.

969 Quaintance, C. W. 1938. Content, meaning, and possible origin of male song in the brown towhee. *Condor* 40:97–101.

970 Queller, D. C. 1994. Extended parental care and the origin of eusociality. *Proceedings of the Royal Society of London B* 256:105–111.

971 Queller, D. C. 1995. The spaniels of St. Marx and the Panglossian paradox: A critique of a rhetorical programme. *Quarterly Review of Biology* 70:485–489.

972 Queller, D. C. 1996. The measurement and meaning of inclusive fitness. *Animal Behaviour* 51:229–232.

973 Queller, D. C., and J. E. Strassmann. 1998. Kin selection and social insects. *BioScience* 48:165–175.

974 Quinn, T. P., and A. H. Dittman. 1990. Pacific salmon migrations and homing: Mechanisms and adaptive significance. *Trends in Ecology and Evolution* 5:174–177.

975 Ralls, K., K. Brugger, and J. Ballou. 1979. Inbreeding and juvenile mortality in small populations of ungulates. *Science* 206:1101–1103.

976 Rand, A. S., M. J. Ryan, and W. Wilczynski. 1992. Signal redundancy and receiver permissiveness in acoustic mate recognition by the tungara frog *Physalaemus pustulosus*. *American Zoologist* 32:81–90.

977 Randall, J. A. 1991. Mating strategies of a nocturnal, desert rodent (*Dipodomys spectabilis*). *Behavioral Ecology and Sociobiology* 28:215–220.

978 Randall, J. A. 1994. Discrimination of foot-drumming signatures by kangaroo rats, *Dipodomys spectabilis*. *Animal Behaviour* 47:45–54.

979 Rätti, O., and P. Siikamäki. 1993. Female attraction behavior of radiotagged polyterritorial pied flycatchers. *Behaviour* 127:279–288.

980 Reboreda, J. C., N. S. Clayton, and A. Kacelnik. 1996. Species and sex differences in hippocampus size in parasitic and non-parasitic cowbirds. *NeuroReport* 7:505–508.

981 Reeve, H. K., and P. Nonacs. 1992. Social contracts in wasp societies. *Nature* 359:823–825.

982 Reeve, H. K., and P. W. Sherman. 1993. Adaptation and the goals of evolutionary research. *Quarterly Review of Biology* 68:1–32.

983 Reeve, H. K., D. F. Westneat, W. A. Noon, P. W. Sherman, and C. F. Aquadro. 1990. DNA "fingerprinting" reveals high levels of inbreeding in colonies of the eusocial naked mole-rat. *Proceedings of the National Academy of Sciences* 87:2496–2500.

984 Reichardt, A. K., and D. E. Wheeler. 1996. Multiple mating in the ant *Acromyrmex versicolor:* A case of female control. *Behavioral Ecology and Sociobiology* 38:219–226.

985 Reyer, H.-U. 1984. Investment and relatedness: A cost/benefit analysis of breeding and helping in the pied kingfisher. *Animal Behaviour* 32:1163–1178.

986 Reynolds, J. D. 1987. Mating system and nesting biology of the red-necked phalarope *Phalaropus lobatus*. *Ibis* 129:225–242.

987 Reynolds, J. D. 1996. Animal breeding systems. *Trends in Ecology and Evolution* 11:68–72.

988 Reynolds, J. D., M. A. Colwell, and F. Cooke. 1986. Sexual selection and spring arrival times of red-necked and Wilson's phalaropes. *Behavioral Ecology and Sociobiology* 18:303–310.

989 Reynolds, J. D., and M. R. Gross. 1990. Costs and benefits of female mate choice: Is there a lek paradox? *American Naturalist* 136:230–243.

990 Reynolds, J. D., M. R. Gross, and M. J. Coombs. 1993. Environmental conditions and male morphology determine alternative mating behaviour in Trinidadian guppies. *Animal Behaviour* 45:145–152.

991 Rice, W. R. 1996. Sexually antagonistic male adaptation triggered by experimental arrest of female evolution. *Nature* 381:232–234.

992 Richardson, H., and N. A. M. Verbeek. 1986. Diet selection and optimization by northwestern crows feeding on Japanese littleneck clams. *Ecology* 67:1219–1226.

993 Ridley, M. 1983. *The Explanation of Organic Diversity*. Oxford University Press, Oxford.

994 Ridley, M. 1988. Mating frequency and fecundity in insects. *Biological Reviews* 63:509–549.

995 Ridley, M. 1993. Clutch size and mating frequency in parasitic Hymenoptera. *American Naturalist* 142:893–910.

996 Riechert, S. E. 1993. Investigation of potential gene flow limitation of behavioral adaptation in an aridlands spider. *Behavioral Ecology and Sociobiology* 32:355–363.

997 Riechert, S. E., and A. V. Hedrick. 1990. Levels of predation and genetically based anti-predator behaviour in the spider, *Agelenopsis aperta*. *Animal Behaviour* 40:679–687.

998 Rintamäki, P. T., R. V. Alatalo, J.Höglund, and A. Lundberg. 1995. Male territoriality and female choice on black grouse leks. *Animal Behaviour* 49:759–767.

999 Rissing, S. W., G. B. Pollock, M. R. Higgins, R. H. Hager, and D. R. Smith. 1989. Foraging specialization without relatedness or dominance among co-founding ant queens. *Nature* 338:420–422.

1000 Robbins, R. K. 1981. The "false head" hypothesis: Predation and wing pattern variation of lycaenid butterflies. *American Naturalist* 118:770–775.

1001 Robert, D., J. Amoroso, and R. R. Hoy. 1992. The evolutionary convergence of hearing in a parasitoid fly and its cricket host. *Science* 258:1135–1137.

1002 Robertson, I. C. 1993. Nest intrusions, infanticide, and parental care in the burying beetle, *Nicrophorus orbicollis* (Coleoptera: Silphidae). *Journal of Zoology* 231:583–593.

1003 Robertson, R. M., and K. G. Pearson. 1982. A preparation for the intracellular analysis of neuronal activity during flight in the locust. *Journal of Comparative Physiology* 146:311–320.

1004 Robertson, R. M., and K. G. Pearson. 1984. Interneuronal organization in the flight system of the locust. *Journal of Insect Physiology* 30:95–101.

1005 Robinson, G. E., R. E. Page, Jr., C. Strambi, and A. Strambi. 1989. Hormonal and genetic control of behavioral integration in honey bee colonies. *Science* 246:109–112.

1006 Rodrigues, M. 1996. Song activity in the chiffchaff: Territorial defence or mate guarding? *Animal Behaviour* 51:709–716.

1007 Rodríguez-Gironés, M. A., P. A. Cotton, and A. Kacelnik. 1996. The evolution of begging: Signaling and sibling competition. *Proceedings of the National Academy of Sciences* 93:14637–14641.

1008 Roeder, K. D. 1963. *Nerve Cells and Insect Behavior.* Harvard University Press, Cambridge, MA.

1009 Roeder, K. D. 1965. Moths and ultrasound. *Scientific American* 212 (Apr):94–102.

1010 Roeder, K. D. 1970. Episodes in insect brains. *American Scientist* 58:378–389.

1011 Roeder, K. D., and A. E. Treat. 1961. The detection and evasion of bats by moths. *American Scientist* 49:135–148.

1012 Rohwer, S., and C. D. Spaw. 1988. Evolutionary lag versus bill-size constraints: A comparative study of the acceptance of cowbird eggs by old hosts. *Evolutionary Ecology* 2:27–36.

1013 Romey, W. L. 1995. Position preferences within groups: Do whirligigs select positions which balance feeding opportunities with predator avoidance? *Behavioral Ecology and Sociobiology* 37:195–200.

1014 Rood, J. P. 1990. Group size, survival, reproduction, and routes to breeding in dwarf mongooses. *Animal Behaviour* 39:566–572.

1015 Rose, N. A., C. J. Deutsch, and B. J. Le Boeuf. 1991. Sexual behavior of male northern elephant seals: III. The mounting of weaned pups. *Behaviour* 119:171–192.

1016 Rosenqvist, G. 1990. Male mate choice and female-female competition for mates in the pipefish *Nerophis ophidion. Animal Behaviour* 39:1110–1116.

1017 Rosenqvist, G., and K. Johansson. 1995. Male avoidance of parasitized females explained by direct benefits in a pipefish. *Animal Behaviour* 49:1039–1045.

1018 Ross, K. G. 1986. Kin selection and the problem of sperm utilization in social insects. *Nature* 323:798–800.

1019 Rothstein, S. I., D. A. Yokel, and R. C. Fleischer. 1988. The agonistic and sexual functions of vocalizations of male brown-headed cowbirds, *Molothrus ater. Animal Behaviour* 36:73–86.

1020 Rowe, L. 1994. The costs of mating and mate choice in water striders. *Animal Behaviour* 48:1049–1056.

1021 Rowe, L., G. Arnqvist, A. Sih, and J. Krupa. 1994. Sexual conflict and the evolutionary ecology of mating patterns: Water striders as a model system. *Trends in Ecology and Evolution* 9:289–293.

1022 Rowe, M. P., R. G. Coss, and D. H. Owings. 1986. Rattlesnake rattles and burrowing owl hisses: A case of acoustic Batesian mimicry. *Ethology* 72:53–71.

1023 Rowe, M. P., and D. H. Owings. 1996. Probing, assessment and management during interactions between ground squirrels (Rodentia: Sciuridae) and rattlesnakes (Squamata: Viperidae). 2: Cues afforded by rattlesnake rattling. *Ethology* 102:856–874.

1024 Rowley, I., and G. Chapman. 1986. Cross-fostering, imprinting, and learning in two sympatric species of cockatoos. *Behaviour* 96:1–16.

1025 Rowley, I., and E. Russell. 1990. "Philandering"—a mixed mating strategy in the splendid fairy-wren *Malurus splendens. Behavioral Ecology and Sociobiology* 27:431–438.

1026 Ruben, J. 1991. Reptilian physiology and the flight capacity of *Archaeopteryx. Evolution* 45:1–17.

1027 Rusak, B., H. A. Robertson, W. Wisden, and S. P. Hunt. 1990. Light pulses that shift rhythms induce gene expression in the suprachiasmatic nucleus. *Science* 238:1237–1240.

1028 Rutowski, R. L. 1997. Sexual dimorphism, mating systems, and ecology in butterflies. In *The Evolution of Mating Systems in Insects and Arachnids*, J. C. Choe and B. J. Crespi (eds.). Cambridge University Press, Cambridge.

1029 Rutowski, R. L., and G. W. Gilchrist. 1986. Copulation in *Colias eurytheme* (Lepidoptera: Pieridae): Patterns and frequency. *Journal of Zoology* 209:115–124.

1030 Ryan, M. J. 1983. Sexual selection and communication in a neotropical frog, *Physalaemus pustulosus. Evolution* 37:261–272.

1031 Ryan, M. J. 1985. *The Túngara Frog.* University of Chicago Press, Chicago.

1032 Ryan, M. J. 1996. Phylogenetics in behavior: Some cautions and expectations. In *Phylogenies and the Comparative Method in Animal Behavior,* E. P. Martins (ed.). Oxford University Press, Oxford.

1033 Ryan, M. J., J. H. Fox, W. Wilczynski, and A. S. Rand. 1990. Sexual selection for sensory exploitation in the frog *Physalaemus pustulosus. Nature* 343:66–67.

1034 Ryan, M. J., and A. Keddy-Hector. 1992. Directional patterns of female mate choice and the role of sensory biases. *American Naturalist* 139:s4–S35.

1035 Ryan, M. J., and A. S. Rand. 1993. Sexual selection and signal evolution: The ghosts of biases past. *Philosophical Transactions of the Royal Society B* 338:187–195.

1036 Ryan, M. J., M. D. Tuttle, and L. K. Taft. 1981. The costs and benefits of frog chorusing behavior. *Behavioral Ecology and Sociobiology* 8:273–278.

1037 Ryan, M. J., and W. Wilczynski. 1991. Evolution of intraspecific variation in the advertisement call of cricket frogs (*Acris crepitans,* Hylidae). *Biological Journal of the Linnean Society* 44:249–271.

1038 Rypstra, A. L., and R. S. Tirey. 1991. Prey size, prey perishability, and group foraging in a social spider. *Oecologia* 86:25–30.

1039 Sætre, G.-P., S. Dale, and T. Slagsvold. 1994. Females pied flycatchers prefer brightly coloured males. *Animal Behaviour* 48:1407–1416.

1040 Sætre, G.-P., T. Fossnes, and T. Slagsvold. 1995. Food provisioning in the pied flycatcher: Do females gain direct benefits from choosing bright-coloured males? *Journal of Animal Ecology* 64:21–30.

1041 Sahlins, M. 1976. *The Use and Abuse of Biology.* University of Michigan Press, Ann Arbor.

1042 Sakaluk, S. K. 1984. Male crickets feed females to ensure complete sperm transfer. *Science* 223:609–610.

1043 Sakaluk, S. K., and A.-K. Eggert. 1996. Female control of sperm transfer and intraspecific variation in sperm precedence: Antecedents to the evolution of a courtship food gift. *Evolution* 50:694–703.

1044 Salzman, C. D., K. H. Britten, and W. T. Newsome. 1990. Cortical microstimulation influences perceptual judgements of motion direction. *Nature* 346:174–177.

1045 Sandberg, R., and F. R. Moore. 1996. Migratory orientation of red-eyed vireos, *Vireo olivaceus,* in relation to energetic condition and ecological context. *Behavioral Ecology and Sociobiology* 39:1–10.

1046 Sargent, R. C. 1989. Allopaternal care in the fathead minnow, *Pimephales promelas:* Stepfathers discriminate against their adopted eggs. *Behavioral Ecology and Sociobiology* 25:379–386.

1047 Sargent, R. C., and M. R. Gross. 1985. Parental investment decision rules and the Concorde fallacy. *Behavioral Ecology and Sociobiology* 17:43–45.

1048 Sargent, T. D. 1976. *Legion of Night—The Underwing Moths.* University of Massachusetts Press, Amherst.

1049 Sato, T. 1994. Active accumulation of spawning substrate: A determinant of extreme polygyny in a shell-brooding cichlid fish. *Animal Behaviour* 48:669–678.

1050 Schaller, G. B. 1964. *The Year of the Gorilla.* University of Chicago Press, Chicago.

1051 Schaller, G. B. 1972. *The Serengeti Lion.* University of Chicago Press, Chicago.

1052 Scheel, D. 1993. Watching for lions in the grass: The usefulness of scanning and its effects during hunts. *Animal Behaviour* 46:695–704.

1053 Scheel, D., and C. Packer. 1991. Group hunting behaviour of lions: A search for cooperation. *Animal Behaviour* 41:711–722.

1054 Schlenoff, D. H. 1985. The startle response of blue jays to *Catocala* (Lepidoptera:

Noctuidae) prey models. *Animal Behaviour* 33:1057–1067.

1055 Schluter, D. 1994. Experimental evidence that competition promotes divergence in adaptive radiation. *Science* 266:798–801.

1056 Schluter, D., and J. D. McPhail. 1993. Character displacement and replicate adaptive radiation. *Trends in Ecology and Evolution* 8:197–200.

1057 Schmidt-Koenig, K., and J. U. Ganzhorn. 1991. On the problem of bird navigation. *Perspectives in Ethology* 9:261–283.

1058 Schneider, J. M., and Y. Lubin. 1996. Infanticidal male eresid spiders. *Nature* 381:655–656.

1059 Schoener, T. W., and D. A. Spiller. 1992. Stabilimenta characteristics of the spider *Argiope argentata* on small islands: Support of the predator-defense hypothesis. *Behavioral Ecology and Sociobiology* 31:309–318.

1060 Schwabl, H. 1983. Auspragung und Bedeutung des Teilzugverhaltnes einer sudwestdeutschen Population der Amsel *Turdus merula*. *Journal für Ornithologie* 124:101–116.

1061 Schwabl, H. 1993. Yolk is a source of maternal testosterone for developing birds. *Proceedings of the National Academy of Sciences* 90:11446–11450.

1062 Schwabl, H., D. W. Mock, and J. A. Gieg. 1997. A hormonal mechanism for parental favouritism. *Nature*.

1063 Schwagmeyer, P. L. 1979. The Bruce effect: An evaluation of male/female advantages. *American Naturalist* 114:932–938.

1064 Schwagmeyer, P. L. 1988. Scramble-competition polygyny in an asocial mammal: Male mobility and mating success. *American Naturalist* 131:885–892.

1065 Schwagmeyer, P. L. 1994. Competitive mate searching in the 13-lined ground squirrel: Potential roles of spatial memory? *Ethology* 98:265–276.

1066 Schwagmeyer, P. L. 1995. Searching today for tomorrow's mates. *Animal Behaviour* 50:759–767.

1067 Schwartz, J. J., S. J. Ressel, and C. R. Bevier. 1995. Carbohydrate and calling: Depletion of muscle glycogen and the chorusing dynamics of the Neotropical treefrog *Hyla microcephala*.

1068 Scott, M. P. 1989. Male parental care and reproductive success in the burying beetle, *Nicrophorus orbicollis*. *Journal of Insect Behavior* 2:133–137.

Behavioral Ecology and Sociobiology 37:125–135.

1069 Scott, M. P. 1990. Brood guarding and the evolution of male parental care in burying beetles. *Behavioral Ecology and Sociobiology* 26:31–40.

1070 Sealy, S. G. 1995. Burial of cowbird eggs by parasitized yellow warblers: An empirical and experimental study. *Animal Behaviour* 49:877–889.

1071 Searcy, W. A. 1992. Measuring response of female birds to male songs. In *Playback and Studies of Animal Communication*, P. K. McGregor (ed.). Plenum Press, New York.

1072 Seeley, T. D. 1977. Measurement of nest cavity volume by the honey bee (*Apis mellifera*). *Behavioral Ecology and Sociobiology* 2:201–227.

1073 Seeley, T. D. 1982. Adaptive significance of the age polyethism schedule in honeybee colonies. *Behavioral Ecology and Sociobiology* 11:287–293.

1074 Seeley, T. D. 1985. *Honeybee Ecology: A Study of Adaptation in Social Life*. Princeton University Press, Princeton, NJ.

1075 Seeley, T. D., R. H. Seeley, and P. Akratanakul. 1982. Colony defense strategies of the honeybees in Thailand. *Ecological Monographs* 52:43–63.

1076 Seeley, T. D., and P. K. Visscher. 1988. Assessing the benefits of cooperation in honeybee foraging: Search costs, forage quality, and competitive ability. *Behavioral Ecology and Sociobiology* 22:229–237.

1077 Seger, J. 1991. Cooperation and conflict in social insects. In *Behavioural Ecology: An Evolutionary Approach* (3rd edition), J. R. Krebs and N. B. Davies (eds.). Blackwell Scientific, Oxford.

1078 Semel, B., and P. W. Sherman. 1986. Dynamics of nest parasitism in wood ducks. *Auk* 103:813–816.

1079 Sereno, P. C., and R. Chenggang. 1992. Early evolution of avian flight and perching: New evidence from the lower Cretaceous of China. *Science* 255:845–848.

1080 Shaywitz, B. A., S. E. Shaywitz, K. R. Pugh, R. T. Constable, P. Skudlarski, R. F. Fulbright, R. A. Bronen, J. M. Fletcher, D. P. Shankweller, L. Katz, and J. C. Gore. 1995. Sex differences in the functional organization of the brain for language. *Nature* 373:607–609.

1081 Sherman, P. M. 1994. The orb-web: An energetic and behavioural estimator of a spider's dynamic foraging and reproductive strategies. *Animal Behaviour* 48:19–34.

1082 Sherman, P. W. 1977. Nepotism and the evolution of alarm calls. *Science* 197:1246–1253.

1083 Sherman, P. W. 1981. Kinship, demography and Belding's ground squirrel nepotism. *Behavioral Ecology and Sociobiology* 8:251–259.

1084 Sherman, P. W. 1985. Alarm calls of Belding's ground squirrels to aerial predators: Nepotism or self-preservation? *Behavioral Ecology and Sociobiology* 17:313–323.

1085 Sherman, P. W. 1988. The levels of analysis. *Animal Behaviour* 36:616–618.

1086 Sherman, P. W. 1989. Mate guarding as paternity insurance in Idaho ground squirrels. *Nature* 338:418–420.

1087 Sherman, P. W., and W. G. Holmes. 1985. Kin recognition: Issues and evidence. In *Experimental Behavioral Ecology and Sociobiology*, B. Hölldobler and M. Lindauer (eds.). G. Fischer Verlag, Stuttgart.

1088 Sherman, P. W., J. U. M. Jarvis, and R. D. Alexander (eds.) 1991. *The Biology of the Naked Mole-Rat*. Princeton University Press, Princeton, NJ.

1089 Sherry, D. F. 1984. Food storage by black-capped chickadees: Memory of the location and contents of caches. *Animal Behaviour* 32:451–464.

1090 Sherry, D. F., M. R. L. Forbes, M. Kjurgel, and G. O. Ivy. 1993. Females have a larger hippocampus than males in the brood-parasitic brown-headed cowbird. *Proceedings of the National Academy of Sciences* 90:7839–7843.

1091 Shields, O. 1967. Hilltopping. *Journal of Research on the Lepidoptera* 6:69–178.

1092 Shields, W. M. 1983. Optimal inbreeding and the evolution of philopatry. In *The Ecology of Animal Movement*, I. R. Swingland and P. J. Greenwood (eds.). Clarendon Press, Oxford.

1093 Shields, W. M. 1984. Barn swallow mobbing: Self-defence, collateral kin defence, group defence, or parental care? *Animal Behaviour* 32:132–148.

1094 Shipman, P. 1997. Birds do it—did dinosaurs? *American Scientist* 153(2067):26–31.

1095 Short, R. V., and E. Balaban (eds.). 1994. *The Differences between the Sexes*. Cambridge University Press, Cambridge.

1096 Shuster, S. M. 1989. Male alternative reproductive strategies in a marine isopod crustacean (*Paracerceis sculpta*): The use of genetic markers to measure differences in the fertilization success among a-, b-, and g-males. *Evolution* 43:1683–1698.

1097 Shuster, S. M. 1992. The reproductive behaviour of a-, b-, and g-male morphs in *Paracerceis sculpta:* A marine isopod crustacean. *Behaviour* 121:231–258.

1098 Shuster, S. M., and C. Sassaman. 1997. Genetic interaction between male mating strategy and sex ratio in a marine isopod. *Nature* 388:373–377.

1099 Shuster, S. M., and M. J. Wade. 1991. Equal mating success among male reproductive strategies in a marine isopod. *Nature* 350:608–610.

1100 Shutler, D., and P. J. Weatherhead. 1991. Owner and floater red-winged blackbirds: Determinants of status. *Behavioral Ecology and Sociobiology* 28:235–242.

1101 Siegel, R. W., J. C. Hall, D. A. Gailey, and C. P. Kyriacou. 1984. Genetic elements of courtship in *Drosophila:* Mosaics and learning mutants. *Behavior Genetics* 14:383–410.

1102 Sikkel, P. C. 1994. Honey, I ate the kids. *Natural History* 103(12):46–51.

1103 Silk, J. B. 1980. Adoption and kinship in Oceania. *American Anthropologist* 82:799–820.

1104 Silk, J. B. 1990. Human adoption in evolutionary perspective. *Human Nature* 1:25–52.

1105 Sillén-Tullberg, B., and A. P. Møller. 1993. The relationship between concealed ovulation and mating systems in anthropoid primates: A phylogenetic analysis. *American Naturalist* 141:1–25.

1106 Silva, A. J., R. Paylor, J. M. Wehner, and S. Tonegawa. 1992. Impaired spatial learning in alpha-calcium-calmodulin kinase II mutant mice. *Science* 257:206–211.

1107 Silverin, B. 1980. Effects of long-acting testosterone treatment on free-living pied flycatchers, *Ficedula hypoleuca*, during the breeding period. *Animal Behaviour* 28:906–912.

1108 Simmons, L. W. 1990. Nuptial feeding in tettigoniids: Male costs and the rates of fecundity increase. *Behavioral Ecology and Sociobiology* 27:43–47.

1109 Simmons, L. W. 1995. Relative parental expenditure, potential reproductive rates, and the control of sexual selection in katydids. *American Naturalist* 145:797–808.

1110 Simmons, L. W., T. Llorens, M. Schinzig, D. Hosken, and M. Craig. 1994. Sperm competition selects for male mate choice and protandry in the bushcricket, *Requena verticalis* (Orthoptera: Tettigoniidae). *Animal Behaviour* 47:117–122.

1111 Simmons, L. W., and M. G. Ritchie. 1996. Symmetry in the songs of crickets. *Proceedings of the Royal Society of London B* 263:305–311.

1112 Simmons, R. E., and L. Scheepers. 1996. Winning by a neck: Sexual selection in the evolution of giraffe. *American Naturalist* 148:771–786.

1113 Singer, P. 1981. *The Expanding Circle: Ethics and Sociobiology.* Farrar, Straus, and Giroux, New York.

1114 Singh, D. 1993. Adaptive significance of female physical attractiveness: Role of waist-to-hip ratio. *Journal of Personality and Social Psychology* 65:293–307.

1115 Singh, D., and R. K. Young. 1995. Body weight, waist-to-hip ratio, breasts, and hips: Role in judgements of female attractiveness and desirability for relationships. *Ethology and Sociobiology* 16:483–508.

1116 Siva-Jothy, M. T., and R. E. Hooper. 1995. The disposition and genetic diversity of stored sperm in females of the damselfly *Calopteryx splendens xanthostoma* (Charpentier). *Proceedings of the Royal Society of London B* 259:313–318.

1117 Skinner, B. F. 1966. Operant behavior. In *Operant Behavior,* W. K. Honig (ed.). Appleton-Century-Crofts, New York.

1118 Skulason, S., and T. B. Smith. 1995. Resource polymorphisms in vertebrates. *Trends in Ecology and Evolution* 10:366–370.

1119 Skutelsky, O. 1996. Predation risk and state-dependent foraging in scorpions: Effects of moonlight on foraging in the scorpion *Buthus occitanus. Animal Behaviour* 52:49–57.

1120 Slagsvold, T., and S. Dale. 1994. Why do female pied flycatchers mate with already mated males: Deception or restricted mate sampling? *Behavioral Ecology and Sociobiology* 34:239–250.

1121 Slobodchikoff, C. N. 1978. Experimental studies of tenebrionid beetle predation by skunks. *Behaviour* 66:313–322.

1122 Slobodchikoff, C. N., J. Kiriazis, C. Fischer, and E. Creef. 1991. Semantic information distinguishing individual predators in the alarm calls of Gunnison's prairie dogs. *Animal Behaviour* 42:713–719.

1123 Smedley, S. R., and T. Eisner. 1996. Sodium: A male moth's gift to its offspring. *Proceedings of the National Academy of Sciences* 93:809–813.

1124 Smith, H. G., and R. Montgomerie. 1991. Nestling American robins compete with siblings by begging. *Behavioral Ecology and Sociobiology* 29:307–312.

1125 Smith, H. G., and R. Montgomerie. 1991. Sexual selection and the tail ornaments of North American barn swallows. *Behavioral Ecology and Sociobiology* 28:195–202.

1126 Smith, M. S., B. J. Kish, and C. B. Crawford. 1987. Inheritance of wealth as human kin investment. *Ethology and Sociobiology* 8:171–182.

1127 Smith, R. L. 1979. Paternity assurance and altered roles in the mating behavior of a giant water bug *Abedus herberti* (Heteroptera: Belostomatidae). *Animal Behaviour* 27:716–728.

1128 Smith, R. L. 1984. Human sperm competition. In *Sperm Competition and the Evolution of Animal Mating Systems,* R. L. Smith (ed.). Academic Press, New York.

1129 Smith, R. L. 1997. Evolution of paternal care in giant water bugs (Heteroptera: Belostomatidae). In *Social Competition and Cooperation among Insects and Arachnids,* II. *Evolution of Sociality,* J. C. Choe and B. J. Crespi (eds.). Cambridge University Press, Cambridge.

1130 Smith, S. M. 1978. The "underworld" in a territorial species: Adaptive strategy for floaters. *American Naturalist* 112:571–582.

1131 Smuts, B. B., and R. W. Smuts. 1993. Male aggression and sexual coercion of females in nonhuman primates and other mammals: Evidence and theoretical implications. *Advances in the Study of Behavior* 22:1–63.

1132 Snow, D. W. 1956. Courtship ritual: The dance of the manakins. *Animal Kingdom* 59:86–91.

1133 Soler, M., J. J. Soler, J. G. Martinez, and A. P. Møller. 1995. Magpie host manipulation by great spotted cuckoos: Evidence for an avian Mafia? *Evolution* 49:770–775.

1134 Sommer, V. 1987. Infanticide among free-ranging langurs (*Presbytis entellus*) at Jodhpur (Rajasthan/India): Recent observations and a reconsideration of hypotheses. *Primates* 28:163–197.

1135 Sommer, V. 1994. Infanticide among the langurs of Jodhpur: Testing the sexual selection hypothesis with a long-term record. In *Infanticide and Parental Care*, S. Parmigiani and F. S. vom Saal (eds.). Harwood Academic Press, Chur, Switzerland.

1136 Srygley, R. B., and P. Chai. 1990. Predation and the elevation of thoracic temperature in brightly colored, Neotropical butterflies. *American Naturalist* 135:766–787.

1137 Stacey, P. B., and J. D. Ligon. 1987. Territory quality and dispersal options in the acorn woodpecker, and a challenge to the habitat-saturation model of cooperative breeding. *American Naturalist* 130:654–676.

1138 Stander, P. E. 1992. Cooperative hunting in lions: The role of the individual. *Behavioral Ecology and Sociobiology* 29:445–454.

1139 Stanford, C. B. 1995. Chimpanzee hunting behavior. *American Scientist* 83:256–261.

1140 St. Clair, C. C., J. R. Waas, R. C. St. Clair, and P. T. Boag. 1995. Unfit mothers? Maternal infanticide in royal penguins. *Animal Behaviour* 50:1177–1185.

1141 Stehle, J. H., N. S. Foulkes, C. A. Molina, V. Simmonneaux, P. Pévet, and P. Sassone-Corsi. 1993. Adrenergic signals direct rhythmic expression of transcriptional repressor CREM in the pineal gland. *Nature* 365:314–320.

1142 Stein, Z., M. Susser, G. Saenger, and F. Marolla. 1972. Nutrition and mental performance. *Science* 178:708–713.

1143 Stenmark, G., T. Slagsvold, and J. T. Lifjeld. 1988. Polygyny in the pied flycatcher, *Ficedula hypoleuca:* A test of the deception hypothesis. *Animal Behaviour* 36:1646–1657.

1144 Stephens, D. W., and J. R. Krebs. 1986. *Foraging Theory.* Princeton University Press, Princeton, NJ.

1145 Stephens, M. 1982. Mate takeover and possible infanticide by a female northern jacana (*Jacana spinosa*). *Animal Behaviour* 30:1253–1254.

1146 Stern, D. L., and W. A. Foster. 1996. The evolution of soldiers in aphids. *Biological Reviews* 71:27–80.

1147 Stern, D. L., and W. A. Foster. 1997. The evolution of sociality in aphids: A clone's-eye-view. In *Social Competition and Cooperation in Insects and Arachnids:* II. *Evolution of Sociality,* J. Choe and B. Crespi (eds.). Princeton University Press, Princeton.

1148 Stewart, K. J. 1987. Spotted hyaenas: The importance of being dominant. *Trends in Ecology and Evolution* 2:88–89.

1149 Stoddard, P. K., M. D. Beecher, C. L. Horning, and M. S. Willis. 1990. Strong neighbor-stranger discrimination in song sparrows. *Condor* 92:1051–1056.

1150 Stone, G. N. 1995. Female foraging responses to sexual harassment in the solitary bee *Anthophora plumipes. Animal Behaviour* 50:405–412.

1151 Stoutamire, W. P. 1974. Australian terrestrial orchids, thynnid wasps and pseudocopulation. *American Orchid Society Bulletin* 43:13–18.

1152 Strassmann, J. E. 1981. Wasp reproduction and kin selection—reproductive competition and dominance hierarchies among *Polistes annularis* foundresses. *Florida Entomologist* 64:74–88.

1153 Strassmann, J. E., C. R. Hughes, D. C. Queller, S. Turillazzi, R. Cervo, S. K. Davis, and K. F. Goodnight. 1989. Genetic relatedness in primitively eusocial wasps. *Nature* 342:268–269.

1154 Strum, S. C. 1987. *Almost Human.* W. W. Norton, New York.

1155 Stumpner, A., and R. Lakes-Harlan. 1996. Auditory interneurons in a hearing fly (*Therobia leonidei*, Ormiini, Tachinidae, Diptera). *Journal of Comparative Physiology A* 178:227–233.

1156 Suga, N. 1990. Biosonar and neural computation in bats. *Scientific American* 262 (June):34–41.

1157 Sugiyama, Y. 1984. Proximate factors of infanticide among langurs at Dharwar: A reply to Boggess. In *Infanticide: Comparative and Evolutionary Perspectives*, G. Hausfater and S. B. Hrdy (eds.). Aldine, Chicago.

1158 Sullivan, B. K. 1983. Sexual selection in Woodhouse's toad (*Bufo woodhousei*). II. Female choice. *Animal Behaviour* 31:1011–1017.

1159 Sullivan, K. A. 1988. Age-specific profitability and prey choice. *Animal Behaviour* 36:613–615.

1160 Sundstrom, L. 1994. Sex ratio bias, relatedness asymmetry and queen mating frequency in ants. *Nature* 367:266–268.

1161 Surlykke, A., and J. H. Fullard. 1989. Hearing of the Australian whistling moth, *Hecatesia thyridion*. *Naturwissenschaften* 76:132–134.

1162 Svärd, L., and J. N. McNeil. 1994. Female benefit, male risk: Polyandry in the true armyworm *Pseudaletia unipuncta*. *Behavioral Ecology and Sociobiology* 35:319–326.

1163 Swaddle, J. P., and I. C. Cuthill. 1994. Preference for symmetric males by female zebra finches. *Nature* 367:165–66.

1164 Swaisgood, R. R., D. H. Owings, and M. P. Rowe. In press. Probing, assessment and management during interactions between ground squirrels (Rodentia: Sciuridae) and rattlesnakes (Squamata: Viperidae). 3: Exploitation of rattling. *Ethology*.

1165 Swan, L. W. 1970. Goose of the Himalayas. *Natural History* 79 (Dec):68–75.

1166 Sweeney, B. W., and R. L. Vannote. 1982. Population synchrony in mayflies: A predator satiation hypothesis. *Evolution* 36:810–821.

1167 Symons, D. 1979. *The Evolution of Human Sexuality*. Oxford University Press, New York.

1168 Székely, T., and J. D. Reynolds. 1995. Evolutionary transitions in parental care in shorebirds. *Proceedings of the Royal Society of London B* 262:57–64.

1169 Takahashi, J. S., and M. Hoffman. 1995. Molecular biological clocks. *American Scientist* 83:158–165.

1170 Temeles, E. J. 1987. The relative importance of prey availability and intruder pressure in feeding territory size regulation by harriers, *Circus cyaneus*. *Oecologia* 74:286–297.

1171 Temeles, E. J. 1994. The role of neighbours in territorial systems: When are they "dear enemies"? *Animal Behaviour* 47:339–350.

1172 Temrin, H. 1989. Female pairing options in polyterritorial wood warblers *Phylloscopus sibilatrix:* Are females deceived? *Animal Behaviour* 37:579–586.

1173 Temrin, H., and A. Arak. 1989. Polyterritoriality and deception in passerine birds. *Trends in Ecology and Evolution* 4:106–108.

1174 Temrin, H., and B. S. Tullberg. 1995. A phylogenetic analysis of the evolution of avian mating systems in relation to altricial and precocial young. *Behavioral Ecology* 6:296–307.

1175 Thielcke, G. 1973. On the origin and divergence of learned signals (songs) in isolated populations. *Ibis* 15:511–516.

1176 Thompson, C. W., N. Hillgarth, M. Leu, and H. E. McClure. 1997. High parasite load in house finches (*Carpodacus mexicanus*) is correlated with reduced expression of a sexually selected trait. *American Naturalist* 149:270–294.

1177 Thompson, D. B. A. 1986. The economics of kleptoparasitism: Optimal foraging, host and prey selection by gulls. *Animal Behaviour* 34:1189–1205.

1178 Thornhill, R. 1976. Sexual selection and nuptial feeding behavior in *Bittacus apicalis* (Insecta: Mecoptera). *American Naturalist* 110:529–548.

1179 Thornhill, R. 1981. *Panorpa* (Mecoptera: Panorpidae) scorpionflies: Systems for understanding resource-defense polygyny and alternative male reproductive efforts. *Annual Review of Ecology and Systematics* 12:355–386.

1180 Thornhill, R. 1992. Female preference for the pheromone of males with low fluctuating asymmetry in the Japanese scorpionfly (*Panorpa japonica*: Mecoptera). *Behavioral Ecology* 3:277–283.

1181 Thornhill, R., and J. Alcock. 1983. *The Evolution of Insect Mating Systems*. Harvard University Press, Cambridge, MA.

1182 Thornhill, R., and S. W. Gangestad. 1996. The evolution of human sexuality. *Trends in Ecology and Evolution* 11:98–102.

1183 Thornhill, R., S. W. Gangestad, and R. Comer. 1995. Human female orgasm and mate fluctuating asymmetry. *Animal Behaviour* 50:1601–1615.

1184 Thornhill, R., and N. W. Thornhill. 1983. Human rape: An evolutionary analysis. *Ethology and Sociobiology* 4:137–173.

1185 Thornhill, R., and N. W. Thornhill. 1992. The evolutionary psychology of men's coercive sexuality. *Behavioral and Brain Sciences* 15:363–421.

1186 Thorpe, W. H. 1979. *The Origins and Rise of Ethology: The Science of the Natural Behaviour of Animals.* Praeger, New York.

1187 Tinbergen, L. 1960. The natural control of insects in pinewoods. 1. Factors influencing the intensity of predation by songbirds. *Archives Neerlandaises de Zoologie* 13:265–343.

1188 Tinbergen, N. 1951. *The Study of Instinct.* Oxford University Press, New York.

1189 Tinbergen, N. 1958. *Curious Naturalists.* Doubleday, Garden City, New York.

1190 Tinbergen, N. 1960. *The Herring Gull's World.* Doubleday, Garden City, New York.

1191 Tinbergen, N. 1963. On aims and methods of ethology. *Zeitschrift für Tierpsychologie* 20:410–433.

1192 Tinbergen, N. 1963. The shell menace. *Natural History* 72 (Aug):28–35.

1193 Tinbergen, N., and A. C. Perdeck. 1950. On the stimulus situations releasing the begging response in the newly hatched herring gull (*Larus argentatus* Pont.). *Behaviour* 3:1–39.

1194 Todd, I. A., and R. J. Cowie. 1990. Measuring the risk of predation in an energy currency: Field experiments with foraging blue tits, *Parus caeruleus. Animal Behaviour* 40:112–117.

1195 Tokarz, R. R., and D. Crews. 1981. Effects of prostaglandins on sexual receptivity in the female lizard, *Anolis carolinensis. Endocrinology* 109:451–457.

1196 Tompkins, L., and J. C. Hall. 1983. Identification of brain sites controlling female receptivity in mosaics of *Drosophila melanogaster. Genetics* 103:179–195.

1197 Townsend, J. M. 1989. Mate selection criteria: A pilot study. *Ethology and Sociobiology* 10:241–253.

1198 Trainer, J. B., and D. B. McDonald. 1995. Singing performance, frequency matching and courtship success of long-tailed manakins (*Chiroxiphia linearis*). *Behavioral Ecology and Sociobiology* 37:249–254.

1199 Trivers, R. L. 1971. The evolution of reciprocal altruism. *Quarterly Review of Biology* 46:35–57.

1200 Trivers, R. L. 1972. Parental investment and sexual selection. In *Sexual Selection and the Descent of Man*, B. Campbell (ed.). Aldine, Chicago.

1201 Trivers, R. L. 1985. *Social Evolution.* Benjamin/Cummings Publishing, Menlo Park, CA.

1202 Trivers, R. L., and H. Hare. 1976. Haplodiploidy and the evolution of the social insects. *Science* 191:249–263.

1203 Truman, J., and L. Riddiford. 1970. Neuroendocrine control of ecdysis in silkmoths. *Science* 167:1624–1626.

1204 Trumbo, S. T. 1990. Reproductive benefits of infanticide in a biparental burying beetle *Nicrophorus orbicollis. Behavioral Ecology and Sociobiology* 27:269–274.

1205 Trumbo, S. T., and A.-K. Eggert. 1994. Beyond monogamy: Territory quality influences sexual advertisement in male burying beetles. *Animal Behaviour* 48:1043–1047.

1206 Tschinkel, W. R., E. S. Adams, and T. Macom. 1995. Territory area and colony size in the fire ant *Solenopsis invicta. Journal of Animal Ecology* 64:473–480.

1207 Tso, I.-M. 1996. Stabilimentum of the garden spider *Argiope trifasciata:* A prey attractant. *Animal Behaviour* 52:183–191.

1208 Tulp, I., S. McChesney, and P. De Goeij. 1994. Migratory departures of waders from north-western Australia: Behaviour, timing, and possible migration routes. *Ardea* 82:201–221.

1209 Tuttle, E. M., S. Pruett-Jones, and M. S. Webster. 1996. Cloacal protuberances and extreme sperm production in Australian fairy-wrens. *Proceedings of the Royal Society of London B* 263:1359–1364.

1210 Tuttle, R. H. 1969. Knuckle-walking and the problem of human origins. *Science* 166:953–961.

1211 Uetz, G. W. 1992. Foraging strategies of spiders. *Trends in Ecology and Evolution* 7:155–159.

1212 Urquhart, F. A. 1960. *The Monarch Butterfly.* University of Toronto Press, Toronto.

1213 van Tets, G. F. 1965. A comparative study of some social communication patterns in the Pelecaniformes. *Ornithological Monographs* 1:1–88.

1214 Veiga, J. P. 1990. Infanticide by male and female house sparrows. *Animal Behaviour* 39:496–502.

1215 Vetter, R. S. 1980. Defensive behavior of the black widow spider *Latrodectus hesperus* (Araneae: Theridiidae). *Behavioral Ecology and Sociobiology* 7:187–193.

1216 Viitala, J., E. Korpimäki, P. Palokangas, and M. Koivula. 1995. Attraction of kestrels to vole scent marks visible in ultraviolet light. *Nature* 373:423–425.

1217 Vincent, A. J. 1994. Operational sex ratios in seahorses. *Behaviour* 128:153–167.

1218 Vincent, A. J. 1994. Seahorses exhibit conventional sex roles in mating competition, despite male pregnancy. *Behaviour* 128:135–151.

1219 Vincent, A. J., I. Ahnesjö, A. Berglund, and G. Rosenqvist. 1992. Pipefishes and seahorses: Are they sex role reversed? *Trends in Ecology and Evolution* 7:237–241.

1220 Vincent, A. J. ., and L. M. Sadler. 1995. Faithful pair bonds in wild seahorses, *Hippocampus whitei*. *Animal Behaviour* 50:1557–1569.

1221 Vines, G. 1992. Obscure origins of desire. *New Scientist* 136(1849):2–8.

1222 Vining, D. R., Jr. 1986. Social vs. reproductive success: The central theoretical problem of human sociobiology. *Behavioral and Brain Science* 9:167–186.

1223 Vitaterna, M. H., D. P. King, A.-M. Chang, J. M. Kornhauser, P. L. Lowerty, J. D. McDonald, W. F. Dove, L. H. Pinto, F. W. Turek, and J. S. Takahashi. 1994. Mutagenesis and mapping of a mouse gene, *Clock*, essential for circadian behavior. *Science* 264:719–725.

1224 vom Saal, F. S., W. M. Grant, C. W. McMullen, and K. S. Laves. 1983. High fetal estrogen concentrations: Correlation with increased adult sexual activity and decreased aggression in male mice. *Science* 220:1306–1309.

1225 von Frisch, K. 1956. *The Dancing Bees.* Harcourt Brace Jovanovich, New York.

1226 von Frisch, K. 1967. *The Dance Language and Orientation of Bees.* Harvard University Press, Cambridge, MA.

1227 Vos, D. R. 1995. The role of sexual imprinting for sex recognition in zebra finches: A difference between males and females. *Animal Behaviour* 50:645–653.

1228 Waage, J. K. 1973. Reproductive behavior and its relation to territoriality in *Calopteryx maculata* (Beauvois) (Odonata: Calopterygidae). *Behaviour* 47:240–256.

1229 Waage, J. K. 1979. Dual function of the damselfly penis: Sperm removal and transfer. *Science* 203:916–918.

1230 Waage, J. K. 1997. Parental investment—minding the kids or keeping control? In *Feminism and Evolutionary Biology: Boundaries, Interactions, and Frontiers*, P. Gowaty (ed.). Chapman and Hall, New York.

1231 Wagner, R. H. 1992. Mate guarding by monogamous female razorbills. *Animal Behaviour* 44:533–538.

1232 Wagner, W. E., Jr. 1992. Deceptive or honest signalling of fighting ability? A test of alternative hypotheses for the function of changes in call dominant frequency by male cricket frogs. *Animal Behaviour* 44:449–462.

1233 Walcott, C. 1972. Bird navigation. *Natural History* 81 (June):32–43.

1234 Wallraff, H. G. 1983. Relevance of atmospheric odours and geomagnetic field to pigeon navigation: What is the "map" basis? *Comparative and Biochemical Physiology* 76A:643–663.

1235 Walter, H. 1979. *Eleonora's Falcon, Adaptations to Prey and Habitat in a Social Raptor.* University of Chicago Press, Chicago.

1236 Ward, P., and A. Zahavi. 1973. The importance of certain assemblages of birds as "information-centres" for food finding. *Ibis* 115:517–534.

1237 Waser, P. M., and W. T. Jones. 1983. Natal philopatry among solitary mammals. *Quarterly Review of Biology* 58:355–390.

1238 Watson, P. J. 1991. Multiple paternity as genetic bet-hedging in female Sierra dome spiders, *Linyphia litigiosa*. *Animal Behaviour* 41:343–360.

1239 Watson, P. J. 1993. Foraging advantage of polyandry for female Sierra dome spiders (*Linyphia litigiosa*: Linyphiidae) and assessment of alternative direct benefit hypotheses. *American Naturalist* 141:440–465.

1240 Watson, P. J., and R. Thornhill. 1994. Fluctuating asymmetry and sexual selection. *Trends in Ecology and Evolution* 9:21–25.

1241 Weatherhead, P. J., R. Montgomerie, H. L. Gibbs, and P. T. Boag. 1994. The cost of extra-pair fertilizations to female red-winged blackbirds. *Proceedings of the Royal Society of London B* 258:315–320.

1242 Weathers, W. W., R. S. Seymour, and R. V. Baudinette. 1993. Energetics of mound-tending behaviour in the malleefowl, *Leipoa ocellata* (Megapodiidae). *Animal Behaviour* 45:33–341.

1243 Weathers, W. W., and K. A. Sullivan. 1989. Juvenile foraging proficiency, parental effort, and avian reproductive success. *Ecological Monographs* 59:223–246.

1244 Webster, M. S. 1994. Female-defence polygyny in a Neotropical bird, the Montezuma oropendula. *Animal Behaviour* 48:779–794.

1245 Wedekind, C. 1994. Mate choice and maternal selection for specific parasite resistances before, during, and after fertilization. *Philosophical Transactions of the Royal Society of London B* 346:303–311.

1246 Wedell, N., and A. Arak. 1989. The wartbiter spermatophore and its effect on female reproductive output (Orthoptera: Tettigoniidae, *Decticus verrucivorus. Behavioral Ecology and Sociobiology* 24:117–125.

1247 Wehner, R. 1983. Celestial and terrestrial navigation: Human strategies—insect strategies. In *Neuroethology and Behavioral Physiology*, F. Huber and H. Markl (eds.). Springer-Verlag, Berlin.

1248 Wehner, R. 1987. "Matched filters"—neural models of the external world. *Journal of Comparative Physiology A* 161:511–531.

1249 Wehner, R., M. Lehrer, and W. R. Harvey (eds.). 1996. Navigation: Migration and homing. *Journal of Experimental Biology* 199:1–261.

1250 Wells, K. D. 1977. Territoriality and male mating success in the green frog (*Rana clamitans*). *Ecology* 58:750–762.

1251 Wells, K. D., and T. L. Taigen. 1986. The effect of social interactions on calling energetics in the gray treefrog (*Hyla versicolor*). *Behavioral Ecology and Sociobiology* 19:9–18.

1252 Welty, J. 1975. *The Life of Birds* (2nd edition). Saunders, Philadelphia.

1253 Wenner, A. M., and P. Wells. 1990. *Anatomy of a Controversy.* Columbia University Press, New York.

1254 Werren, J. H., M. R. Gross, and R. Shine. 1980. Paternity and the evolution of male parental care. *Journal of Theoretical Biology* 82:619–631.

1255 West, M. J., and A. P. King. 1985. Social guidance of vocal learning by female cowbirds: Validating its functional significance. *Zeitschrift für Tierpsychologie* 70:225–235.

1256 West, M. J., and A. P. King. 1990. Mozart's starling. *American Scientist* 78:106–114.

1257 West, M. J., A. P. King, and D. H. Eastzer. 1981. The cowbird: Reflections on development from an unlikely source. *American Scientist* 69:56–66.

1258 Westcott, D. 1992. Inter- and intra-sexual selection: The role of song in a lek mating system. *Animal Behaviour* 44:695–703.

1259 Westcott, D. 1994. Leks of leks: A role for hotspots in lek evolution? *Proceedings of the Royal Society of London B* 258:281–286.

1260 West-Eberhard, M. J. 1975. The evolution of social behavior by kin selection. *Quarterly Review of Biology* 50:1–33.

1261 West-Eberhard, M. J. 1979. Sexual selection, social competition, and evolution. *Proceedings of the American Philosophical Society* 123:222–234.

1262 Westneat, D. F., and R. C. Sargent. 1996. Sex and parenting: The effects of sexual conflict and parentage on parental strategies. *Trends in Ecology and Evolution* 11:87–91.

1263 Westneat, D. F., and P. W. Sherman. 1993. Parentage and the evolution of parental behavior. *Behavioral Ecology* 4:66–77.

1264 Westneat, D. F., P. W. Sherman, and M. L. Morton. 1990. The ecology and evolution of extra-pair copulations in birds. *Current Ornithology* 7:330–369.

1265 Wheeler, D. A., C. P. Kyriacou, M. L. Greenacre, Q. Yu, J. E. Rutila, M. Rosbash, and J. C. Hall. 1991. Molecular transfer of a species-specific behavior from *Drosophila simulans* to *Drosophila melanogaster. Science* 251:1082–1085.

1266 Wheeler, P. E. 1991. The influence of bipedalism on the energy and water budgets of early hominids. *Journal of Human Evolution* 21:117–136.

1267 Wheeler, P. E. 1991. The thermoregulatory advantages of hominid bipedalism in open equatorial environments: The contribution of increased convective heat loss and cutaneous evaporative cooling. *Journal of Human Evolution* 21:107–115.

1268 Whitham, T. G. 1979. Habitat selection by *Pemphigus* aphids in response to resource limitation and competition. *Ecology* 59:1164–1176.

1269 Whitham, T. G. 1979. Territorial defense in a gall aphid. *Nature* 279:324–325.

1270 Whitham, T. G. 1980. The theory of habitat selection examined and extended using *Pemphigus* aphids. *American Naturalist* 115:449–466.

1271 Whitham, T. G. 1986. Costs and benefits of territoriality: Behavioral and reproductive release by competing aphids. *Ecology* 67:139–147.

1272 Wickler, W. 1968. *Mimicry in Plants and Animals.* World University Library, London.

1273 Wickler, W., and U. Seibt. 1981. Monogamy in Crustacea and man. *Zeitschrift für Tierpsychologie* 57:215–234.

1274 Wickman, P.-O., and C. Wiklund. 1983. Territorial defence and its seasonal decline in the speckled wood butterfly (*Pararge aegeria*). *Animal Behaviour* 31:1206–1216.

1275 Widemo, F., and I. P. F. Owens. 1995. Lek size, male mating skew and the evolution of lekking. *Nature* 373:148–151.

1276 Wiederman, M. W., and E. R. Allgeier. 1992. Gender differences in mate selection criteria: Sociobiological or socioeconomic explanation? *Ethology and Sociobiology* 13:115–124.

1277 Wiklund, C. G., and M. Andersson. 1994. Natural selection of colony size in a passerine bird. *Journal of Animal Ecology* 63:765–774.

1278 Wiklund, C., A. Kaitala, V. Lindfors, and J. Abenius. 1993. Polyandry and its effect on female reproduction in the green-veined white butterfly (*Pieris napi* L.). *Behavioral Ecology and Sociobiology* 33:25–33.

1279 Wiklund, C., and B. Sillén-Tullberg. 1985. Why distasteful butterflies have aposematic larvae and adults, but cryptic pupae: Evidence from predation experiments on the monarch and the European swallowtail. *Evolution* 39:1155–1158.

1280 Wilcox, R. S. 1979. Sex discrimination in *Gerris remigis:* Role of a surface wave signal. *Science* 206:1325–1327.

1281 Wilcox, R. S. 1984. Male copulatory guarding enhances female foraging success in a water strider. *Behavioral Ecology and Sociobiology* 15:171–174.

1282 Wiley, R. H., B. J. Hatchwell, and N. B. Davies. 1991. Recognition of individual males' songs by female dunnocks: A mechanism increasing the number of copulatory partners and reproductive success. *Ethology* 88:145–153.

1283 Wilkinson, G. S. 1984. Reciprocal food sharing in the vampire bat. *Nature* 308:181–184.

1284 Wilkinson, G. S. 1992. Information transfer at evening bat colonies. *Animal Behaviour* 44:501–518.

1285 Wilkinson, G. S., and G. N. Dodson. 1997. Function and evolution of antlers and eye stalks in flies. In *The Evolution of Mating Systems in Insects and Arachnids*, J. C. Choe and B. J. Crespi (eds.). Cambridge University Press, Cambridge.

1286 Will, M. W., and S. K. Sakaluk. 1994. Courtship feeding in decorated crickets: Is the spermatophore a sham? *Animal Behaviour* 48:1309–1315.

1287 Williams, G. C. 1966. *Adaptation and Natural Selection.* Princeton University Press, Princeton, New Jersey.

1288 Williams, G. C. 1975. *Sex and Evolution.* Princeton University Press, Princeton, New Jersey.

1289 Williams, G. C. 1985. A defense of reductionism in evolutionary biology. *Oxford Surveys in Evolutionary Biology* 2:1–27.

1290 Williams, T. C., and J. M. Williams. 1978. An oceanic mass migration of land birds. *Scientific American* 239 (Oct):166–176.

1291 Willows, A. O. D. 1971. Giant brain cells in mollusks. *Scientific American* 224 (Feb):68–75.

1292 Wilson, D. S., and E. Sober. 1994. Reintroducing group selection to the human behavioral sciences. *Behavioral and Brain Sciences* 17:585–654.

1293 Wilson, E. O. 1971. *The Insect Societies.* Harvard University Press, Cambridge, MA.

1294 Wilson, E. O. 1975. *Sociobiology: The New Synthesis.* Harvard University Press, Cambridge, MA.

1295 Wilson, E. O. 1976. Academic vigilantism and the political significance of sociobiology. *BioScience* 26(183):187–190.

1296 Wilson, E. O. 1980. Caste and division of labor in leaf-cutter ants (Hymenoptera: Formicidae: *Atta*), II. The ergonomic optimization of leaf cutting. *Behavioral Ecology and Sociobiology* 7:157–165.

1297 Wilson, E. O. 1983. Caste and division of labor in leaf-cutter ants (Hymenoptera: Formicidae: *Atta*), III. Ergonomic resiliency inforaging by *A. cephalotes*. *Behavioral Ecology and Sociobiology* 14:47–54.

1298 Wiltschko, R., D. Nohr, and W. Wiltschko. 1981. Pigeons with a deficient sun compass use the magnetic compass. *Science* 214:343–345.

1299 Wiltschko, W., R. Wiltschko, and C. Walcott. 1987. Pigeon homing: Different effects of olfactory deprivation in different countries. *Behavioral Ecology and Sociobiology* 21:333–342.

1300 Wingfield, J. C., and T. P. Hahn. 1994. Testosterone and territorial behaviour in sedentary and migratory sparrows. *Animal Behaviour* 47:77–89.

1301 Wingfield, J. C., and M. C. Moore. 1987. Hormonal, social and environmental factors in the reproductive biology of free-living male birds. In *Psychobiology of Reproductive Behavior: An Evolutionary Perspective*, D. Crews (ed.). Prentice-Hall, Englewood Cliffs, NJ.

1302 Withers, G. S., S. E. Fahrbach, and G. E. Robinson. 1995. Effects of experience and juvenile hormone on the organization of the mushroom bodies of honey bees. *Journal of Neurobiology* 26:130–144.

1303 Wittenberger, J. F. 1981. *Animal Social Behavior*. Duxbury Press, Boston.

1304 Wong, Z., V. Wilson, A. J. Jeffreys, and S. L. Thein. 1986. Cloning a selected fragment from a human DNA "fingerprint": Isolation of an extremely polymorphic minisatellite. *Nucleic Acids Research* 14:4605–4616.

1305 Woodroffe, R., and A. Vincent. 1994. Mother's little helpers: Patterns of male care in mammals. *Trends in Ecology and Evolution* 9:294–297.

1306 Woodward, J., and D. Goodstein. 1996. Conduct, misconduct and the structure of science. *American Scientist* 84:479–490.

1307 Woolfenden, G. E., and J. W. Fitzpatrick. 1984. *The Florida Scrub Jay: Demography of a Cooperative-Breeding Bird*. Princeton University Press, Princeton, NJ.

1308 Wourms, M. K., and F. E. Wasserman. 1985. Butterfly wing markings are more advantageous during handling than during the initial strike of an avian predator. *Evolution* 39:845–851.

1309 Woyciechowski, M., L. Kabat, and E. Król. 1994. The function of the mating sign in honey bees, *Apis mellifera* L.: New evidence. *Animal Behaviour* 47:733–735.

1310 Wright, R. 1994. *The Moral Animal: Evolutionary Psychology and Everyday Life*. Pantheon, New York.

1311 Wynne-Edwards, V. C. 1962. *Animal Dispersion in Relation to Social Behaviour*. Oliver & Boyd, Edinburgh.

1312 Yack, J. E. 1992. A multiterminal stretch receptor, chordotonal organ, and hair plate at the wing-hinge of *Manduca sexta*: Unravelling the mystery of the noctuid moth ear B cell. *Journal of Comparative Neurology* 324:500–508.

1313 Yack, J. E., and J. H. Fullard. 1990. The mechanoreceptive origin of insect tympanal organs: A comparative study of similar nerves in tympanate and atympanate moths. *Journal of Comparative Neurology* 300:523–534.

1314 Yager, D. D., and M. L. May. 1990. Ultrasound-triggered, flight-gated evasive maneuvers in the flying praying mantis, *Parasphendale agrionina*. II. Tethered flight. *Journal of Experimental Biology* 152:41–58.

1315 Yalden, D. W. 1985. Forelimb function in *Archaeopteryx*. In *The Beginnings of Birds*, M. K. Hecht, J. H. Ostrom, G. Viohl, and P. Wellnhofer (eds.). Freunde des Jura-Museums Eichstatt, Willibaldsburg, West Germany.

1316 Ydenberg, R. C., and L. M. Dill. 1986. The economics of fleeing from predators. *Advances in the Study of Behavior* 16:229–249.

1317 Ydenberg, R. C., L. A. Giraldeau, and J. B. Falls. 1988. Neighbours, strangers, and the asymmetric war of attrition. *Animal Behaviour* 36:343–347.

1318 Yosef, R., and D. Whitman. 1993. Imperfect defense. *Living Bird* 12(4):27–29.

1319 Young, D. 1989. *Nerve Cells and Animal Behaviour.* Cambridge University Press, Cambridge.

1320 Yu, Q., A. C. Jacquier, Y. Citri, M. Hamblen, J. C. Hall, and M. Rosbash. 1987. Molecular mapping of point mutations in the *period* gene that stop or speed up biological clocks in *Drosophila melanogaster. Proceedings of the National Academy of Sciences* 84:784–788.

1321 Zach, R. 1979. Shell-dropping: Decision-making and optimal foraging in northwestern crows. *Behaviour* 68:106–117.

1322 Zahavi, A. 1975. Mate selection—A selection for a handicap. *Journal of Theoretical Biology* 53:205–214.

1323 Zahavi, A. 1977. Reliability in communication systems and the evolution of altruism. In *Evolutionary Ecology*, B. Stonehouse and C. M. Perrins (eds.). Macmillan, London.

1324 Zeh, D. W., and R. L. Smith. 1985. Paternal investment in terrestrial arthropods. *American Zoologist* 25:785–805.

1325 Zeh, D. W., J. A. Zeh, and E. Bermingham. 1997. Polyandrous, sperm-storing females: Carriers of male genotypes through episodes of adverse selection. *Proceedings of the Royal Society of London B* 264:119–125.

1326 Zeh, J. A. 1997. Polyandry and enhanced reproductive success in the harlequin beetle-riding pseudoscorpion. *Behavioral Ecology and Sociobiology* 40:111–118.

1327 Zeh, J. A., and D. W. Zeh. 1996. The evolution of polyandry I: Intragenomic conflict and genetic incompatibility. *Proceedings of the Royal Society of London B* 263:1711–1717.

1328 Zeh, J. A., and D. W. Zeh. 1997. The evolution of polyandry II: Post-copulatory defenses against genetic incompatibility. *Proceedings of the Royal Society of London B* 264:69–75.

1329 Zehring, W. A., D. A. Wheeler, P. Reddy, R. J. Konopka, C. P. Kyriacou, M. Rosbach, and J. C. Hall. 1984. P-element transformation with *period* locus DNA restores rhythmicity to mutant, arrhythmic *Drosophila melanogaster. Cell* 39:369–376.

1330 Zeil, J., A. Kolber, and R. Voss. 1996. Structure and function of learning flights in bees and wasps. *Journal of Experimental Biology* 199:245–252.

1331 Zielinski, W. J., F. S. vom Saal, and J. G. Vandenbergh. 1992. The effect of intrauterine position on the survival, reproduction and home range of female house mice (*Mus musculus*). *Behavioral Ecology and Sociobiology* 30:185–192.

1332 Zippelius, H. 1972. Die Karawanenbildung bie Feld- und Hausspitzmaus. *Zeitschrift für Tierpsychologie* 30:305–320.

1333 Zucker, I. 1983. Motivation, biological clocks and temporal organization of behavior. In *Handbook of Behavioral Neurobiology: Motivation*, E. Satinoff and P. Teitelbaum (eds.). Plenum Press, New York.

1334 Zuk, M., R. Thornhill, J. D. Ligon, and K. Johnson. 1990. Parasites and mate choice in red jungle fowl. *American Zoologist* 30:235–244.

Illustration Credits

Unless cited below, full bibliographic information for all copyrighted illustrations, tables, or photographs can be found in the Bibliography. We are grateful to the following publishers for their permission to adapt or reprint copyrighted material.

Page 26: From "The Far Side" by Gary Larson. Reproduced by permission of Chronicle Features, San Francisco; Figure 2-12: From K. Iizuka, 1987. *Kuro the Starling.* © 1987 by Thomson Canada, Ltd; Figure 7-23: © 1987 by Prentice-Hall, Inc., Englewood Cliffs, NJ.; Figure 9-10: © 1976 by the University of Massachusetts Press, Amherst, MA.

Figures 15-7, 15-8 © 1996 by the American Institute of Biological Sciences.

Figures 3-6 © 1992; 3-14 © 1993; 6-8 © 1994; 5-15 © 1992; 15-19 © 1994; 7-8 © 1992; 7-7 © 1993; 4-7 © 1989; 2-6 © 1991; 3-12 © 1981; 4-2 © 1983; 4-28 © 1972; 2-8 © 1991; 6-19 © 1975; 6-12 © 1990; 14-9 © 1992; 10-35 © 1991; 7-20 © 1991; 8-21 © 1978; 9-15B © 1970; 9-18 © 1987; 9-43 © 1979; 12-8 © 1981; 4-23 © 1974; 5-1 © 1991; 10-13 © 1977; Tables 3.1 © 1981; 9-3 © 1981 by the AAAS.

Figures 5-8 © 1963; 10-8 © 1961; 15-11 © 1975 by Harvard University Press, Cambridge, MA.

Figures 1-8 © 1978; 3-1 © 1975; 5-21 © 1974; 5-37 © 1971 by Scientific American, Inc. All rights reserved.

Figures 4-17 © 1995; 16-7 © 1993 by the American Psychological Association.

Figures 6-16 © 1995; 7-12 © 1978; 7-11 © 1990, 1992; 4-5 © 1995 by Wiley-Liss, Inc., a subsidiary of John Wiley & Sons, Inc.

Figures 4-9 © 1995; 7-4 © 1996; 7-15 © 1991; 7-16 © 1992; 8-5 © 1992; 8-7 © 1996; 8-8 © 1994; 9-14 © 1982, 1993; 9-26 © 1994; 9-27 © 1994; 10-3 © 1995; 10-30 © 1990; 10-31 © 1990; 11-6 © 1993; 11-8 © 1997; 12-10 © 1993; 12-12 © 1994; 12-20 © 1993; 12-30 © 1994; 12-34 © 1995; 13-2 © 1995; 13-16 © 1994; 13-19 © 1995; 13-23 © 1995; 13-28 © 1996; 14-4 © 1995; 14-7 © 1993; 15-13 © 1995; 15-17 © 1995; 16-11 © 1993 by Academic Press Limited/W. B. Saunders Limited, London.

Figure 9-1: From N. Tinbergen and H. Falkus, 1970. *Signals for Survival.* © 1970 by Oxford University Press, New York; Figures 12-9 © 1994; 12-19 © 1995; by Oxford University Press.

Figure 7-3 © 1995; 7-18B © 1985; 9-39 © 1996 by The University of Chicago Press.

Figures 4-6 © 1992; 4-12 © 1994; 4-14 © 1993; 4-19 © 1993 by the National Academy of Sciences, U.S.A.

Figure 11-17 © 1996 by the Company of Biologists Ltd.

Figures 4-3 © 1992; 4-4 © 1982; 8-3 © 1996; 10-9 © 1988; 10-10 © 1988; 10-22 © 1992; 11-13 © 1996; 11-21 © 1992; 11-30 © 1992; 12-23 © 1990; 12-28 © 1994; 12-32 © 1994; 13-3 © 1995; 13-24 © 1993; 13-25 © 1994; 15-12 © 1992 by Springer-Verlag.

Figures 2-3 © 1987; 3-3 © 1992; 4-24 © 1979; 6-21 © 1991; 7-2 © 1995; 7-13 © 1997; 7-14 © 1997; 8-6 © 1987; 8-14 © 1955; 8-16 © 1978; 8-22 © 1986; 10-12 © 1987; 10-19 © 1991; 10-25 © 1986; 11-3 © 1979; 12-31 © 1994; 12-33 © 1996; 13-6 © 1989; 13-7 © 1992; 13-26 © 1995; 14-13 © 1994; 15-26 © 1992; 15-27 © 1989; Tables 2-1 © 1983; 13-1 © 1990 by Macmillan Magazines, Ltd.

Index

About the Book

Editor: Peter Farley

Project Editor: Kerry Falvey

Copy Editor: Norma Roche

Production Manager: Christopher Small

Book Layout and Production: Jefferson Johnson

Illustrations: Precision Graphics and Nancy Haver

Book Design: Susan Brown Schmidler

Cover Design: MBDesign

Cover Manufacturer: Henry N. Sawyer Company, Inc.

Book Manufacturer: The Courier Companies, Inc.